Advances in
Usability Evaluation
Part I

Advances in Human Factors and Ergonomics Series

Series Editors

Gavriel Salvendy
Professor Emeritus
School of Industrial Engineering
Purdue University

Chair Professor & Head
Dept. of Industrial Engineering
Tsinghua Univ., P.R. China

Waldemar Karwowski
Professor & Chair
Industrial Engineering and
Management Systems
University of Central Florida
Orlando, Florida, U.S.A.

3rd International Conference on Applied Human Factors and Ergonomics (AHFE) 2010

Advances in Applied Digital Human Modeling
Vincent G. Duffy

Advances in Cognitive Ergonomics
David Kaber and Guy Boy

Advances in Cross-Cultural Decision Making
Dylan D. Schmorrow and Denise M. Nicholson

Advances in Ergonomics Modeling and Usability Evaluation
Halimahtun Khalid, Alan Hedge, and Tareq Z. Ahram

Advances in Human Factors and Ergonomics in Healthcare
Vincent G. Duffy

Advances in Human Factors, Ergonomics, and Safety in Manufacturing and Service Industries
Waldemar Karwowski and Gavriel Salvendy

Advances in Occupational, Social, and Organizational Ergonomics
Peter Vink and Jussi Kantola

Advances in Understanding Human Performance: Neuroergonomics, Human Factors Design, and Special Populations
Tadeusz Marek, Waldemar Karwowski, and Valerie Rice

4th International Conference on Applied Human Factors and Ergonomics (AHFE) 2012

Advances in Affective and Pleasurable Design
Yong Gu Ji

Advances in Applied Human Modeling and Simulation
Vincent G. Duffy

Advances in Cognitive Engineering and Neuroergonomics
Kay M. Stanney and Kelly S. Hale

Advances in Design for Cross-Cultural Activities Part I
Dylan D. Schmorrow and Denise M. Nicholson

Advances in Design for Cross-Cultural Activities Part II
Denise M. Nicholson and Dylan D. Schmorrow

Advances in Ergonomics in Manufacturing
Stefan Trzcielinski and Waldemar Karwowski

Advances in Human Aspects of Aviation
Steven J. Landry

Advances in Human Aspects of Healthcare
Vincent G. Duffy

Advances in Human Aspects of Road and Rail Transportation
Neville A. Stanton

Advances in Human Factors and Ergonomics, 2012-14 Volume Set:
Proceedings of the 4th AHFE Conference 21-25 July 2012
Gavriel Salvendy and Waldemar Karwowski

Advances in the Human Side of Service Engineering
James C. Spohrer and Louis E. Freund

Advances in Physical Ergonomics and Safety
Tareq Z. Ahram and Waldemar Karwowski

Advances in Social and Organizational Factors
Peter Vink

Advances in Usability Evaluation Part I
Marcelo M. Soares and Francisco Rebelo

Advances in Usability Evaluation Part II
Francisco Rebelo and Marcelo M. Soares

Advances in
Usability Evaluation
Part I

Edited by

Marcelo M. Soares
and
Francisco Rebelo

CRC Press
Taylor & Francis Group
Boca Raton London New York

CRC Press is an imprint of the
Taylor & Francis Group, an **informa** business

CRC Press
Taylor & Francis Group
6000 Broken Sound Parkway NW, Suite 300
Boca Raton, FL 33487-2742

First issued in paperback 2019

© 2013 by Taylor & Francis Group, LLC
CRC Press is an imprint of Taylor & Francis Group, an Informa business

No claim to original U.S. Government works

ISBN-13: 978-1-4398-7024-2 (hbk)
ISBN-13: 978-0-367-38113-4 (pbk)

Visit the Taylor & Francis Web site at
http://www.taylorandfrancis.com

and the CRC Press Web site at
http://www.crcpress.com

Table of Contents

Section III: Product Design and Evaluation

Preface

Successful interaction with products, tools and technologies depends on usable designs, accommodating the needs of potential users and does not require costly training. In this context, this book is concerned with emerging concepts, theories and applications of human factors knowledge focusing on the discovery and understanding of human interaction with products and systems for their improvement.

The book is organized into three sections that focus on the following subject matters:

- Devices and their user interfaces
- User Studies
- Product Design and Evaluation

In the section "Devices and their interfaces," the focus is on optimization of user devices, with emphasis on visual and haptic feedback.

In the section "User studies," the focus goes to the limits and capabilities of special populations, particularly the elderly, which can influence the design. Generally, the effect of changes in force and kinematics, physiology, cognitive performance, in the design of consumer products, tools and workplaces is discussed.

The section "Product and design evaluation" employs a variety of user-centered evaluation approaches, for developing products that can improve safety and human performance and at same time, the efficiency of the system. Usability evaluations are reported for different kinds of products and technologies, particularly for cellular phones, earphone controls, mattresses and pillows, package and professional tools (i.e. platform lifts, pruning shears) and service systems (i.e. office and communication).

We would like to thank the Editorial Board members for their contributions.

This book will be of special value to a large variety of professionals, researchers and students in the broad field of human performance who are interested in feedback of devices' interfaces (visual and haptic), user-centered design, and design for special populations, particularly the elderly. We hope this book is informative, but even more - that it is thought provoking. We hope it inspires, leading the reader to contemplate other questions, applications, and potential solutions in creating good designs for all.

April 2012

Marcelo Soares
Federal University of Pernambuco
Brazil

Francisco Rebelo
Ergonomics Laboratory
Interdisciplinary Centre for the Study of Human Performance
Technical University of Lisbon
Portugal

Editors

Section I

Devices and Their User Interfaces

CHAPTER 1

Improving Small Visual Displays for Low Vision Users

Morgan Blubaugh and Mark Uslan

AFB Tech, American Foundation for the Blind
Huntington, WV
MUslan@afb.net
MBlubaugh@afb.net

Caesar Eghtesadi, Ph.D.

Tech For All, Inc.
KEghtesadi@TFAConsulting.com

ABSTRACT

The use of small visual displays (SVDs) embedded in electronic devices presents a growing usability issue for the 25 million adult Americans with vision loss. This paper provides an overview of the usability barriers presented by SVDs for the expanding demographic of people with vision loss. Efforts to improve usability include the development of the American Foundation for the Blind's Optics Lab to measure SVD optical parameters, and research conducted in partnership with the Atlanta VA Rehabilitation R&D Center of Excellence to test the effectiveness of an SVD research-based guideline to predict SVD character recognition by people with vision loss.

Keywords: Small visual display, vision loss, accessibility, usability, contrast ratio, image quality, low vision.

1   INTRODUCTION

The use of electronic small visual displays (SVDs) embedded in home, office, and medical devices has become increasingly commonplace and presents a growing

3

usability issue for people with vision loss. The quality of these displays can vary extensively from product to product, and in many cases the text and images produced present significant interpretation challenges for people with low vision.

This problem affects millions of Americans with low vision. In 2008, the National Health Interview Survey (NHIS) established that there were over 25 million Americans aged 18 and over who reported having vision loss (Pleis, J.R., & Lucas, J.W., 2009). In the survey, the term "vision loss" referred to individuals who reported that they had trouble seeing, even when wearing glasses or contact lenses, as well as to individuals who reported that they were blind or unable to see at all. During the same year, it was reported by the World Health Organization (WHO) that of the total world population reporting visual impairment (both blind and low vision), 86% had low vision (Foster, A., Gilbert, C., & Johnson, G., 2008).

The incidence of low vision within the United States is expected to rise in the future. Although the risk of vision loss increases with age, the Baby-Boom Generation—those between 46 and 66 years of age—currently comprise the largest population of individuals with vision loss. Experts predict that by 2030, rates of vision loss will double as this demographic begins to experience the full extent of age-related eye conditions (Prevent Blindness America, 2008).

The accessibility needs of SVD users with low vision vary depending on the type and stage of eye condition. In order to improve the usability of these ubiquitous devices, manufacturers need to take the usability requirements of low vision users into consideration when researching, developing, designing, and updating their products.

Perhaps the most serious usability issue caused by SVDs is the barrier to independently monitoring and caring for one's own well being. Many home medical devices, including scales, thermometers, blood pressure monitors, and diabetes monitoring equipment such as blood glucose meters, insulin pumps, and insulin pens, have embedded SVDs that make it difficult for a low-vision user to distinguish the characters on the display (Uslan, M.M., Burton, D.M., Wilson, T.E., Taylor, S., Chertow, B.S., Terry, J.E., 2007).

Despite the widespread use of SVDs, standards governing their design and establishing usability conventions for users with vision loss have not been published. Several conventional standards that address the usability of large displays, such as computer monitors and televisions have been published (Video Electronics Standards Association, 2001; International Organization for Standardization, 1993). Moreover, a large amount of research (Legge, G. E., Pelli, D. G., Rubin, G. S., Schleske, M. M., 1985) has been devoted to understanding how individuals with or without vision loss process the types of information found on visual displays, such as alphanumeric characters, objects, and icons (Patching, G.R., & Jordan, T.R., 2005; Avidan et al, 2002). Yet there is a distinct lack of information specifically addressing the usability of SVDs for users with low vision.

Over the past six years AFB Tech, has been addressing the problem of SVD accessibility by:

Establishing an optics lab to measure the display characteristics of SVDs that have a direct relationship to SVD usability for people with low vision. These display characteristics will be published in an online database, which will allow consumers and manufactures to research the usability of SVDs.

Partnering with the VA Palo Alto Health Care System to develop a research-based design guideline for SVD usability. The basis for the guideline is the Barten Square Root Integral (SQRI), which is an image quality metric that combines measurements of display quality and the user's vision level to predict display performance (a measure of usability).

This paper provides: 1) an overview of relevant SVD display characteristics, 2) descriptions of common and advanced SVD technologies, 3) a discussion of the optical instrumentation system used by AFB to measure SVDs, along with sample results from our findings, 4) a description of the research being conducted to predict SVD character recognition by people with low vision.

2   RELEVANT OPTICAL CHARACTERISTICS OF SVDS FOR USERS WITH VISION LOSS

Whether or not a person can reasonably use a given SVD is contingent upon multiple display characteristics, including resolution, luminance, contrast ratio, reflection, spatial frequency, font size, font style, and the physical size of the display itself.

Contrast ratio is defined as the relationship in luminance between the foreground and background; a higher contrast ratio is generally easier to read than a lower contrast ratio (Rubin, G.S., & Legge, G.E., 1989). Spatial frequency relates to the number of line pairs (e.g. a dark line against a light background) in a specified area. The higher the number of line pairs in the area under assessment, the higher the spatial frequency measurement. Figure 1 shows a simple graphical representation of high and low contrast and spatial frequency.

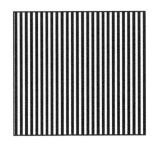  

(a) High contrast, high spatial frequency square-wave grating

b) High contrast, low spatial frequency square-wave grating

(c) Low contrast, low spatial frequency square-wave grating

Figure 1: Examples of Different Contrast Ratio and Spatial Frequency

Contrast can be affected by a number of factors, including the display technology itself and the reflectivity of the display's surface. Reflection is an important factor for the usability of SVDs; the clear protective covers on SVDs reflect a minor to excessive quantity of ambient light depending on the type of covering material (glossy or matte) used. We are primarily concerned with two types of reflective condition: specular and diffuse. Specular reflection is the reflection of light off a surface at the angle opposite the incoming one. Specular reflection can be demonstrated by looking into a mirror and seeing a distinct reflected image. This type of reflection poses the biggest visibility problem for all users of displays, and particularly for users with vision loss. Diffuse reflection is the reflection of light off a surface at many different angles. Diffuse reflection can be demonstrated by looking at a plain sheet of white paper in bright light: a distinct but consistent reflection is perceived.

High levels of reflection can significantly reduce the perceived contrast of a display (such as when reading a copier display in a bright office), and extremely reflective conditions (when trying to read a cell phone display outside on a sunny day, for example) can render a display nearly unreadable. People with ocular and macular diseases are particularly susceptible to visual disturbances caused by reflection. Reverse contrast polarity seems to be especially beneficial to individuals who have particular difficulties with glare.

Regardless of polarity, the contrast ratio of a display significantly affects a person's ability to discern the displayed information and is perhaps the single most important characteristic of any visual display.

In addition to contrast ratio and spatial frequency, font size and spatial resolution also affect the usability of SVDs for people with vision loss. As a general rule, the larger the size of the displayed information, the easier it is to read; the higher the spatial resolution (which measures the ability of the display to produce detail), the clearer the display.

3   CHALLENGES PRESENTED BY SVD TECHNOLOGIES TO USERS WITH VISION LOSS

Several common SVD technologies present visibility and legibility challenges for people with low vision. Basic device technologies include: LCDs, light emitting diodes (LEDs), and e-paper. The majority of currently available devices with SVDs use LCD technology. LCDs can be grouped into two categories: "backlit," which provide illumination and can be viewed in the dark or in ambient light, or "reflective," which do not provide illumination and therefore can only be viewed in ambient light conditions. LCDs can present information in color or monochrome. By definition, LEDs are always backlit, and are typically found only on simple segmented displays, which use icons and/or segments to display information instead of pixels, such as home appliances and digital clocks. The spatial resolution of SVDs is dependent on the quality of the specific device, and is not consistent within technology types.

Significant number of home medical devices currently on the market use reflective, monochrome LCD screens. In many cases, particularly with blood glucose meters and thermometers, the displays are also limited to seven-segment output. In the course of our research we found that the measured contrast ratio of these products generally ranged from 2–25, well below that of backlit color displays, and low enough to cause significant usability issues for low-vision users. When a specular reflection was introduced, the contrast of these displays fell by 50–90%. In some of these devices, the size of the displayed text tended to be relatively large—up to 40 mm—particularly in home blood pressure monitors and some of the blood glucose meters.

Cell phones and personal audio players use a range of display technologies, although LCD is still the most prevalent display type. These SVDs provide high levels of brightness as well as flexibility in color and resolution. Nearly all of the cell phones we examined featured full-color, backlit pixelated displays. The contrast ratio of these screens ranged from 250–400 making them the highest contrast SVDs we measured. To achieve these high contrasts, the majority of these displays use high-gloss screen coverings, which can create significant amounts of reflection in specular reflection conditions. Two of the emissive displays we measured and whose results are included below were the Samsung Intensity and the Apple iPod Touch. Both of these displays have high contrast compared to any other display type tested with contrast ratios of 279 and 350 respectively. The cell phone exhibited poor reflection properties, dropping to a contrast ratio of 1.97 without a specular reflection and 1.25 with. According to our research, the iPod Touch has one of the best displays available on a mobile device. In sunlight conditions with no specular reflection, the contrast ratio measured 134, dropping to 3.6 with a specular reflection. While this drop is significant, it is smaller than that seen in most other comparable displays. Many of these devices allow the user adjust the size of the font; font height can range anywhere from 1–9 mm.

LEDs can provide a bright reverse-polarity display that produces high contrast but can suffer from reflection issues largely because many LED displays have a convex, plastic cover that reflects light. These displays are not generally found on handheld or mobile devices, so it's rare that they are used in sunlight conditions. However, since they often are installed on stationary devices, such as ovens and microwaves, it is more difficult for users to adjust the display to a preferred angle to reduce reflection. LEDs are usually simple seven-segment displays that present large characters ranging from 10–30 mm in height.

E-paper is a relatively new low-power technology used on most book readers that replicates the visual effect of printed text on paper. This effect comes at the cost of a non-illuminated display and relatively poor contrast. Some e-paper displays have a matte surface that does not create specular reflection, so the displays have relatively high contrast in sunlight, and in some cases are more usable by people with low vision than other display types in sunlight conditions. Our research found that contrast ratios of e-paper displays range from 2–6 and only drop 5–10% with a specular reflection in sunlight conditions. E-paper displays usually let the user adjust the font size; display height can range from 2–10 mm.

Printers, copiers, fax machines, and other pieces of large office equipment typically use backlit LCD screens that provide high levels of contrast. These products are usually not used in direct sunlight, so reflection conditions for these displays are not a major concern. Typically, office equipment displays use character sizes of 5–15 mm in height.

Smaller SVD devices, such as home thermostats, calculators, postage scales, and digital clocks typically use reflective LCD displays that exhibit a lower level of contrast compared to other reflective LCD displays, ranging from 1.5–5 and dropping by as much as 30% with a specular reflection. The font size of these devices can range anywhere from 2–30 mm in height, depending on the device type and function.

4   OPTICAL INSTRUMENTATION SYSTEM

To measure the display characteristics of SVDs, a replicable optical instrumentation system based on established metrology standards (Video Electronics Standards Association Display Metrology Committee, 2001) was established. The system was used to measure main display characteristics including the most important parameters to SVD accessibility: contrast, reflection, resolution, and font size. The instrumentation and processes are intended to measure a wide range of devices, from home medical devices to digital audio players.

The instrumentation has two separate setups that allow measurements of contrast and reflection. One setup is designed for measuring the display in a bright setting (e.g., office or sunlight conditions), the other for measuring the display in a dark setting (only for backlit displays). Figure 2 provides an image of the optical instrumentation system taking measurements in both setups. The main components of the system are:

- A positioning stage with translating platforms necessary to align the optics equipment and the target SVD device, regardless of device size.
- A computer-controlled luminance meter with close-up lens.
- A video magnifier with an adjustable flex-arm camera attached to the table to provide accessibility for low-vision operators.
- A Canon EOS digital camera outfitted with a 100 mm macro lens and extension tubes to provide a detailed image of the target screen.
- A 12-inch sampling sphere with measurement ports at 0, ±8, ±20, and ±30 degrees to provide carefully controlled illumination of the target screen.

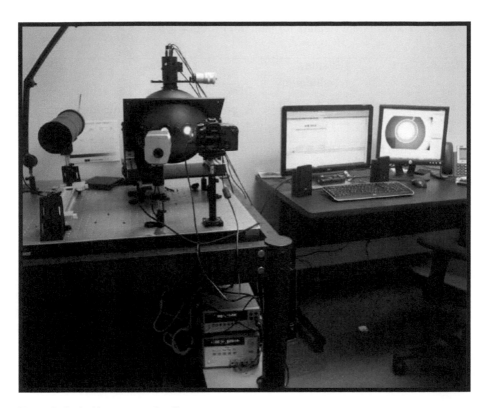

Figure 2: Optical Instrumentation Set-up

The above system is complemented with a dedicated computing environment running software that controls the remote operation of all electronic instrumentation on the positioning table. The camera interfaces with the system using the Canon EOS Digital Software Development Kit (EDSDK). A software algorithm was developed in MATLAB to compute image quality parameters. The image analysis software was written using the Microsoft .NET framework coupled with the Canon EDSDK to analyze raw image data and calculate display parameters. These parameters are uploaded to the online database, which currently has data on over 150 devices in a number of categories, including: cell phones, personal audio players, home medical devices, office equipment, and home appliances.

4.2 Sample Results

Results of the sample devices are presented in Table 1. This selection of devices was intended to capture a sampling of some of the most popular devices on the market for a variety of functions. The devices sampled in the areas of: blood glucose meter (OneTouch Ultra 2), home appliances (Honeywell RTH7600D Thermostat), mobile phones (Samsung Intensity), mp3 players (Apple iPod Touch), and e-book readers (Amazon Kindle 2).

Table 1 Measurement Results for Multiple Product Types

Product Type	Display Technology	Contrast			Font Height (mm)
		Dark Room	Specular Excluded (Sunlight)	Specular Included (Sunlight)	
Blood Glucose Meter (OneTouch Ultra 2)	Reflective LCD	N/A	16.9	2.1	6-14
Thermostat (Honeywell RTH7600D)	Reflective LCD	279	2.0	1.3	2.5-3.5
Cell Phone (Samsung Intensity)	Backlit LCD	350	134	3.6	1.5-9.0
Mp3 Player (iPod Touch 3G)	Backlit LCD	N/A	3.8	3.4	2-10
eBook Reader (Amazon Kindle 2)	E-paper	N/A	2.3	1.6	2.5-14

The methodology for determining contrast was to first measure the reflection properties of each display and then calculate contrast for the two light conditions under assessment. The Contrast columns present the results for the two light levels (Dark Room [0 lux] and Sunlight [20,000 lux]) and two reflection conditions (Specular-Excluded/Diffuse [SE] and Specular-Included [SI]).

Purely reflective displays (blood glucose meter, Kindle, and thermostat) exhibit contrast only when external illumination is introduced; therefore there are no Dark Room condition results for these devices. Level of contrast for these displays is dependent upon the type of light being reflected.

Font height describes the size of the font in millimeters. We reported the smallest and largest sizes presented on the displays. Some displays like the iPod Touch, Kindle, and cell phone have adjustable font heights, a very helpful feature to people with vision loss. The iPod Touch also has a built in screen magnifier, which gives it a very large dynamic range of text. The seven-segment displays like the blood glucose monitor and thermostat typically use a large but fixed-size font.

5   PERFORMANCE TESTING

As part of its efforts to improve SVD usability, AFB has partnered with the VA Palo Alto Health Care System to test the effectiveness the Barten SQRI as a research-based SVD guideline. The Barten SQRI is computed from two measurements: the Modulation Transfer Function (MTF) of the display, which measures both the contrast and spatial frequencies of the display, and the Contrast Sensitivity Function (CSF) of the user, which measures the minimum contrast required for the user to view various spatial frequencies (Lovegrove, W.J., Bowling, A., Badcock, D., and Blackwood, M., 1980; Barten, P.G.J., 1990).

The objective of performance testing in individuals with central vision loss is to investigate the relationship between the contrast threshold required for successful recognition of a single digit/ character and the Barten SQRI. Performance testing was conducted with single-digit numerals (0–9) presented in six heights (0.25, 0.50, 1.0, 2.0, 4.0, and 8.0 degrees). The study showed that the Barten SQRI significantly predicts digit/character recognition performance ability ($r^2 = 0.81$; $p < 0.01$) in people with central vision loss (Schuchard, R.A., Uslan, M., Wilson, T., and LiKamWa, W., 2009). Combined contrast measures of individual observers' contrast sensitivity characteristics and an image display's contrast characteristics may provide effective performance measures of the effect of central vision loss on everyday tasks with SVDs.

6   DISCUSSIONS AND FUTURE DIRECTION

The usability of SVDs among people with vision loss is a significant issue for millions of Americans. The large aging population, who possess decreased contrast sensitivity due to normal age-related vision changes, adds to the urgency of this problem. Prevalent usage of SVDs in daily life has increased dramatically with the introduction of cell phones, home medical devices, e-book readers, personal audio devices, and more. While these devices have revolutionized daily activities, their prevalence has also created a major usability issue for people with low vision, regardless of age. Currently, no guidelines exist to help manufacturers to develop SVDs with optimum optical characteristics for people with vision loss.

Further performance testing is being conducted on the common SVDs (like those mentioned in this article) that people use in their everyday lives. This performance testing will also test participants with a more representative sampling of eye conditions. The main study objective remains the same: to validate the Barten SQRI image quality matrix as a means to predict performance using SVDs for people with low vision, which is in essence a measure of usability.

The final goal of this project will be to publish guidelines and best practices to help increase industry awareness of the relationship between customers' visual abilities and the usability of manufactured devices. Eventually universal design principals should be used to develop SVD devices using these guidelines and best practices.

ACKNOWLEDGEMENTS

This paper summarizes the ongoing accessibility research and outcomes of a field-initiated development project (#H133G090026) led by Mark Uslan, Director of AFB Tech, and funded by the U.S. Department of Education's National Institute on Disability and Rehabilitation Research (NIDRR) to research small visual displays for people with vision loss.

REFERENCES

Avidan, G., Harel, M., Hendler, T., Ben-Bashat, D., Zohary, E., & Malach, R. (2002). Contrast sensitivity in human visual areas and its relationship to object recognition. *Journal of Neurophysiology, 87,* 3102-3116.

Barten, P.G.J. (1990). Evaluation of subjective image quality with the square-root integral method, *Journal of the Optical Society of America, 7,* 2024-2031.

Foster, A., Gilbert, C., & Johnson, G. (2008). Changing patterns in global blindness: 1998-2008. *Community Eye Health Journal, 21,* 37-39.

International Organization for Standardization. (1993). *ISO 9241-3:1992 Ergonomic requirements for office work with display terminals (VDTs), Part 3: Visual display requirements.*

Legge, G.E., Pelli, D.G., Rubin, G.S., Schleske, M.M. (1985). Psychophysics of Reading, II: Low vision. *Vision Research, 25,* 253-266.

Lovegrove, W.J., Bowling, A., Badcock, D. & Blackwood, M. (Oct 1980). Specific reading disability: differences in contrast sensitivity as a function of spatial frequency. *Science,* 210(4468), 439-440.

Patching, G. R., & Jordan, T. R. (2005). Spatial frequency sensitivity differences between adults of good and poor reading ability. *Investigative Ophthalmology & Visual Science,* 46(6), 2219-24.

Pleis, J.R., & Lucas, J.W. (2009). Summary health statistics for U.S. adults: National Health Interview Survey, 2008. National Center for Health Statistics. *Vital Health Stat 10,* 242.

Prevent Blindness America & National Eye Institute. (2008). *Vision Problems in the U.S.: Prevalence of Adult Vision Impairment and Age-Related Eye Disease in America 2008 Update.* Retrieved from: http://www.preventblindness.net/site/DocServer/VPUS_2008_update.pdf?docID=1561.

Rubin, G.S. & Legge, G. E. (1989). Psychophysics of reading, VI: The role of contrast in low vision. *Vision Research, 29,* 79-91.

Schuchard, R.A., Uslan, M., Wilson, T., LiKamWa, W. Evaluating the Barten SQRI for Predicting Small Visual Display Character Recognition. ARVO Abstracts online and on CDROM (2009). #4725.

Uslan, M.M., Burton, D.M., Wilson, T.E., Taylor, S., Chertow, B.S., Terry, J.E. (2007). Accessibility of home blood pressure monitors for blind and visually impaired people. *Journal of Diabetes Science and Technology, 1,* 218-227.

Video Electronics Standards Association Display Metrology Committee. (2001). Flat-panel display measurements standard. Version 2.0. Milpitas, CA.

CHAPTER 2

Design of Steering Torque Feedback in a Fixed-base Driving Simulator and Its Validation

De-Yu WANG, Liang MA

Department of Industrial Engineering, Tsinghua University, Beijing, China, 100084
wangdy11@mails.tsinghua.edu.cn

ABSTRACT

A driving simulator provides versatile, controllable, economical, and safe platform for researchers. However, one aspect worthy of special attention is the issue of validity. Next to visual feedback, steering wheel torque feedback is a very important feedback channel in simulators. This paper discussed how the fidelity of a driving simulator was improved by adding a steering torque feedback system to ensure a high level of simulator validity. We used subjective evaluation and objective vehicle handling tests to assess the effectiveness of the feedback system. The results show that the addition of torque feedback increased driver's concentration on handling vehicle and enhanced the fidelity of the simulator.

Keywords: virtual reality, driving simulator, force feedback, fidelity

1   INTRODUCTION

In the studies concerning vehicle and transportation facilities, driving simulators with virtual reality environment has been commonly used. Popular topics include driving distraction and impairment, novice drivers, traffic sign design, vehicle ergonomics, adaptive cruise control, head-up display etc. (Sung, Arup Dutta, Marc Wittmann, David Uzzell, and M. Hoedemaeker). In comparison to instrumented cars used in field studies, simulators have many advantages (Blana, 1996). They provide an economic yet versatile platform. Researchers are free to create desired replicable scenarios and control experiment variables. It is easy to deeply customize the virtual environment. Moreover, vehicle parameters and status are easy to monitor and be precisely recorded for further analysis. Driver behaviors are also easy to monitor, since complicated and expensive monitoring instruments are easier to use in laboratory than in vehicles used in field tests. Last but not least, simulators are safer experiment device. They eliminate potential risks to the driver and other road users. This is especially emphasized in researches where the participants are required to complete risky tasks, such as answering cellphone while driving.

One major aspect worthy of special attention in modern simulators is their validity, since it is very hard to depict the real world to its full complexity. The validity of the system is in close connection with the simulation fidelity, which determines the condition of the participants doing experiments. In simulator studies, a popular and very meaningful research area is the influence of distraction and impairment. In these studies, the effect of the impairment factor should be separated from other undesired factors. Thus, the driving feeling in the simulator should be as close to the reality as possible. Similarly, in studies involving vehicle upholstery design and traffic infrastructure design, mental concentration of the driver is an important reference for evaluation. Simulator itself should not bring any distraction to the driver in the experiments. To achieve higher fidelity, we should first provide sufficient feedback channels and then improve the feedback quality. Major feedback channels include visual, auditory, haptic, and motional ones. Rockwell (1972) estimates that over 90% of information input to a driver come via vision. Gordon (1966) pointed out that steering feel is the second most important source of input after visual feedback. During the interviews with the participants in previous projects, some of them reported that the steering wheel lacked torque feedback, which had a negative effect on system fidelity. Consequently, the experiments would have the problem of validity. Thus, it is necessary to add an effective steering torque feedback system. In system evaluation, we used both subjective assessment and objective measurements. Post-test questionnaire was used as an effective way to collect subjective evaluation (Ronald R. Mourant, 2002). For objective measurements, both vehicle speed and vehicle moving path was recorded for further analysis (Stuart T. Godley, 2002). In data analysis, we examined relative validity of the simulator instead of its absolute validity (Harms, 1994). This is because the variables in researches are usually independent (Törnros, 1998) and relative validity focuses more on predicting the tendency of controlled variables.

The simulator discussed in this study is located in the virtual reality and driving simulation lab in the Department of Industrial Engineering, Tsinghua University. It is a fixed-base driving simulator with a full car cabin. The simulation environment is created by VEGA and is displayed via projectors on a panoramic screen. The sounds from the engine and from the ambient environment are also delivered. The acceleration and braking pedals are reconnected to sensors. The steering rack and pinion system is disconnected from the steering wheel and is replaced by an optical encoder to take the steering action of the driver. Previously, several research projects were carried out on this platform (Wang, 2010 and Wang, 2009).

In this paper, we discussed how we improved the fidelity of the simulator in our laboratory by adding steering torque feedback system to it. First, a simplified torque feedback model is built. Then the design and installation of a steering wheel torque feedback system in a driving simulator is briefly discussed. At last the subjective assessment and objective evaluation for the feedback system is also presented and analyzed.

2   STEERING WHEEL TORQUE FEEDBACK SYSTEM

In this section, the torque feedback model should first be determined. Then the upgrade scheme will be introduced, including the hardware design and software realization. At last, the evaluation test and its results are discussed. Expectedly, the driver should feel the torque feedback on the steering wheel after modification, and their driving experience should be more realistic.

2.1 Feedback torque model

In our torque feedback system, the relative feedback torque $P_{feedback}$ is given by the function

$$P_{feedback} = \begin{cases} 0.515|\theta|\sqrt{v} & |\theta| \leq 20° \\ (10 + 0.015|\theta|)\sqrt{v} & |\theta| > 20° \end{cases}$$

where θ represents the angular position of the steering wheel and v represents the vehicle velocity.

In general, the steering torque feedback will always serve in the direction that brings the steering wheel back to its initial position. The torque increases with the steering angle. When the steering wheel is steered away from the initial point within ±20°, the feedback torque quickly increases. Beyond 20°, the feedback increases linearly with the steering angle. The slope is not as large as that near the initial point of the steering wheel. The relationship between torque and steering wheel angle is shown in Figure 1. In terms of speed of the vehicle, the feedback torque is in linear relationship with the square root of the vehicle velocity. This is consistent with the fact that the steer wheel is heavier under high speed. The relationship between torque and speed of the vehicle is shown in Figure 1.

In reality, the torque feedback on a steering wheel is complicated. However, it could also be defined by the dynamic condition of the vehicle. The torque could be a function of speed, steering wheel angle, as previous studies suggest (Zhang, 2009). Expectedly, our driving simulator will be used in studies concerning general driving conditions. So a simplified model will provide sufficient fidelity.

To establish this model, we referred to a previous literature (Zhao Xue-Ping, 2009) and obtained key values for regression. The key values are a series of torque levels under different driving conditions. These values are representative of general driving scenarios, so that the model produced from them are expected to be close to the reality.

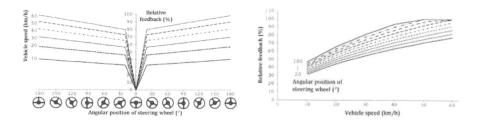

Figure 1: The relationship between torque feedback and steering wheel position. The relationship between torque feedback and speed of the vehicle

It is noteworthy that the model only defined the relative level of torque feedback. Thus, the unit is the percentage of maximum torque. For general use in laboratory studies, 5N·m is a suitable level of maximum torque.

2.2 Realization

The structure of the system is shown in Figure 2. The torque feedback was generated from three major components. 1) Sensors were used to measure the input of the driver. They detected the positions of the pedals and the steering wheel, and then sent the readings to the computer. The simulation program that ran on the computer read the input of the sensors and then calculated the output signal for 2) the control module. The control module received the signal and modulated the power output that drove 3) the torque motor. Under different levels of power supply, the torque motor would deliver different levels of torque to the steering wheel. The whole process was repeated at the refresh rate of 60Hz, determined by the virtual scenario. Thus, the torque feedback was delivered to the driver in real time without any perceivable delay.

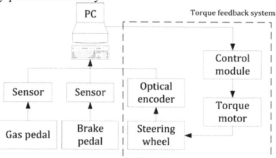

Figure 2: The structure of the simulator system

2.3 Validation

Comparative experiments were conducted on the driving simulator to test the effect of steering wheel torque feedback on the fidelity of the virtual environment.

The independent variable of the experiments was the setting of the steering system. The two levels of the variable were a) nonfeedback and b) feedback. The participants were required to complete two driving tasks under both settings. As dependent variables, the participants' performance and their evaluations were collected. It was expected that the driver behavior would be more natural when steering feedback exists. The vehicle handling dynamics should also be improved.

2.3.1 Method

Procedures

There were seven steps in the validity test: 1) introduction; 2) randomizing the order of two settings in each test session; 3) city free roam with/without feedback, or test session 1; 4) double lane-change with/without feedback, or test session 2; 5) subjective assessment, or respond session; 6) showing the result; 7) rewarding.

In the first session, the subjects drove the vehicle through the city following instructions and then free roam at will. The participants changed lanes and made various types of turns. This session lasted for 5 minutes so that the participant got familiar with the simulation system (Daniel V. McGehee, 2004). Meanwhile, the participants drove the simulator with or without the feedback torque, and compare in a general basis the steering feelings under each setting to that on a real car.

The second test session was handling test, which was performed in a double lane-change scenario. We used ISO 3888-1: 1999 standard for simulator handling test (ISO, 1999). In the test, the driver completes two successive lane-change in a test track defined by traffic cones (see Figure 3). The vehicle speed during the maneuver was set to 15, 20, 25, 30, 40 km/h for each time respectively. The driving paths were recorded for further investigation. Expectedly, introducing the steering feedback would enhance the driving experience, thus improving the driver's control over the vehicle during intense driving.

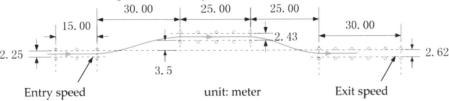

Figure 3: ISO double lane-change test track

Virtual scenarios

The virtual scenario was designed to depict real road scenes to help the driver understand the driving tasks. Two scenarios were designed for the tests. In the city roam session, we created a typical city area with multi-lane road network and various kinds of buildings (see Figure 4). To reduce the complexity of the task, there were no public traffic in the city. The second scenario, double lane-change, was based on the ISO double lane-change test track. In this study, were produced the lane-change scenario by placing cars and buses of the equivalent size in the

18

corresponding position (see Figure 5). This would help the participant to connect the virtual driving with their daily driving conditions.

Figure 4: Test scenario 1, the virtual city

Figure 5: Test scenario 2, double lane-change

Participants

There were four participants in the validation tests including three males and one female. All the participants were licensed drivers in China. They were all in good health conditions. No participant had driving simulator experience. Their information is listed in Table 1.

Table 1: Information about validation test participants

Subject	Gender	Age (year)	Driving Experience (year)	Mileage (km)	Preferred vehicle
A	Male	23	5	12,000	Land Rover
B	Male	22	1	3,000	Chery
C	Male	53	14	80,000	Volkswagen
D	Female	22	1	8,000	Toyota

Measurements

The validity of the simulation system was evaluated by objective performance and by subjective assessment. The vehicle path was recorded as objective measurement of the driving performance. In a realistic simulation system, the driver makes the minimum adjustment in lane-change. We counted the times the driver adjusts the vehicle moving direction, which is represented by the local apex. For subjective measurement, participants' evaluations were collected by post-test questionnaires. After driving under feedback and non-feedback settings, the driver described their feelings through five seven-point scale questions about the similarity and fidelity of the steering system from a subjective perspective (see Table 3).

2.3.2 Results

Objective measurement

All the participants completed the double lane-change test within three repli-cations under each speed limit. After lane change, the driver would line up with the new lane by adjusting the vehicle from side to side. In this study, the number of adjustments after each lane-change would be an indicator of the vehicle stability. In Table 2 listed the mean vehicle adjustments among all participants under each speed level. When torque feedback exists, the number of adjustments is smaller than that when torque feedback is absent.

Table 2: Mean vehicle adjustments

Speed	15	20	25	30	40
Non-feedback (times, $\bar{x} \pm s$)	2.75±2.38	3.50±0.50	3.25±0.83	2.00±1.41	1.75±0.83
Feedback (times, $\bar{x} \pm s$)	0.75±1.30	1.75±1.79	0.75±0.83	1.25±1.30	1.00±0.71

Subjective assessment

The result is shown in Table 3. For each question, higher score reflects the system has higher fidelity in a certain aspect. The result of the after-test questionnaire shows an overall tendency that feedback steering has a higher fidelity than the non-feedback one, which is consistent to the assumption.

Table 3: Mean ratings for seven-point scale post-test questionnaire

Question	Non-feedback	Feedback
1. Rate the similarity of the steering feeling between the simulator and a real car.	2.5±1.5	5.5±0.9
2. How easy is it to handle the virtual vehicle at will.	3.8±1.5	6.0±0.0
3. Rate the accuracy of vehicle handling.	3.0±1.0	5.3±0.4
4. How easy is it to align with the straight after changing lanes.	2.0±1.0	4.0±1.6
5. Overall similarity between the simulator and a real car.	2.5±0.5	5.3±0.8

3 DISCUSSION

From the result we found that after lane-change, the driver may have difficulties aligning with the straight (see Figure 6). It is more likely to happen when torque feedback was absent. Under feedback setting, the number of cases when driver need to make adjustment decreased from 16 to 7. This phenomenon is in accordance with the observation made by Ronald R. Mourant (2002). The lack of steering wheel torque feedback is the major reason for impaired vehicle control. The torque feedback system effectively established an important feedback channel and improved the drivers' control over the vehicle.

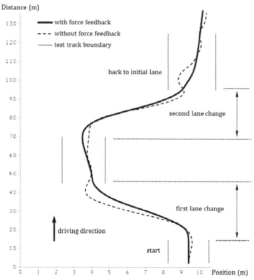

Figure 6: Path of the vehicle in double lane-change test, participant 03, speed of 15km/h

In this study, the feedback model includes speed and angular position as variables. In reality, the feedback on a steering wheel is more complex. In modern passenger cars and commercial cars, the steering wheel is the connection between a driver and a car. The feedback that a driver feels from the steering wheel includes vibration and torque resistance, the former of which reflects the road surface condition and the latter contains more information than normally understood. The driver comprehends the information from steering wheel and other channels and instantly makes driving maneuver decisions. As modern cars are usually equipped with power steering unit, the torque feedback characteristic becomes more complicated. In order to find the relationship between the feedback torque and its determining factors, we could either build a steering system model and perform calculation or use key values to build up a regression model. The former approach would produce a precise function but takes much effort. The latter one allows us to quickly establish an approximate relationship. As the objective of this study is to provide an effective torque feedback to the users of the simulator, the regression method is adopted.

In daily driving, a driver could feel feedback of various characteristics when the vehicle is in different conditions. Driving tasks could be generalized as two types: low speed turning and high speed lane-change. The former usually happens in parking lots and when drivers make turns at crossings. Under low speed condition, the driver completes the turning maneuver by setting an angular position of the steering wheel. Thus, the feedback torque should be small so that smaller force is required. On the contrary, when the vehicle is travelling at high speed, any slight changes in steering angle would change the motion of the vehicle significantly. In such conditions, the torque feedback should be large so that the steering wheel is more stable. This variation in steering torque feedback discussed here is consistent with the design of power steering system in modern passenger cars.

As for the actual effect of the steering feedback, the subjective ratings indicate that this system enhanced the driver experience. However, for the second question in the questionnaire, the easiness to handle the virtual vehicle, one participant gave the same rating on two settings. A short interview revealed that her daily ride was a Toyota, which is easier to steer than European or American cars. Therefore, she was more familiar with lighter steering wheel and be less sensitive about the feedback force. In reality, steering characteristic vary significantly from vehicle to vehicle. Consequently, participants for future study are expected to have various preferences for steering feeling. In general, male drivers and those who drive European or American brand cars prefer to have greater steering feedback than female drivers and those who drive Asian brand cars. Furthermore, in some cases, the driver would be distracted from the unexpected feedback. To provide better simulation of real world driving, the algorithm of the steering feedback should be redesigned to fit more driving conditions and be flexible to satisfy different preferences.

4   CONCLUSION AND FUTURE PROSPECT

The results of the tests are in general agreement with the assumption that the addition of torque feedback had a positive effect on driver experience especially in overtaking maneuvers. The major benefit of the torque feedback is that it provides another feedback channel for the driver to judge the position of the steering wheel, thus making decisions in intense driving conditions. Without torque feedback, the driver would need to make such judgment through vision, which is very distracting. The torque feedback system eliminates the undesired distraction of the simulator operation and enhances the immersion of the system.

For further improvement in simulator fidelity, there are still many approaches to be done. On the steering wheel, more complex and flexible feedback algorithm is expected. Taking more variables into the model will be a proper goal. Enriching the sounding will also enhance the overall immersion for the driver. The dashboard and other in-vehicle control devices are not as important but still effective way to keep driver in the virtual reality.

ACKNOWLEDGEMENT

We appreciate the support for this study from the National Natural Science Foundation of China (NSFC, grant number: 71001092).

REFERENCES

Arup Dutta, Donald L. Fisher, D. A. N., 2004. Use of a driving simulator to evaluate and optimize factors affecting understandability of variable message signs. Transportation Research Part F 7, 209227.

Blana, E., 1996. A survey of driving simulators around the world. Working paper, Institute of Transport Studies, University of Leeds.

Daniel V. McGehee, John D. Lee, M. R. J. D. K. B., 2004. Quantitative analysis of steering adaptation on a high performance fixed-base driving simulator. Transportation Research Part F 7, 181–196.

David Uzzell, R. M., 2005. Simulating traffic engineering solutions to predict changes in driving behavior. Transportation Research Part F 8, 311–329.

Gordon, D. A., 1966. Experimental isolation of drivers visual input. Human Factors: The Journal of the Human Factors and Ergonomics Society vol. 8 no. 2, 129–138.

Harms, L., 1994. Driving performance on a real road and in a driving simulator: results of a validation study. Vision in Vehicles Vol. V.

ISO, 1999. ISO 3888-1:1999 passenger cars -test track for a severe lane-change maneuver - part1: Double lane-change. (accessed March 11, 2011). URL http://www.iso.org/iso/iso catalogue/catalogue tc/catalogue detail.htm?csnumber=31317

M. Hoedemaeker, K. B., 1998. Behavioral adaptation to driving with an adaptive cruise control (ACC). Transportation Research Part F1, 95–106.

Marc Wittmann, Mikls Kiss, P. G. A. S. M. F. E. P. H. K., 2006. Effects of display position of a visual in-vehicle task on simulated driving. Applied Ergonomics 37, 187–199.

Rockwell, T. H., 1972. Skills, judgment and information acquisition in driving. Human Factors in Highway Traffic Safety Research, 133–164.

Ronald R. Mourant, P. S., 2002. Evaluation of force feedback steering in a fixed based driving simulator. In: Proceedings of the Human Factors and Ergonomics Society 46th Annual Meeting. pp. 2202–2205.

Stuart T. Godley, Thomas J. Triggs, B. N. F., 2002. Driving simulator validation for speed research. Accident Analysis and Prevention 34, 589–600.

Sung, Eun-Jung; Min, B.-C. K. S.-C. K. C.-J., 2005. Effects of oxygen concentrations on driver fatigue during simulated driving. Applied Ergonomics 36, 25–31.

Törnros, J., 1998. Driving behavior in a real and a simulated road tunnel -a validation study. Accident Analysis and Prevention Vol. 30, No.4, 497503.

Wang, Y., 2009. In-vehicle secondary task study based on human-machine interactive simulation. Ph.D. thesis, Department of Industrial Engineering, Tsinghua University.

Wang, Y., 2010. Studies of hazard perception training and intervention for novice drivers. Ph.D. thesis, Department of Industrial Engineering, Tsinghua University.

Zhang, Y., 2009.Research and development of steering wheel return simulation system. Master's thesis, School of Automation, Wuhan University of Technology.

Zhao Xue-Ping, Li Xin, C. J. M. J.-L., 2009. Parametric design and application of steering characteristic curve in control for electric power steering. Mechatronics 19, 905–911.

CHAPTER 3

Design of a Touch-based User Interface for Naval Command & Control and Comparison with a Current Onboard System in a Scenario-based Usability Test

Jessica Schwarz, Oliver Witt

Fraunhofer Institute for Communication, Information Processing and Ergonomics (FKIE)
Wachtberg, Germany
jessica.schwarz@fkie.fraunhofer.de

ABSTRACT

Operators in the command center of naval ships are assisted in their assessment of the tactical situation around the ship and their subsequent actions by Command & Control Systems (C2-Systems). As the development of present C2-Systems has been focused on increased automation of working processes rather than on the ergonomic design of the user interface, Fraunhofer FKIE has been commissioned by the German Navy with the design of an ergonomically improved user interface for naval C2-systems. A touch-based design concept was developed according to ergonomic guidelines and using ecological interface design as a theoretical framework. The new touch-based design concept was then evaluated in an experimental setting in comparison with the interface of a current onboard system. The sample consisted of 12 operators of the German Navy who completed scenario-based tests related to the tasks of tactical picture compilation and engagement using both user interfaces. The measured data included accuracy, situation awareness, workload and subjective ratings. The results revealed that the touch-based user interface can reduce the workload and enhance the situation awareness of operators. Moreover the touch-based interface was rated significantly more positive by the operators in comparison to the current user interface. The results suggest the touch-based interface can be regarded as a promising alternative to conventional interfaces.

Keywords: Touch-based user interfaces, Command & Control Systems, Usability-test, Situation awareness

1   INTRODUCTION

The domain of Naval Warfare contains many time and safety critical tasks. These tasks include rapidly identifying objects in the vicinity of the ship (e.g. as friendly or hostile) and planning the engagement response for objects that show suspect or hostile intent. Operators in the Command Center of naval ships are assisted in these tasks by Command & Control Systems (C2-Systems).

As an effect of a continuous technology push many steps in the identification and engagement process run semi-automatically. For example, based on an identification algorithm the system provides the operator proposals on the identity (ID) of tracks that have not yet been identified. These automation strategies can relieve the operator on the one hand, but on the other hand they might also lead to a loss of situation awareness (Endsley & Kiris, 1995). It is critical though that the operator stays in the loop as he always remains responsible for the decisions and actions performed in interaction with the system. Especially in ambiguous situations when the algorithm of the system is not able to produce a reliable result it is the responsibility of the operator to correctly interpret the situation based on his or her existing knowledge and the information provided by the system. Hence, there is a pressing need to produce an ergonomic design of the C2 user interface that supports the operator in his decision making, provides a sufficient level of situation awareness, and allows for fast and accurate interaction.

In a study commissioned by the German Navy, Fraunhofer FKIE addressed these issues by designing and evaluating a touch-based user interface for naval C2 systems. The study focused on the tasks of "tactical picture compilation" and "engagement" in the domain of Anti-Air-Warfare (AAW) as these tasks impose high cognitive demands on the operator, especially for combat situations.

2   DESIGN OF A TOUCH-BASED USER INTERFACE

According to the ISO-standards for ergonomics, DIN EN ISO 9241-210 and DIN EN ISO 9241-110, a touch-based user interface for naval C2 systems was developed. The development process included the following main activities:

1. Understanding and specifying the context of use
2. Specifying the user and organisational requirements
3. Producing design solutions
4. Evaluating designs against requirements.

Ecological Interface Design (Burns & Hajdukiewicz, 2004) was used as a theoretical framework for the specification of the context of use and the specification of user and organizational requirements. The analysis included the modeling of the identification and engagement process by the use of a decision ladder (described in Witt, Ley & Schwarz, 2010). The process specified which subtasks are involved in the decision making process and generated ideas concerning how the operator could be properly assisted in these tasks by the user interface.

This paper focuses on the description of the final design solution and the evaluation of this design concept in comparison with the user interface of a current onboard system. As Figure 1 shows the user interface consists of a 24" multitouch monitor which is used as the sole display and input device. As all inputs are completed by touch the user interface is equipped with touch-specific layouts and functionalities.

Figure 1. Touch-based interface consisting of a 24" multitouch monitor.

An important component of the system is constituted by the Tactical Display Area (TDA) – a radar screen that shows the objects (also called tracks) which have been detected by sensors in the area surrounding the ship. On current onboard systems the TDA is typically displayed on one screen, while additional information about the tracks' properties (e.g. velocity, altitude, bearing) or own ship information (e.g. gun status) is displayed in different windows on a separate screen. In order to enhance situation awareness, we followed the approach to integrate most of the task-relevant information into the TDA. Thus, the operator can keep his focus on the TDA and does not have to switch between different screens and windows.

Figure 2 shows a screenshot of the TDA. Directly beside the chart area there are bars for altitude (on the left) and velocity (on the right) in which all tracks are displayed according to their current altitude and velocity. These bars should make it easier for the operator to recognize if a track suddenly looses height or increases its speed which can be indicators of hostile intent.

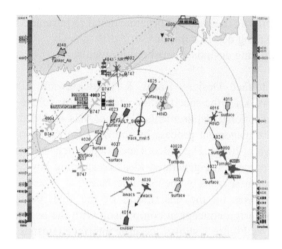

Figure 2. Tactical Display Area (TDA) of the touch-based user interface.

The interface supports the use of well-known touch specific gestures in order to provide an intuitive interaction with the system. For example, the zooming in and out of the TDA is accomplished by the spreading apart or pinching together of two fingers on the screen. A short finger tap on a track symbol causes a selection of the track and additional information on the track is displayed right beside the symbol (see Figure 3a). A long finger press induces different actions depending on the chosen work domain. For the task of tactical picture compilation a long finger press opens a radial menu where the operator can, for example, change the ID of the track (see Figure 3b). In contrast, for the engagement task a long finger press opens a panel for weapon availability which is used to select weapons for an engagement (see Figure 3c).

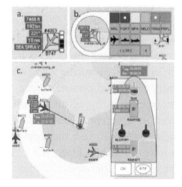

Figure 3. a. selected track, b. radial menu for tactical picture compilation, c. display showing weapon availability for engagement

3   EVALUATION STUDY

After the concept development was completed, an experimental evaluation was conducted in order to examine if the touch-based design concept really constitutes an improvement over the user interfaces of current C2-Systems. Thanks to the German Navy, the opportunity was provided to take the most recent C2-System of a navy ship as reference system and to directly compare these two interfaces in scenario-based usability tests.

3.1   Method

3.1.1   Participants

The participant group consisted of 12 operators of the German Navy (11 men and 1 woman). Half of them had good knowledge of the C2-System used in the evaluation. The other half was unfamiliar with the system. The age of the operators ranged from 22 to 48, $M = 26$.

3.1.2   Task description

The evaluation consisted of a scenario-based test with respect to the tasks of "tactical picture compilation" and "engagement" and a subjective assessment of the user interfaces. For the task of tactical picture compilation the test was divided into two phases which differed in the degrees of freedom the operators were given for completing their tasks.

Tactical picture compilation Phase 1

Phase 1 is characterised by a rather high standardisation of task execution which should ensure comparable outcomes and high internal validity. The task was to identify a certain track on the TDA every 30 seconds according to previously announced ID-Criteria. As an example the ID neutral was assigned to air tracks that were travelling on an air route with an altitude above 250hft and a velocity below 400kts. Participants' situation awareness was measured in this phase by certain questions the operator had to answer during the scenario. According to the technique SPAM (Durso & Dattel, 2004) the time required to answer the questions was taken as a measure of situation awareness. These questions included: "How many air tracks are heading towards the own ship?", "What is the distance of the track who is nearest to the own ship?", "Select the hostile track whose velocity is highest".

Tactical picture compilation Phase 2

In phase 2 task execution corresponded more to realistic working conditions and was characterised by a low standardization of task execution which should lead to a higher external validity. The operator was commissioned to observe the tracks on the TDA and to indicate when tracks show changes in their behaviour that call for a change of their ID or if new tracks appear. The events that had to be recognized are listed in the following:

1. Two unknown tracks appear (ID suspect)
2. Suspect track is heading towards the own ship (change ID to hostile)
3. One unknown track appears (ID friendly)
4. Two unknown tracks appear (ID suspect)
5. Track leaves airroute and is heading towards the own ship (change ID to hostile)
6. Hostile track is firing a missile
7. Suspect track is heading towards the own ship (change ID to hostile)
8. Hostile track is firing a missile
9. Two hostile tracks are firing missiles
10. Neutral track is firing a missile

Engagement

For the task of engagement, only a test with high standardisation of task execution was conducted due to technical restrictions. The test was based on an artificial scenario where tracks were heading towards the own ship at two-minute intervals. For each approaching track, operators had to initiate an engagement in a predefined way and answer questions about the current engagement process e.g. "with which weapon can the target be engaged at the earliest possible time?".

3.1.3 Variables and experimental design

The evaluation was conducted as a within-subject design with N=12 participants. The experimental design consisted of three independent variables:

- User interface (touch-based interface vs. interface of C2-System)
- Task (tactical picture compilation vs. engagement)
- Experience (operators with high and low experience with respect to the C2-System)

All participants completed all task scenarios twice, once with each user interface. To avoid sequence and learning effects the scenarios differed between the two user interfaces with respect to track positions and the direction of engagement. Additionaly, the order in which the user interfaces were tested was varied. For each task and each user interface the following dependent variables were measured:

- Task performance (accuracy and response time)
- Situation Awareness
- Subjective Workload (assessed by NASA-TLX, Hart & Staveland, 1988)
- Subjective assessment of the interfaces with questionnaires

3.1.4 Procedure

The participants completed all tasks and assessments first using one interface and then the procedure was repeated using the other interface. For each user interface the evaluation started with the task of tactical picture compilation. After an introduction and a training phase the test for tactical picture compilation Phase 1 was conducted. Subsequently subjects had to rate their workload for that task. Tactical picture compilation Phase 2 started without a training phase and was followed again by a workload rating. After completing Phase 2 the user interface was assessed using questionnaires with respect to the task of tactical picture compilation. After an introduction and a training phase the test for the engagement task was conducted followed by a task-specific workload rating and interface assessment questionnaire. Overall the assessment of both interfaces took one day for each participant.

3.2   Results

3.2.1   Task performance regarding tactical picture compilation

As a measure of accuracy the error rate was used. That is the percentage of times a track was assigned with either a wrong or no ID within a 30 seconds timeslot. For the touch-based user interface the error rate was a bit higher (17%) than for the user interface of the current C2-System (14%) but the difference was not significant.

Regarding the response time for the identification the average identification time for the touch-based user interface was $M = 14,4$ with $SD = 3,4$ and for the interface of the current system $M = 16,2$ with $SD = 3,0$. So it took a bit more time to set the ID in the current system, even though the difference was not significant.

3.2.2.   Task performance regarding engagement

For the task of engagement, all tasks could be accomplished without errors with both interfaces except one task where the error rate was 17% for the user interface of the current C2-system. The task was to select the track with the least time to impact. With the touch-based user interface no errors occurred.

With respect to response time, the user interfaces only differed significantly on two of the five tasks. The tasks were to select the track with the least time of impact and to initiate an engagement with a specific weapon. For these tasks significantly more time was required with the current C2-System than with the touch-based user interface, $t(11) = 6,09$, $p<.001$ and $t(11) = 3,71$, $p<.01$, respectively.

3.2.3   Situation Awareness

In phase 1 of the task of tactical picture compilation ,situation awareness was measured by questions on the current situation. The time to answer these questions did not produce significant differences but a significant difference was found regarding the error rate, $t(11) = 3,53$; $p<.01$. The error rate was 13% for the touch-based user interface and 22% for the current C2-System.

An indication of the situation awareness can also be inferred from the events that were recognised in the second phase of tactical picture compilation. The user interfaces did not differ significantly in the total percentage of recognised events (92% for the touch-based interface, 84% for the user interface of the C2-System) but a separate examination for each event shows that almost every event was recognised by more subjects on the touch-based interface. As shown in Figure 4, event 5 was recognised by all subjects on the touch-based interface but by only five subjects on the interface of the C2-System.

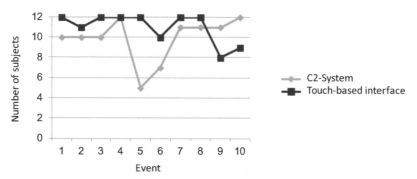

Figure 4. Number of subjects who recognised the respective event.

3.2.4   Workload

For the analysis of workload a repeated measures ANOVA was conducted to evaluate the impact of the user interface and task (tactical picture compilation phase 1, phase 2 and engagement) on perceived workload. The analysis revealed significant main effects for both factors ($F(1,11) = 23.37$, $p<.01$ for user interface; $F(2,22) = 27.11$, $p <.001$ for task) but no significant interaction. As figure 5 shows, the workload was rated higher for every task on the user interface of the C2-System. For both interfaces Phase 2 of the tactical picture compilation task led to the highest workload level.

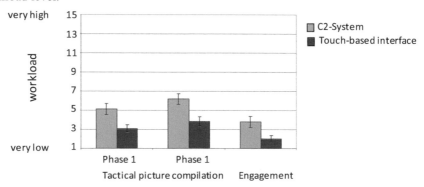

Figure 5. Mean subjective workload levels.

Subjective Assessment

In the subjective assessment of the user interfaces participants had to rate the interfaces according to some criteria which were essential for the tasks of tactical picture compilation and engagement. At the end subjects were also asked to rate general aspects of the interfaces like the input technique and the visualisation of information. Figure 6 shows the mean values of the ratings summarised for the questions related to tactical picture compilation, engagement and the general rating. In all three categories the touch-based user interface was rated significantly more positive than the interface of the C2-System.

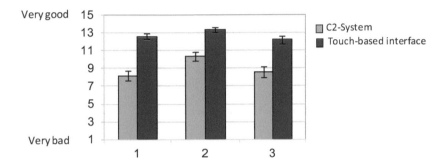

Figure 6. Mean values of the subjective rating (1=tactical picture compilation, 2=engagement, 3=general rating)

3.3   Discussion

The results of this evaluation reveal that the newly developed touch-based interface has some advantages over the user interface of the current C2-System. Even though the results did not show consistent, significant differences favouring either interface with respect to task performance, it was shown that the touch-based interface lead to a lower perceived workload level and was rated more positively by the operators. The touch-based interface also seems to enhance situation awareness as can be seen, particularly in the case of event 5 in phase 2 of the task tactical picture compilation. Event 5 consisted of an air track that left the airroute and was heading towards the own ship. The results suggest that this event is better detected on the TDA of the touch-based interface.

For the interpretation of the results it has to be taken into account that some of the operators already knew the current C2-System but had to get to know the touch-based interface. However, when investigating if the experience level of the operators with the current C2 system has an effect on the outcome no significant results were found. This may be explained by the fact that other C2-Systems are similar in layout and functionality and operators could profit from their general knowledge of C2-systems.

4   CONCLUSION

The results indicate that a touch-based application could represent a promising alternative to existing interfaces for naval Command & Control systems. However, further tests are required to assure that the touch-based interface also enables a proper and fast interaction in underway conditions, as well as under stress and in prolonged interaction periods.

ACKNOWLEDGMENTS

The authors would like to acknowledge the German Navy for the support during all phases of the study.

REFERENCES

Burns, C. M. & Hajdukiewicz, J. R. (2004). *Ecological Interface Design*. Boca Raton, FL: CRC Press.

DIN EN ISO 9241-210: Ergonomics of human-system interaction - Part 210: Human-centred design for interactive systems. *International Organization for Standardization*, 40(4), 759-768.

DIN EN ISO 9241-110: Anforderungen an die Gebrauchstauglichkeit – Leitsätze (ISO 9241-11). Beuth: Berlin 1998

Durso, F.T. & Dattel, A.R. (2004). SPAM: The real-time assessment of SA. In S. Banbury & S. Tremblay (Eds.), A cognitive approach to situation awareness: Theory and application Aldershot, UK: Ashgate pp. 137-154 2010

Endsley, M. R., & Kiris, E. O. (1995). The out-of-the-loop performance problem and level of control in automation. Human Factors, 37, 381–39

Hart, S.G. & Staveland, L. E. (1988). Development of NASA-TLX (Task Load Index): Results of empirical and theoretical research. In *Human Mental Workload* (pp. 139-183).

Witt, O., Ley, D. & Schwarz, J. (2010). Entscheidungsleitern als Gestaltungsgrundlage für Benutzungsoberflächen eines Marine-Einsatzsystems. In VDI/VDE-Gesellschaft für Messung Automatisierungstechnik (Hrsg.), USEWARE 2010 (VDI-Bericht, S. 315-324). Düsseldorf: VDI-Verlag.

CHAPTER 4

Study of the Relation "Action and Command" to the Interior of an Automobile

Caio Márcio Almeida e Silva 1, Maria Lúcia Leite Ribeiro Okimoto 2 and Luis Carlos Paschoarelli 3 e

[1] Federal University of Paraná |
Instituto de Tecnologia para o Desenvolvimento - Lactec
Curitiba, Brasil
caiomarcio1001@yahoo.com.br

[2] Federal University of Paraná
Curitiba, Brasil
lucia.demec@ufpr.br

[3] UNESP - Univ Estadual Paulista
Bauru, Brasil
paschoarelli@faac.unesp.br

ABSTRACT

This paper had as objective investigates the stereotyped expectations involved in the relationship between the actions and commands, in the interior of an automobile. An experiment was accomplished with two groups of participants: habitual drivers of vehicles, and no drivers of vehicles. The results confirm the hypothesis of confirmation of the stereotyped expectations, related to the respective actions.

Keywords: design, intuitive use, commands, stereotyped expectations, automobile

1. INTRODUCTION

The studies about the interaction between user and product are fundamental for the development and application of the industrial design. Among these studies, they stand out those that try understand how it happens the intuitive use of a product. In agreement with Sudjic (2010), "the design is used to mold the perceptions of as the objects they should be understood. Per times, it is treated of a subject of direct communication: to work a machine it is necessary to understand intuitively what she is, and how to do her to execute the wanted operation." It is in that perspective that the intuitive use appears.

The intersection of the intuition in the design has been worked starting from different approaches. Some more theoretical ones (Norman, 2010; and Bürdek, 2006); associating projects of products to the intuition (Rutter, Becka and Jenkins, 1997; and Frank and Cushcieri, 1997) and starting from usability tests (Blackler, Popovic and Mahar, 2003). there are still those approaches on the use intuitito, starting from: No-intentional design [Bürdek (2006), Brandes (2000), Suri and Ideo (2005)]; Natural Interaction [Norman (2010)]; Identification starting from archetypes [Sudjic (2010)]; the Intuitive Use [Blackler, Popovic and Mahar (2003)]; Expectativas Extereotipadas [Smith (1981); Iida (2005); Kroemer and Grandjean (2005)] and Product Experience based the intuitive use [Silva(2012)].

The intuitive use can be made present in the interaction with technological interfaces, so much for the actuation of products, as in their current unfoldings of the use. Any actuation happens starting from controls, then characterized by own sub-systems, that you/they are to feed the system "man-machine". They are worked, he/she saw of rule, with the movement of the hands and/or of the fingers (IIDA, 2005). Para this author, they can be: steering wheels, cranks, buttons, keyboards, mouse, joysticks, remote controls, among others. And, in this sense, they can be contained in agreement with his/her typology.

In the automotive context, the application of the commands is considered of extreme importance, contemplating in the usability of the interface and in the user's safety. In if treating of safety, Iida (2005) it alerts for the prevention of accidents with controls starting from cares that they should be considered from the project, with prominence for: location, orientation, I lower, covering type, canalization, windowsill, resistance, blockade, lights and codes. However, only the development of the control cannot be shown enough, having the need of some unfolding's.

One of those unfoldings is his/her discrimination. Iida (2005) it presents some discriminations of controls in the intention of differentiating them, when the same ones one find inserted in a same context. For so much, some should be considered varied: he/she forms, size, colors, texture, operational way, location and signs. Important other unfolding refers to the combination of codes. Iida (2005) it points that different ways exist of codifying the display cases and that these can be combined in some commands. The author still highlights that, in critical situations, those codes can be used in a redundant way, as for instance, associating colors and forms.

There is still the unfolding of the relationship between controls and display cases. Kroemer and Grandjean (2005) they present five aspects to be considered: relative speeds of movement, stereotyped expectations, ethnic differences, controls and corresponding display cases, and control panels.

The stereotyped expectations present special importance in the studies on the intuitive use, once the demands are wide. Like this, it can be considered stereotype as being it "... reflex conditioned that if it turns subconscious and automatic", in other words, a "... behavior without originality and of adaptation to the present situation, and characterized by the automatic repetition of a previous model, anonym or impersonal" (KROEMER AND GRANDJEAN, 2005).

In agreement with Iida (2005), some of those stereotypes are natural or innate, for they be marked by the own organism. In that sense, different expectations can happen when the users come across interfaces of products in that there are movement stereotypes, of operation, of result, among other, characterizing, like this, the stereotyped expectations.

In the mark of the intuitive use, the stereotyped expectations can reveal indications of the possibility of a certain effect in certain situations, on the part of the population (IIDA, 2005). That situation happens when the experience recorded the corresponding patterns in the brain and theses are shown capable of they be applied, again, in other experiences (KROEMER AND GRANDJEAN, 2005).

That association between stereotypes of the controls and movements was worked in an experimental way by Smith (1981), which evaluated the movements and knob controls, key, the, faucets (taps) and of the keyboard for calculator and telephone. The evaluation was made starting from tests with three groups of participants: 92 engineers, 80 women and 55 specialists in ergonomics. The result of the tests demonstrates that in some cases, there was significant difference among the participants' of the experiment three groups. Starting from that, the author suggests that that difference can be influenced so much by the experience, as for the training of the equipment; offered for one of the groups. Another identified result was that the people can be trained to do incompatible movements. However, the forecast of time for that training is larger. Finally, the author appears that in emergency situations and/or panic, a tendency exists of doing the movement compatible, same having done training of incompatible movements

In general, it is observed that the stereotyped expectations can be influenced by the users' perception and of no-users, when they need to relate actions to commands of a product (or interface) specific.

In this sense, the objective of the present study to analyze the applied intuitive use inside an automobile. Particularly, it was evaluated the stereotyped expectations involved in the relationship between the actions and commands, among user (drivers) and no users of automobiles, verifying like this, the influence of popular stereotypes.

2. MATERIALS AND METHODS

They participated in the study 18 subjects, being eleven experienced drivers and seven individuals that never drove an automobile. All the participants were voluntary and they signed a Term of Free and Illustrious Consent, assisting the Code of Deontology of the Certified Ergonomist - Norma ERG BR 1002 (ABERGO, 2003). Analysis protocols of controls and ballpoint pen were used.

Most of the tests happened in the Laboratory of Ergonomics and Usability of the Federal University of Paraná. The others were accomplished in the participants' residences, because that scenery change would not influence in the answers.

The procedure for the accomplishment of the test was systematized in the following way: initially, it was explained the objectives of the study and the interest in the participation. After they accept and they sign TCLE, the subject was questioned was conductive or not of vehicles. In the sequence, he/she felt beginning to the "block" of statements that you/they were read by the mediator of the experiment. Those statements correspond to the routine actions that a driver has to execute inside an automobile. The statements began in the same way and, soon afterwards, the actions were specified, as demonstrated to proceed: - In his/her opinion, which the most appropriate command for the accomplishment of the task of: to signal a curve for the right or left; to work him/it "blinks - it alerts"; to work the horn; to control the intensity of exit of air of the air conditioning; to control the volume of the sound apparel; to adjust the direction of the rear-view mirror manually; to adjust the direction of the rear-view mirror automatically; to work the brake using hand; to work the brake using the foot; to accelerate the car using the foot; to work the clutch; to control (to arise and to go down) manually the glass of the window; to work the electric glass of the window; to open the door of the car being inside the same. When some participant didn't know about any the meaning term, the mediator read the extracted meaning of dictionary of the Portuguese Language.

Starting from the statement, the participants had to identify in the protocol (Figure 01), which of the five verbs, the action could be appropriately related. After the selection of the verb, it was asked the subject that it selected a control in the corresponding line. As example, to choose the button of alternate control, the participant first should make the association with the verb to press, for then to choose the control of number 14 among the four only possibilities (13, 14, 15 and 16).

As she went enunciating the actions, the participants just answered indicating the number in the table. Like this, the researcher wrote the number of the control in the respective answer. The same ones took on average about 20 minutes to answer to the questionnaire. As research technique, the interview was used. As metric, it was used to solemnity-report "evaluation of specific attributes of the product."

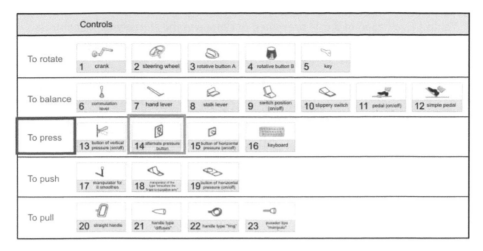

Figure 1 It fixes of controls indicating the relationship between the verb and a control in a same line. Source: The authors (2012).

3. RESULTS

3.1 Commands and automobiles

Before tabulating the results, it was accomplished an evaluation of the 14 presented actions and the stereotypes of controls for accomplishment of each one of them. That evaluation was accomplished by one of the authors of the present article and with a sample of two popular automobiles of each one of the four main marks of automobiles marketed in Brazil (Fiat, Ford, GM and Volkswagen). We Tried to identify which the command type used in the automobiles for then, to identify the stereotype. The result of that evaluation was synthesized in the Table 1.

3.2 Commands and participants

The data were organized in four tables. Two of them correspond to the participants' relationships with the types of controls (verbs); one of the participants that drive and another of the participants that you/they don't drive. The Table 2 correspond to the relationships between the actions and the specific controls; an also refers to the participants that drive and other to the participants that don't drive.

Starting from the most frequent answers (verbs or commands with a larger frequency than 50%), it was analyzed the relationship that these possess with the popular stereotypes of the presented actions. Soon afterwards, a new table was built indicating in which actions existed, in fact, that influence of the stereotypes in the participants' answers. The confirmation was made starting from a larger frequency.

Table 1 Evaluation of the 14 presented actions and the stereotypes of controls.

N	Action	Control stereotype
1	to signal a curve for the right or left	Lever
2	to work the "pisca - alerta"	Button of horizontal pressure (on/off)
3	to work the horn	Button of horizontal pressure
4	to control the intensity of exit of air of the air conditioning	Rotative button A
5	to control the volume of the sound apparel	------------
6	to adjust the direction of the rear-view mirror manually	Commutation lever
7	to adjust the direction of the rear-view mirror automatically	Slippery switch (4 directions)
8	to work the brake using hand	Hand lever
9	to work the brake using the foot	Simple pedal
10	to accelerate the car using the foot	Simple pedal
11	to work the clutch	Simple pedal
12	to control (to arise and to go down) manually the glass of the window	Crank
13	to work the electric glass of the window	Slippery switch
14	to open the door of the car being inside the same	Straight handle

Table 2 Confirmation of the it influences of the popular stereotypes in the relationship "action and command" for an automobile.

N	Action	They drive (Specific commands)	They don't drive (Specific commands)	They drive (Command typology)	They don't drive (Command typology)
1	to signal a curve for the right or left	----	----	----	----
2	to work the "pisca - alerta"	----	----	Confirmed	Confirmed
3	to work the horn	Confirmed	----	Confirmed	Confirmed
4	to control the intensity of exit of air of the air conditioning	Confirmed	Confirmed	Confirmed	Confirmed
5	to control the volume of the sound apparel	----	----	----	----
6	to adjust the direction of the rear-view mirror manually	Confirmed	----	Confirmed	Confirmed
7	to adjust the direction of the rear-view mirror automatically	----	----	----	----
8	to work the brake using hand	Confirmed	Confirmed	Confirmed	Confirmed
9	to work the brake using the foot	Confirmed	----	Confirmed	Confirmed
10	to accelerate the car using the foot	Confirmed	Confirmed	Confirmed	Confirmed
11	to work the clutch	Confirmed	----	Confirmed	Confirmed
12	to control (to arise and to go down) manually the glass of the window	Confirmed	Confirmed	Confirmed	Confirmed
13	to work the electric glass of the window	----	----	----	----
14	to open the door of the car being inside the same	Confirmed	----	Confirmed	Confirmed

Starting from the results synthesized in the table 3, we identified the confirmation of the it influences of the stereotypes in the perception of the different publics' controls. When we considered the whole sample of the research (participants that drive and they don't drive), we have four actions with stereotyped expectations. They are them: to control the intensity of exit of air of the air conditioning, to work the brake using hand, to accelerate the car using the foot, and to control (to arise and to go down) manually the glass of the window.

To the we just consider the participants that drive, we have eight actions to generate stereotyped expectations. They are them: to work the horn, to control the intensity of exit of air of the air conditioning, to adjust the direction of the rear-view mirror manually, to work the brake using hand, to work the brake using the foot, to accelerate the car using the foot, to work the clutch, to control (to arise and to go down) manually the glass of the window, and to open the door of the car being inside the same. Already if we isolate the participants' group that you/they don't drive, we have four actions to generate stereotyped expectations. They are them: to control the intensity of exit of air of the air conditioning, to work the brake using hand, to accelerate the car using the foot, and to control (to arise and to go down) manually the glass of the window.

4. DISCUSSION

In that work the intuitive use was approached starting from stereotyped expectations related to the present commands in the interface of the interior of an automobile. The stereotypes were not related to the display cases, as it was proposed by Kroemer and Grandjean (2005), but, related with the mental models that the participants possessed of each action, as well as, with the control type.

Stereotypes depend on previous models. When we treated of previous models for the participants of that experiment, we should consider that the group that doesn't drive doesn't possess the same experience related to the interior of an automobile that the group of the ones that drives. However, the one that don't drive possess an experience that cannot be disrespected (it is while ride, be observing in a film, in advertising materials, others).

Starting from the results, he/she identified that the previous models influenced in the answers. When if they took care of people that already drove, 61,5% of the stereotyped expectations related to the controls were confirmed. In the people's group that you/they didn't drive, that percentage reduced for 30,7%. it is Worth to stand out that all of the public's stereotyped expectations that it doesn't drive if they made presents in the public's expectations that it drives.

Of the four stereotypes confirmed by the participants' group that you/they don't drive (to Control the intensity of exit of air of the air conditioning, to work the brake using hand, to accelerate the car using the foot, and to control (to arise and to go down) manually the glass of the window), three are visible actions for people that are not as drivers. They are them: to control the intensity of exit of air of the air conditioning, to work the brake using hand, and to control (to arise and to go down) manually the glass of the window. The only that cannot be shown visible it is the actuation of the brake with pedal. However, the mental model of actuation of the brake using the foot is made present culturally.

Another aspect to be considered is that for the participants' group that you/they drive, all of the indications of specific commands almost coincided with the indication of the typology of the commands (verbs). Already for the participants' group that you/they don't drive, per times, there was a difference between the specific commands and their typologies. That difference, it can also be a contribution in the development or I redraw of commands for the interior of automobiles. That contribution can be executed starting from the people's perception that you/they don't possess a consolidated mental model. Like this, they can contribute without larger restrictions, as well as, without larger functional restrictions and of previous experiences related to the automobile. They can yes, to bring the repertoire of other experiences.

Concerning the popular stereotypes, Iida (2005) it pointed that the stereotyped expectations could reveal indications of the expectation of a certain effect in certain situations, on the part of the population. That expectation was confirmed for the participants that had direct experience of interaction with the product (drivers) and partially confirmed with the participants that had experience with the product, but no while drivers.

In Smith's experiment (1981) it was identified that the stereotyped expectations of the movement of Knob, of the box's lock, of the movement of the lever and of the faucet of the sink they were confirmed. Already the expectation regarding the positioning of the numbers in the calculator, it was not confirmed. Like this, we identified that the operations related with the action of to "rotate" and to push, they were confirmed. The same didn't happen with the operations related to the action of "organizing". In the Experiment, four expectations were confirmed. Each one of them relative to a typology different from command: to rotate, to pull, to push and to press.

5. CONCLUSION

This paper had as objective investigates the stereotyped expectations involved in the relationship between the actions and commands. For so much, an experiment was accomplished with two groups of participants: the ones that drive, and the ones that don't drive. Finally, the confirmation of the stereotyped expectations was verified, related to the actions.

For the participants' group that you/they drove, it was verified that of the 14 actions, eight had their expectations stereotyped confirmed. Already for the participants' group that you/they don't drive, only four stereotyped expectations were confirmed. Of those four, three are applicable to the visible controls for people that are not as drivers.

It is recommended that future works are segmented the participants' group better. For instance, to select the participants that drive a certain category of cars (sporting, SUV, utilitarian, show off, etc). Other aspect that can be perfected in another study is the type of visual representation of the commands and the statement of the actions. Per times, the same ones were shown confused or ambiguous. He/she can she, then, represent the actions through videos, and the commands starting from three-dimensional models (or prototypes), and not of illustrations.

ACKNOWLEDGMENTS

This study was supported by CAPES - Coordination for the Improvement of Higher Level Personnel [Coordenação de Aperfeiçoamento de Pessoal de Nível Superior] and Fundação Araucária.

REFERENCES

Bullinger, Hans-Jörg., 1994. Ergonomie Produkt- und Arbeitsplatzgestaltung. Unter Mitarb. von Rolf Ilg und Martin Schmauder. – Stuttgart: Teubner.

Bürdek, Bernhard E., 2006. História, teoria e prática do design de produtos. São Paulo: Edgard Blücher, 2ª edição.

Chapanis, A. ,Lindenbaum, L. E., 1959. A reaction time study of control-display linkages. Human Factors, 1(4): 1-7.

Cybis, Walter de Abreu; Bertiol, Adriana Holtz; Faust, Richard., 2007. Ergonomia e usabilidade: conhecimentos, métodos e aplicações. São Paulo: Novatec Editora.

Frank, T. e Cushcieri, A., 1997. 'Prehensile atraumatic grasper with intuitive ergonomics' Surgical Endoscopy Vol 11 (1997) 1036–1039

Grandjean, E., 1983. Précis d'ergonomie. Paris: Les Éditions d'Organisation.

Iida, Itiro, 2005. Ergonomia: projeto e produção. 2ª edição. São Paulo: Editora Edgard Blücher.

Kroemer, K. H. E., Grandjean, E., 2005. Manual de Ergonomia: adaptando o trabalho ao homem. Trad. Lia Buarque de Macedo Guimarães. Porto Alegre: Editora Bookman.

Norman, Donald A., 2010. O design do futuro. Rio de janeiro: Rocco.

Nornam, Donald A., 2006. O Design do dia-a-dia. Rio de janeiro: Rocco.

Oliveira, F.I.; Rodriguez, S.T., 2006. Affordances: a relação entre agente e ambiente. In: Ciencias & Cognição. vol.09, p.120-130.

Paulheim, H., Döweling, S., Tso-Sutter, K.H.L., Probst, F. e Ziegert, T., 2009. Improving Usability of Integrated Emergency Response Systems: The SoKNOS Approach. In: Proceedings of GI Jahrestagung, 1435-1349.

Popovic, V.; Blackler, A; Mahar, D., 2003. The nature of intuitive use of products: an experimental approach. In: Design Studies 24. Grã-Bretanha: Elsevier, p. 491-509.

Rutter, B. G.; Becka, A. M. e Jenkins, D. A., 1997. 'User-centered approach to ergonomic seating: a case study' In: Design Management Journal Vol Spring 27–33.

Silva, C. M. A., Okimoto, M. L., 2011. Diretrizes para utilização dos aspectos para o uso intuitivo no desenvolvimento de interfaces de produtos tridimensionais. In: Anais do 5º Congresso Internacional de Design da Informação, Florianópolis.

Silva, C. M. A. e., 2012. Experiência com o produto a partir do uso intuitivo. Dissertação de Mestrado – Programa de Pós-graduação em Design, Universidade Federal do Paraná, Curitiba.

Smith, S. L., 1981. Exploring compatibility with words and pictures. In: Human Factors. V. 23, n. 3, p. 305-315.

Sudjic, Deyan., 2010. A linguagem das coisas. Tradução de Adalgisa Campos da Silva. Rio de Janeiro: Intrínseca.

Tullis, T., Albert, B., 2008. Measuring the user experience - Collecting, analyzing and presenting usability metrics. Burlington: Morgan Kaufmann.

CHAPTER 5

Empirical Investigation of Conflict and Interference within Haptic Controlled Human-Excavator Interface

Benjamin Osafo-Yeboah, Steven Jiang

North Carolina A&T State University
Greensboro, NC, USA
xjiang@ncat.edu

ABSTRACT

Traditional human-excavator interface rely mainly on visual modality and to some lesser extent on auditory modality to communicate between human and the excavator. However, increased operator workload has resulted in visual overload resulting operator fatigue and high error rate. To address these shortcomings, a haptic controlled human-excavator interface that incorporates visual, auditory and haptic cues as means of communication between human and excavator is proposed. However, the application of multiple sensing modalities in a human-excavator interface presents a set of issues and complications that need to be resolved before the full potential of multimodality in human-excavator system could be realized. In this work, an empirical study is conducted to investigate how visual, audio and haptic modalities could be used efficiently and effectively in human-excavator interface, potential conflicts that may arise from using multiple modalities and how the effects of these conflicts could be minimized. Results show that conflicts do exist between the visual and haptic modalities in the haptic control excavator interface, and that this interference does impact the operation of the haptic control excavator. Further, the results show that performance of novice operators were impacted more by the interference between the sensory cues than the performance of expert operators.

Keywords: haptic, excavator, conflict, interference

1   INTRODUCTION

Haptic refers to sensing and manipulation through touch. The origin of the word haptic can be traced back to the Greek words: *haptikos* meaning "able to touch" and *haptesthai* which translates to "able to lay hold of" (Katz & Krueger, 1989; Révész, 1950). However, today it is used broadly to encompass the study of touch and the human interaction with external environment via touch. More commonly, the word "haptic" or "haptics" refers to the capability to sense a natural or synthetic mechanical environment through touch and includes kinesthesia (or proprioception), the ability to perceive one's body position, movement and weight (Hayward, Astley, Cruz-Hernandez, Grant, and Robles-De-La-Torre, 2004). The recent explosion in computer technology and the need for better and intuitive ways for humans to interact with machines and computer-generated virtual environments has led to increased interest in haptics.

A haptic user interface is an interface that uses computer-controlled mechanism to allow users to interact with systems/machines through the sense of touch. Haptic user interfaces have wide range of applications that range from surgical devices in medicine to aviation, gaming and virtual reality industries, though their use in commercial products is low due to the technical challenges of their implementation. Recently, there is interest incorporating haptic into the user interface of an excavator. A haptic-controlled excavator will provide a simultaneous exchange of information between the operator and the excavator, thus, enabling the operator to experience an "immersed" interaction in the environment in which the task is being performed. By making use of the haptic control interface instead of the traditional levers and pedals, the excavator operators will be freed from solving the inverse kinematic relationship, and result in a more efficient and effective task performance and shorten training time for novice operators (Kontz and Book, 2007). In combination with visual display, haptic interface can be used to train operators to better perform digging tasks that require hand-eye coordination, and provide valuable help to novice operators to improve their task performance. Further, since human cognitive processes and perception build largely upon multimodality, a proper combination of haptic, visual and auditory modalities will result in a flow of information on several parallel channels which has been shown to enhance effectiveness of interaction (Krapichler, Haubner, Lösch, Schuhmann, Seemann, and Englmeier, 1999). Although the haptic interface promises reduced mental workload and improved operator performance over the traditional lever/pedal interface, its use as a control interface for the excavator has not been fully explored because the technology is still being developed. Currently, the concept of haptic-controlled excavator interface is under development at Georgia Institute of Technology.

One of the common concern in a multimodal interface design is the conflict. As stated by Oviatt, Cohen, Wu, Vergo, Duncan, Suhm, and Landay (2000), making a system multimodal by just adding a further modality to the system may not necessarily lead to improvement in the system and may instead increase operator's cognitive load due to the increased number of processing channels. In some cases the different modalities may actually interfere with each other and may have a negative impact on system performance.

The goal of this research is to conduct an empirical study to investigate the conflict and interference in a haptic-controlled excavator.

2   METHODOLOGY

2.1   Participants

Twenty-four students were recruited from the North Carolina Agricultural & Technical State University to take part in this empirical study. A between group design was used in this experiment. The participants were grouped into two groups: *novices* and *experts*. The novices group consisted of 20 volunteers with no prior knowledge of the haptic controlled excavator simulator, while the experts group consisted of 4 volunteers made up of members of Center for Compact and Efficient Fluid Power Systems (CCEFP) research team at North Carolina A&T State University who have had the experience of interacting and manipulating the haptic-controlled excavator.

2.2 Equipment

The equipment for this experiment consisted of two Gateway computers, a Tobii® Eye Tracker T60, and a Phantom Omni 5.3 Haptic device. The two Gateway computers ran the excavator simulation program and were connected with the Tobii® Eye Tracker via a local network. Computer number one interfaced with the Phantom Omni and ran the excavator dynamics simulation, while computer number two ran the xPC-target simulation. The excavator simulation graphics were displayed on the Tobii® Eye Tracker connected via a local network to the two Gateway computers. The schematic layout of the equipment setup is shown in Figure 1. The Phantom Omni device sat next to the Tobii® Eye Tracker on the right hand side of participants and had 6 degrees of freedom in total: up-down, left-right, front-back, and a rotating stylus with 3 degrees of freedom. The C++ and Mat Lab programming that ran the simulation was developed by Mark Elton of Georgia Institute of Technology.

2.3 Task

Participants were asked to perform a series of tasks that required them to move the boom/bucket assembly to the desired location using the stylus of the Phantom Omni device. Next, they had to position the bucket at the work area (trench), then scoop/dig soil, move content to the desired location (bin), and rotate anticlockwise to open bucket and unload its content. Participants performed two tasks, and the order of the tasks was randomized among all participants.

Task #1: Dig soil from the marked area to fill *bin #1* (bin to the left of trench). Accomplish this by using the stylus of Phantom Omni device to control and manipulate the boom/bucket assembly of the simulated excavator. When the bin is full, there will be an audio alert and the content of the bin turns green.

Task #2: Dig soil from the marked area to fill *bin #2* (bin to the right of trench). Accomplish this by using the stylus of the Phantom Omni device to control and manipulate the boom/bucket assembly of the simulated excavator. When the bin is full, there will be an audio alert and the content of the bin turns green.

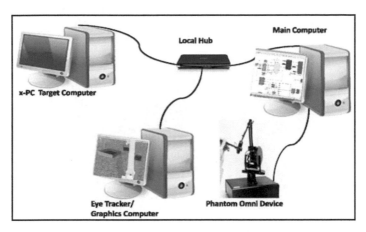

Figure 1 Schematic equipment setup of simulation

2.4 Procedure

Participants were briefed on the purpose of the study upon arrival, and then asked to read and sign a consent form. They were briefed on how to complete a computer-based NASA TLX workload assessment after which a pre-test questionnaire was administered to collect demographic information. Participants were informed that their eye movements would be recorded with a remote desktop Tobii® Eye Tracker T60 and that they should maintain a steady head position as much as possible during the test. A short demo of the simulation was given, and participants were given a few minutes to familiarize themselves with the simulator. Questions about the simulator and controls from participants were answered by the experimenter after which actual testing started.

To take the test, participants were seated in front of Tobii® Eye Tracker T60 with their heads about 60cm from the monitor. Participants' head positions were adjusted so that their head was in the middle of the monitor when viewed from behind. Once the appropriate head position is found, participants' eyes were calibrated. This was done by asking the participant to follow the red calibration dot/ball with their eyes as it moved randomly across the screen, briefly stopping at each of the four diagonals and the center of the screen. After calibration, the excavator simulation was initiated and participants were asked to carry out the assigned digging tasks with the Phantom device while their eye movements were recorded with the Tobii® Eye Tracker. The experimental set up for this study is shown in Figure 2. Upon completion, participants were thanked, debriefed, and asked to complete the NASA TLX workload assessment and a post-test questionnaire. They were also asked for comments about their experience of using the haptic control excavator interface. Overall, the test took about one hour to complete.

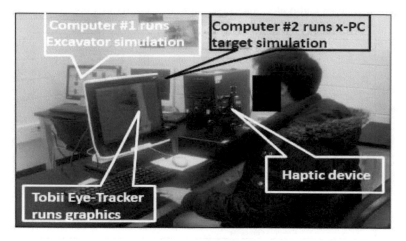

Figure 2 Experimental set-up

3 RESULTS

Objective performance measures of task completion time (s), number of scoops to fill a bin, and number of scoops dropped outside of bin for both expert and novice operators were collected during the experiment. The descriptive statistics of the objective performance measures can be seen in Table 1. Table 2 provides the descriptive statistics of the subjective workload assessment results obtained from the participants. In addition, eye tracking data were also captured during the experiment. For this study, the fixation count, fixation length and the 1st fixation duration within the area of interest (AOI), and the fixation count and fixation length outside the AOI were used. In order to investigate the conflicts among multiple modalities, tasks that depend on auditory, visual and haptic cues were analyzed for both expert and novice operators. The descriptive statistics for these measures can be shown in Tables 3 and 4.

Prior to conduct any inferential statistical analysis, normality test, test for independence and test for homogeneity of variance (HOV) were performed on each of the data sets and compared to $\alpha=0.05$ significance level. For Within-AOI measures, Shapiro-Wilk's test indicated a lack of normality for fixation length *(w=0.9954* and *p=0.0009)*, fixation count *(w=0.8304* and *p=0.0001)*, and the first fixation duration $(w=0.2885$ and $p=0.001)$. Further, Levene's test for homogeneity of variance showed there is a significant variation for fixation length $(F_{(1, 46)} =4.55,$ and $p=0.0383)$, fixation count $(F_{(1, 46)} =2.52,$ and $p=0.01192)$ except for fixation duration $(F_{(1, 46)} =0.47,$ and $p=0.4958)$. Similarly, Outside-AOI measure data also violated normality assumption. Consequently, a non-parametric statistical analysis, the Mann-Whitney-Wilcoxon test was used to analyze the data.

Table 1 Descriptive statistics for task completion time, number of scoops and number of drops

Expertise	Statistic	Objective Performance Measure		
		Completion Time (seconds)	Number of Scoops	Number of Drops
Novices	Mean	216.1	7.6	0.9
	SD	67.6	1.4	1.0
Experts	Mean	138.1	6.4	0.3
	SD	6.4	0.3	0.5

Table 2 NASA TLX subjective workload assessment for experts and novices

Workload Metric	Expertise	
	Novices	Experts
Mental	12.30	6.00
Physical	12.13	5.67
Temporal	5.38	1.92
Performance	6.49	6.92
Effort	13.11	11.83
Frustration	9.78	5.50
Total Workload	59.19	37.83

Results from the non-parametric Mann-Whitney-Wilcoxon test showed that there was no statistically significant difference in fixation count outside AOI between experts and novices, $(z=48.0,$ and $p=0.9054)$. However, there was a statistically significant difference in fixation count within AOI between experts and novices $(z=104.50,$ and $p=0.018)$. Further, the results showed that there was a statistically significant difference in fixation length within AOI between experts and novices $(z=23.00,$ and $p=0.0398)$. However, outside the AOI, there was no statistically significant difference in fixation length between experts $(\mu=20.626)$ and novices $(\mu=25.275)$, $(z =165.00,$ and $p=0.3988)$.

The higher number of fixation count and fixation length within the AOI by expert operators, a difference which the results above show are significant, may be due to the fact that novice operators had harder time keeping their eyes focused in the task area compared to expert operators. In fact, the results show that, novices were nearly twice as likely (3.78 vs. 2.0) to look outside the area of interest while performing the task than experts (Table 5).

Table 3 Descriptive statistics for fixation count, fixation length and 1st fixation duration on AOI for experts and novices

Expertise	Statistic	Performance Measure		
		Fixation Count	Fixation Length (seconds)	1st Fixation Duration (seconds)
Novices	Mean	281.35	0.671	0.195
	SD	120.979	0.075	0.416
Experts	Mean	190.125	0.752	0.147
	SD	43.525	0.247	0.068

Table 4 Fixation count & length not on area of interest (AOI)

Expertise	Statistic	Performance Measure	
		Fixation count (seconds)	Fixation length (seconds)
Novice	Mean	7.975	0.164
	SD	22.56	0.1
Experts	Mean	1.625	0.095
	SD	1.061	0.048

Table 5 Mean number of scan paths outside the area of interest (AOI)

Expertise	Mean # of Scan Paths Outside AOI
Experts	2.0
Novices	3.78

This may be due to the interference between the sensory cues that are required for successful execution of the excavation task. The fact that fixation count and fixation length values were higher for novices than for experts may be due to the fact that novices had more difficulty in extracting useful information necessary for task performance compared to experts. The high fixation count also might be an indication that, novices were less efficient in performing the assigned task compared to experts. Thus while experts focused most of their attention on the task area (the screen), novices were unable to focus their full attention on the task area, but alternated between looking at the screen and their hands. The results from this study, therefore, show the existence of possible interference between visual, haptic and auditory cues in the haptic control excavator interface.

Further, to gain an understanding of operators' mental processes as they carried out the excavation task, the gaze plots and scan path data obtained using eye-tracking were analyzed. Since what the human eye looks at is usually reflects what goes on mentally, the gaze plot data was used to gauge operators' mental processes. The gaze plots for expert and novice operators are shown in Figures 3 and 4 respectively.

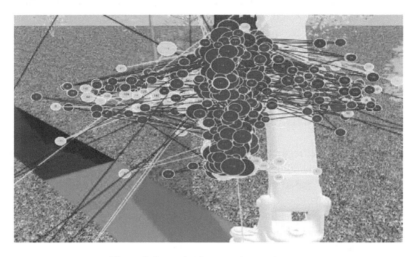

Figure 3 Gaze plot for expert operators

Figure 4 Gaze plot for novice operators.

From the gaze plot data, it was observed that experts' attention were focused on the environment where the task was performed, this is demonstrated by the fact that, most gaze lines of experts operators were within the work area as seen in Figure 3. Novice operators on the other hand, were unable to fully focus or limit their eye movements to the work area as demonstrated by numerous gaze lines that go off the screen as seen in Figure 4. The off screen gaze lines are indication of novices attempting to look at their hands as they performed the excavation task with the phantom device. This may be due to the fact that, novices struggle to keep their eyes focused on the screen where the actual excavation task takes place, but rather keep their eyes from looking on the screen to looking at their hands. A situation similar to an experienced driver having the ability to accelerate and brake a vehicle without having to look at the accelerator or brake pedals, an involuntary action. On the other hand, an inexperienced driver might be tempted to look at the accelerator or brake pedal in order to move or stop a vehicle, a voluntary action.

4 DISCUSSION AND CONCLUSION

In summary, the results from this study show that operators had difficulty coordinating their hand-eye movement while operating the haptic-controlled excavator. Further, the results show that, novice operators had more difficulty coordinating their hand-eye movement than did expert operators. This may be due to the fact that experts were able to retrieve information from memory to help them accomplish the task while novices did not.

In order to design an effective, intuitive and easy to use interface, it is important that the complimentary sensing cues are integrated in a way that capitalizes on the strengths of each mode in order to overcome the weakness in each other. To design a robust and easy to use haptic-controlled excavator interface, it is important that issues of conflict and interference between the multiple sensing cues used in the design are well understood. To accomplish this, an empirical study was conducted to assess whether conflict exists between visual, haptic and auditory cues that are necessary for the smooth operation of the haptic control excavator interface.

The goal of the empirical study was to identify if there were conflict between visual, haptic and auditory cues in the haptic interface, and whether these conflicts had an impact on the performance of the operator. Results from the empirical study show that conflicts do exist between the visual and haptic modalities in the haptic control excavator interface, and this interference does impact the operation of the haptic control excavator. From the results, performance of novice operators was impacted more by the interference between the sensory cues than the performance of expert operators. Finally, results from the empirical study show that novice operators had a harder time coordinating their hand-eye movement than expert operators. Overall, the results from the empirical study provided an understanding of the interference between the sensory modalities and their effects on operator performance.

ACKNOWLEDGMENTS

This research is partially funded by the National Science Foundation Engineering Research Center for Compact Efficient Fluid Power (CCEFP).

REFERENCES

Hayward, V., Astley, O. R., Cruz-Hernandez, M., Grant, D., and Robles-De-La-Torre, G. 2004. Haptic interfaces and devices. *Sensor Review, 24*(1), 16-29.

Katz, D. and Krueger, L. E. 1989. *The world of touch*: Edited by Krueger, LE. Hillsdale, New Jersey. Erlbaum Associates, Inc.

Kontz, M., and Book, W. 2007. Electronic control of pump pressure for a small haptic backhoe. *International Journal of Fluid Power, 8*(2), 5.

Krapichler, C., Haubner, M., Lösch, A., Schuhmann, D., Seemann, M., and Englmeier, K. H. 1999. Physicians in virtual environments multimodal human-computer interaction. *Interacting with Computers, 11*(4), 427-452.

Oviatt, S., Cohen, P., Wu, L., Vergo, J., Duncan, L., Suhm, B., and Landay, J. 2000. Designing the user interface for multimodal speech and pen-based gesture applications: state-of-the-art systems and future research directions. *Human-Computer Interaction, 15*(4), 263-322.

Révész, G. 1950. Psychology and art of the blind. Longmans Green, New York.

Section II

User Studies

Relationship among Fall-Related Self-Efficacy, Activities of Daily Living and Fall Risk in the Elderly

Chien-Lung Chan[a], Shao-Sung Huang[b,c], Nan-Ping Yang[d], Wan-Yu Chen[a]

[a]Dept. of Information Management, Yuan Ze University,
Taoyuan, Taiwan, clchan@im.yzu.edu.tw
[b]Dept. of Medicine, National Yang Ming University, Taipei, Taiwan
[c]Dept. of Internal Medicine, Taoyuan Veterans Hospital, sshuang2@vghtpe.gov.tw
[d]Community Health Research Center & Institute of Public Health,
National Yang-Ming University, yang.nanping@gmail.com

ABSTRACT

Elderly fall is regarded as one of the most important issues in the area of public health research. Studies have revealed that people who are 65 years old and above have a one-third chance of falling annually in average. Falls not only inflict physical injuries and incur medical costs, but also cause psychological problems and difficulties in the daily activities of the elderly. This study examined the relationship between the falls self-efficacy, daily activities, and fall risk of the elderly to help healthcare organizations and home caregivers reduce falls. Between January and May 2011, surveys and interviews were conducted with 129 people aged 65 years and above in Taoyuan county, Taiwan. Results indicate that falls self-efficacy, daily activities, and marital status were associated with the fall risk among the elderly.

Keywords: elderly, falls, fall self-efficacy, daily activities, fall risk

1   INTRODUCTION

World Health Organization (WHO) defined the ageing society as the population above 65 years old higher than all population 7% in the area. The ageing society reminds us that the issue of health care and elderly medical treatment ever more important. Therefore, elderly fall and other fall-related issues should not be overlooked. In Taiwan, fall resulting death is the second cause of deaths in accidental deaths for the elderly (Department of Health, 2009). According to the Taiwan Joint Commission on Hospital Accreditation 2009 report, there are 10,094 fall events in medical institutions across Taiwan. In all the fall events, 47% elderly above 65 years old have the highest incidence of fall events (Taiwan Joint Commission on Hospital Accreditation, 2009).

Fall and fall-related complication could cause other disease deterioration and the medical costs of falling are considerable. In 2000, there were a total of 10,300 fatal falls and 2.6 million non-fatal falls in the U.S., and the medical costs are approximately USD19.2 billion. Meanwhile, the United Kingdom has spent 1 billion pounds of medical costs on fall events (Stevens et al., 2006). Falls bring 0.85% to 1.5% disbursement of medical health insurance (Heinrich et al., 2010); an enormous financial burden to the society.

However, fall risk factors are very complex. Physical factors include mobility, visual disabilities, arthritis, walking gait, body balance, muscular endurance, cognitive impairment, response time, diabetes, and incontinence (Sherrington and Finch, 2004; McMichael et al., 2008; Delbaere et al., 2010). Physical factors of fall can be summarized into three categories: (1) degeneration or pathological changes of perceptive function, (2) diseases, and (3) shortage of muscular endurance and balance due to lack of exercise (Chu, Chi and Chiu, 2005). For the psychological factors, previous research have shown that melancholy, anxious, and fear of fall are all related to elderly fall (McMichael et al., 2008; Delbaere et al., 2010). Other external fall risk factors, such as slippery surfaces, safeguard providing, wearing supportive shoes and bright environment are related to fall in the elderly (Huang and Acton, 2004; Letts et al., 2010).

Psychological factor is one of the fall factors to cause the occurrence of fall. Elderly's movements or fall-related self-efficacy might influence their behavior. Bandura considers that personal behavior approach is different from their belief, and self-efficacy is a belief of a person having the capability to finish his goal (Bandura, 1977). Fortinsky discovered that most of the elderly having over confidence of balance are below 75 years old (Fortinsky et al., 2009). Therefore, our study presumes it might be due to the relationship on the interaction of both the physiological and psychological factors to fall risk in the elderly.

2   FALL RISK AND FACTORS

2.1   Fall-Related Self-efficacy

The concept of fall-related self-efficacy is derived from Bandura's concept of self-efficacy. The definition of self-efficacy is that people believe themselves have the abilities to complete a job (Bandura, 1977). In previous studies, fall-related self-efficacy is defined as the confidence of not falling, when people are carrying out activities (Huang and Acton, 2004, Pang and Eng, 2008). Fall-related self-efficacy can be used to estimate the fall risk of people carrying out activities (Hellstro¨m et al., 2009). Vellas also found sex and age are related to fall-related self-efficacy; and the relationship with falls, fall-related self-efficacy is stronger than balance and gait. Besides, those elderly with lower fall-related self-efficacy reduced the number of movement times (Vellas et al., 1997).

The fear of falling is a long-term status associated with experience, and fall-related self-efficacy is a short-time status because it would change according to vicarious experience, verbal persuasion, and emotional arousals (Bandura, 1977). In this study, the definition of fall-related self-efficacy is the confidence of maintaining balance or not falling when people are doing their daily activities.

2.2   Activities of Daily Living (ADL)

Activities of daily living(ADL) means the most basic abilities of people take care of themselves to keep living, maintaining personal hygiene and most basic cleaning, repeatability and commonality activities in their daily lives. Joshua defined activities of daily living as doing basic work on a daily basis, such as eating, moving, bathing, dressing, and toileting (Wiener et al., 1990).

There are many scales to survey the activities of daily living, and they are widely used in any kind of research, such as disease, mortality, insurance, nursing, and home care. In our study, the definition of activities of daily living is the abilities to take care self eating, dressing, living, moving, and keeping clean on a daily basis (Toba et al., 2009; Cumming et al., 2000; Lord et al., 2003; Tinetti and Williams, 1998).

2.3   Fall Risk

Fall is defined as people losing their balance, causing their heads, arms, knees, or body to touch or hit the floor (Huang and Acton, 2004; Zecevic et al., 2006). Yardly defined fall as an instant event causing somebody to fall to floor or lowlands (Yardley et al., 2005). And the fall risk is the possibility of falling. The factors of fall are classified into: (1) physiological factors, (2) psychological factors, (3) drug factors, and (4) others factors.

Nevitt discussed that among those elderly who have fallen in the past year, 55% of them would fall again with minor injuries, and 6% would produce fracture, dislocation or serious injuries. Besides, the degree of injury would be affected by age, neuromuscular, and cognitive impairment (Nevitt et al., 1989; Nevitt, Cummings and Hudes, 1991). Swanenburg mentioned that relative to normal people, people who take multiple or high-fall risk drugs is prone to falling with the odds ratio of 2.3 (Swanenburg et al., 2010). Vitry also said taking mental drugs might increase the fall risk (Vitry et al., 2010).

On the other hand, Delbaere used data mining techniques and found that people having visual defects, mobility barriers, balance disorders lower reflection time, or depression are more prone to falling than others (Delbaere et al., 2010). Graafmans also concurred that people who have mobility barriers, dizziness, postural hypotension, bad mental states, or apoplexy histories are more prone to falling than others (Graafmans et al., 1996). Finally, in other factors, Letts highlighted that those elderly who rely on hand tools for walking might increase the fall risk (Letts et al., 2010).

3   METHODS

We surveyed and interviewed those elderly aged 65 years and above in Taoyuan county, Taiwan. The questionnaire consists of four parts: (1) falls related self-efficacy, (2) activities of daily living, (3) fall risk, and (4) personal information.

Falls Efficacy Scale International(FES-I) is used to measure fall-related self-efficacy, and the internal reliability(Cronbach's α) of the scale is 0.96 (Yardley et al., 2005). The scale has 16 questions about activities of daily living. The grading is the degree of concern for falling while the subjects are carrying out those activities. Higher score means the degree of concern for falling is higher, resulting in lower fall-related self-efficacy.

For the activities of daily living, Barthel Index is used to measure activities of daily living and the scale reliability is above 0.80 (Minosso et al., 2010). Barthel Index consists of 10 questions on the activities of daily living, such as eating, walking, taking a bath, defecation, and urination. The grading is the degree of assistance from others or time spent while the subjects are carrying out those activities. Higher score means higher independence of daily living.

In the past studies, fall risk is measured by using the numbers of falls in a period of time. However, fall-related self-efficacy is measured for the degree of concern for falling at a particular moment or recently. Therefore, using the number of falls within a period of time to measure the fall risk is not logical (Graafmans et al., 1996; Delbaere et al., 2010; Fortinsky et al., 2009; Hendrich, Bender and Nyhuis, 2003).

The Hendrich II Fall Risk Model was used to measure the fall risk of those subjects. The scale has 8 questions with weight score. Higher score means higher fall risk. The Cronbach's α of the scale is 0.87 and the internal reliability of scale is good (Hendrich, Bender and Nyhuis, 2003).

4   RESULT

4.1   Participants

Surveys and interviews were conducted with 129 people aged 65 years and above between January 2011 and May 2011 in Taoyuan county, Taiwan. As shown in Table 1, out of the 129 subjects, 69 (53.5%) are males and 60 (46.5%) are females. The average age is 78.31 (s.d. = 8.11) years old, most of them are 75 to 79 years old, and 23 (20.2%) subjects are between 85 and 89 years old. As for the subjects' marital status, 46 (35.7%) subjects are widowed, 42 (32.6%) subjects are married, and 40 (31.0%) subjects are single. In the surveying the living conditions, 89 (69.0%) subjects are living in long term care agencies, and 36 (27.9%) subjects are living at home. Based on the walk status' findings, 74 (57.4%) subjects are walking normally, 34 (26.4%) subjects require a hand tool to walk, and 19 (14.7%) subjects are using wheelchair to move around.

Table 1 Characteristics (number and percent) of the participants(n= 129)

Variables	Number (%)
Sex	
Male	69 (53.5)
Female	60 (46.5)
Marital Status	
Single	40 (31.0)
Widowed	46 (35.7)
Missing	1 (0.8)
Living conditions	
Alone	3 (2.3)
With friends	1 (0.8)
With family	36 (27.9)
Long term care agencies	89 (69.0)
Walk status	
Normally	74 (57.4)
Hand tool	34 (26.4)
Wheelchair	19 (14.7)
Can't walk	1 (0.8)
Missing	1 (0.8)

4.2   Falls Efficacy Scale-International(FES-I), Barthel Index and Fall Risk

In the Falls Efficacy Scale International(FES-I), the average score of subjects is 34.7 (s.d. =14.356). This number is very close to the study of Yardly et al(2005). As shown in Table 2, most of the subjects are more concerned of falling while "Walking on a slippery surface" in which the average score of this question is 2.66 (s.d. =1.138). The second highest score, 2.58 (s.d. =1.214), came from the question, "Going up or down stairs." And the third highest score, 2.55 (s.d. =1.196), came from the question, "Walking on an uneven surface." It means that these three activities have a certain degree of challenge for these subjects.

Table 2   Fall related self-efficacy of participants measured by FES-I(n=129)

Falls Self-Efficacy Scale (FES-I)	Max	Min	Mean	S.d.
Total Score of FES-I	64	16	34.7	14.356
Cleaning the house	4	1	2.04	1.071
Getting dressed or undressed	4	1	1.69	0.983
Preparing simple meals	4	1	1.69	1.022
Taking a bath or shower	4	1	2.29	1.182
Going to the shop	4	1	1.96	1.104
Getting in or out of a chair	4	1	2.09	1.075
Going up or down stairs	4	1	2.58	1.214
Walking around in the neighborhood	4	1	2.02	1.087
Reaching for something above your head or on the ground	4	1	2.22	1.136
Going to answer the telephone before it stops ringing	4	1	1.88	1.024
Walking on a slippery surface	4	1	2.66	1.138
Visiting a friend or relative	4	1	2.00	1.136
Walking in a place with crowds	4	1	2.34	1.160
Walking on an uneven surface	4	1	2.55	1.196
Walking up or down a slope	4	1	2.51	1.261
Going out to a social event	4	1	2.14	1.099

& 4 point scale, 1=not at all concerned, 4=very concerned.

In the Barthel Index, the average score of subjects is 86.05 (s.d. =21.872). Most of subjects are independent completely in the daily living, and it has 69 (53.5%) subjects. Besides, moderate dependence of subjects is 40 (31%) subjects. Otherwise, heavy dependence of subjects is 10 (7.8%) subjects. Finally, mild dependence of subjects is 6 (4.7%) subjects.

In the Hendrich II Fall Risk Model, the average score of subjects is 2.64 (s.d. =2.046). The subjects who having altered elimination are 58 (45%) subjects, 34 (26.4%) said they felt dizziness or vertigo, and 19 (14.7%) subjects have taken sedative hypnotics (Benzodiazepin). In Get-Up-And-Go test of the Hendrich II Fall Risk Model, 62 (48.1%) subjects can get up in a single movement, 47 (36.4%) subjects need pushes up in one attempt, 7 (5.4%) subjects are multiple attempts then successful, and 13 (10.1%) subjects can't rises by themselves. Finally, according to the scale, 22 (17.1%) subjects are classified having high fall risk (score $\geq$ 5) and 107 (82.9%) subjects are classified without high fall risk (score <5).

4.3   Correlation among activity of daily living, fall-related self-efficacy and fall risk

Table 3 Hendrich II Fall risk score of participants

Hendrich II Fall Risk Model	Number (%)
Confusion/Disorientation/Impulsiveness	7 (5.4)
Depression	6 (4.7)
Altered Elimination	58 (45.0)
Dizziness/Vertigo	34 (26.4)
Gender (Male)	68 (53.5)
Any antiepileptics	0 (0.0)
Any benzodiazepines	19 (14.7)
Up-And-Go Test	
Rises in a single movement	62 (48.1)
Pushes up in one attempt	47 (36.4)
Multiple Attempts, successful	7 (5.4)
Unable to rise without assist	4 (3.1)
Overall Fall risk	
Having high fall risk	22 (17.1)
Without high fall risk	107 (82.9)

Table 4   Logistic regression of fall risk with FES-I and Barthel Index

	β	p value	OR	95%CI
FES-I	0.075	0.000[***]	1.078	1.038-1.118
Constant	-4.518	0.000[***]	0.011	
Barthel Index	-0.049	0.000[***]	0.952	0.934-0.970
Constant	0.926	0.041[*]	2.524	
Barthel Index	-0.087	0.019[*]	0.917	0.853-0.986
FES-I	-0.015	0.648	0.985	0.926-1.049
Barthel Index*FES-I	0.001	0.217	1.001	0.999-1.003
Constant	1.376	0.380	3.961	

Barthel Index and fall risk have a strong negative correlation (r=-0.616, p<0.001). The correlation coefficient of fall risk and FES-I is 0.478 (p <0.001), and Barthel Index and FES-I is -0.508 (p <0.001). Fall risk and FES-I have positive correlation. Besides, Barthel Index and FES-I are negative correlation. The Barthel Index score is higher, and then the score of FES-I is lower(means higher fall related self-efficacy). Barthel Index is a kind of protection factor where by the fall risk is lower when the Barthel Index score is higher.

In the binary logistic regression of FES-I and fall risk, the FES-I has a positive influence on fall risk, β =0.075 (p value <0.001), odds ratio =1.087. Barthel Index has a negative influence on fall risk, β =-0.049 (p value <0.001), odds ratio =0.952. n the binary logistic regression of fall risk with independent variables FES-I, Barthel Index and interaction term of FES-I * Barthel Index. We found only Barthel Index has a negative influence on fall risk, β = -0.087 (p =0.019), odds ratio =0.917. Both FES-I and interaction term are not significant. Barthel index turns out to be a better predictor of fall risk than FES-I.

5   DISCUSSION

Subjects with higher Barthel Index have lower FES-I score. Lower FES-I means higher fall-related self-efficacy. Fall-related self-efficacy has a positive relationship with activities of daily living. The result confirms the findings from previous studies (Pang and Eng, 2008; Cumming et al., 2000). and it also explained the concept of information sources of Bandura's self-efficacy. Elderly with higher activities of daily living have fewer falls in the past, hence, they have higher falls-related self-efficacy (Bandura, 1977).

Subjects with higher score of FES-I (lower fall-related self-efficacy), has higher fall risk. Fall-related self-efficacy has a negative relationship with fall risk. Elderly with lower fall-related self-efficacy is more prone to falling, and once again, the result confirms previous studies' findings (Tinetti and Williams, 1998).

Subjects with higher Barthel Index score have lower fall risk. Activity of daily living has a negative relationship with fall risk. Elderly with lower activities of daily living is more prone to falling, and the result is also consistent with previous study (Hellstro¨m et al., 2009). Falls are related to the deterioration of activities of daily living.

The interaction term of fall-related self-efficacy, and activities of daily living is not significant. Since activities of daily living might represent most of the fall risk factors, the influence of the interaction of fall-related self-efficacy and activities of daily living is not significant.

6   CONCLUSIONS

Previous studies stated that the main reason of falls is deterioration of activities of daily living (Tinetti and Williams, 1998). Bandura's research, suggested that part of the information sources to measure self-efficacy originates from personal previous experiences. Therefore, activities of daily living has a positive relationship with fall-related self-efficacy (Bandura, 1977).

Previous studies highlighted that activities of daily living has a negative relationship with fall risk (Hellstro"m et al., 2009; Toba et al., 2009; Cumming et al., 2000; Lord et al., 2003; Tinetti and Williams, 1998). Our study supports those previous studies and found that activities of daily living are the most important factor to estimate the fall risk, because activities of daily living is directly related to the physiological state.

Finally, we found that fall-related self-efficacy has a negative relationship with fall risk and the result concurs with the previous studies. When compared with activities of daily living, the fall-related self-efficacy has weaker linkage with fall risk.

ACKNOWLEDGMENTS

This study was supported by a grant from National Science Council, Taiwan(NSC 99-2221-E-155-045-MY3 & NSC 99-2221-E-155-041-MY3). This study was approved by the Institutional Review Board(IRB) of Taoyuan General Hospital(Approval number: TYGH99050). The authors want to thank the staff in Nursing department, Taoyuan Veteran Hospital; Taoyuan Ren-Ai Senior Citizens' home, Bade Veterans' home for helping collection of data.

REFERENCES

Bandura, A. 1977. Analysis of self-efficacy theory of behavioral change. *Cognitive Therapy and Research* 1(4):287-310.

Chu, L., I. Chi, and A. Chiu. 2005. Incidence and predictors of falls in the Chinese elderly. *Annals of the Academy of Medicine* 34(1):60-72.

Cumming, R.G., G. Salkeld, and M. Thomas, et al. 2000. Prospective study of the impact of fear of falling on activities of daily living, SF-36 scores, and nursing home admission. *The Journals of Gerontology. Series A, Biological Sciences and Medical Sciences* 55(5):M299-M305.

Delbaere, K., J.C.T. Close, and J.r. Heim, et al. 2010. A multifactorial approach to understanding fall risk in older people. *Journal of the American Geriatrics Society* 58(9):1679-1685.

Department of Health, Executive Yuan, R.O.C. (Taiwan). 2009. *Cause of Death Statistics 2009*.

Fortinsky, R.H., V. Panzer, and D. Wakefield, et al. 2009. Alignment between balance confidence and fall risk in later life: Has over-confidence been overlooked? *Health, Risk & Society* 11(4):341-352.

Heinrich, S., K. Rapp, and U. Rissmann, et al. 2010. Cost of falls in old age- A systematic review. *Osteoporosis International* 21(6):891-902.

Hellstro"m, K., B. Vahlberg, and C. Urell, et al. 2009. Fear of falling, fall-related self-efficacy, anxiety and depression in individuals with chronic obstructive pulmonary disease. *Clinical Rehabilitation* 23(12):1136-1144.

Hendrich, A.L., P.S. Bender, and A. Nyhuis. 2003. Validation of the Hendrich II Fall Risk Model: A large concurrent case/control study of hospitalized patients. *Applied Nursing*

Research 16(1): 9-21.

Huang, T.-T. and G.J. Acton. 2004. Effectiveness of Home Visit Falls Prevention Strategy for Taiwanese Community-Dwelling Elders- Randomized Trial. *Public Health Nursing* 21(3):247-256.

Letts, L., J. Moreland, and J. Richardson, et al. 2010. The physical environment as a fall risk factor in older adults: Systematic review and meta-analysis of cross-sectional and cohort studies. *Australian Occupational Therapy Journal* 57(1):51-64.

Lord, S.R., S. Castell, and J. Corcoran, et al. 2003. The effect of group exercise on physical functioning and falls in frail older people living in retirement villages- a randomized, controlled trial. *Journal of the American Geriatrics Society* 51(12):1685-1692.

McMichael, K.A., J.V. Bilt, and L. Lavery, et al. 2008. Simple balance and mobility tests can assess falls risk when cognition is impaired. *Geriatric nursing (New York)* 29(5):311-23.

Minosso, J. S. M., F. Amendola, and M. R. M. Alvarenga, et al. 2010. Validation of the Barthel Index in elderly patients attended in outpatient clinics in Brazil. *Acta Paulista De Enfermagem* 23(2):218-223.

Nevitt, M. C., S. R. Cummings and E. S. Hudes. 1991. Risk Factors for Injurious Falls: A Prospective Study. *Journal of Gerontology* 46(5):164-170.

Nevitt, M. C., S. R. Cummings, and S. Kidd, et al. 1989. Risk Factors for Recurrent Non-syncopal Falls. *The Journal of the American Medical Association* 261(18) :2663-2668.

Pang, M.Y.C. and J.J. Eng. 2008. Fall-related self-efficacy, not balance and mobility performance, is related to accidental falls in chronic stroke survivors with low bone mineral density. *Osteoporosis International* 19(7):919-927.

Sherrington, C. and C. Finch. 2004. Physical activity interventions to prevent falls among older people: Update of the evidence. *Journal of Science and Medicine in Sport* 7(1):43-51.

Stevens, J. A., P. S. Corso, and E. A. Finkelstein, et al. 2006. The costs of fatal and non-fatal falls among older adults. *Injury Prevention* 12(5):290-295.

Swanenburg, J., E.D.d. Bruin, and D. Uebelhart, et al. 2010. Falls prediction in elderly people: A 1-year prospective study. *Gait & Posture* 31(3): 317-321.

Taiwan Joint Commission on Hospital Accreditation. 2009. *Report: Patient safety reporting system in Taiwan 2009.*

Tinetti, M.E. and C.S. Williams. 1998. The effect of falls and fall injuries on functioning in community-dwelling older persons. *The Journals of Gerontology. Series A, Biological Sciences and Medical Sciences* 53(2):112-119.

Toba, K., R. Kikuchi, and A. Iwata, et al. 2009. "Fall Risk Index" helps clinicians identify high-risk individuals. *Japan Medical Association Journal* 52(4):237-242.

Vellas, B. J., S. J. Wayne, and L. J. Romero, et al. 1997. Fear of falling and restriction of mobility in elderly fallers. *Age Ageing* 26(3):189-193.

Vitry, A.I., A.P. Hoile, and A.L. Gilbert, et al. 2010. The risk of falls and fractures associated with persistent use of psychotropic medications in elderly people. *Archives of Gerontology and Geriatrics* 50(3):e1-e4.

Wiener, J. M., R. J. Hanley, and R. Clark, et al. 1990. Measuring the activities of daily living: Comparisons across national surveys. *The Gerontological Society of America* 45(6):s299-s237.

Yardley, L., N. Beyer, and K. Hauer, et al. 2005. Development and initial validation of the Falls Efficacy Scale-International (FES-I). *Age Ageing* 34(6):614-619.

Zecevic, A.A., A.W. Salmoni, and M. Speechley, et al. 2006. Defining a fall and reasons for falling: comparisons among the views of seniors, health care providers, and the research literature. *The Gerontologist* 46(3):367-376.

CHAPTER 7

Accident Characteristics and Safety Measures for Aging Workers in Korea

Chan-O Kim, Hyoung-Jun Choi, Dong-Won Choi

Department of Safety Engineering,
Seoul National University of Science & Technology
Seoul, Republic of Korea

ABSTRACT

The Korean average life span is increasing due to the growth of economy and the improvement of quality of life. So, Korea is rapidly approaching to the aging society. In addition, the age of core industrial workers has been changed from 20s to 40s, and also advanced-age workers are increasing in industrial field every year.

However, Korea does not have a sufficient interest about the advanced-age workers, yet. And there is no definition of advanced-age worker. And so, first of all, it is needed to define the advanced-age worker.

Usually, aged person could be defined the person whose age is over 60 or 65 years. However, the National Pensions Act provides that the requirement of old-age pension is over the age of 60 years. And the Elderly Welfare Act & the Magna Charta provide that the senior citizen is over the age of 65 years. And the Aged Employment Promotion Act defines that the senior citizen is over the age of 55 years. Considering these facts, in this thesis, it is defined that the advanced-age workers is over the age of 55 years.

The results of statistical analysis of industrial accidents of advanced-age workers are as follows: In 2007 the number of industrial accidents of advanced-age workers is 19,133 that is 21.22% of whole industrial accidents. And in 2008 the number of industrial accidents of advanced-age workers is 21,083 that is 22.00% of whole industrial accidents. And in 2009 the number of industrial accidents of advanced-age workers is 24,996 that is 25.55% of whole industrial accidents. And in 2010 the number of industrial accidents of advanced-age workers is 26,622 that is 26.98% of whole industrial accidents. It shows that the number of industrial accidents of advanced-age workers is increasing continuously.

Physical characteristics of advanced-age workers are as follows: Endurable power of advanced-age workers is 86% of the 30s. And quickness ability of advanced-age workers is 80% of the 30s. And instantaneous reacting ability of advanced-age workers is 68% of the 30s. And muscular strength of advanced-age workers is 18% of the 30s. Because of these characteristics, advanced-age worker is easier to get the industrial accidents in comparison with younger workers.

Types of industrial accidents of advanced-age workers are as follows: The upset is important accident because of the physical characteristics of advanced-age worker.

In addition, the characteristics of industrial accidents of advanced-age workers appear in part of the loss day. The average loss day increases corresponding with the ages. The average loss day of major industrial accidents are as follow: In case of occupational disease, the average loss day has been sharply increased corresponding with the ages. And accident aftereffects of advanced-age workers are prolonged to longer term, especially, in case of electric shock accident, traffic accident, fall accident, etc.

So, in this thesis, it is analyzed the characteristics of industrial accidents of advanced-age workers and found out the problems about the safety management for the advanced-age workers. As a result, it is presented the improving measures of the safety management for the advanced-age workers.

Keywords: Advanced-age worker, Aging worker, Industrial accidents of advanced-age worker, Safety measures for advanced-age worker

1. INTRODUCTION

1.1 Background and Purpose of the Study

In 2000, Korean society has already become an aging society which, according to UN, is a society where aged people older than 65 years comprise 7 percent of society. Even after then, while the birth rate is decreasing, the average life expectancy is increasing. In other words, Korean society is quickly becoming an aging society. Also, it is expected that, in 2026, Korean society will become a super aging society which the aged people comprise 20.0 percent of the society.

Table 1 Composition Rate Changes of People of Different Ages

	1980	1994	2000	2003	2004	2010	2020	2030
Total	38,124	44,6423	47,008	47,925	48,199	49,594	50,650	50,296
0~14 years	12,951	10,653	9,911	9,719	9,633	8,552	7,034	6,217
15~64years	23,717	31,446	33,702	34,238	34,396	35,741	35,948	32,475
older than 65	1,456	2,542	3,395	3,969	4,171	5,302	7,667	11,604
Composition ratio	100.0	100.0	100.0	100.0	100.0	100.0	100.0	100.0
0~14 years	34.0	23.9	21.1	20.3	20.0	17.2	13.9	12.4
15~64 years	62.2	70.4	71.7	71.4	71.4	72.1	71.0	64.6
older than 65	3.8	5.7	7.2	8.3	8.7	10.7	15.1	23.1

Source: National Statistical Office, 「Changes of Future Population」, 2001

In addition, average age of the majority of workers in industrial fields is changing from 20s to 40s. In other words, the working field is also becoming aging. Despite that, there is not sufficient interest in advanced-age workers. As a result, industrial accidents of advanced-age workers are increasing. However, there is no definition of advance-age workers yet and there is no standard for advanced-age workers by the Occupational Safety and Health Acts.

Therefore, in this thesis, the definition of advanced-age workers will be suggested. In addition, analysis of the accident characteristics of advanced-age workers and the differences from accident characteristics of general workers will be provided. Finally, safety measures will be given to prevent accidents of advanced-age workers.

2. DEFINITION OF ADVANCED-AGE WORKERS

2.1 Definition of Advanced-age Workers

There are many discussions with the age range of advanced-age workers internationally. According to WHO(World Health Organization), the workers older than 45 years are defined as "the aging workers". So, in general, the age related to aging is regarded as 45 years.

However, in Korea, there is no clear definition of aging workers. The followings are the definitions of advanced-age workers obtained through the analysis of the existing reports and thesis. Usually, the person whose ages are older than 60 or 65 years are designated the aged people. However, according to the article 61 of the National Pensions Act, old-age pension is given to a policyholder who holds a policy for more than 20 years, and the pension is provided from the point which he or she becomes 60 years old (55 years old for special type workers) to his or her death. And, according to the Welfare Act for the Aged or National Basic Living Security Act, people older than 65 years are the aged people. Also, according to the Act for Anti-discrimination of Age and Encouragement of the Aged Employment, the person older than 55 years are defined the aged people. In this thesis, considering the youngest age of former definitions, the advanced-age worker are defined the workers who are older than 55 years and do economic activity.

2.2 Physical Characters of Advanced-age Workers

In doing industrial activities, physical characteristics of workers are very important. In carry out the work, physical strength is important as much as motivation to work and technique. However, advanced-age workers have less physical ability than young adults and middle-aged workers. The physical characteristics of advanced-age workers are as follows: Advanced-age workers have 14% of endurance of the workers of 30s. Their agility, quickness and minuteness are 20%, 32% and 48% of those of the young workers, respectively. Because advanced-age workers have less ability of looking, hearing and thinking, they are more vulnerable to industrial accidents.

The followings are functional characteristics of advanced-age workers. First, their physiological ability such as sensory ability and balance decreases quickly. Because their elasticity of crystalline lens decreases, their vision becomes blurry. Also, as their hearing ability decreases, they make mistakes in communicating and understanding danger signs and often forget things. Second, their muscular strength decreases. Third, they are well aware of knowledge and skill obtained during training, but they cannot adapt themselves to new techniques well. Fourth, there are big individual differences in functional ability among advanced-age workers. In other words, due to aging, resistance to disease and strength recovery decrease, and the possibility of accidents increases compared to the workers of other ages.

3 ANALYSIS OF CASES OF INDUSTRIAL ACCIDENTS

Before finding out accident characteristics of advanced-age workers, general trend industrial accidents of Korea were analyzed. The data were obtained after analyzing "trend of industrial accidents" from 2007 to 2010 published by KOSHA (Korea Occupational Safety & Health Agency). Then the trend of industrial accidents is analyzed for different fields and ages.

3.1 Trend of Overall Industrial Accidents

The table 2 shows an increase of the workers in whole industries, as shown in 2007 (12,528,879 workers), in 2008 (13,489,986 workers), in 2009 (13,884,927 workers), and in 2010 (14,198,748 workers). The number of injured workers also increased, as shown in 2007 (90,147 workers), in 2008 (95,806 workers), in 2009 (97,821 workers), and in 2010 (98,645 workers). On the contrary, the industrial accident rate is decreasing by 0.1% every year, as shown in 2007 (0.72%), in 2009 (0.70%), and in 2010(0.69%).

Table 2 Trend of Industrial Accidents of Korea (2007~2010)

Category	2007	2008	2009	2010
A	12,528,879	13,489,986	13,884,927	14,198,748
B	90,147	95,806	97,821	98,645
C	2,406	2,422	2,181	2,200
D	0.72	0.71	0.70	0.69
E	1.92	1.80	1.57	1.55
F	63,934,071	70,087,376	51.900.074	56,707,886

A: Number of workers(people), B : Number of injured workers(people)
C: Number of dead workers(people), D: Accident Rate(%), E: Total Death Rate
F: Loss Days

The death accidents decreased from 2007 (2,406 people) and 2008 (2,422 people) to 2009 (2,181 people) and increased in 2010 (2,200people). Total death rate decreased, as shown in 2007 (1.92), in 2008 (1.80), in 2009 (1.57), in 2010 (1.55) .

3.2 Trend of Industrial Accidents with Industrial Fields

The table 3 shows the trend of industrial accident injuries corresponding with industrial fields in Korea. There were the most injured workers 34,117 in 2007 and 35,819 in 2008 from the manufacturing industry. While, there were the most injured workers 38,222 in 2009 and 36,538 in 2010 from others industries.

Table 3 Trend of Industrial Accidents with Industrial Fields in Korea (2007~2010)

A	B	C	D	E	F	G	H	I
2007	1	1,429,885	1,239	247,460	193,993	997	41,671	944,525
	2	12,528,879	16,105	3,095,377	2,887,634	53,984	697,833	5,777,946
	3	90,147	1,593	34,117	19,050	121	4,736	30,530
2008	1	1,594,793	1,246	255,073	297,521	1,077	43,408	996,468
	2	13,489,986	15,275	3,103,942	3,248,508	54,479	703,249	6,364,533
	3	95,806	1,326	35,819	20,473	99	4,739	33,350
2009	1	1,560,949	1,146	257,686	236,747	1,110	44,356	1,019,904
	2	13,884,927	13,732	3,182,262	3,206,526	52,952	708,584	6,720,871
	3	97,821	1,118	32,997	20,998	114	4,372	38,222
2010	1	1,608,361	1,117	269,630	221,617	1,173	46,264	1,068,560
	2	14,198,748	12,548	3,196,182	3,200,645	54,080	711,094	7,024,199
	3	98,645	1,084	34,069	22,504	85	4,365	36,538

A: Year, B: Category, C: Overall Industry, D: Mining E: Manufacturing
F: Construction, G: Electricity, Gas and Water
H: Transportation, warehousing and communication service, I: Others
Category 1: Number of Business, 2: Number of Workers, 3: Number of Injured Workers
※Others industries include forestry, fishing, farming and financial insurance

The trend of industrial accidents corresponding with industrial fields as follows: In the mining industry, the number of overall workers and injured workers decreased altogether. In the manufacturing industry, from 2007 to 2010, the number of workers increased. The number of injured workers increased from 2007 to 2008 and decreased in 2009, and then increased again in 2010. In construction industry, the number of workers increased from 2007 to 2008 and started to decrease from 2009 while the injured workers increased. In the electricity, gas and water industry, workers decreased from 2007 to 2009 and increased again in 2010. The injured workers decreased from 2007 to 2008, increased in 2009, and decreased again in 2010. In the transportation, warehousing and communication service, the number of workers increased steadily. The number of injured workers also increased from 2007 to 2008, but decreased later. In the others industries, the number of workers steadily increases. The injured workers increased from 2007 to 2009 and then decreased from 2010.

4. ANALYSIS OF INDUSTRIAL ACCIDENTS OF ADVANCED-AGE WORKERS

4.1 Trend of Advanced-age Workers Employment

The Table 4 shows the chronological list of labor population from 2007 to 2010 published by national statistical office. It shows that aged workers who are older than 50 years comprise from 28.22 percent to 31.22 percent of the whole workers, and the rate is growing greater.

Table 4 Trend of Advanced-age workers Employment (2007~2010) (unit: 1000 people)

Category	2007	2008	2009	2010
A	24,216	24,347	24,394	24,748
B	6,834	7,055	7,346	7,728
C	28.22	28.97	30.11	31.22

A: Total (older than 15 years), B: Older than 50 years,
C: Rate of the employed workers older than 50 years (%)

4.2 Trend of Advanced-age Workers' Industrial Accidents

The Table 5 shows the trend of the advanced-age workers' industrial accidents. The rate of the injuries of advanced-age workers older than 55 years is increasing, as shown in 2007(19,133 workers, 21.22%), in 2008(21,083 workers, 22.00%), in 2009(24,996 workers, 25.55%), and in 2010(26,622 workers, 26.99%).

Table 5 Trend of Advance-age Workers' Industrial Accidents

Category	2007	2008	2009	2010
A	90,147	95,806	97,821	98,645
B	19,133	21,083	24,996	26,622
C	21.22	22.00	25.55	26.99

A: Total injured workers, B: Older than 55 years,
C: Rate of the injured workers older than 55 years (%)

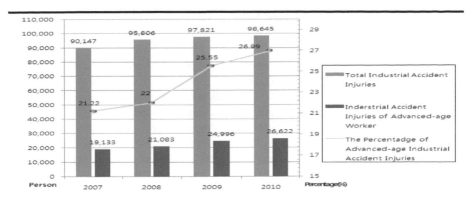

Figure 1 Trend of Advanced-age Workers Industrial Accidents

4.3 Trend of Industrial Accidents of Advanced-age Workers with Industrial Fields

The Table 6 shows the trend of industrial accident injuries of advanced-age workers corresponding with industrial fields. In mining industry, the rate of injured advanced-age workers is very high, comprising from 65.86% to 68.98%. It was the highest rate. In the manufacturing industry, the rate shows from 13.45% to 17.72%. From 2007 to 2010, the rate increased. In the construction industry, the rate shows an increase from 27.80% to 34.66%. In the industry of electricity, gas, and water, transportation, warehousing and communication service, and the other industries show increase too.

Table 6 Trend of Industrial Accidents of Advanced-age Workers with Industrial Fields

A	B	C	D	E	F	G	H	I
2007	1	90,147	1,593	34,117	19,050	121	4,736	28,137
	2	19,133	1,099	4,590	5,333	15	869	6,359
		(21.22%)	(68.98%)	(13.45%)	(27.80%)	(12.40%)	(18.35%)	(22.60%)
2008	1	95,806	1,326	35,819	20,473	99	4,739	30,522
	2	21,083	893	5,044	6,042	15	889	7,060
		(22.00%)	(67.34%)	(14.08%)	(29.51%)	(15.15%)	(18.76%)	(23.13%)
2009	1	97,821	1,118	32,997	20,998	114	4,372	33,961
	2	24,996	763	5,273	6,751	16	913	9,320
		(25.55%)	(68.25%)	(15.98%)	(32.15%)	(14.03%)	(20.88%)	(27.44%)
2010	1	98,645	1,084	34,069	22,504	85	4,365	33170
	2	26,622	714	6,038	7,799	14	976	9,576
		(26.99%)	(65.86%)	(17.72%)	(34.66%)	(16.47%)	(22.36%)	(28.87%)

A: Year, B: Category, C: Whole Industry, D: Mining, E: Manufacturing, F: Construction,
G: Electricity, gas and water, H: Transportation, warehousing and communication service, I: Other
Category 1: Total Injured Workers, 2: Number of Injured Advanced-age Workers (rate)

4.4 Trend of Advanced-age Workers' Industrial Accident Types

The table 7 shows the trend of advanced-age workers' industrial accident types in 2008. Several kinds of industrial accidents took place the most in the following order, which are upset, fall, stricture, fall-flying, collision, cutting, collapse and electric shock accident. Among them, upset accident comprised 28.04% of overall industrial accidents. Especially, the rate of advanced-age workers' upset accidents comprised 31.91% which is the highest rate of the overall upset accidents. The major cause is analyzed that advanced-age workers lack physical ability compared to young and middle-aged workers.

Table 7 Trend of Advanced-age Workers' Industrial Accident Types in Korea

	Total	Fall	Upset	Collision	Fall-Flying	Collapse	Stricture	Cutting	Electric Shock
A	95,806	14,027	18,527	7,279	8,670	942	15,250	6,615	448
B	21,083	3,655	5,912	1,440	1,955	274	2,331	1,203	57
	(100.00)	(17.34)	(28.04)	(6.83)	(9.27)	(1.30)	(11.06)	(5.71)	(0.27)
C	22.01	26.06	31.91	19.78	22.55	29.09	15.29	18.19	12.72

A: Whole Injured Workers (peoples), B: Injured Advanced-age Workers (rate %)
C: Rate of Injured Advanced-age Workers (%)

5. PROBLEMS AND IMPROVEMENT MEASURES OF SAFETY MANAGEMENT FOR ADVANCED-AGE WORKERS

5.1 Problems of Safety Management for Advanced-age Workers

Based on the trend of advanced-age workers' industrial accidents and their physical characteristics, the problems of safety management for advanced-age workers in Korea as follows:

1. In industrial fields, the rate of advanced-age workers is increasing.
2. In industrial fields, there are no differences in safety management for advanced-age workers compare to other general workers.
3. Advanced-age workers learn more slowly than general workers do.
4. Advanced-age workers have less ability of looking and hearing.
5. The rate of upset and fall accident of advanced-age workers' is high.

5.2 Improvement Measure of Advanced-age Workers' Industrial Accidents

5.2.1 Implementation and Improvement of Safety & Health Education for Advanced-age Workers

Currently, in industrial fields, advanced-age workers and young and middle-aged workers are getting the same safety & health education altogether. As mentioned before, advanced-age workers learn very slowly and have little memory ability compared to the general workers. When getting education with the same material used for the general workers, advanced-age workers show lack of understanding the material due to the decreased vision. Therefore, advanced-age workers need to get more frequent education. Also, the texts in the books need to be bigger and pictures need to be used more to help their understanding.

5.2.2 Safety Measures Using Multiple Senses for Advanced-age Workers

In general, advanced-age workers have less ability of hearing and looking compared to general workers. Therefore, advanced-age workers cannot recognize and react to general danger signs swiftly compared to the general workers. Thus, at the work places where advanced-age workers work, redundant signs that integrate visual and auditory signs are needed. In carrying out general works, supports for their decreased visual ability are also needed. First, intensity of illumination needs to be appropriate for advanced-age workers to see easily. Second, the size of the text need to be bigger and magnifying glasses need to be offered. Third, implications using scales need to be changed into digital implications so that they can see easily. Auditory supports are also needed because advanced-age workers lack hearing ability. First, noise should be minimized. Second, in sending information to them, visual way should be preferred rather than phonetic way. Third, information needs to be provided in a way that integrates visual and auditory senses. Advanced-age workers need to get additional support of sense.

5.2.3 Exclusion of Advanced-age Workers in Dangerous Works and Safety Measures Against Upset and Fall Accidents

For the safety management of advanced-age workers, they need to be excluded in carry out dangerous works. Advanced-age workers lack endurance and quickness compared to general workers. However, in industrial fields, advanced-age workers and general workers are working altogether. Especially, due to the decreased physical ability, advanced-age workers experience upset accident more often. And advanced-age workers tend to fall often, so the appropriate safety measures need to be taken. First, they need to wear non-slippery shoes when working. Second, they need to work on the ground more than on the high places because they cannot balance themselves well. If it is inevitable for them to work in the high places, safety handrails need to be built for their safety.

6. CONCLUSION

Since the Korean society became an aging society in 2000, aging is quickly going on in Korea due to the development of economy and society. Also, an increase in life expectancy due to low birth rate and the breakthrough in science and medical area in contributing to the rate of advanced-age workers in industrial fields.

In this thesis, physical characteristics of advanced-age workers were analyzed. And the rate of injured advanced-age workers in the whole industries was high. After seeing the trend of advanced-age workers' employment, the trend of industrial accident types of different fields was analyzed.

The followings are the types and characteristics of advanced-age workers' accidents. The employment rate of advanced-age workers was increasing, as shown in 2007 (28.22%), and in 2010 (31.22%). The rate of injured advanced-age workers was also increasing as shown in 2007 (21.22%) and in 2010 (26.99%). Because there are no special safety measures for advanced-age workers, it needs to be improved. After the analysis of industrial accidents of advanced-age workers, safety measures for them were not appropriate enough especially in the mining and construction industries.

Finally, after the analysis of industrial accident types, it was found that advanced-age workers experienced upset and fall accidents the most. It was resulted from their lack of physical ability. Considering industrial accident types of advanced-age workers, the following safety measures need to be taken. First, advanced-age workers need to get special safety measures separated from general workers. In other words, considering their decreased memory and visual ability, special safety education needs to be given for them. Second, safety support using multiple senses such as danger signs and visual aids need to be given. In other words, warning and sign that uses visual and auditory aid need to be used for advanced-age workers that lack visual and hearing ability. Third, advanced-age workers need to be excluded from dangerous works, and the safety measures preventing upset and fall accidents need to be taken. Because they lack in physical ability compared to general workers, they need to be excluded from dangerous works of high places and wear non-slippery shoes.

In this thesis, physical characteristics of advanced-age workers and their industrial accidents are analyzed, and appropriate safety measures were proposed.

If appropriate safety management is given in industrial fields, industrial accidents of advanced-age workers will be lessened.

REFERENCES

National Statistical Office, "Changes of Future Population", 2011

Jung-Chul Lee, Chan-Sik Lee, "Industrial Accident Injury Types of Middle/Advanced-age Construction Workers", Thesis of Architectural Institute of Korea, Vol. 22, No. 5, P. 201~ 208, 2008

Sung-Rok Jang, Eun-Ah Kim, "Safety Measures for an Aging Society–Focusing on Busan-", Korea Occupational Study, Vol. 17, No. 4, P. 184 ~ 188, 2002

Yu-Chang Kim, "Analysis of Industrial Accident Injury Types of Advanced-age Workers in Small-sized Works", Korea Occupational Study, Vol. 14, No. 3, P. 163~167, 1999

Ministry of Labor, Analysis of Industrial Accident Injuries, 2007~2010

National Statistical Office, Chronological Listing of Labor Force Population, 2011

Korea Occupational Safety & Health Agency, "Safety and Welfare of Advanced-age Workers", 2006

CHAPTER 8

Multi-scale Entropy Analysis for Evaluating the Balance of the Flatfeet

Tsui-Chiao Chao, Bernard C. Jiang

Department of Industrial Engineering and Management
Yuan Ze University
Chungli, Taoyuan, Taiwan
aatcchao@saturn.yzu.edu.tw

ABSTRACT

The foot can be regarded as a reference for the human aging process. A lack of walking or exercise results in poor development of the foot muscles, which increases the risk of falling and also increases the flatfooted population. Analyzing the complexity of postural dynamics at different time points is helpful when assessing the balancing ability of participants. This study measured the center of pressure (COP) signals from standing participant, to distinguish differences in balancing ability for flatfooted people versus those with normal feet.

An experiment was conducted to determine changes in the COP. Participants were asked to stand on a platform for 65 s with their eyes open or closed. We measured changes in the COP distribution from front-to-back (anteroposterior, AP) and side-to-side (mediolateral, ML), respectively. The method of empirical mode decomposition (EMD) was used to reconstruct short-term COP signals, and the concepts of sample entropy (SampEn) and multiscale entropy (MSE) were applied to analyze the complexity of balance. Balance was represented by a complexity index value, C_I.

Nineteen participants (11 men and 8 women) aged 22 to 42 years (mean age 29.74 ± 6.04 years) were assessed. The results showed that the C_I for the AP direction was lower for flatfooted participants with open eyes, compared with participants with normal feet (4.92±0.55 and 5.26±0.58, respectively). Flatfooted participants with closed eyes also showed lower C_I compared with the comparison participants (4.33 ±0.57 and 5.11±0.58, respectively). We also compared the results of our MSE analysis with nine traditional stabilogram metrics. The *P*-value for the eyes-open condition was .203 (not significant) and that for eyes-closed was .01. This result indicated a significantly lower C_I among flatfooted participants compared with normal-foot participants for the eyes-closed condition.

The MSE method of analysis can clearly distinguish between normal feet and flatfeet. Further experiments will be conducted to identify various symptoms of flatfeet. To help elderly avoid falling down and to alleviate their symptoms, strategies will be developed to improve their balance control.

Key words: Balance, Center of pressure (COP), Empirical mode decomposition (EMD), Flat feet, Multi-scale entropy (MSE), Sample entropy (SampEn)

1 INTRODUCTION

The effects of aging on the feet include physical and psychosocial concomitants. The contour of the foot widens with age and with lack of walking or exercise, resulting in poor development of the foot muscles, which easily leads to foot deformities and falls. Conditions such as metatarsalgia and hallux valgus are most often seen in geriatric patients (Edelstein et al, 1988). Scott et al (2007) found ageing to be associated with significant changes in foot characteristics, particularly foot posture and the severity of hallux valgus. Age-related changes in the feet have resulted in an increase in the flatfooted population (Staheli et al, 1987).

Historically, physiological signal analysis needed huge amount of time series data that were difficult and complicated to analyze. By contrast, multiscale entropy (MSE) analysis lends itself to physiologic time series analysis (Costa et al, 2002), does not assume any particular mechanism, and requires far less sample data. MSE analysis of biological systems has been applied to time series data of human gait dynamics (Costa et al, 2003) and has been used to assess differences in participants' balancing abilities (Costa et al, 2007). The MSE method allows for comparison of the degree of complexity at different time series, and shows that healthy dynamics are the most complex. Specific conditions are associated with distinct MSE curve profiles result of complexity could be used for diagnostic suggestions. Our study used MSE analysis to assess participants' balance and found that flatfooted people showed increased sway and less postural stability than people with normal feet.

A person's balancing ability is difficult to quantify. Postural balancing indicators may be used for assessment. Changes in postural sway are measured at the center of pressure (COP) at various time points of quiet standing (Collins et al, 1994). Experiments can help to determine changes in the COP during different time periods of standing. Sway data collected from a force plate placed underneath the participant are analyzed according to time-series changes in COP sway. Costa et al (2007) used this method to study the complexity of postural dynamics, by obtaining short-term data and to analyze differences in the balance of young versus elderly people. Optimum parameters include empirical mode decomposition (EMD) used to de-trend the signals.

The current study measured the COP signals from standing participants, and then used MSE analysis to distinguish postural stability for flatfooted and normal-footed adults. The reconstructed time series signals for COP data revealed differing complexities for balance associated with each foot type. Furthermore, we compared the results of our MSE analysis with nine traditional stabilogram metrics reported in the literature for flatfooted and normal-footed individuals. The *P*-values were examined to determine whether the differences between the two groups in balance ability were statistically significant.

2   METHOD

2.1   Center of Pressure (COP)

The pressure point of the body weight delivered to the focus of the ground will cause the center of the body to sway when standing, thereby providing trajectory data. The COP measurement is used to quantify the complex dynamics of postural control and enables comparison of those dynamics (Pyykko et al, 1990; Hirabayashi et al, 1995).

Priplata et al (2002) proposed two directional indicators of change in COP distribution through the feet during postural sway, namely antero-posterior (AP) and medio-lateral (ML). These measurements are obtained from front-to-back (AP) and side-to-side (ML). The complex variability of the COP position across time series are quantified for both the AP (represented on the Y axis) and ML (X axis) directions, as shown in Figure1.

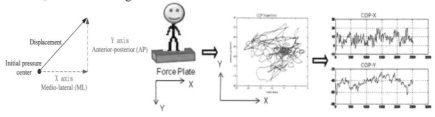

Figure 1. Initial pressure center is on the platform, with the AP position shown on the X axis and the ML position shown on the Y axis. A complexity index diagram of the postural stability of the participant is obtained from COP data during periods of quiet standing.

2.2   Stabilogram Index of Quiet Standing

Changes in the quiet standing index over time could reflect impaired balance and an increased tendency to fall (Schilling et al, 2009). Numerous metrics or features can be used to characterize a stabilogram, with measurement being based on the displacement of the COP on a force platform. Several COP measures are based on time domain parameters (Souvik and Tibarewala, 2011). Most references mention nine traditional stabilogram metrics to characterize the COP stabilogram (Prieto, 1996; Jiang et al, 2011), as follows:

1) Total excursion (TOTEX)
2) Mean displacement velocity (MVELO)
3) Planar deviation (PD)
4) Sway area (AREA-SW)
5) Mean resultant distance (MDIST)
6) 95% confidence circle area (AREA-CC)
7) 95% confidence ellipse area (AREA-CE)
8) Rfa (range fore-aft)
9) Rsw (range sideways)

The AREA-CC, AREA-CE, and AREA-SW provide estimates of the area of stabilogram. The MDIST, MVELO, PD, Rfa, Rsw, and TOTEX are time-domain distance measures used to estimate either AP or ML displacement of the COP data.

2.3   Complexity and Multiscale Entropy Analysis Methods

Scholars have analyzed the complexity of entropy to assess disorder, uncertainty, or randomness, as proposed by Kolmogorov (1965). This method was introduced to measure the complexity of physical and physiologic time series with a limited number of data points. The basic theory is derived from approximate entropy (ApEn) (Pincus, 1991) and modified sample entropy (SampEn) (Richman and Moorman, 2000). Costa et al (2002) introduced related applications using mulit-scale entropy(MSE) analysis to quantify the information content of a signal across multiple scales. Costa et al (2003) used MSE analysis to examine human gait dynamics and postural stability. Yang (2010) analyzed the influence of attention-demanding tasks using MSE analysis of postural stability COP signals.

The EMD method (Huang et al, 1998) de-trends the non-stationary time series signals to minimize confounding influences. Costa et al (2007) analyzed the new time series signals summation from high frequency intrinsic mode functions (IMFs) by EMD. Yang (2010) applied a signal reconstruction stage to choose appropriate IMFs.

Two input parameters, a run length m and a tolerance window r, must be specified for *SampEn* to be computed. The equation of sample entropy, *SampEn(m,r,N)*, where N represents the length of a time series, is the negative logarithm of the conditional probability that two sequences similar for m points remain similar at the next point, with self-matches not being included in the calculation for probability. The most widely established values for m and r are $m=1$ or $m=2$, and r between 0.1 and 0.25 times the standard deviation of the original time series (Costa et al, 2007). The original data from one-dimensional discrete time series are divided into non-overlapping windows of length n, corresponding to scale factor n. The data points inside each window are averaged and an entropy measure is calculated for each coarse-grained time series, plotted as a function of the scale. The method analyzes the profile of the obtained MSE curves, as represented by the area under the curve $\sum_{i=1}^{6} \text{SampEn(i)}$. The result, known as the complexity index (C_I), is shown in Figure 2:

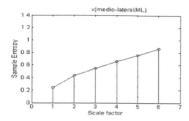

Figure 2. The C_I is the area under the curve shown, representing the area from a scale factor of 1 to 6.

3   EXPERIMENT

3.1   Flat Foot Types

Several methods are used to classify flatfeet (Razeghia and Batt, 2002). The footprint method is the most widely used and is similar to the arch index. Staheli et al. (1987) obtained static footprints from participants standing in a relaxed position and used the results to classify foot types for which the support in the central zone of the foot was greater than the width of the metatarsal support. Hui et al (2007) used the clearer and simpler "three-line" method to classify the footprint type. The current study used both the Staheli and Hui methods of classification to determine flatfoot and normal foot.

3.2   Equipment

In this study, COP signals were collected from a force plate CATASYS 2000® (Danish Product Development Ltd. 1994, Denmark) with a sampling rate at 40 Hz (Figure 3).

Figure 3. Force plate CATASYS 2000® (Danish Product Development Ltd. 1994, Denmark).

3.3   Procedure

Participants were required to understand the processes of our trial before the actual experiment. The underlying conceptual framework of our study included the following criteria: 1) participants were not mentally ill; 2) participants were asked to look straight ahead and then open or close their eyes; 3) participants stood alone quietly and in a natural erect posture; and 4) no disturbances were permitted during a trial.

The analyzed COP sway time series were derived from signals continuously recorded for 65 s. The time series length was approximately 2516 for each participant per trail, obtained from the COP plate.

One experiment per participant was constructed with five steps: 1) classify foot type of participant; 2) capture the consecutive coarse-grained time series data of COP position; 3) de-trend the raw data of dynamic behavior by EMD; 4) apply entropy (MSE) measurement to quantify the information content of all coarse-grained time series on a given time scale; and 5) plot entropy values as a function scale and analyze the profiles of the MSE curves for C_I. The procedure is shown in Figure 4.

Figure 4. Experimental process used for assessing postural stability of flatfooted participants using MSE analysis.

4   RESULTS AND DISCUSSION

The following parameters were defined for MSE analysis: run length $m=2$; tolerance window $r=0.2$; run time 65 s, thus data length for the time series $N=2516$; scale number n=7; reconstructed time series signals were suggested by summing the IMF2 and IMF3 of intrinsic mode functions of higher frequencies reconstructed by EMD; and C_I, was defined as the area under the curve for scales 1 to 7. We divided the time series averages for complexity from 19 participants into two groups, namely nine people with normal feet, and 10 flatfooted people.

We then plotted the entropy values as a function of scale to analyze the profile of the obtained MSE curves. To observe the area of the curve we plotted the results as shown in Figure 5, representing COP sway time series in AP and ML directions for the six different scale curves, with the average area summed. The complexity of flatfeet was clearly lower than that of normal feet in the AP direction, for the eyes-open and eyes-closed conditions.

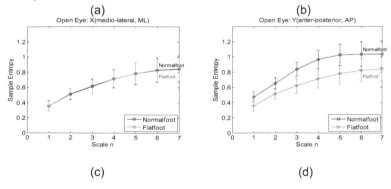

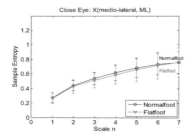
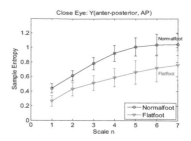

Figure 5. The MSE analysis for reconstructed COP signal by IMF2 and IMF3, comparing C_I values for flatfeet and normal feet. (a) Complexity of $ML_{(2,3)}$ direction with eyes open. (b) Complexity of $AP_{(2,3)}$ direction with eyes open. (c) Complexity of $ML_{(2,3)}$ direction with eyes closed. (d) Complexity of $AP_{(2,3)}$ direction with eyes closed.

Note: $MSE\text{-}ML_{(2,3)}$: complexity of the signal reconstructed by IMF2 and IMF3 in ML direction.

$MSE\text{-}AP_{(2,3)}$: complexity of the signal reconstructed by IMF2 and IMF3 in AP direction.

Table 1 shows the complexity result for the eyes-open condition, and Table 2 show the result for eyes-closed, compared with nine traditional stabilogram metrics. The mean C_I of AP was lower for flatfooted patients in both the eyes-open (4.92 ± 0.55) and eyes-closed (4.33 ± 0.57) conditions compared with normal-footed participants (5.26 ± 0.58 and 5.11 ± 0.58, respectively). Thus the results revealed that the COP data obtained from flatfooted patients were less complex than that of normal foot participants.

Furthermore, the results were compared with nine traditional stabilogram metrics, using the two sample t-test with the significance level (P-value) set at .05. Postural stability for AP differed significantly between the normal and flatfooted groups for the eyes closed condition ($P = .01$), but the difference between the groups for the eyes-open condition ($P = .203$) was still significant than nine traditional stabiliogram metrics. Thus the measurement of complexity could distinguish between flat feet and normal feet. For C_I values, the differences between the groups were statistically more significant for MSE analysis than for the nine comparative traditional stabilogram metrics. This finding revealed that MSE analysis could be applied to distinguish flatfeet from normal feet, and that the results were clearer than those yielded by the other nine traditional stabilogram metrics.

The appropriate frequency signals must be selected for MSE analysis of postural ability. The high frequency IMF's were used in this study; however, various combinations of IMF's may be evaluated for eyes-open and eyes-closed, respectively. Electromyography (EMG) analysis is widely used to analyze postural dynamics in flatfooted individuals. The next important consideration was whether IMF signals of MSE implied a specific muscle activity. In elderly people, flatfeet are a sign of age-related degradation and should be monitored. We thus observed the changes in stability for each participant using the MSE method, and compared the degree of postural stability at different time.

One observation in this study was that emotion could affect a participant's postural ability. Participants were therefore asked to keep calm. If they felt impatient in a happy, sad, or unpleasant mood, the standard deviation (SD) values were large. We thus asked participants keep remain emotionally stable during the trials and we did not allow distractions during the experiment.

Table 1. The comparison of COP C_l values for flatfeet and normal feet with open-eyes participants

Stabilometric parameters	19 adults (age: mean±SD, 29.74±6.04 years)		
	Normal feet	Flat feet	p value
AREA-CC (95%) (mm^2)	267.17±153.33	311.69±157.11	0.541
AREA-CE (95%) (mm^2)	254.48±154.32	275.11±117.75	0.75
AREA-SW (mm^2/s)	14.33±7.90	15.54±6.75	0.725
MDIST (mm)	4.79±1.79	5.10±1.18	0.665
PD	5.36±1.93	5.76±1.38	0.6
Rfa (mm)	20.40±6.36	23.16±6.28	0.356
Rsw (mm)	15.98±5.26	17.29±5.15	0.592
TOTEX (mm)	549.88±116.60	568.53±124.99	0.741
MVELO (mm/s)	9.16±1.94	9.48±2.08	0.741
OpenEye MSE-ML$_{(2,3)}$ (complexity)	4.05±0.69	4.05±0.68	0.989
OpenEye MSE-AP$_{(2,3)}$ (complexity)	5.26±0.58	4.92±0.55	0.203

Note: The values shown are mean $\pm$ SD, and the p value is for statistically significant level.
MSE-ML$_{(2,3)}$: complexity of the signal reconstructed by IMF2 and IMF3 in ML direction.
MSE-AP$_{(2,3)}$: complexity of the signal reconstructed by IMF2 and IMF3 in AP direction.

Table 2. The comparison of COP C_l values for flatfeet and normal feet with close-eyes participants

Stabilometric parameters	19 adults (age: mean±SD, 29.74±6.04 years)		
	Normal feet	Flat feet	p value
AREA-CC (95%) (mm2)	315.30 ±203.60	453.62 ±252.24	0.205
AREA-CE (95%) (mm2)	317.58±212.20	430.08 ±240.76	0.295
AREA-SW (mm2/s)	20.16±10.04	27.94 ±15.39	0.207
MDIST (mm)	4.99±1.76	5.91 ±1.77	0.272
PD	5.68±1.95	6.81 ±2.05	0.237
Rfa (mm)	22.04±5.65	30.54 ±9.94	0.036
Rsw (mm)	22.49±8.36	24.18 ±8.71	0.672
TOTEX (mm)	713.50±146.25	805.73 ±254.49	0.344
MVELO (mm/s)	11.89±2.44	13.43 ±4.24	0.344
CloseEye MSE-ML$_{(2,3)}$ (complexity)	3.53±0.63	3.43 ±0.80	0.769
CloseEye MSE-AP$_{(2,3)}$ (complexity)	5.11±0.58	4.33 ±0.57	0.010

5 CONCLUSION AND FUTURE WORK

MSE analysis is an efficient method to compare the postural stability of participants with normal feet versus flatfeet. This study analyzed the C_I results of the features of COP signal in the AP direction, and found that 1) higher frequency signals IMF2 and IMF3 as extracted by EMD exhibited clear differences in postural stability for participants in the two groups, in the eyes-closed condition. The statistical analysis for C_I indicated that flatfooted people have a lower ability to balance than do people with normal feet. 2) The C_I analysis demonstrated a greater ability to distinguish between the two types of feet compared with any of the nine traditional stabilogram metrics, in both the eyes-closed and eyes-open conditions. This conclusion was supported by the greater statistical significance for differences in C_I compared with the other metrics.

The MSE method thus showed a clear ability to distinguish between flatfooted and normal-footed participants in this study. The differences were more pronounced for the eyes-closed condition relative to eyes-open, for the summation frequency of IMF2 and IMF3 using sway dynamics in the time series of the COP during quiet standing. It possible exists that different combinations of IMFs would result in an even more significant distinction of postural stability between the flatfooted and the normal-footed individuals. Our future research will include additional participants, so that we can more clearly establish and confirm the meaning of each IMF in conjunction with each of the two eye conditions. To improve the control and balance ability of flatfooted people, we hope to develop strategies to help them prevent falling. We also intend to use MSE analysis to identify useful strategies of physical exercise, podiatry, orthotics, splints, or shoe modifications to improve the symptoms of geriatric foot deformities.

ACKNOWLEDGEMENTS

This research was supported by the National Science Council (NSC 100-2221-E-155 -065 -MY3), Taiwan.

REFERENCES

Collins, J. J. and C. J. De Luca. 1994. Random walking during quiet standing. *Physical Review Letters* 73(5): 764–767.

Costa, M., A. L. Goldberger, and C. K. Peng. 2002. Multiscale entropy analysis of complex physiologic time series. *Physical Review Letters* 89(6): 068102-1-4.

Costa, M., C. K. Peng, and A. L. Goldberger, et al. 2003. Multiscale entropy analysis of human gait dynamics. *Physica A* 330(1):53-60.

Costa, M., A. A. Priplata, and L. A. Lipsitz, et al. 2007. Noise and poise: enhancement of postural complexity in the elderly with a stochastic-resonance–based therapy. *A Letters Journal Exploding the Frontiers or Physics* 77:68008-1-5.

Danish Product Development Ltd., 1994. CATSYS 2000 force plate, Stolbjergrej 19, Dk-3070 Snekkersten, Denmark.

Edelstein, J. E. 1988. Foot care for the aging. *Physical Therapy* 68(12):1882-1886.

Hirabayashi, S. I. and I. Yuuji. 1995. Developmental perspective of sensory organization on postural control. *Brain and Development* 17(2):111–113.

Huang, N. E., Z. Shen, and S. R. Long, et al. 1998. The empirical mode decomposition method and the Hilbert spectrum for non-stationary time series analysis. *Proceedings A of Royal Society* 454: 903-995.

Hui, J., S. Chen, and F. Xia, et al. 2007. Comparison among two footprint-based approaches, radiograph and X-ray measurement for flatfoot. *Chinese Journal of Anatomy* 30(2):233-234.

Jiang, B. C., W. H. Yang, and J. S. Shieh, et al. 2011. Entropy-based method for COP data analysis. *Theoretical Issues of Ergonomics Science*, DOI:10.1080/1463922X.2011.61710.

Kolmogorov, A. N. 1965. Three approaches to the quantitative definition of information. *Problems of Information Transmission* 1(1): 1-7.

Razeghi, M. and M. E. Batt. 2002. Foot type classification: A critical review of current methods. *Gait & Posture* 15(3): 282-291.

Pincus, S. M. 1991. Approximate entropy as a measure of system complexity. *Proceedings of the National Academy of Sciences of the United States of America* 88:2297-2301.

Prieto, T. E., J. B. Myklebust, and R. G.Hoffmann, et al. 1996. Measures of postural steadiness: Differences between healthy young and elderly adults. *IEEE Transactions on Biomedical Engineering* 43(9): 956-966.

Priplata, A., J. Niemi, and M. Salen, et al. 2002. Noise-enhanced human balance control. *Physical Review Letters* 89:238101-1-4.

Pyykko, I., P. Jantti, and H. Aalto. 1990. Postural control in elderly subjects. *Age Ageing* 19: 215-221.

Richman, J. S. and J. R. Moorman. 2000. Physiological time-series analysis using approximate entropy and sample entropy. *American Journal of Physiologic Heart and Circulatory Physiology* 278: H2039-H2049.

Scott, G., H. B. Menzb, and L. Newcomb. 2007. Age-related differences in foot structure and function. *Gait & Posture* 26(1):68–75.

Staheli, L. T., D. E. Chew, and M. Corbett. 1987. The longitudinal arch: A survey of eight hundred and eighty-two feet in normal children and adults. *J Bone Joint Surg Am.* 69(3): 426-428.

Schilling, R. J., E. M. Bollt, and G. D. Fulk, et al. 2009. A quiet standing index for testing postural sway of healthy and diabetic adults across a range of ages. *IEEE Transactions on Biomedical Engineering* 56(2):292–302.

Souvik, D. and D. N. Tibarewala. 2011. Stabilometric postural steadiness analysis of poststroke hemiplegic patients. *International Journal of Engineering Science and Technology* 3(6):4627-4637.

Yang, W. H. 2010. Data mining on physiological single-Integrate dissimilarity approach and signal reconstruction for complexity analysis, Ph.D. Dissertation, Industrial Engineering and Management Department, Yuan Ze University, Chung-Li, Taiwan.

CHAPTER 9

Academics, Life Experiences, and Symptoms of ADHD and ODD

Valerie J. Rice, Louis Banderet*, Donna Merullo*, Gary Boykin

Army Research Laboratory-Army Medical Department Field Element
Ft. Sam Houston, San Antonio, TX
Valerie.Rice1@US.Army.Mil
*U.S. Army Research Institute of Environmental Medicine, Natick, MA

ABSTRACT

Men and women diagnosed with Attention Deficit and Hyperactivity Disorder (ADHD) as children can enter the U.S. Army, within certain constraints. The purpose of this research was to identify service members attending Army Health Care Specialist (HCS) Advanced Individual Training (AIT) who scored above, and below, adult norms on ADHD and Oppositional Defiant Disorder (ODD) self-rating scales, and to examine each group's academic history, past life events, and grade point average (GPA) during HCS training. Results revealed those scoring above adult norms on current ADHD symptoms of inattention, hyperactive/impulsive, and combined types reported a lower science orientation. The inattentive and combined types reported experiencing greater hardships prior to active duty ($p < 0.5$). Those scoring above adult norms for ODD reported a lower science orientation ($p < .05$). No differences for above or below norm groups were found for HCS final GPA for any of the symptom classifications, indicating that service members with symptoms of ADHD or ODD performed as well academically as those with fewer symptoms.

Keywords: attention deficit and hyperactivity disorder, hyperactive/impulsive, inattentive, oppositional defiant disorder,

1 INTRODUCTION

Approximately 6% to 9% of children in the United States are diagnosed with Attention Deficit and Hyperactivity Disorder (ADHD), with one to two thirds continuing to experience symptoms as adults (Barkley, Fischer, Smallish, and Fletcher, 2002; Wender, Wolf, and Wasserstein, 2001). Young men and women can enter active duty with a diagnosis of ADHD if they meet the "Physical Standards for Appointment, Enlistment, or Induction" (DOD Directive 6130.3), achieve an acceptable score on the Armed Services Vocational Aptitude Battery (administered without special accommodations), and have not taken medications for the disorder within the previous year. Oppositional Defiant Disorder (ODD) occurs during childhood, although the same symptoms can afflict adults (Harpold, et. al., 2007). Although ODD is not mentioned directly regarding military entrance standards, a current record or history of misconduct such as occurs with behavioral disorders is considered disqualifying due to concerns regarding individual adaptability to service demands (Sackett and Mavor, 2006).

Once on active duty, US Army soldiers attend Basic Training followed by Advanced Individual Training (AIT). Soldiers receive their military occupational specialty (MOS), i.e. the job they will perform during their tenure on active duty during AIT. Reducing academic attrition, while maintaining high training standards is a business case, cost-effective goal of all AIT programs. Approximately 34% of service members attending 68W, Health Care Specialist AIT reported symptoms above the adult norm on the Barkley Adult ADHD Rating Scale (Barkley, 2011a; Barkley and Murphy, 1998 and 2006; Rice, 2006). Therefore, it is important for faculty members to understand the unique characteristics of this sub-population, so training can be designed to meet the needs of all students.

ADHD is a disorder of attention. Three types of ADHD exist: ADHD Combined (ADHD-C); ADHD Predominantly Inattentive (ADHD-I); and ADHD Predominantly Hyperactive-Impulsive (ADHD-H/I) (APA, 2000). To be diagnosed with ADHD, symptoms must significantly interfere with normal functioning (be disruptive and inappropriate for one's developmental level), and have been present for at least six months (Ibid). Individuals with ADHD experience impairments in academic functioning (Hechtman, 2000), including High School (Barkley and Murphy, 1998, Barkely, et al. 2002), college (Heiligenstein, Guenther, Levy, Savino & Fulwiler, 1999), and with executive functioning as adults (Biederman, et al., 2007). Yet other research suggests those with ADHD think and approach problems differently, and that their unique perspective can be an advantage (Grossberg, 2005; Hallowell and Ratey, 2006).

The criteria a diagnosis of ODD includes a pattern of behavior lasting for at least six months, displayed more often than typical for children of the same age and peer group. ODD does not have a readily apparent adult diagnosis (Harpold, et al., 2007), yet some adults meet the diagnostic criteria (Barkley, Edwards, Laneri, Fletcher, Metevia, 2001; Harpold, et al., 2007; May and Bos, 2000; Murphy and Barkley, 1996a; Murphy, Barkley and Bush., 2002) and have more difficulties academically (Harpold and colleagues (2007)

The purpose of this paper is to compare the academic background, life experiences, and grade point averages (GPAs) of service members attending 68W Health Care Specialist (HCS) training who report current symptoms of ADHD or ODD that are above the adult norms on the Barkley Adult ADHD Rating Scale (Barkley and Murphy, 1998; Barkley and Murphy, 2006), with service members attending HSC training whose ratings are below the adult norms. This information can assist instructors in understanding the individuals attending their classes, so they can better assist their learning.

2 METHODS

2.1 Procedure. A total of 579 service members attending 68W AIT completed a demographic survey, the Banderet Life Experience Questionnaire (BLEQ), and the Barkley Adult ADHD Rating Scale (Barkley and Murphy, 1998) during their first two weeks of class. Final grades for the course were attained from the Department of Combat Medical Training, after their 16 weeks of training. All service members were briefed on the nature of the study, their freedom to choose to participate, and their ability to withdraw their participation at any time without penalty. Those who agreed to participate completed a Volunteer Agreement Affidavit prior to completing the study questionnaires.

2.2 Instrumentation

Demographics. Demographics included age, gender, and level of academic achievement (GED, High School, Some College/Associates Degree, College Degree/Graduate School). Volunteers also reported their high school science course grades and cumulative high school grades (A, B, C, D, F).

Banderet Life Experiences Questionnaire (BLEQ). The BLEQ examines life experiences (before basic training) and has 26 statements rated with a Likert Scale with discrete anchor points of Strongly Disagree, Disagree, Neutral, Agree, and Strongly Agree (Rice et al., 2006). The questionnaire has 6 factors of physical activity, accountability, achievement, hardship, science orientation, and responsibility. Development of the questionnaire was guided by Levine's (2005) research which emphasizes that autonomy, self-initiated action, learning from experience, and dealing with occasional failure are essential during child development to produce a competent and happy adult.

Barkley's Adult ADHD Rating Scale. This early version of the rating scale was originally developed by Barkley and Murphy (1998). It measures symptoms of ADHD among adult populations using self-rating checklists to examine the existence, prevalence, and impact of ADHD symptoms. Based on the DSM-IV for ADHD (APA, 2000), the scale includes a self-assessment of ADHD symptoms, symptom interference with everyday life activities, and an 8-question scale on

Oppositional Defiant Disorder (ODD). Although these rating scales cannot take the place of a professional clinical diagnosis, they provide information on the presence of ADHD symptoms and offer a method to evaluate the impact of those symptoms on academic performance.

2.3 Statistics. An alpha level of 0.05 was used with the Statistical Package for Social Sciences® (SPSS for Windows, Rel. 12.0.0. 2003 Chicago: SPSS, Inc.). The Fisher's Exact Test was used to examine gender differences, while Student's t-tests, with Levene's Test for Equality of Variance, were used to determine differences between those who scored above versus below adult normative values. For the descriptive information for age, gender, and ADHD and ODD classifications, data from all 579 volunteers were used. For subsequent analyses (including an additional analysis of gender), comparisons were made between those who scored above the adult norms for ADHD-C, ADHD-I, ADHD-H/I, or ODD, compared with those who scored below adult norms for all four categories (n = 412.

3 RESULTS

Participants were mostly male (male=350, female=228) with an age range of 17-40 years and a mean of 20.84 years (Male: $M = 21.33 \pm 4.34$; Female: $M = 20.15 \pm 3.79$). One participant did not provide gender information.

Of the total population, the percentage scoring above adult norms were 7.25% for ADHD-I, 8.80% for ADHD-H/I, 27.24%for ADHD-C, and 10.19% for ODD. Significant differences were found on mean ADHD scores for those scoring above and below the adult norms for each sub-category (Table 1).

3.1 ADHD-I, Inattention. For the ADHD-I category, 90.75% (412) scored below the adult norm and 9.25% (42) scored above the adult norm. There were no differences for gender (Fishers Exact (1, N=453) = .74, with 8.88% of men and 9.84% of women scoring above adult norms. There were no differences for education level (Fishers Exact (1, $N = 453$) = .87), high school science grade (χ^2 (2, $N = 410$) = 2.94, p = .23), or final high school grade (χ^2 (2, $N = 410$) = 2.61, p = .27). Results of t-tests with the six BLEQ factors and with students final GPA in the 68W Health Care Specialist (HCS) Course are shown in Table 2.

3.2 ADHD-H/I, Hyperactive/Impulsive. In the hyperactive/impulsive category, 88.98% (412) scored below the adult norm and 11.02% (51) scored above the adult norm. There were no differences for gender (Fishers Exact (1, N=462) = 1.00) or education level (Fishers Exact (1, N=462) = .66). No differences were seen for High School science grade ($\chi^2(1, N = 462) = .33, p = .57$) or High School final grade ($\chi^2(2, N = 461) = 3.45, p = .18$). Table 3 shows the results of t-tests with the six factors of the BLEQ and with students final GPA in the HCS Course.

Table 1. T-test Results for Scores Above and Below Adult Norms.

	N	Mean (SD)	t	p-value
Inattention				
Above Adult Norm	42	16.19 (3.00)		
Below Adult Norm	412	4.36 (2.99)	24.40	.00
Hyperactive/Impulsive				
Above Adult Norm	59	17.76 (2.57)		
Below Adult Norm	412	6.57 (3.11)	28.59	.00
Combined ADHD Score				
Above Adult Norm	158	36.14 (8.28)		
Below Adult Norm	412	15.05 (6.96)	26.63	.00
Opposition Defiant Disorder				
Above Adult Norm	51	16.92 (3.73)	29.70	.00
Below Adult Norm	412	4.11 (3.00)		

Table 2. Inattention T-test on BLEQ.

	Threshold	Mean (SD)	t	df	p-value
Physical Activity	Below	3.11 (.89)	-.36	452	.72
	Above	3.05 (.96)			
Accountability	Below	3.75 (.83)	-1.49	452	.14
	Above	3.55 (.89)			
Achievement	Below	2.61 (.72)	-.87	452	.38
	Above	2.51 (.65)			
Hardship	Below	2.71 (.67)	2.05	452	.05*
	Above	2.98 (.85)			
Science Orientation	Below	3.38 (.94)	-2.32	452	.03
	Above	2.93 (1.23)			
Responsibility	Below	3.21 (1.01)	-1.50	452	.13
	Above	3.01 (1.14)			
Final HCS GPA	Below	83.59 (8.48)	-1.85	452	.07
	Above	81.05 (8.73)			

*actual p-value = .046

Table 3. Hyperactive/Impulsive: T-test on BLEQ.

	Threshold	Mean (SD)	t	df	p-value
Physical Activity	Below	3.11 (.89)	-.94	463	.18
	Above	2.98 (.98)			
Accountability	Below	3.75 (.83)	-.03	463	.77
	Above	3.75 (.90)			
Achievement	Below	2.61 (.72)	-1.77	463	.13
	Above	2.42 (.65)			
Hardship	Below	2.71 (.67)	1.88	463	.06
	Above	2.93 (.80)			
Science Orientation	Below	3.38 (.94)	-2.13	463	.04
	Above	3.03 (1.16)			
Responsibility	Below	3.22 (1.00)	.75	463	.46
	Above	3.34 (1.09)			
Final HCS GPA	Below	83.59 (8.48)	-.15	463	.88
	Above	83.41			

3.3 ADHD-C, Combined. For the ADHD Combined Type, 72.28% (412) scored below the adult norm and 27.72% (158) scored above the adult norm. There were no differences for gender (Fishers Exact (1, N=469) = .70), education level (Fishers Exact (1, N=469) = .40), HS science grade ($\chi2(2, N = 468) = .07, p = .96$), or final HS grade ($\chi2(2, N = 468) = .29, p < .86$). Table 4 contains t-test results for ADHD-C and the six factors of the BLEQ and with students final HCS GPA.

3.4 ODD. For Oppositional Defiant Disorder, 87.67% (412) scored below the adult norm on all forms of ADD, while 12.53% (59) scored above the adult norm for ODD. There were no differences for gender (Fishers Exact (1, N=470) = .20), education level (Fishers Exact (1, N=469) = .40), or HS science grade ($\chi2(2, N = 460) = .13, p = .94$). A difference was shown for final HS grade ($\chi2(2, N = 460) = 19.3, p < .00$). Table 5 shows the t-tests results for ODD with the six factors of the BLEQ and with students final HCS GPA.

4 DISCUSSION

While the prevalence of ADHD among adults is not well documented, one finding from a study of 720 individuals who were renewing their driver's licenses found that when participants were required to meet both the symptom thresholds for both current and childhood functioning, the prevalence for all types of ADHD was

Table 4. ADHD-C: T-test on BLEQ.

	Threshold	Mean (SD)	t	df	p-value
Physical Activity	Below	3.11 (.89)	-1.57	569	.12
	Above	2.97 (.98)			
Accountability	Below	3.75 (.83)	.18	569	.86
	Above	3.77(.88)			
Achievement	Below	2.61 (.72)	-1.82	569	.07
	Above	2.49 (.68)			
Hardship	Below	2.71 (.67)	2.27	569	.02
	Above	2.87 (.78)			
Science Orientation	Below	3.38 (.94)	-3.10	569	.00
	Above	3.08 (1.09)			
Responsibility	Below	3.22 (1.00)	-.88	569	.38
	Above	3.13 (1.08)			
HCS Final GPA	Below	83.59 (8.48)	-.74	569	.46
	Above	83.02 (7.80)			

Table 5. ODD: T-test on BLEQ.

	Threshold	Mean (SD)	t	df	p-value
Physical Activity	Below	3.11 (.89)	-.68	469	.50
	Above	3.00 (1.10)			
Accountability	Below	3.75 (.83)	1.26	469	.21
	Above	3.90 (.81)			
Achievement	Below	2.61 (.72)	-1.62	469	.11
	Above	2.45 (.64)			
Hardship	Below	2.71 (.67)	1.56	469	.12
	Above	2.89 (.86)			
Science Orientation	Below	3.38 (.94)	-3.14	469	.01
	Above	2.96 (1.07)			
Responsibility	Below	3.23 (1.00)	1.43	469	.15
	Above	3.41 (1.05)			
HCS Final GPA	Below	83.59 (8.48)	-1.52	469	.13
	Above	81.80 (8.75)			

4.7% (Murphy and Barkley, 1996b). In a second, much larger study (n=3199), the prevalence of adult ADHD was 4.4% (Kessler, Adler, Barkley et al., 2006). In our study, the interest was primarily in current symptoms that might interfere with a student's ability to perform well academically; therefore, volunteers were required to meet only the criteria for current symptoms. The prevalence of ADHD was higher in this study, possibly due to the less stringent criteria used and perhaps, partly due to the stressors faced by the studied population. For example, reactions to stress or sleep loss may be similar to the symptoms associated with ADHD, such as difficulty paying attention and forgetting things. The scores for our volunteers were similar to those found in Barkley and Murphy's adult study (1998, 2006) for all categories except the ADHD-C type (Table 6).

Table 6. Comparison of Means and Standard Deviations from this study with initial normative data by Barkley and Murphy (1998, 2006)*.

	Barkley's Study	This Study Mean (SD)
Inattention		
17-29	6.3 (4.7)	6.17 (4.46)
Hyperactive/Impulsive		
17-29	8.5 (4.7)	8.46 (4.54)
ADHD Combined		
17-29	14.7 (8.7)	20.79 (11.82)
Opposition Defiant Disorder		
17-29	6.1 (4.7)	6.17 (5.01)

*For 30-49 age range N=20, none scored above adult norms

No gender or academic achievement level differences were found for those scoring above adult norms for the three types of ADHD. These findings are contrary to other findings showing ADHD occurs more often among boys than girls (Pliszka, 2007), and lower academic achievement occurs among those with diagnosed ADHD (Biederman et al., 2006; Hechtman, 2000). While High School science grades were not lower among those with ODD symptoms, their science orientation was lower, however, and poorer academic and arithmetic achievement has been found among those with a lifetime history of ODD (Harpold, et al., 2007).

Science orientation was consistently lower for all types of ADHD, as well as for ODD. The finding that those with symptoms of ADHD were less interested in science than their counterparts concurs with Barkley and colleagues findings that youth with ADHD perform more poorly on arithmetic (among other subjects) (Barkley, 2011b) and executive functioning (Biederman et al., 2006). In addition to difficulties with self regulation, poorer executive functioning translates to difficulties organizing, prioritizing, and initiating tasks; focusing, sustaining and shifting attention; regulating alertness and sustaining effort, and using working memory and accessing recall (Brown, 2005), all of which are important for success in science-focused fields.

The BLEQ hardship score was higher among those scoring above adult norms on ADHD-I and ADHD-C. The questions about hardship include "Before basic training, sometimes I had to go without enough food, clothing, or shelter for a few days; As a child growing up I was responsible for many tasks and duties; Before basic training I performed heavy construction or farm labor regularly; Before I joined the military service, I never failed at anytime that mattered; Growing up I had a challenging survival experience, e.g. lost in the woods; I witnessed a long-term illness in one of my relatives; and Before basic training I had few hardships in my life." One possible explanation may involve hardship experiences with 'treatments' such wilderness or 'survival experiences'. A hereditary component of ADHD (Alberts-Corush, Firestone, and Goodman, 1986) may also explain the relationship, if parents with ADHD have difficulties parenting their own children with ADHD. Another possible explanation is that coping with ADHD symptoms may be considered as challenging. Whatever the cause, experiencing hardships is not necessarily a negative, as BLEQ Hardships have previously been found positively predictive of academic performance in the 68W HCS program (Rice et al., 2006).

Finally, and perhaps most important, those with symptoms above adult norms for each of the three sub-types of ADHD did not have lower final 68W HCS GPA's. Nor was the final GPA achieved by those with high symptoms of ODD significantly different than those with fewer or less intense symptoms. While certainly there are a multitude of potential issues affecting those with symptoms commensurate with ADHD and ODD, in this study they did not negatively impact academic performance. Mitigating factors might include their past experiences with overcoming personal hardships. Levine (2005) claims those who have faced hardships and overcome them, more easily enter the workforce and transition into adulthood. The individuals reporting symptoms of ADHD and ODD in this study have also completed the process of applying for entry onto active duty and passed Basic Training, demonstrating their ability to set goals and achieve them. Finally, while the military environment is challenging, it is also supportive, with instructors providing tutoring, personal counseling in handling life skills such as time management or financial management, and the opportunity to re-take the training should they fail the first time.

5 LIMITATIONS

One limitation of self-report data is the possibility of participants answering to appear competent and capable, however volunteers were assured of non-attribution of responses. A second limitation of this study is the subject population, which is all military. It is unknown whether these results can be generalized to other, non-military populations. Third, this study examined only current symptoms corresponding to ADHD and ODD, thus comparisons with other research studies that used different ADHD criteria (such as a childhood history) are difficult.

6 CONCLUSION

Service members attending 68W HCS training score above adult norms on self-ratings of current symptoms commensurate with ADHD and ODD at rates ranging from 7.25% for ADHD-I to 27.24% for ADHD-C. Those scoring above adult norms report less interest in science and having experienced greater hardships prior to entering active duty, however their academic achievement during 68W HCS training parallels that of service members who score below adult norms.

DISCLAIMER

The views expressed in this article are those of the authors and do not reflect the official policy or position of the Department of the Army, Department of Defense, or the US Government.

REFERENCES

Alberts-Corush, J., Firestone, P., & Goodman, J. T. 1986. Attention and impulsivity characteristics of the biological and adoptive parents of hyperactive and normal control children. *American Journal of Orthopsychiatry, 56*: 413-423.

American Psychiatric Association (APA). 2000. *Diagnostic and Statistical Manual of Mental Disorders, Fourth Edition*, Washington, DC.

Barkley, R. 2002. Major Life Activity and Health Outcomes Associated With Attention-Deficit/ Hyperactivity Disorder, *Journal of Clinical Psychiatry, 63* (suppl 12): 10-15.

Barkley, R. A. 2011a. *Barkley Adult ADHD Rating Scale*. New York: Guilford Press.

Barkley, R.A. 2011b. "ADHD in adults: History, Diagnosis, and Impairments". Retrieved 25 Jan 2011 from www.continuingedcourses.net.

Barkley, R.A., Edwards, G., Laneri, M., Fletcher, K., Metevia, L. 2001. Executive functioning, temporal discounting and sense of time in adolescents with attention deficit hyperactivity disorder (ADHD) and oppositional defiant disorder (ODD). *J Abnorm Child Psychol.* 29:541–556.

Barkley, R.A., Fischer, M., Smallish, L., Fletcher, K. 2002. The persistence of attention-deficit/hyperactivity disorder into young adulthood as a function of reporting source and definition of disorder. *J Abnorm Psychol.* 111: 279–89.

Barkley, R. and Murphy, K. 1998. *Attention-Deficit/Hyperactivity Disorder: A Clinical Workbook*. 2nd ed. New York, NY: Guilford Press.

Barkley, R. A. & Murphy, K. 2006. *Attention deficit hyperactivity disorder: A clinical workbook (3rd ed.)*. New York: Guilford.

Biederman J, Petty C, Fried R., Fontanella, J., Doyle, A.E., Seidman, L.J., and Faraone, S.V. 2006. Impact of psychometrically defined executive function deficits in adults with ADHD. *Am J Psychiatry* 163:1730–1738.

Biederman, J., Petty, C.R., Fried, R., Doyle, A.E., Spencer, T., Seidman, L.J., Gross, L., Poetzl, K., and Faraone, S.V. 2007. Stability of executive function deficits into young adult years: a prospective longitudinal follow-up study of grown up males with ADHD. *Acta Psychiatrica Scandinavica* 116 (2): 129-136.

Brown, T.E. (2005). *Attention Deficit Disorder: The unfocused mind in children and adults* (pp 20-58). New Haven, CT: Yale University Press Health and Wellness.

Grossberg, B. 2005. *Making ADD work: On-the-job strategies for coping with attention deficit disorder.* NY, NY: Berkley Publishing.

Hallowell, E.M. and Ratey, J.J. 2006. *Delivered from Distraction: Getting the Most out of Life with Attention Deficit Disorder.* New York: Ballantine.

Harpold, T., Biederman, J., Gignac, M., Hammerness, P., Surman, C., Potter, A., Mick, E. 2007. Is Oppositional Defiant Disorder a Meaningful Diagnosis in Adults? *The Journal of Nervous and Mental Disease*, 195 (7): 601- 605.

Hechtman, L. 2000. Assessment and diagnosis of attention-deficit/hyperactivity disorder. *Child Adolesc Psychiatr Clin N Am* 9:481–498

Heiligenstein, E., Guenther, G., Levy, A., Savino, F. and Fulwiler, J. 1999. Psychological and Academic Functioning in College Students with Attention Deficit Hyperactivity Disorder. *Journal of American College Health*, 47(4): 181-185.

Kessler, R. C., Adler, L., Barkley, R. A., Biederman, J., Conners, C. K., Demler, O., Faraone, S. V., Greenhill, L. L., Howes, M. J., Secnik, K., Spencer, T., Ustun, T. B., Walters, E. E., & Zaslavsky, A. M. 2006. The prevalence and correlates of adult ADHD in the United States: Results from the National Comorbidity Survey Replication. *American Journal of Psychiatry* 163(4): 716-23.

Levine, M. *Ready or Not, Here Life Comes.* New York: Simon & Schuster, 2005.

May, B, and Bos, J. 2000. Personality characteristics of ADHD adults assessed with the Millon Clinical Multiaxial Inventory-II: Evidence of four distinct subtypes. *J Pers Assess* 75:237–248.

Murphy, K. and Barkley, R.A. 1996a. Attention deficit hyperactivity disorder adults: comorbidities and adaptive impairments. *Compr Psychiatry* 37: 393–401.

Murphy, K., and Barkley, R. A. 1996b. Prevalence of DSM-IV ADHD symptoms in adult licensed drivers. *Journal of Attention Disorder,* 1: 147-161.

Murphy, K.R., Barkley, R.A., and Bush, T. (2002) Young adults with attention deficit hyperactivity disorder: Subtype differences in comorbidity, educational and clinical history. *J Nerv Ment Dis.* 190:147–157.

Pliszka S. 2007. AACAP Work Group on Quality Issues. Practice parameter for the assessment and treatment of children and adolescents with attention-deficit/hyperactivity disorder. *J Am Acad Child Adolesc Psychiatry.* 46(7): 894-921.

Rice, V.J. 2006. Symptoms of ADHD among trainees attending health care specialist training. Presented at the Joint Accessions Research & Best Practices Symposium. San Antonio Hilton, San Antonio, TX.

Rice, V.J., Butler, J., Marra, D., DeVilbiss, C., Bundy, M., Headley, D., Dixon, M., Patton, D., Rose, P., and Banderet, L. 2006. A prediction model for the personal academic strategies for success (PASS) and academic class composite (AC^{2}T) tool. ARL AMEDD internal report for the Army Medical Department Center and School, Ft. Sam Houston, TX.

Sackett, P.R. and Mavor, A.S. 2006. *Assessing Fitness for Military Enlistment. National Research Council.* The National Academies Press: Washington, DC.

Wender, P.H., Wolf, L.E., and Wasserstein, J. 2001. Adults with ADHD: an Overview. *Ann N Y Acad Sci.* 931:1-16.

CHAPTER 10

Age Differences in Postural Strategies for Low Forward Reach

Mahiyar F. Nasarwanji, Victor L. Paquet, Edward Steinfeld

University at Buffalo, State University of New York,
Amherst, New York 14260 USA
mfn3@buffalo.edu

ABSTRACT

Age related declines in ability and function are well documented in the literature. Adults over the age of 65 have reduced strength and range of motion (ROM) as compared to the younger population. Aging therefore can lead to difficulties completing tasks of daily living. Older adults cope with excessive physical demands by modifying the way they perform a task or by modifying the environment. Difference in behaviors between older and younger adult have been identified for seated reach tasks. These differences in behavior may illustrate how older adults cope with tasks that are not optimally designed for their abilities. This pilot study investigates the differences in the postural strategies adopted for a low forward reach task between older and younger adults to further the understanding of coping strategies adopted by individuals with reduced ability.

Maximum unrestricted low forward reach was assessed for 18 older adults and 25 younger adults. Maximum low forward reach was defined as the maximum reach distance from the foot to the center of a can (0.5 kg) placed onto a shelf at knee height. Posture was not restricted; however, participants had to keep their feet together and could not kneel when performing the task. Anthropometry and active ROM data were recoded for each individual. Interaction strategy or postural joint angles including knee flexion angle, back flexion angle and shoulder flexion angle were coded using still images for the maximum low forward reaching task.

There were significant differences between young and older adults in the maximum low forward reach distance. After controlling for stature and gender, there were significant differences in the multivariate effect of interaction strategy adopted between groups. Univariate analysis indicated differences in back flexion and shoulder flexion angles between groups. Results indicate that shoulder flexion ROM had a greater influence on reach as compared to back flexion ROM.

The findings of the study suggested that older adults were more likely to stand with their feet straighter and not flex their back or shoulder as much as younger adults. This could be due limitations in balance, stability and strength in the lower limbs. The differences in behavior might suggest that different design affordances may be required to aid the different groups in completing forward reaching tasks. Further research is needed to understand why individuals with limited ability adopt different interaction strategies. In addition, the results demonstrate that it is important to consider differences in interaction strategies employed by user groups when evaluating human task performance and in design of products and environments.

Keywords: age difference, interaction strategy, coping, reach

1   INTRODUCTION

As individuals age there is a documented decrease in both ability and function. Adults over the age of 65 have reduced strength and range of motion (ROM) as compared to the younger population (Bohannon, Bear-Lehman, Desrosiers, Massy-Westropp, & Mathiowetz, 2007; Doriot & Wang, 2006; Reese & Bandy, 2010; Roach & Miles, 1991; Wright, Govindaraju, & Mital, 1997). Decrements in ability and function are often associated with challenges when completing activities of daily living (ADL) (Czaja, Weber, & Nair, 1993; Rogers, Meyer, Walker, & Fisk, 1998). Studies have also reported challenges interacting with consumer products. One such study indicated that 94% of the older adult population sampled encountered some usability problems with household products (Hancock, Fisk, & Rogers, 2001).

Although some older adults delegate challenging tasks to others or avoid such tasks, others persevere with ADL tasks, often with difficulty (Rogers et al., 1998). When investigating ADL, up to 19% of the sampled older adult population reported using compensatory or coping behaviors or adopting modifications to the layout of the environment (Rogers et al., 1998). Other studies, that investigate interactions between individuals with limited function and products, have also identified coping strategies (Kanis, 1993; Wahl, Oswald, & Zimprich, 1999). However, research studies on this topic have emphasized measuring human capabilities and not the study of actual user actions (Kanis, 1998). Coping strategies are often investigated post-hoc to explain unexpected results. Coping strategies illustrate how older adults adapt to tasks that are not optimally designed for their abilities. In addition, changes in interaction strategy create challenges when developing and evaluating predictive models of performance. As an example, when evaluating performance for high reach tasks, predictive models (such as the Human Anthropometrics and Data Requirements Investigation and Analysis) could not accurately assess the percent of population capable of completing the task since they were not programmed to account for individuals who would stand on their toes when reaching (Gyi, Sims,

Porter, Marshall, & Case, 2004; Marshall, Case, Oliver, Gyi, & Porter, 2002; Porter, Case, Marshall, Gyi, & Oliver, 2004).

The increasing older adult population (Anderson & Hussey, 2000), suggests that functional reach and its relevance to the design of home environments is of great importance to maintaining independence and a safe living environment. Older adults have been shown to face challenges with upper and lower cabinets, especially in the kitchen (Gyi et al., 2004; Lenker, Feathers, Nasarwanji, & Paquet, 2008). However reaching within the lower cabinet, especially to the lower shelves and to the back of deep cabinets had not received much attention as it is often assumed to be within the reach envelope of individuals. Although information on reach envelopes is available for older adults (Moelnbroek, 1998), there is limited knowledge on functional maximum low forward reach with objects in the hand.

When investigating the design of car cabins, both stature and age were shown to have an influence on the seated reach postures (Chaffin, Faraway, Zhang, & Woolley, 2000). Similarly, postural differences were found between older and younger adults when evaluating seated ROM (Chateauroux & Wang, 2008; Doriot & Wang, 2006). Differences in posture have also been identified for force application tasks (Chaffin, Andres, & Garg, 1983; Daams, 1990). There is no information on difference between older and younger adults in the postural strategy adopted for unrestricted standing low reach tasks. This information could help identify limiting ability measures and improve predictive performance of biomechanical models to accommodate the aging population.

The overall goal of this study was to better understanding physical coping strategies adopted by individuals with reduced ability. The specific goals were to 1) measure the functional maximum low forward reach distance for younger and older adults, 2) identify if there are differences in the postural strategies adopted for a low forward reach between older and younger adults and 3) identify which personal ability characteristics, if any, limited low forward reach.

2   METHODS

Forty three participants volunteered to participate in the study. Demographic information including sample size and age for the younger adults and older adults who participated in the study is summarized in Table 1. Most participants were right handed; four older adults were left handed, one of whom was initially right handed but had to use her left hand due to a cerebrovascular accident (stroke). In addition to basic demographic information, anthropometric measures that would influence forward reach were collected including stature, acromion height, upper arm length, lower arm length, patella height and trochanter height. All anthropometric measures were collected with the participant standing using previously published standardized landmarks (Harrison & Robinette, 2002; Smith, Norris, & Peebles, 2000). For the patella height the base of the patella (proximal point) was palpated and for the trochanter height the greater trochanter was used. Active ROM for the back (lumbar flexion) and shoulder flexion was also collected using manual goniometers based on standardized procedures to evaluate limiting factors for reach (Reese & Bandy,

2010). To evaluate lower limb function, the standardized sit-to-stand time for 5 trials was used based on standardized procedures (Bohannon, 1995, 2006). The only modification made to the sit-to-stand test was that older adults were allowed to use their hands for support if the trial could not be completed otherwise. Table 1 summarizes the anthropometric measures, ROM measures and sit-to-stand time collected for younger and older adults.

Maximum low forward reach was defined as the maximum horizontal reach distance achievable in the sagittal plane from the lateral malleolus of the fibula to the center of a can placed onto a shelf at patella height when the can was held with a power grip (Figure 1). All participants had adequate grip strength to complete the task.

Table 1 Demographic data, anthropometric measures, active range of motion measures, and sit-to-stand time measured for younger and older adults

	Younger adults		Older adults	
Demographic information	Sample size (n)	Age (SD) (years)	Sample size (n)	Age (years)
Males	16	23.75 (5.40)	3	81.67 (11.95)
Females	9	23.33 (5.00)	15	82.00 (8.75)
Overall	25	23.60 (5.13)	18	81.94 (8.79)
Anthropometric measures (mean (SD)) (m)				
Stature	1.736 (0.084)		1.545 (0.068)	
Acromion Height	1.425 (0.075)		1.291 (0.064)	
Upper arm length	0.300 (0.031)		0.289 (0.023)	
Lower arm length	0.258 (0.030)		0.242 (0.027)	
Patella height	0.549 (0.038)		0.505 (0.025)	
Trochanter Height	0.909 (0.049)		0.839 (0.045)	
Active range of motion measures (mean (SD)) (degrees)				
Back flexion	113.33 (7.90)		74.69 (25.06)	
Shoulder flexion	160.20 (10.85)		100.26 (28.98)	
Sit-to-stand time (mean (SD)) (s)				
No support	7.72 (1.59) (n = 25)		14.67 (2.70) (n = 10)	
With support	-		19.12 (4.72) (n = 8)	
Overall	7.72 (1.59)		16.65 (4.27)	

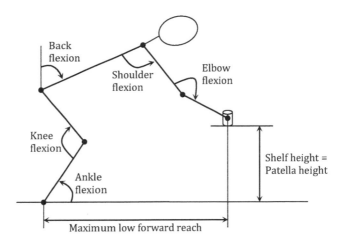

Figure 1 Schematic representation of the human, shelf height set, joint angles measured and maximum low forward reach distance measured

A vertically and horizontally adjustable shelving system was set to the patella height for each participant. The participant was then instructed to grasp the can (height = 82.75 mm, diameter = 73.5 mm, 0.36 kg) using a power grip and place the can on the shelf as far as comfortably possible without losing balance. The postural strategy was not restricted and participants were allowed to bend their knees or squat when reaching; however, they had to keep their feet together and could not kneel during the task. There were no external supports present in the environment that the participant could use, other than the shelf itself. At the instant the participant placed the can into the shelf a still digital image was collected of the sagittal plane to extract information on the postural strategy adopted. The horizontal distance from the front of the foot to the center of the can was recorded. To account for the measurement offset from the front of the foot to the lateral malleolus of the fibula, 20 mm was added to the recorded distance based on published anthropometric data (Friendly Systems Ltd., 1995). Participants were given an initial trial run, where gross adjustments were made to the shelf in the horizontal direction to accommodate participant stature and reach capability. The recorded trial was repeated if the participant did not place the can straight down, pushed the can further or used support from the shelf in any way. A spotter was always present, especially for older adults, as a precautionary measure in case the participant lost balance during the trial. This study was carried out as part of the larger study in which participants were paid.

The postural interaction strategy for reach represented by joint angles adopted by participants were measured for the ankle, knee and back; and shoulder and elbow for the hand used to reach from the still image recorded using the software Kinovea v0.8.15 (Figure 1). The hand used to reach was also recorded. Postural interaction strategy could not be extracted for four older adults due to occlusion and or corrupt images. Figure 1 shows a schematic representation of the human as a simple five-joint kinematic chain, the shelf height set, joint angles measured and maximum low forward reach measured.

Differences between groups in the maximum low forward reach distance were evaluated using a univariate ANOVA procedure using stature and gender as a covariate. Difference between groups in the postural strategy adopted (joint angles) were evaluated using a multivariate ANOVA procedure using stature and gender as a covariate followed by univariate ANOVAs.

3   RESULTS

Most of the older adults cohort were residents of an assistive or independent living facility and most exceeded the reported life expectancy value of 77.1 years for the U.S. population (Kinsella & Velkoff, 2001). In addition, 83% of the older adults (n = 15) reported using a walking aid (walker or cane) most of the time. In comparison, the younger adult cohort represented a majority of college students from the local university. All anthropometric measures were significantly correlated with each other except lower arm length with stature and acromion height (Table 1). Hence, only stature was used as a covariate in all the statistical models. In addition, active ROM for back flexion and shoulder flexion were correlated.

There were significant differences between groups ($F_{1,37}$ = 14.795, p > 0.001) in the maximum low forward reach distance when controlling for stature and gender. Younger adults were able to reach 1.085 m as compared to 0.808 m for older adults (Figure 2). Older adults had a greater variation in maximum low reach distance with a coefficient of variation of 0.218 as compared to only 0.062 for younger adults. All individuals, except one, used their dominant hand to reach.

Figure 2 Cumulative probability distribution of maximum low forward reach for younger and older adults in meters

After controlling for stature and gender there were significant differences between younger and older adults ($F_{6,30} = 4.01$, $p = 0.005$) in the multivariate effect of interaction strategy (joint angles) adopted for the maximum low forward reach task. Univariate analysis indicated significant differences between younger and older adults in the postural strategy adopted for back flexion angles and shoulder flexion angles for the maximum low forward reach task (Figure 3). Younger adults adopted back flexion and shoulder flexion angles of 10.5° and 27.8° higher than those of older adults respectively. In addition, on average, younger adults had lower knee and ankle flexion angles as compared to older adults. Greater variation was found for younger adults in the ankle and knee flexion angles which can be attributed to some younger adults who opted to squat.

There was a low positive correlation ($r = 0.22$) between sit-to-stand time and knee flexion angles adopted, indicating that individuals with limited lower limb function (represented by high sit-to-stand time) adopted greater knee flexion angles or stood with their legs straighter. As there were no differences between groups, ankle and knee flexion were not investigated further.

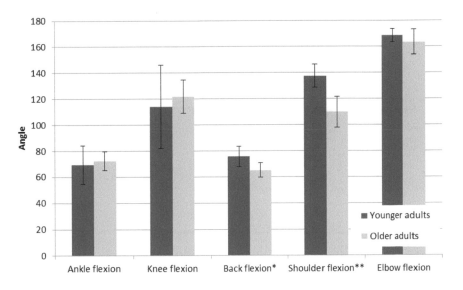

Figure 3 Specified joint angles adopted by younger and older adults when performing maximum low forward reach in degrees; error bars indicate standard deviation (Difference between groups; *: p > 0.005, **: p > 0.001)

Back flexion angles and shoulder flexion angles adopted were significantly correlated with the active ROM of the back ($r = 0.607$) and shoulder ($r = 0.808$) respectively. Some older adults were capable of adopting shoulder and back flexion angles greater than what was expected based on the active ROM during the reaching task (Figure 4). This indicates that function measured during a task may vary from that measured with standardized performance tests due to motivational and contextual factors, e.g. purposeful activity and a familiar environment. In addition, both younger and older adults adopted shoulder flexion angles closer to the maximum active ROM limit as compared to back flexion angles as seen by the groupings closer to the diagonal line in Figure 4 for shoulder flexion. Hence, for maximum low forward reach at patella height shoulder flexion was probably a greater limiting factor as compared to back flexion for both groups.

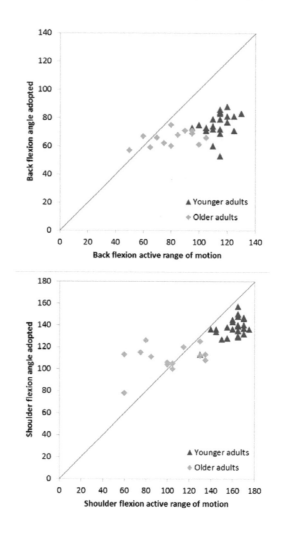

Figure 4 Correlation between postural angle adopted and active range of motion for back (left) and shoulder (right)

4   DISCUSSION

The goals of this study were to measure functional maximum low forward reach for older and younger adults, and investigate differences in the postural interaction strategy adopted between groups for the reach task. Younger adults had a maximum low functional reach approx. 0.2 m greater than that of older adults at patella height. Reach distances for older adults matched published findings for reach envelopes, but were approximately 0.05 m shorter due to the functional nature of the task selected in the current study (Moelnbroek, 1998). The increased variability in reach distances for the older adult population can be attributed to greater variation in the ability measures of ROM and lower limb function.

Lower cabinets have been shown to pose a challenge to older adults (Lenker et al., 2008). However, based on the sampled data all older adults should be able to reach the back of lower cabinets with a designed depth of up to 24 inches (0.61 m) when the shelf is at knee height. The current study only evaluated functional reach in an unrestricted environment and does not account for challenges associated with environmental factors such as obstructions created by the countertop (top of the cabinet) to line of sight and reach, or the influence of using the counter top or shelf as a support to increase functional reach distances. Hence, although the current study indicates that the lower cabinet depth was acceptable, it is only the first step towards creating functional low reach envelopes; the influence of environmental factors needs to be investigated to arrive at design relevant recommendations for functional reach distances.

There were differences in the biomechanical and externally visible postural interaction strategy for low forward reach when standing, especially for shoulder and back flexion angles. These findings are similar to those found for seated task where age had a significant effect of the posture adopted when reaching (Chaffin et al., 2000; Chateauroux & Wang, 2008; Doriot & Wang, 2006). For low reach tasks shoulder flexion ROM may be a greater liming factor compared to trunk flexion ROM for both older and younger adults. In addition, during the functional reach task participants adopt angles greater than what would be predicted based on ROM alone. Indicating that individual's function during a task may be different from what is evaluated during standardized performance tests. Task motivation and context could explain some of the difference in standardized and task based performance measures.

In general, older adults adopted lower back flexion and shoulder flexion angles while standing with their legs straighter, as compared to younger adults; indicating that older adults did not lean as far forward as younger adults. Increased trunk flexion angles and shoulder flexion angles during a forward reach task would move the center of gravity further towards the toes increasing instability. Since the older adult sample had low lower limb function, with sit-to-stand times that were comparable to those with balance disorders (Whitney et al., 2005), balance in addition to ROM could be a limiting factor for reach. For individuals with low lower limb function, and those who use walking aids commonly, balance should be considered when evaluating reach tasks to better understand difference in the biomechanical strategy adopted.

The variation in strategies within group and difference in strategies between groups indicate that individuals may not interact similarly and instead adopt strategies from a repertoire of interaction strategies based on their ability (Park, Martin, Choe, Chaffin, & Reed, 2005; Rosenbaum, Meulenbroek, Vaughen, & Jansen, 2001). Due to this self-optimization of strategy based on ability and the results of the current study it would be foolhardy to assume that older and younger adults interact similarly. Furthermore, variation in interaction strategies between individuals, especially those with limited function, should be considered in design. Different types of design affordances may be required to aid the different groups in completing forward reaching tasks. Information on the interaction strategies employed by different user groups seems essential when evaluating human task performance. Data for studies such as this could be incorporated into existing predictive biomechanical models to account for variation in interaction strategy and the influence of age and ability, which would improve the predictive capability of such models, particularly when applied to diverse populations and consumer product use.

5 CONCLUSIONS

The study found difference in the low maximum functional reach between older and younger adults. Limitations in shoulder flexion seemed to be the most critical factor that limits reach, however other factors such as balance need to be considered especially for the older adult population. Difference in the postural interaction strategy were also identified for the reach task between younger and older adults, indicating that older adults may be using different or alternate interaction strategies to accommodate for limited function. Additional theoretical and empirical research is needed to better understand why older adults adopt different strategies and what the limiting ability characteristics for such tasks are. In addition, considering the influence of the environment and the influence of the object in the hand such as weight and size need to be considered. It is important that the interaction strategies employed by user groups is also considered when evaluating human task performance to accurately describe results.

ACKNOWLEDGEMENTS

The research was supported in part by Grant #H133E050004-07 from the U.S. Department of Education, National Institute on Disability and Rehabilitation Research (NIDRR) as part of the Rehabilitation Engineering Research center on Universal Design at the Center for Inclusive Design and Environmental Access at the University at Buffalo. The contents of this paper reflect the views of the authors and do not necessarily reflect the views of the U.S. Department of Education or NIDRR.

REFERENCES

Anderson, G. F., & Hussey, P. S. (2000). Population aging: A comparison among industrialized countries. Health Affairs, 19(3), 191-203.

Bohannon, R. W. (1995). Sit-to-Stand test for measuring performance of lower extremity muscles. Perceptual and Motor Skills, 80, 163-166.

Bohannon, R. W. (2006). Reference values for the five-repetition sit-to-stand test: A descriptive meta-analysis of data from elders. Perceptual and Motor Skills, 103, 215-222.

Bohannon, R. W., Bear-Lehman, J., Desrosiers, J., Massy-Westropp, N., & Mathiowetz, V. (2007). Average Grip Strength: A Meta-Analysis of Data Obtained with a Jamar Dynamometer from Individuals 75 Years or More of Age. Journal of Geriatric Physical Therapy, 30(1), 28-30.

Chaffin, D. B., Andres, R. O., & Garg, A. (1983). Volitional Postures during Maximal Push/Pull Exertions in the Sagittal Plane. Human Factors: The Journal of the Human Factors and Ergonomics Society, 25(5), 541-550.

Chaffin, D. B., Faraway, J. J., Zhang, X., & Woolley, C. (2000). Stature, Age, and Gender Effects on Reach Motion Postures. Human Factors: The Journal of the Human Factors and Ergonomics Society, 42(3), 408-420.

Chateauroux, E., & Wang, X. (2008). Effects of Age, Gender, and Target Location on Seated Reach Capacity and Posture. Human Factors: The Journal of the Human Factors and Ergonomics Society, 50(2), 211-226.

Czaja, S. J., Weber, R. A., & Nair, S. N. (1993). A human factors analysis of ADL activities: A capacity-demand approach. The Journals of Gerontology, 48(Special Issue), 44-48.

Daams, B. J. (1990). Static Force Exertion in Standardized, Functional and Free Postures. Proceedings of the Human Factors and Ergonomics Society Annual Meeting, 34(10), 724-728.

Doriot, N., & Wang, X. (2006). Effects of age and gender on maximum voluntary range of motion of the upper body joints. Ergonomics, 49(3), 269-281.

Friendly Systems Ltd. (1995). People size v1.4: Friendly Systems Ltd.

Gyi, D. E., Sims, R. E., Porter, J. M., Marshall, R., & Case, K. (2004). Representing older and disabled people in virtual user trials: data collection methods. Applied Ergonomics, 35, 443-451.

Hancock, H. E., Fisk, A. D., & Rogers, W. A. (2001). Everyday products: Easy to use ... or not? Ergonomics in Design 12-18.

Harrison, C. R., & Robinette, K. M. (2002). CAESAR: Summary statistics for the adult population (Ages 18-65) of the United States of America: United States Air Force Research Laboratory.

Kanis, H. (1993). Operation of Controls on Consumer Products by Physically Impaired Users. Human Factors: The Journal of the Human Factors and Ergonomics Society, 35, 305-328.

Kanis, H. (1998). Usage centered research for everyday product design Applied Ergonomics, 29(1), 75-82.

Kinsella, K., & Velkoff, V. A. (2001). An aging world: 2001. (P95/01-1). Washington, DC: U.S Government Printing Office Retrieved from http://www.nia.nih.gov/NR/rdonlyres/A69F17DC-3B7D-4C3C-A90A-96CACE6E7EDF/0/agingworld2001.pdf.

Lenker, J. A., Feathers, D., Nasarwanji, M. F., & Paquet, V. L. (2008). Perspective of elders on product usability in the home. Paper presented at the International Conference of Aging, Disability and Independence, St Petersburg, FL.

Marshall, R., Case, K., Oliver, R., Gyi, D. E., & Porter, J. M. (2002). A task based 'design for all' support tool. Robotics and Computer-Integrated Manufacturing, 18, 297-303.

Moelnbroek, J. F. (1998). Reach envelopes of older adults. Paper presented at the Proceedings of the Human Factros and Ergonomics Society 42nd Annual Meeting, Chicago.

Park, W., Martin, B. J., Choe, S., Chaffin, D. B., & Reed, M. P. (2005). Representing and identifying alternative movement techniques for goal-directed manual tasks. Journal of Biomechanics, 38(3), 519-527.

Porter, J. M., Case, K., Marshall, R., Gyi, D. E., & Oliver, R. (2004). 'Beyond Jack and Jill': designing for individuals using HADRIAN. International Journal of Industrial Ergonomics, 33(3), 249-264.

Reese, N. B., & Bandy, W. D. (2010). Joint range of motion and muscle length testing. Missouri, St. Louis: Saunders Elsevier.

Roach, K. E., & Miles, T. P. (1991). Normal Hip and Knee Active Range of Motion: The Relationship to Age. PHYS THER, 71(9), 656-665.

Rogers, W. A., Meyer, B., Walker, N., & Fisk, A. D. (1998). Functional limitations to daily living tasks in the aged: A focus group analysis Human Factors, 40(1), 111-125.

Rosenbaum, D., Meulenbroek, R. J., Vaughen, J., & Jansen, C. (2001). Posture-based motion planing: Applications to grasping Psychological Review, 108(4), 709-734.

Smith, S., Norris, B., & Peebles, L. (2000). OLDER ADULTDATA: The handbook of measurements of capability of the older adults - Data for design safety Nottingham, U.K.: University of Nottingham, Product Safety and Testing Group.

Wahl, H.-W., Oswald, F., & Zimprich, D. (1999). Everyday Competence in Visually Impaired Older Adults: A Case for Person-Environment Perspectives. Gerontologist, 39(2), 140-149.

Whitney, S. L., Wrisley, D. M., Marchetti, G. F., Gee, M. A., Redfern, M. S., & Furman, J. M. (2005). Clinical Measurement of Sit-to-Stand Performance in People With Balance Disorders: Validity of Data for the Five-Times-Sit-to-Stand Test. Physical Therapy, 85(10), 1034-1045.

Wright, U., Govindaraju, M., & Mital, A. (1997). Reach profiles of mean and women 65-89 years of age. Experimental Aging Research, 23, 369-395.

CHAPTER 11

An Application of Ballistic Movement Models for Comparing Ageing Differences While Interacting with a Touchscreen

Ray F. Lin, Shin-Wen Shih, & Bernard C. Jiang

Department of Industrial Engineering & Management, Yuan Ze University
Chung-Li, Taiwan 32003
juifeng@saturn.yzu.edu.tw

ABSTRACT

To make sure innovative touchscreen techniques can benefit well the elderly population, this study utilized ballistic movement models to evaluate the differences of movement speed and accuracy between elder and young adults while interacting with a touchscreen. Six elder and six young participants conducted ballistic movements on a touchscreen monitor. The measured data of movement time and endpoint error were utilized to test the ballistic movement models. Our results showed that ballistic movement models fitted well young participants' data. However, elder participants performed movements with a conservative manner; their movement time and variable errors did not increase with increased movement distance as much as that of young participants'. Future research will focus on enhancing experimental designs to obtain solid conclusions so that the results could help developing touchscreen design guidelines.

Keywords: ageing, ballistic movements, ballistic movement models, aiming movement, touchscreens, interactive displays

1   INTRODUCTION

Touchscreen displays are quickly become an important interactive device for a variety of electronic products, such as information kiosks, global positioning systems, MP3 players, smart phones, and tablet computers. By attaching touchscreens to electronic devices, users can directly interact with what is displayed on screens with their fingers, rather than indirectly with additional input devices. Because of zero displacement between input and output, control and feedback, and hand action and eye gaze, touchscreens have been considered as the most intuitive way for human computer interaction.

To enhance touchscreen usability, as traditional input devices, researchers study user's task performance while using touchscreen devices. Common hand-control movements for interacting with touchscreens include click, drag, drop, multi-finger zoom in and zoon out. All of these movements begin by a pointing movement, such as pointing the finger to a certain location on the touchscreen surface. The time of pointing movement influences user's motivation and satisfaction with touchscreen devices. Hence, studying pointing movements is one of effective ways to evaluate and improve touchscreen devices.

1.1   Age-associated Changes and Relevant Impacts of Touchscreen Device Usage

Because the world's population is ageing dramatically, device designers and developers pay more and more attention to the elderly population. Compared to young adults, the elderly's physical, cognitive, and sensory abilities are relatively diminished (Craik and Salthouse 2000). The elderly have a variety of age-associated changes, such as reduced muscle strength (Ranganathan et al. 2001, Carmeli et al. 2003), reduced range of motion (Carmeli et al. 2003, Gajdosik et al. 2009), reduced manual dexterity (Carmeli et al. 2003, Hourcade and Berkel 2008) and greater difficulty of executing hand-control movements (Shiffman 1992, Maki and McIlroy 2006). These changes could generate obstacles while executing target acquisition task (e.g., selecting an icon or a menu item). Hence, to produce elder-friendly products, understanding of these changes and developing compensatory strategies are imperative.

Several studies showed that the elderly have greater difficulty executing pointing movements on touchscreen devices. For example, Fezzani et al. (2010) tested ageing differences while performing numerous sequences of serial pointing movements with a stylus on a digitizer tablet and found that elder participants performed movements with a longer duration. Hourcade and Berkel (2008) also evaluated ageing differences when using pens to interact with handheld computers and concluded that elder participants achieved lower accuracy rates. Furthermore, Moffatt and McGrenere (2007) compared ageing effects on pen-based target acquisition on a tablet computer and found that elder participants required longer movement time and achieved lower movement accuracy.

1.2   Fitts' Law as a Common Method

To compare ageing differences, Fitts' law (Fitts 1954) has been the most common method. As shown in Equation 1, Fitts' law describes the speed-accuracy tradeoff relationship while performing target acquisition tasks in which a user uses a stylus or finger to quickly reach a target by moving certain distance.

$$MT = a + b \times \log_2 \frac{2A}{W} \tag{1}$$

where MT is movement time; a and b are experimentally determined variables; the logarithmic term is called "index of difficulty"; A is movement amplitude; W is target width.

Fitts' law (1954) is a useful method to measure age differences while using touchscreen devices, but the results obtained by this method do not give us adequate information to know why elder people execute movements with a longer duration. Specifically, Fitts' law allows researchers to obtain the result of pointing movement time that is an overall performance of movement speed, movement accuracy and sensory ability. Therefore, by applying Fitts' law as methodology, one has difficulty clarifying the extent to which speed, accuracy, and sensory individually contribute to long-duration movements. To develop effective techniques for helping elder users of touchscreen devices, it is necessary to understand the factors that cause the differences between the elderly and young adults in more detail.

1.3   Ballistic Movement Models as a Potential Method

To evaluate the performances of movement speed and accuracy individually, Lin and Drury (2011) suggested a method of using ballistic movement models. Lin et al. (2009) proposed a general model that stated that an aiming movement described by Fitts' law is consisted of several subsequent ballistic movements. The understanding of ballistic movements can predict the performance of Fitts-type aiming movements. To understand how one performs ballistic movements, Lin and Drury (2011) verified two models, originally proposed by Hoffmann and Gan (Hoffmann 1981, Gan and Hoffmann 1988) and Howarth et al. (Howarth *et al.* 1971), to predict ballistic movement time and endpoint variability. As shown in Equation 2, the ballistic movement time ($t_{ballistic}$) is linearly related to the square root of ballistic movement distance ($\sqrt{d_u}$).

$$t_{ballistic} = e + f\sqrt{d_u} \tag{2}$$

where e and f are experimentally determined constants. Furthermore, Equation 3 shown as below shows that the endpoint variability of a ballistic movement is linearly related to the square of ballistic movement distance (d_u^2).

$$\sigma^2 = g + h \times d_u^2 \tag{3}$$

where g and h are experimentally determined constants. These two ballistic movement models was validated with the hand-control movements performed on a drawing table and were further validated for the hand-control movements performed in a true three-dimensional environment (Lin and Ho 2011).

1.4   Research Objective

This research aimed at testing ballistic movement models for comparing ageing differences of movement speed and endpoint variability while interacting with a touchscreen. Compared to Fitts' tapping movements, ballistic movements are relatively essential to describe one's capability of movement control. The speed-accuracy tradeoff relationship described by Fitts' law is a combination resulted from the properties of ballistic movement time and ballistic movement variability. To enhance the usability of touchscreen designs, ballistic movements could be a more representative experimental task to measure.

2   METHOD

2.1   Participants and Apparatus

Six colleague students and six health elder adults participated in this study. These colleague students, aged from 23 to 27 years, were familiar with the use of computers. The elder adults, aged from 62 to 79 years, were all healthy and had a regular exercise habit in the morning. Both sexes were equally distributed in these two groups. All of them were right-handed with normal or corrected-to-normal vision.

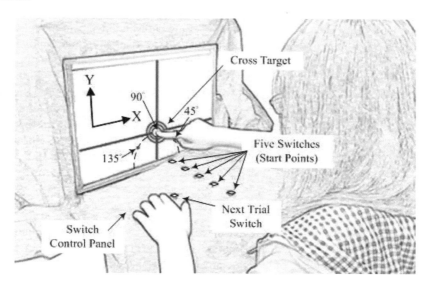

Figure 1 Executions of ballistic movement on a modified touchscreen

Experimental apparatus included a personal computer, a 19" modified LED monitor, and a switch control panel. The computer was used to run Visual Basic 2010 using a self-developed program that both displayed the experimental tasks and measured task performance. The LED monitor was attached with a touch panel on which participants could perform ballistic movements by using their pointing fingers. The monitor was modified to a condition where its backlight could be turned on and off rapidly by the program to make movements ballistically. As shown in Figure 1, on the center of switch control panel, five push switches were mounted linearly toward the monitor to generate five start points. The intervals between the centers of these switches were 50 mm, providing five different distances (50, 100, 150, 200, and 250 mm) between the touch panel and these switches. An additional switch, called the "Next Trial Switch", was mounted on the left side of the control plane that was used to continue the next trial.

2.2 Experimental Setting and Procedures

The experiments were conducted in a darkened room in which the modified LED monitor was the only illumination source. As shown in Figure 1, the participant sat on a chair in a distance about 500 mm from the LED monitor while performing experimental tasks. To perform experimental tasks, participants executed ballistic movements from one of the five push switches to the center of a cross target showed on the monitor. To start tasks, the participants first used their pointing fingers to press down a switch indicated by the program and then moved quickly toward the cross target displayed right after the switch was pressed. Once the pointing finger were moved away the switch, the backlight of the monitor immediately turned off and the movement time started to record. When the finger touched the screen, the backlight turned on and the information about the cross target and the endpoint of that movement were recorded and displayed on the screen. By clicking the "Next Trial Switch", participants could continue on the next trial.

2.3 Experimental Variables

Independent variables were different age group of participant (Age Group) and movement distance (Distance). To make sure that executing movements were not obstructed by any switches and to eliminate learned kinesthetic feedback, cross targets were programmed to appear at three different locations on a virtual circle (radius = 57 mm) at the angles of 45°, 90° and 135°, respectively. The virtual extension line of five push switches hit the center of the circle. Hence, five values of ballistic movement distance (d_u) tested in this study were 76, 115, 160, 208, and 256 mm. Every experimental combination was replicated 8 times, resulting in a total 120 trials (3 angles × 5 distances × 8 replications). All the trials were randomly conducted by each participant, taking about 20 to 30 minutes to finish. Each participant performed two formal measurements in different days. There was a one-hour practice and ten minutes practice before the first and the second formal measurements, respectively.

Three dependent variables were ballistic movement time, horizontal endpoint error (X error), and vertical endpoint error (Y error). As shown in Figure 1, X error was measured in the horizontal direction and Y error was measured in the vertical direction of the monitor.

3   RESULTS

3.1   Ballistic Movement Duration

Analysis of variance was performed on the movement time, using a mixed model with Distance and Age Group as fixed effects and Participant as a random effect nested within Age Group. The results showed significant the main effect of Distance ($F_{4,40} = 32.11$, $p < 0.001$), implying that the increase of ballistic movement distance resulted in increased ballistic movement time. However, the main effect of Age Group was not significant ($F_{1,10} = 1.22, p > 0.05$).

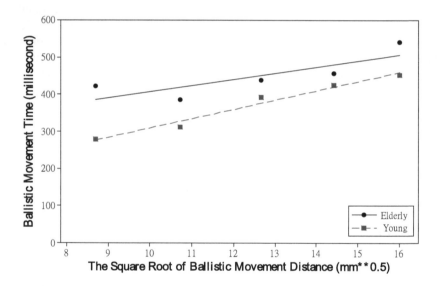

Figure 2 Relationship between ballistic movement time and the square root of distance

Because the significant main effect of Distance was found, the application of Equation 2 was tested. The means of ballistic movement time ($t_{balllitic}$) were calculated for elder and young participants respectively and were then regressed on to the square root of ballistic movement distance ($\sqrt{d_u}$). As shown in Figure 2, the model fitted young participants' data better than elder participants. Equation 2 accounted for 96.4% variance of the overall young participants' data. However, the model did not predict well the overall elder participants' data ($p > 0.05$). Although the main effect of Age Group was not significant, there was a small trend for young participants to have shorter movement time compared to elder participants.

3.2   Ballistic Movement Accuracy

Endpoint errors are consisted of constant error and variable error. To analyze whether Distance and Age Group had significant effects on the two types of errors, eight replications of each experimental combination were calculated as the constant error and the variable error (measured by variance). Since the constant errors were small (less than 2 mm), only the results of variable error were discussed in this article.

Analysis of variance, as movement time, was performed on X- and Y-variable errors using a mixed model with Distance and Age Group as fixed effects and Participant as a random effect nested within Age Group. The results only showed significant main effect of Distance on Y-variable error ($F_{4,40} = 7.59, p < 0.001$), indicating that the increase of ballistic movement distance resulted in increased Y-variable error.

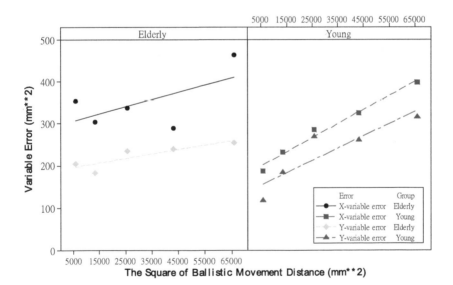

Figure 3 Relationships between two types of variable errors and the square of distance for elder adults and young adults

Although the main effect of Distance had no statistically significant effect on X-variable error and the main effect of Age Group had no significant effect on both types of variable errors, Equation 3 was tested for two groups' both axes of errors. The two error variances, calculated from the raw data for each movement distance, were regressed on to $d_u{}^2$. The results showed that Equation 3 was only applicable for young participants' data ($p < 0.05$), in which the model accounted for 97.4 % variance of X-variable error and 74.0 % variance of Y-variable error. As shown in Figure 3, two groups of participants committed smaller Y-variable error compared to X-variable error. Furthermore, there was a trend for young participants to have smaller variable errors compared to elder participants.

4   DISCUSSION

This preliminary study showed the potential of ballistic movement models for comparing ageing differences while interacting with touchscreens. With a self-modified LED monitor, we successfully measured ballistic movement time and ballistic movement endpoint variability when participants performed ballistic movements on the touchscreen by using their pointing fingers. Although statistically significant differences between the elderly and young adults were not found and two ballistic movement models did not fit well elder participants' data, we revealed the differences of movement speed and accuracy by plotting measured data against certain formats of movement distance. Relevant explanations of statistical issues and the findings related to ageing are discussed below.

There are two potential explanations for no significant difference between two groups of participants. The results of analysis of variance showed that the main effect of Age Group had no significant effect on both movement time and two axes of variable errors. The first explanation of this finding might be the small number of participants. In this study, only 12 participants were recruited and they were nested within two different age groups. While large variance was found among participants, it was difficult to show statistically significant difference between age groups. The second explanation could be that the elder participants recruited in this study were much healthier than normal elder people. As mentioned in the method section, all the elder participants had a regular exercise habit and all of them were healthy without any issues of executing hand-control movement. This might lessen the discrepancy of motor skills between two groups of participants.

Furthermore, two reasons could explain why ballistic movement models did not fit the data well. The results of model validation showed that both ballistic movement time model and ballistic movement variability model fitted well young participants' data, but not elder participants' data. The first potential reason to explain the results could be that the ballistic movements designed in this study were inappropriate. The ballistic movements were not executed in an exactly vertical direction toward the surface of screen, especially for short distance movements. This inappropriate design might impact the predictions of models, especially the ballistic movement variability model. Secondly, the elderly might perform ballistic movement in a different manner. As shown in Figure 2 and Figure 3, the elder participants' data of the movement time and two axes of variable errors were not affected by movement distance too much; they did not increase dramatically with increased movement distance. Several studies (e.g., Chaparro *et al.* 1999, Bakaev 2008, Fezzani *et al.* 2010) that utilized Fitts' law reported similar findings and found that the elderly tend to be more conservative while executing target-acquisition tasks.

Except for the statistical issues, several differences between two groups of participants were found. As just discussed, elder participants tended to be more conservative; the measured data were not significantly affected by movement distance. Compared to young participants, they took relatively a long duration while performing short distance movements (see Figure 2) and they produced relatively high variable errors of short distance movements and low variable errors of long distance movements (see Figure 3).

Suggestions for future research and design suggestions for the elderly were made after this preliminary study. To obtain solid conclusions, the number of participants should be increased and the experimental issue of non-vertical ballistic movement direction should be solved. While ballistic movements were measured, instead of Fitts-type movements, we found detail characteristics between the elderly and young adults. As expected, the elderly performed movements with a slower speed. However, the elderly had relatively high variable errors even in short distance movements. Furthermore, the discrepancy between vertical variable error and horizontal error on a touchscreen was larger for elder participants, compared to young participants. These findings can be utilized to develop design guidelines for touchscreen devices.

5   CONCLUSIONS

This study utilized ballistic movement models to compare the differences of speed and accuracy between the elderly and young adults while interacting with a touchscreen. The results showed that (1) because of few experimental issues, there was no statistically significant difference found between two groups of participants, (2) two ballistic movement models fitted well young participants' data, but not elder participants' data, and (3) elder participants performed movements with a conservative manner that could be utilized to develop touchscreen design guidelines. To provide solid conclusions, few suggestions were made for future research.

ACKNOWLEDGMENTS

The authors would like to acknowledge the grant support from Taiwan National Science Council (NSC 100-2221-E-155-063) for funding the paper submission and presentation.

REFERENCES

Bakaev, M., 2008. Fitts' law for older adults: Considering a factor of age. *Proceedings of the VIII Brazilian Symposium on Human Factors in Computing Systems.* Porto Alegre, Brazil, 260-263.

Carmeli, E., Patish, H. & Coleman, R., 2003. The aging hand. *Journal of Gerontology: Medical Sciences,* 58 (2), 146-152.

Chaparro, A., Bohan, M., Fernandez, J., Choi, S.D. & Kattel, B., 1999. The impact of age on computer input device use: Psychophysical and physiological measures. *International Journal of Industrial Ergonomics,* 24, 503-513.

Craik, F.I.M. & Salthouse, T.A., 2000. *The handbook of aging and cognition*: Mahwah, NJ: Erlbaum.

Fezzani, K., Albinet, C., Thon, B. & Marquie, J.-C., 2010. The effect of motor difficulty on the acquistion of a computer task: A comparison between young and older adults. *Behavior & Information Technology,* 29 (2), 115-124.

Fitts, P.M., 1954. The information capacity of the human motor system in controlling the amplitude of movement. *Journal of Experimental Psychology,* 47, 381-391.

Gajdosik, R.L., Linden, D.W.V., Mcnair, P.J., Riggin, T.J., Albertson, J.S., Mattick, D.J. & Wegley, J.C., 2009. Slow passive stretch and release characteristics of the calf muscles of older women with limited dorsiflexion range of motion. *Clinical Biomechanics,* 19 (4), 398-406.

Gan, K.-C. & Hoffmann, E.R., 1988. Geometrical conditions for ballistic and visually controlled movements. *Ergonomics,* 31, 829-839.

Hoffmann, E.R., 1981. An ergonomics approach to predetermined motion time systems. *Proceedings from the 9th National Conference (Institute of Industrial Engineers, Australia),* 30-47.

Hourcade, J.P. & Berkel, T.R., 2008. Simple pen interaction performance of young and older adults using handheld computers. *Interacting with Computers,* 20, 166-183.

Howarth, C.I., Beggs, W.D.A. & Bowden, J.M., 1971. The relationship between speed and accuracy of movement aimed at a target. *Acta Psychologica,* 35, 207-218.

Lin, J.-F., Drury, C., Karwan, M. & Paquet, V., Year. A general model that accounts for fitts' law and drury's modeled.^eds. *Proceedings of the 17th Congress of the International Ergonomics Association,* Beijing, China.

Lin, J.-F. & Drury, C.G., 2011. Verification of two models of ballistic movements. *Lecture Notes in Computer Science,* 6762, 275-284.

Lin, R.F. & Ho, Y.-C., 2011. Verification of ballistic movement models in a true 3d environment. *The 2nd East Asian Ergonomics Federation Symposium.* National Tsing Hua University, Hsinchu, Taiwan.

Maki, B.E. & Mcllroy, W.E., 2006. Control of rapid limb movements for balance recovery: Age-related changes and imp.lcations for fall prevention. *Age and Ageing,* 35 (2), 12-18.

Moffatt, K. & Mcgrenere, J., Year. Slipping and drifting: Using older users to uncover pen-based target acquistion dfficultiesed.^eds. *Proceedings of the 9th international ACM SIGACCESS conference on computers and accessibility.*

Ranganathan, V.K., Siemionow, V., Sahgal, V. & Yue, G.H., 2001. Effects of aging on hand function. *Journal of American Geriatrics Society,* 49 (11), 1478-1484.

Shiffman, L.M., 1992. Effects of aging on adult hand function. *The American Journal of Occupational Therapy,* 46, 785-792.

CHAPTER 12

The Impact of Soldiers' Age on Balance

Gary L. Boykin Sr., Valerie J. Rice and Petra E. Alfred

Army Research Laboratory-Army Medical Department Field Element
Fort Sam Houston, San Antonio, Texas
gary.boykin@amedd.army.mil

ABSTRACT

Equilibrium and balance can deteriorate with age, yet there are few reports examining this phenomenon within the military community. The Army Research Laboratory-Army Medical Department Field Element conducted research to examine balance differences between a younger (≤ 38 yrs, n=14) and an older (≥ 39 yrs, n=14) group of Soldiers using the Sharpened Romberg Test (SRT). Participants performed the SRT with eyes open and eyes closed while standing on the Advanced Mechanical Technology Inc. (AMTI) OR6-7 Biomechanics Force Platform. Average Radial Displacement (ARD), ARD-Standard Deviation (ARD-SD), and Path Length (PL) were computed over 22 seconds. Independent t-tests results indicated younger Soldiers had less sway ($p < .05$) than older Soldiers in two of three balance parameters: ARD—eyes open (t(26) = -2.2, p=.033) and Path Length—eyes open (t(26) = -2.30, p=.037).

Keywords: age, balance, Sharpened Romberg Test, tandem stance

1   INTRODUCTION

Postural control is essential for normal activities and crucial for athletics (Pendergrass, Moore & Gerber, 2003). The physical demands required of Soldiers lifting, carrying, running and their complex dynamic visual-motor tasks, shooting while quickly moving through variable terrain are similar to the demands placed on athletes. The load carried by the Soldier repeated intense physical demands add to their requirements. For example, Soldiers often complete tasks while carrying rucksacks that impact postural sway (Schiffman, Bensel, Hasselquist, Norton & Piscitelle, 2009) and intense physical exertion (e.g., a two-mile maximum run) affects postural control immediately following exertion (Pendergrass, Moore &

Gerber, 2003). A Soldier's job requires excellent balance, flexibility, and the ability to recover quickly.

Balance decreases with age (Briggs, Gossman, Birch, Drews, & Shaddeau, 1989; Fowler and Inman, 2006; Iverson, Gossman, Shaddeau, & Turner, 1990; Steffen & Mollinger, 2005; Sumi et al., 1988). Most studies investigating aging effects have been conducted with volunteers aged 50 and older (Briggs, Gossman, Birch, Drews, & Shaddeau, 1989; Fowler and Inman, 2006; Iverson, Gossman, Shaddeau, & Turner, 1990; Steffen & Mollinger, 2005). However, studies have identified that age-related declines in balance occur as early as age 30-40 in men and 30 in women (Fregly & Graybriel, 1968; Fregly, Smith, & Graybiel, 1973).

Potential reasons for age-related declines in balance include peripheral nervous system aging and accompanying decreases in exercise, fitness, strength, proprioception and muscle mass, and increases in medication use (Briggs, Gossman, Birch, Drews, & Shaddeau, 1989; Iverson, Gossman, Shaddeau, Sand Turner, 1990; Notermans, van Dijk, vander Graff, van Gijn, & Wokke, 1994).

While the majority of active duty military are relatively young, ages range from 17 to 62 years of age (Army Policy Message, 2006). As Soldiers age, physical training requirements are amended accordingly (Headquarters Department of the Army, 2010); however, job tasks remain rigorous and demanding, particularly during deployment into active war zones. Should age-related declines in balance exist, military duty assignments and post-injury assessments that include measures of balance and equilibrium may need to consider age. In the case of post-injury assessments, balance is considered when evaluating Soldiers who have been exposed to Improvised Explosive Devices (IEDs) and may have incurred a minimal to moderate Traumatic Brain Injury. In the absence of baseline balance data for deployed Soldiers, age related normative data may need to be used.

The Sharpened Romberg Test (SRT) is an adaptation of the original Romberg Test (Lanska & Goetz, 2000) The SRT is also known as the tandem, augmented, modified, or quantitative Romberg (Fregly & Grabiel, 1966; Fregly & Grabiel, 1968; Fitzgerald, 1996). The SRT is considered to be a reliable assessment of vestibular function (Fregly & Grabiel, 1966; Fregly & Grabiel, 1968; Notermans, van Dijk, vander Graff, van Gijn, & Wokke, 1994), with high test re-test reliability ($r = 0.91$) (Fowler & Inman; Graybiel & Fregly, 1966). Typically, the assessment criterion is the length of time an individual can stand in a static position with one foot directly in front of the other. Researchers use various protocols to define arm and head positions (Lee, 1998). Placement of the dominant foot (front or back) does not appear to impact the results (Briggs, Grossman, Birch, Drews, & Shaddeau, 1989; Pandaretaki, Kostadakos, Hatzitaki & Grouios, 2004). The SRT could be an expeditious, low-cost, field-expedient balance evaluation procedure.

The purpose of this research is to determine whether age impacts balance as measured by the AMTI OR6-7 force platform among a group of military service members. Force platform outcome measures examined include Average Radial Displacement (ARD), ARD-standard deviation (ARD-SD), and Path Length (PL).

2   METHODS

2.1 Procedure

Volunteers consisted of 28 healthy service members. Volunteers were briefed on the study and signed a Volunteer Agreement Affidavit prior to participating. All volunteers underwent screening by a health care practitioner and were identified as not having any medical condition or illness which could impact their balance. Volunteers were not taking any medications that could impact their balance. Volunteers completed a demographic questionnaire and a series of balance evaluations, however only select measures are reported in this paper.

Volunteers wore socks and stood on the AMTI OR6-7 force platform (a 3x3" biomechanical balance platform) with the dominant foot forward, arms crossed against the chest, and each palm resting on the opposing shoulder. The platform was clearly marked with reflective tape identifying proper foot placement and alignment. The SRT position was demonstrated to the volunteer, who then demonstrated it back to the researchers. As part of the eyes open instruction, it was suggested that volunteers visually fix their gaze on a stationary object or reference point in front of them. Volunteers performed the SRT with their eyes open and then with their eyes closed (Figure 1). Each participant was flanked by a spotter for safety. Each position was held for 22 seconds, as technical literature suggests 20 seconds for standard Romberg evaluations (RSscan International, 2001; Sumi et al., 1988). If the volunteer began to sway excessively (beyond an estimated 30 degrees from the participant's midline), the test was stopped and data from that session were truncated. Volunteers were offered rest (5 minutes) between evaluations. The balance platform (described below) was re-calibrated between volunteers.

Figure 1. Sharpened Romberg with eyes open and closed (courtesy of ARL AMEDD Field Element Research, 2009)

2.2 Instrumentation

Sharpened Romberg Test. The SRT measures a person's ability to maintain balance during tandem standing (one foot in front of the other, heel to toe), with eyes open and eyes closed. It has been shown to have good intra- and inter-rater reliability and good test-retest reliability among healthy women and men, as well as significant correlations with other balance tests r = 0.95 to .99 intraclass and r = 0.73 to 0.93 within rater (Franchignoni, Tesio, Martino, & Ricupero, 1998), r = 0.88 test-retest reliability (Hamilton, Kantor, & Magee, 1989).

Advanced Mechanical Technology, Inc. (AMTI). The AMTI OR6-7 Biomechanics Force Platform was used for this study. This balance platform was designed for the precise measurement of three orthogonal force components along the X, Y, and Z axes, and the movements about the three axes. It uses strain gauges mounted on four precision strain elements. Data are automatically recorded and stored on BioAnalysis with NetForce software.

2.3 Independent and Dependent Variables

Independent variables. The independent variables are age and SRT positions with eyes open and closed. The two age groups, based on median age, were young (38 and younger) and older (39 and older), with 14 volunteers per group.

Dependent variables. The balance parameters of Average Radial Displacement (ARD), Average Radial Displacement-Standard Deviation (ARD-SD), and Path Length (PL) were selected as dependent variables (measured in centimeter (cm) units).These measures are considered to be clinically relevant and to have mathematical integrity (Rose, et al, 2002; Wolff et al., 1998). Each parameter was calculated using standard statistical equations offered by AMTI (Table 1, AMTI, 2000*).* Center of Pressure (COP) is the center point of force in the x and y directions exerted on a force plate during static posture testing; ARD is the mean radial distance of the COP from the center point, calculated for each frame. ARD-SD is the standard deviation PL is the total distance the center of pressure travels. Smaller values for ARD and PL indicate greater postural control, and a lower standard deviation indicates less variability. Independent samples t-tests were used for the analysis with a p-value of $<.05$. Figure two illustrates scatter plots with correlation constants for balance parameters ARD ($R^2 = 0.043$) and PL ($R^2 = 0.033$).

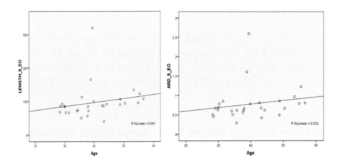

Figure 2. Linear regression for Path Length and Average Radial Displacement vs. age—eyes open.

3   RESULTS

Participants

A total of 28 volunteers, 23 men (82%) and 5 women (18%) participated in this study. Their mean age was 40.46 ± 8.69 years with a range of 28.5 to 56.67 years, and a median age of 38.8 years. The majority of volunteers were Caucasian males (n=20, 71%) and were on active duty (n=26, 93%).

Table 1. Balance Parameter Statistical Equation Formulas

Dependent Measure	Formula
ARD	$$RDavg = \sum_{i=1}^{N} \frac{ri}{N}$$ where: $r_{i=\sqrt{x_i^2+y_1^2}}$
ARD-SD	$$SDrad = \sqrt{\frac{1}{N}\sum_{i=1}^{N}(r_{1-RDavg})^2}$$ where: $r_{1=\sqrt{x_i^2+y_1^2}}$
PL	$$L = \sum_{i=2}^{n}\sqrt{(x_i-x_i)^2+(y_i-y_{i-1})^2}$$

Key: ARD = Average Radial Displacement, ARD-SD = Average Radial Displacement – Standard Deviation, PL = Path Length

T-Tests

Eyes Open vs. Eyes Closed.   Participants swayed more with their eyes closed, as indicated by each of the three measures, ARD (t(13) =-4.02, p=.001), (t(13) =-3.70, p=.003); ARD-SD (t(13) =-2.85, p=.014), (t(13) =-3.27, p=.006); PL (t(13) =-4.81, p=.000), (t(13) =-3.96, p=.002).   Results are shown in Tables 2 through 4, respectively.

Table 2. Eyes Open vs. Eyes Closed for Average Radial Displacement

Age Group	Eyes	Mean (cm) (SD)	t	p-value
Younger ≤38	Open	0.60 (.138)	-4.02	.001
	Closed	1.54 (.915)		
Older ≥39	Open	0.95(.561)	-3.70	.003
	Closed	1.60 (.711)		

Table 3. Eyes Open vs. Eyes Closed for Average Radial Displacement-Standard Deviation

Age Group	Eyes	Mean (cm) (SD)	t	p-value
Younger ≤38	Open	0.33 (.071)	-2.85	.014
	Closed	1.11 (1.04)		
Older ≥39	Open	0.61 (.610)	-3.27	.006
	Closed	1.11 (.710)		

Table 4. Eyes Open vs. Eyes Closed for Path Length

Age Group	Eyes	Mean (cm) (SD)	t	p-value
Younger ≤38	Open	76.45 (15.71)	-4.81	.000
	Closed	192.01 (94.27)		
Older ≥39	Open	117.40 (64.65)	-3.96	.002
	Closed	214.80 (101.25)		

Age comparison. For ARD there was a significant difference with eyes open ($t(26) = -2.2$, $p=.033$), but not with eyes closed ($t(26) = -.171$, $p=.866$) (Table 5 and Figure 2). For ARD-SD, no significant differences were found during balance testing with eyes open ($t(26)=-1.70$, $p=.100$) or with eyes closed ($t(26)=.015$, $p=.988$) (Table 6). For PL, a significant difference was seen with eyes open ($t(26)= -2.30$, $p=.037$), but not with eyes closed ($t(26)= -.617$, $p=.543$) (Table 7 and Figure 3).

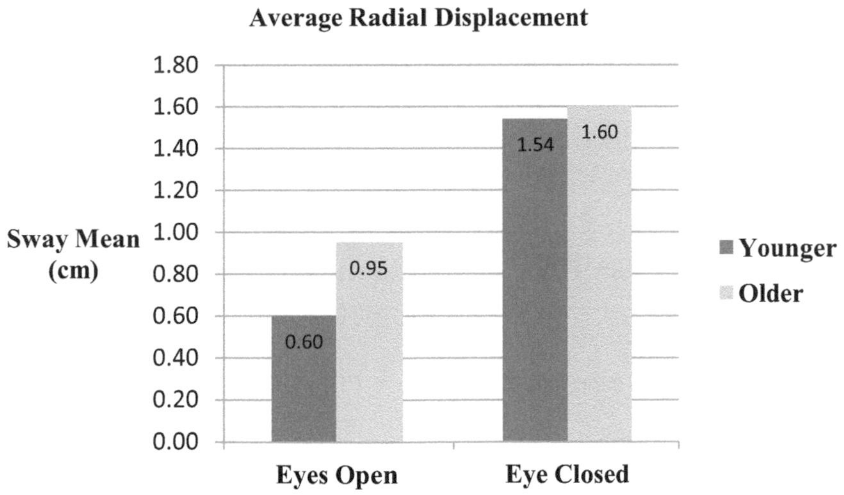

Figure 2. Eyes Open vs. Eyes Closed ARD Sway (cm) by Age Group (statistically significant)

Table 5. Average Radial Displacement (ARD) - Younger ≤ 38 vs. Older ≥39

Age Group	Age Group	Mean (cm) (SD)	t	p-value
Eyes Open				
	Young ≤38	.60 (.138)	2.45	.033
	Older ≥39	.95 (.561)		
Eyes Closed				
	Young ≤38	1.54 (.915)	.171	.866
	Older ≥39	1.60 (.711)		

Table 6. ARD-Standard Deviation (ARD-SD) - Younger ≤ 38 vs. Older ≥39

Age Group	Age Group	Mean (cm) (SD)	t	p-value
Eyes Open				
	Young ≤38	0.33 (.071)	1.705	.100
	Older ≥39	0.61 (.610)		
Eyes Closed				
	Young ≤38	1.11 (1.04)	.015	.988
	Older ≥39	1.11 (.710)		

Table 7. Path Length (PL) - Younger ≤ 38 vs. Older ≥39

Eyes	Age Group	Mean (cm) (SD)	t	p-value
Eyes Open				
	Young ≤38	76.45 (15.7)	-2.30	0.37
	Older ≥39	117.40 (64.6)		
Eyes Closed				
	Young ≤38	192.01 (94.2)	-.617	.543
	Older ≥39	214.82 (101.2)		

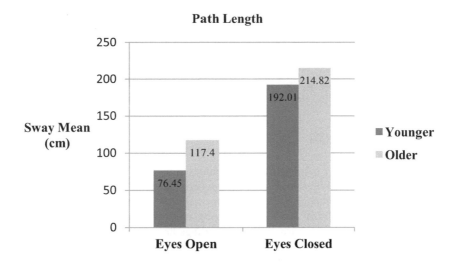

Figure 3. Eyes Open vs. Eyes Closed Sway by Age Group (statistically significant)

4   DISCUSSION

Greater postural control was seen with eyes open than with eyes closed, indicating vision as a stabilizing influence on postural control. This finding is supported by other researchers (Black, Wall, Rockette, and Kitch, 1982; Sumi et al., 1988).While younger individuals demonstrated less sway, according to each dependent measure, significance was found only for two measures during the eyes open condition (ARD = 36% difference and PL = 35% difference). That is, younger volunteers had greater postural control as measured by smaller sway excursion average radial displacement (ARD) and smaller path length (PL) when they performed the SRT with their eyes open. It appears that younger individuals were more affected by performing the task with their eyes closed than were older individuals, as illustrated in Figures 2 and 3. In other words, vision appears to have been a more stabilizing factor for younger individuals. These findings are counter to findings in which older participants have more postural sway on all tests, including both eyes open and eyes closed tasks (Steffen & Mollinger, 2005). The need and use of corrective lenses was not recorded during this study, and while volunteers could leave their glasses on, most chose to remove them for the duration of the balance testing. If more of the older volunteers had myopia (nearsightedness) then their ability to focus on a far-away object or reference point may have been compromised. While presbyopia (farsightedness) is an age-related visual difficulty with focusing on nearby objects, many older individuals have elements of both and may use trifocal lenses.

5   LIMITATIONS

The limitations of this study are the small sample size and specific sample population (military) which restrict generalizability of results. Other limitations were not accounting for conditions that may have assisted with explaining the results such as recording glasses use and vision, and body shape and composition, nor did we record whether the volunteer normally wore eyeglasses.

6   CONCLUSION

The results of this study demonstrated differences in balance performances between younger and older Soldiers using the Sharpened Romberg Stance. Less postural sway was seen with younger individuals, especially in the eyes open condition. It is suggested that additional studies on balance and age be conducted, accounting for visual input to determine whether such findings have functional implications. It is important to ascertain whether these findings are accurate among a larger population for several reasons. First and foremost, it is important to know whether older Soldiers can continue to perform their wartime tasks acceptably. Second, the U.S. Army physical fitness standards are being overhauled requiring the addition of balance performance, functional testing, and both age and gender-related normative data will need to be established (Schloesser, 2011; Army Readiness Training TC 3-22-20, 2010). Finally, age-related normative balance data would be useful during post-injury testing.

REFERENCES

AMTI. (2000). BioAnalysis Technical Manual.

Army Physical Readiness Training. (2010). TC 3-22-20.

Army Policy Message 06-06. Change to Maximum Age Criteria. 5 April 2006. Retrieved January 30, 2012 from: http://www.armyreenlistment.com /Messages/Policy/PM_06_06_ age.pdf

Black, F., Wall, C., Rockette, H., and Kitch, R. (1982). Normal subject postural sway during the Romberg Test. Am J *Otolaryngol, 3*, 309-318.

Briggs, R.C., Gossman, M., Birch, R., Drews, J.E. and Shaddeau, S.A. (1989). Balance performance among non-institutionalized elderly women). *Physical Therapy, 69*, 748-756.

Fitzgerald, B. (1996). A review of the Sharpened Romberg Test in diving medicine. *SPUMS Journal, 26*(3), 142-146.

Fowler, N. and Inman, V. (2006). Sharpened Rhomberg vs. Sway Magnetometry in assessing postural instability before and after driving simulation participation. Undergraduate Research Journal for the Human Sciences, 5, retrieved November 2011 from www.kon.org/urc/v5/fowler.html

Franchignoni, F., Tesio, L., Martino, M., and Ricupero C. (1998). Reliability of four simple, quantitative tests of balance and mobility in healthy elderly females. *Aging Clin Exp Res, 10*, 26-31.

Fregly, A.R. and Grabiel, A. (1966). A new quantitative ataxia test battery. Acta Oto-laryngologia (Stockholm, 61, 292-312.

Fregly, A.R. and Grabiel, A. (1966). An ataxia test battery not requiring rails. *Aerospace Medicine, 39*, 277-282).

Fregly, A.R., Smith, M.J., and Graybiel, A. (1973). A revised normative standards of performance of men on a quantitative ataxia test battery. *Acta Otolaryng, 75*: 10-16.

Hamilton, K., Kantor, L., and Magee, L. Limitations of postural equilibrium tests for examining simulator sickness. *Aviat Space Environ Med, 60*, 246-251.

Headquarters Department of the Army. (2010). Army Physical Readiness Training, TC 3-22.20.

Heitmann, D.K., Gossman, M.R., Shaddeau, S.A., and Jackson, J.R. (1989). Balance performance and step width in noninstitutionalized, elderly, female fallers and nonfallers. *Physical Therapy, 69*, 923-931.

Iverson, B.D., Gossman, M.R., Shaddeau, S.A., and Turner, M.E. Jr. (1990). Balance performance, force production and activity levels in non-institutionalized men 60 to 90 years of age. *Physical Therapy, 70*, 348-355.

Lafiandra, M and Harman, E. (2003). The Distribution of Force between the Upper and Lower Back during Load Carriage. *Medicine & Science in Sports and Exercise.* American College of Sports Medicine.

Lanska, D, Goetz, G. (2000). "Romberg's sign: development, adoption, and adaptation in the 19th century". *Neurology* 55 (8): 1201–6.

Lee, C.T. (1998). Sharpening the Sharpened Romberg. *SPUMS Journal, 28* (3), 125-131.

Notermans, N.C., van Dijk, G.W., van der Graff, Y., van Gijn, J., and Wokke, J.H.J. (1994). Measuring ataxia: Quantification based on the standard neurological examination. *J Neurology, Neurosurgery and Psychiatry, 57*, 2-26.

Panaretaki, E., Kostadakos, S. S., Hatzitaki, V., & Grouios, G. (2004). Standing with one foot in front of the other (sharpened romberg position): footedness effects. (Poster Session).

Pendergrass, TL., Moore, JH., Gerber, JP. Postural control after a 2-mile run. Mil Med. Nov 2003;168(11):896-903.

Rose J, Wolff D, Jones V, Bloch D, Oehlert J, Gamble J. Postural balance in children with cerebral palsy. Dev Med Child Neurol 2002;44:58-63.

RSscan International. (2007). Users Manual, Romberg.

Schiffman, J., Bensel C., Hasselquist, L., Gregorczyk, K., Piscitelle L. Effects of Carried Weight on Random Motion and Traditional Measures of Postural Sway. Med Sport. 2009;13:108–124. 130.

Schloesser, K. (2011, February 28). TRADOC revises Army Physical Fitness Test. Retrieved February 12, 2012, from http://www.army.mil/article/52548/TRADOC.

SPSS Inc. (2007). SPSS Base 16 Reference Guide. SPSS Inc., Chicago IL.

Steffen, T.M. and Mollinger, L.A. (2005). Age and gender related test performance in community-dwelling adults. Journal of Neurological Physical Therapy, 29 (4), 181-188.

Sumi, K. Watanabe, T., Kobayashi, F., Takeshima, N., Suzukim M., Tsuneji, M. Maed, K., and Kata, T. (1988). A study on age-related changes in postural sway. Jpn J Geriat, 25, 296-300.

Thessaloniki, Greece, Aristotle University of Thessaloniki, Department of Physical Education & Sport Science, [2004], p.284-285 Greece:

Thompson, J. Sebastianelli, W. & Slobounov, S. (2005) EEG and postural correlates of mil traumatic brain injury in athletes. Neuroscience Letters, 377, 158-163.

Wolff, D., Rose, J., Jones, V., Bloch, D., Oehlert, J., and Gamble, J. (Postural balance measurements for normal children and adolescents. J Orthop Res 1998;16:271-5.

CHAPTER 13

Cure Performance and Usability Evaluation of Electro-conductive Textiles Based on Healthcare Clothing System with TENS for the Elderly

Ryang-Hee Kim and Gilsoo Cho

Dept. of Fashion Design, University of Hanseo,
360 Daegok-ri, Haemi-myeon,
Seosan-si, Chungcheongnam-do, ZIP: 356-706, South Korea,
yanghee1003@naver.com
Dept. of Clothing and Textiles, Yonsei University, 262 Seongsan-no,
Seodaemun-Gu, Seoul, ZIP: 120-749, South Korea
gscho@yonsei.ac.kr

ABSTRACT

Modern people are interested in healthier elderly life and preventing chronic diseases, and more innovative portable devices for healthcare with a more comfortable and interactive ubiquitous computing environment. There have been integrated and converged studies among medicine, textiles, and electronic technology. The electro-conductive yarn and fabric may be applied to the development of more innovative and high-tech portable smart textile systems. This research aimed to develop the E-textile based Smart Medical Glove Systems (SMGS) with Transcutaneous Electrical Nerve Stimulator (TENS) as healthcare devices for the Elderly chronic diseases as the hypertensive, and to investigate the cure performance of the SMGS. The SMGS with TENS is used for the cure of hypertension by electrically stimulating meridian points on the palm based on the principles of acupuncture in oriental medicine. Thirty-two female subjects (Age: 40-

70's) were divided into two groups: Acute hypertension group (N: 16) and Chronic hypertension group (N: 16), had been suffering from hypertension participated in the cure performance test. Blood pressure and pulse rate were measured before and after wearing the SMGS with TENS for 15min. The subjects filled out a post-study questionnaire consisting of usability questions for each of the SMGS. The blood pressure of all 32 subjects decreased from 140.19±5.69 mmHg to 123.16±6.16 mmHg in systolic blood pressure after wearing SMGS. The paired T-test showed that the changes were statistically significant ***p<.001 for both systolic and diastolic blood pressure in both groups. On the result of the usability, using Principle Components Analysis (PCA), Varimax rotated PCA was applied to the measured usability. The sixteen usability questionnaires were categorized into 4 factors according to their correlations, and the results of usability were shown as wear-comfort: 3.1875, user-cognition: 4.1250, user-interaction: 3.8516, use-information: 3.6719. This research found that the blood pressure of all of the subjects after wearing SMGS decreased significantly, and we can suggest that the SMGS with TENS could be an easy-to-use and comfortable high-tech healthcare device based on E-textiles for the hypertensive Elderly.

Keywords: High-tech Healthcare Clothing, Smart Medical Glove System, Transcutaneous Electrical Nerve Stimulator, Cure Performance, Usability of High-tech Healthcare

1   INTRODUCTION

Since an evolution in the medical environment has led to the development of increasingly user-friendly and comfortable wearable medical devices. Demand of wearable medical clothing has increased for more efficient and effective health services.

Hypertension is typical chronic disease which frequently occurs under stressful Conditions and Habit of eating situations that may cause obesity. Currently, it is estimated that 100 billion people worldwide have hypertension, comprising almost 24% of the elderly population, and the rate of hypertension in the United States is increasing and reached 29% in 2004, whereas in case of Korea, about 30,2% male, 25.6%. For the treatment of hypertension, preventative lifestyle changes such as a regularly physical exercise, and weight loss for significant reduction of blood pressure, and antihypertensive drugs are currently available. However, medications could cause medical poisoning and side effects over time, leading to additional complications, and also side effects of these medications are not negligible

E-textile based on smart technologies will give the textile and wearable functional industry a high-tech value of making daily life healthier and more comfortable. Smart clothing can be achieved in electro-active textiles by combining multifunctional electronic device (Ross, D. D. and Tao, X., 2003). There is a need for research on medical clothing which is centered on the concepts of more personalized healthcare and greater empowerment of the patients that will enable them to manage their health conditions anytime and anywhere.

Therefore, this research is to design the portable Smart Medical Glove System (SMGS) with Transcutaneous Electrical Nerve Stimulation (TENS), which may be able to cure for the hypertensive patients in daily life, and to estimate the effectiveness of the Smart Medical Glove System with a cure performance and usability evaluation.

2   SMART MEDICAL CLOTHING FOR HEALTHCARE

2.1   CONCEPT OF SMART MEDICAL CLOTHING

Strese et al. (2004) defines smart textiles as 'textiles with integrated electronics and micro-systems which could be in clothes and in technical textiles'. Also, they define three levels of integration, but without specific names: solutions adapted to clothes, e.g., mobile phone in a pocket, electronics and micro systems integrated into clothes or textiles with connectable modules, e.g., with textile conductors, functions integrated into the textile via direct insertion into textile fibers, e.g., woven displays (Strese, et al. 2004).

The main functions may be expected of 'intelligent textiles' and 'smart clothes'. Whatever the nuances interpreted, the fundamental property expected of smart textiles is a measurable, reliable and useful responsiveness to environmental conditions and stimuli, such as heat, light and moisture. Another functions that may be expected of 'intelligent textiles' and 'smart clothes' in information, and actuator technologies, switches, transponders and touch pads, sensors for pressure and temperature.

The initial forays into wearable computing applications for outdoor and sports clothing, such as Levi's and Philips' ICD + jacket (2000), Burton's snowboarding jacket (2001) and France Telecom's create wear (2004) were relatively short-lived due to their niche appeal, rather cumbersome nature and expense, although new designs have recently been released. Imperatives provided by medical needs have given stimulus to the resolution of these problems in new ways.

The potential benefits of smart textiles in medical use can be summarized as follows: integration of functionality into textile interface, flexible materials which conform to the body, enable patient mobility whilst undergoing monitoring, continuous monitoring of vital signs for post-operative chronically ill babies and elderly patients, reduction of invasive procedures, inclusive design solutions for all users, low power needs linked to communication network, cost-effective solutions appropriate for disposable usage, facilitate healthcare, enable integration of feedback and therapies into monitoring.

2.2   High-Tech Healthcare Device with Smart Materials

Effective development of smart textile solutions to medical problems and procedures can be achieved only through a combination of several areas of expertise and research: several research groups now have combined medical knowledge and requirements with those of material scientists, textile developers and clothing designers and manufacturers. With the advent of ubiquitous (or pervasive) computing, sensors, antennae, miniature processor chips, radio frequency identification (RFID) tags and readers embedded in the domestic or hospital environment will be able to interact with sensors on the body or clothing, the 'smart home' for the elderly then becomes a reality-enabling a greater degree of independence and dignity whilst still being cared for. The union of the electronics industry with textiles has bred the new field of electronic textiles or e-textiles to provide the next generation of wearable computing. This research process starts from user requirements and needs, and takes a human centered product development approach focused on design solutions.

A number of technology platforms have now been established, based on different underlying technologies, by pioneering companies such as Softswitch and Eleksen, which integrate electronic functionality into textiles and clothing. Yet, primarily research for sportswear, portable entertainment and lifestyle products, and power supply for all electronic solutions still remains a fundamental problem when attempting to impart electronic functionality into clothing whilst simultaneously remaining completely portable.

2.3   Industrial Design and Usability

The origin of inclusive industrial design goes back to the design for the aged or the handicapped and mostly the focus is on providing the means of usability and accessibility. Usability can be a software application, website, book, tool, machine, process, or anything a human interacts with. The primary notion of usability is that an object designed with a generalized users' psychology and physiology in mind is, ISO defines usability as "The extent to which a product can be used by specified users to achieve specified goals with effectiveness, efficiency, and satisfaction in a specified context of use." Usability consultant Jakob Nielsen and computer science professor Ben Shneiderman have written (separately) about a framework of system acceptability, where usability is a part of "usefulness" and is composed of: 1)Learnability: How easy is it for users to accomplish basic tasks the first time they encounter the design?, 2)Efficiency: Once users have learned the design, how quickly can they perform tasks?, 3)Memorability: When users return to the design after a period of not using it, how easily can they establish proficiency?,4)Errors:

How many errors do users make, how severe are these errors, and how easily can they recover from the errors?, 5)Satisfaction: How pleasant is it to use the design?

In regard of medical smart products in medical care need to be more context-specific and the examples which follow indicate the wide range of approaches currently being undertaken in the several previous researches. Textiles for use in performance will be determined by the type of smart fiber and yarn characteristics, textile structures and finishes that may be engineered and placed in relation to zones on the body. Closer fit and appropriate protection can be enhanced by stretch fibers, molding, bonded seams or components and reinforced areas of support essential in effective performance related to health and illness. Technical fibers and yarn developments are key characteristics of all fabric innovation but will not be covered in detail (R. H. Kim, S. M. Cho, and G. Cho, 2010, R. H. Kim and G. Cho, 2011).

Therefore, we aimed to develop E-textile based Smart Medical Glove System (SMGS) with TENS for hypertension, and SMGS was designed glove system stimulating the meridian points on the palm for hypertension, to determine effectiveness of cure performance by clinical field test experiment and usability survey.

3   METHODOLGY

3.1   Materials of SMGS

E-textiles for SMGS

The design of Smart Medical Glove System based on E-textile sensors electrically stimulating four meridian points (P1, P2, P3, P4) for hypertension, supervised by an oriental doctor (Ha, K. S., 1998, Shin, I. S., 2007, Shin, M.S., et al, 2008, R. H. Kim and G. Cho, 2011). In the SMGS, the sensors were stitched in a ring shape using stainless steel thread for four meridian points on the left palm designated for hypertension treatment. The distance between two rings is 5mm, the gap between two stitches is 2mm, the length of stitches is 5mm, and the diameter is 20mm. The conductive snap in the middle connects to the transmission line.

We used the textile-based sensor which was made with stainless steel yarn and fabric to convey the electrical stimulation, electrical resistivity of stainless steel yarn and fabric specimens was measured according to 2000 20K 20M 2000M, for Linear Resistance (Ω/m) of conductive thread and sheet Resistance (Ω/sq) of conductive fabric using multi-o-meter is an electronic measuring instrument by using DMM4050® digital multi-meter (Hitachi).

The results of conductive properties of stainless steel yarn and fabric specimens were shown in Table 1, and they have excellent flexible and durable functionalities, commercially available specimens.

Transcutaneous Electrical Nerve Stimulator (TENS) and Transmission line

Transcutaneous Electrical Nerve Stimulator (TENS) were adapted the low frequency stimulator MB-430 (Hubidic Co. LTD., Korea) as the TENS model, with a stimulating frequency range of 1~1200Hz. The user can control the device manually by using set programs. The transmission line with a cable connector of Hubidic Co., LTD. was used to be inserted into the stimulator of TENS and to be put the conductive snap of sensor (Figure 1). It has a characteristic impedance of 60Ω at low frequency.

Table 1 Characteristics of conductive thread and fabric

Property ⟍ Type of E-textiles	Composition	Yarn Type	Electrical Linear resistance (Ω/m)	Electrical Sheet resistance (Ω/sq)	Finish
E-yarn	Stainless -steel / Polyester	Multi-filament	14±7		-
E-fabric	Polyester	Multi-filament		less than 0.04 Min.:50`~ Max:0.005	Electrol-ess plating; NiCo

3.2   Design of SMGS

Figure 1 shows the design of E-textile based sensors electrically stimulating four meridian points (P1, P2, P3, P4) for hypertension, were consulted to an oriental doctor (Ha, K. S., 1998, Shin, I. S., 2007, Shin, M.S., et al, 2008). Then, it was embedded the conductive snap is combined with bottom of conductive snap to transfer electrical wave from module TENS to the meridian site on palm in the middle to connect with the transmission line (Figure 1). In the of SMGS , the sensors were stitched in a ring shape using stainless steel thread (E-yarn) for four meridian points on the left palm designated for hypertension treatment.

3.3   Study Participants

Study participants were enrolled in a clinical study conducted by the previous studies. Study participants ranged in age from 30 to 70's years old. 32 female subjects in their 30~70's who were suffering from high blood pressure 130mmHg as systolic and 90mmHg as diastolic participated in this study. They were divided into two groups in accordance with their hypertension history and Age (Shin, 2007), (Van, 2007); occurring over the past three years is chronic hypertension, occurring within three years is incipient hypertension

Control participant exclusion criteria were any diagnosis of vascular disease, diabetes mellitus (DM), cancer (clinically or by anamnesis), renal disease, liver disease, thyroid disease, and acute or chronic inflammatory disease. Almost participant had been taking hypertension drugs or supplements. They were divided into two group: Acute group (female; N: 16), Chronic group (female; N: 16).

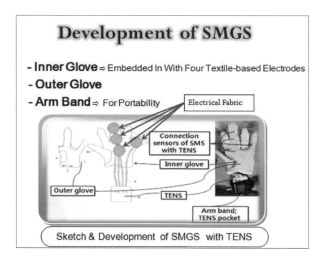

Figure 1 E-textile based sensors electrically stimulating four meridian points

3.4   Experimental Procedure

The subjects were rested for 15 min and then were measured their physiological signals of blood pressure and pulse rate before wearing the SMGS. Then they were stimulated on the meridian points of hypertension on their left hand by the SMGS with TENS during 15 min. Then, we measured the blood pressure and pulse rate after wearing SMGS. The last step was the usability evaluation that the subjects assessed wear-comfort, user-cognition, and user-satisfaction of the SMGS by 5 Likert scale.

3.5   Statistical Analysis

The evaluation and analysis of the results of before-after their physiological signals of blood pressure and pulse rate were statistically assessed by using the paired T-test with SAS 9.1. The selected usability principles are classified systematically using Principle Components Analysis (PCA) using Factor Analysis.

4   RESULTS AND DISCUSSION

4.1   Cure Performance

The results of the cure performance test of the Smart Medical Glove System with Transcutaneous Electrical Nerve Stimulator (TENS) for hypertension treatment are presented below. A system that performs palm stimulation as a mean of lowering blood pressure and pulse rate was presented, with the results in terms of effect presented in Table 2 and Figure 3.

The blood pressure of all 32 subjects decreased from 140.19±5.69 mmHg to 123.16±6.16 mmHg in systolic blood pressure after wearing SMGS. The paired T-test showed that the changes were statistically significant ***$p<.001$ for both systolic and diastolic blood pressure in both groups.

Table 2 Changes of blood pressure within 16 Acute and 16 Chronic hypertension groups before and after wearing SMGS by Wilcoxon rank test

Mean ± S.D	Systolic blood Pressure (mmHg)		
	Before Wearing SMGM	After Wearing SMGS	Difference
Acute; N(16)	135.06 ± 2.43	118.38 ± 3.95	-16.69 ± 4.25
Chronic; N(16)	145.31 ± 2.24	127.94 ± 3.77	-17.38 ± 3.79
p value 2)	<.0001***	<.0001***	

* $p<.05$, ** $p<.01$, *** $p<.001$
1) p-value for paired t-test
2) p-value for Wilcoxon rank sum test

4.2   Usability Evaluation of SMGS

On the result of the usability, using Principle Components Analysis (PCA), Varimax rotated PCA was applied to the measured usability. The sixteen usability questionnaires were categorized into 4 factors according to their correlations, and the results of usability were shown as wear-comfort: 3.1875, user-cognition: 4.1250, user-interaction: 3.8516, use-information: 3.6719. PCA reduced the 14 usability m easurements to four factors (Table3). The varimax rotation more clearly disting uished the usability components. The first factor (User-cognition) showed Reco gnition; .854, Predictability; .854, Learnability; .814, and Consistency; .642. Th e second factor (Use-information) showed Enjoyment; .854, Familiarity; .833, Visibility; .703, Defaults; .617. The third factor (User-interaction) show Direct-manipulation; .742, Feedback; .711, Low physical effort; .697, Responsiveness; .694. The fourth factor (Wear comfort) showed Tactile Sensation; .869, Hetero geneity; .783. Reliability within these factors were calculated by Cronbach's alpha, Factor1; 0.763, Factor2; 0.782, Factor3; 0.846, Factor4; 0.821. All of the reliability within these factors was very a high correlation respectively.

Table 3 Descriptive statistics of usability (N: 32)

	glove	Mean	S.D.	S.D. of Mean
Wear comport	SMGS1	3.1875	.47093	.08325
User- cognition	SMGS1	4.1250	.62540	.11056
User- interaction	SMGS1	3.8516	.84924	.15013
User- information	SMGS1	3.6719	.66429	.11743

* p<.05, ** p<.01, *** p<.001
1) p-value for independent two-sample t-test
2) p-value for paired t-test

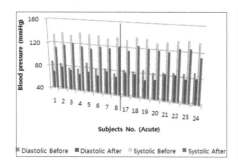 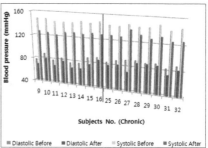

(A) (B)

Figure 2 Changes of blood pressure within Acute (A) and Chronic (B) hypertension group (N=16) before and after wearing SMGS

5   CONCLUSIONS

The Smart Medical Glove System (SMGS) using Transcutaneous Electrical Nerve Stimulation (TENS) have been successfully developed by using E-textiles and using embedded process. The Smart Medical Glove System with Transcutaneous Electrical Nerve Stimulation (TENS) might be able to care for hypertensive patients. It consisted of the E-textile based sensor for stimulating the four points.

This study found that subject's blood pressure dropped significantly after wearing the SMGS for 15 minutes. The paired T-test between two groups showed that both systolic and diastolic blood pressure were statistically significant decreased (***p<.001). 16 usability principles are categorized into 4 factors: (1) User-cognition, (2) Use-information, (3) User-interaction, (4) Wear comfort, usability principles for each group is developed as a checklist items. More specially, the results of this research program are summarized in Predictability is SMGS: 4.25±0.67 SMGS is considered effective device (p<0.0037***). Overall assessment of usability is found that SMGS is evaluated usable device. Although this investigation on the usability is estimated statistically significant, SMGS is evaluated ease-to-use device. Finally, this research found that the blood pressure of all of the subjects after wearing SMGS decreased significantly, and we can suggest that the SMGS with TENS could be an easy-to-use and comfortable high-tech healthcare device based on E-textiles for the hypertensive Elderly.

REFERENCES

Anderson, J. A., Wangner, J., Bessesen, M., and Williams, L. C. 2011, Usability Testing in the Hospital, *Journal of Human Factors and Ergonomics in Manufacturing & Service Industries*, 00 (0) 1-12 DOI: 10. 1002/hfm, Wiley Periodicals, Inc.

ISO 9241-11, 1998, Ergonomic Requirements for Office Work with Visual Display Terminals (VDTs)-Part II: Guidance on usability, ISO, Geneva.

Kim, R. H., Cho, S. M. and Cho, G. "Smart medical glove system for hypertension", *Proceedings of Applied Human Factors and Ergonomics*, paper ID : 1091, 2010.

Kim, R. H., Cho, S. M. and Cho, G. "Care performance and utility of Smart Glove System", *Proceedings of The 14th annual IEEE International Symposium on Wearable Computers*, pp. 208-214, 2010.

Kim, R. H. and Cho, G. 2011, Effectiveness of the Smart Healthcare Glove System for the Elderly with Hypertension, *Journal of Human Factors and Ergonomics in Manufacturing & Service Industries*, ID HFM-10-0159 1002/hfm, Wiley Periodicals, Inc

Shin, M. S., Han, C. H., Kim, B. Y., Kim, K. J., Park, S. H., Choi, S. M. 2008, A study on the recognition and actual condition of clinic oriental doctors' treatment for hypertension, *Journal of Korean Acupuncture Society*, 6(25), 23-33.

Tao, X 2003, Wearable Electronics and Photonic, pp. 8-12, Cambridge: WOODHEAD Publishing Limited and CRC Press LLC.

W. S. Jang, and Ji, Y. G., 2011, Usability Evaluation for Smartphone Augmented Reality Application User Interface, Thesis, University of Yonsei.

CHAPTER 14

Investigating the Primary Factors of Post-Sale Service to Enhance Service Quality for Mobile Communication Devices

Ping-Yu Chang

Department of Industrial Engineering and Management, MingChi University of Technology,
#84, GungJuan Rd., Taishan District, New Taipei City, Taiwan, 243,
pchang@mail.mcut.edu.tw

ABSTRACT

With the rapid increasing competitive environment and the rising consumers' awareness, enhancing post-sale service becomes a pertinent strategy in strengthening customer satisfaction. Hence, other than focusing on improving product quality, industries have devoted many resources to improve post-sale service quality. A good post-sale service will not be achieved without constructing appropriate strategies. If an effective and integrated post-sale strategy can be developed, customer satisfaction will be enhanced, the overall operational cost can be reduced, and the brand image will be broadened and promoted. Although post-sale service has been realized to have effect on the service quality, its value has not yet been explored in both academia and industries. Hence, this research analyzes the characteristics of mobile communication devices, the post-sales service flow, reverse logistics network design, and the influence of the flow design on the service quality. Primal factors that have significant effect on the service quality are identified and some strategies are proposed for managerial prospects in designing post-sale service.

Keywords: Mobile communication device, Post-sale, Service quality, Reverse logistics

1   INTRODUCTION

With the rapid increasing competitive environment and the rising consumers' awareness, enhancing post-sales service becomes a pertinent strategy in strengthening customer satisfaction. Also, the development of Customer Relationship Management (CRM) shows that the price of reserving old customers is much less than the price of attracting new customers. Juran (1974) thought that quality is not only for sales but to satisfy customers' specification. Hence, other than focusing on improving product quality, industries have devoted many resources to improve the quality of post-sale service. Although post-sale service becomes important in achieving higher service quality, its effectiveness and factors that affect post-sales service is not yet realized. Post-sale service is considered as the trigger point of the reverse logistics that will have effect on the retailing process. Furthermore, companies should consider both retailing sales and reverse logistics so that their service quality can be improved and sales can be increased.

Figure 1 demonstrates the reverse logistics of product. In Figure 1, customers return product directly to regional distribution center and the product will be checked for possibility of repairing. If there is a possibility of repairing, the product is sent to manufacturing plants for repairing, remanufacturing or reuse. Otherwise, the product will be disposed. With well management of retailing and reverse logistics, it is believed that cost and service quality can be improved.

Although reverse logistics can bring vantages to companies, only a few research have considered the entire aspect of reverse logistics. Similarly, fewer industries have contributed to the integration of post-sale service and reverse logistics due to the cost associated with conducting such experiments. However, mobile communication devices such as mobile phone and pad are usually with high price and customers intends to repair product if they are damaged. Hence, post-sale service quality and efficiency becomes an important strategy to increase customer satisfaction for mobile communication devices. To improve post-sale service quality, some factors in reverse logistics should be identified to develop reverse logistics systems so that customer satisfaction will be enhanced.

This research analyzes the characteristics of mobile communication devices, the post-sales service flow, and reverse logistics network design to identify main factors of post-sale service of mobile communication devices. Furthermore, QFD (Quality Function Deployment) is adapted to transform experts' opinions in deciding the main factors. This research is organized as follows. Section 1 is the introduction while Section 2 discusses the literature of reverse logistics. Section 3 demonstrates the simulation system and setup of reverse logistics and service quality and Section 4 discusses the conclusion.

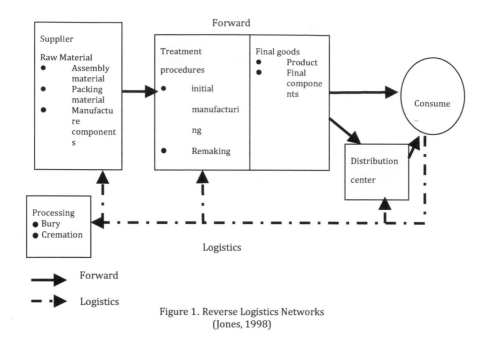

Figure 1. Reverse Logistics Networks
(Jones, 1998)

2   LITERATURE REVIEW

Slater (1994) pointed out that the maintenance of competitive advantage is concerned as an increasingly difficult job. The face of the global market and post-sales service quality can be regarded as maintaining a competitive factor. The focus of post-sales service is implemented to satisfy customers with the resource within the enterprises. With the consumer's heightened awareness, customer satisfaction is no longer limited to product prices and features, post-sales service should also be considered to strengthen the service quality so that customer satisfaction can be improved. Davidow and Uttal (1988) created a new definition of customer service, that is, customer service means that customers can better understand the core product or service the potential value of various characteristics, behavior or information. This definition covers the traditional customer service activities, such as: order processing, complaining about treatment. Also, a lot of new services such as product performance tracking, taking the initiative to inform the repair, and fault diagnosis are included.

Lalonde and Zinszer (1976) divided customer service into three categories: pre-sales services, order processing, post-sales service. Regan (1963) considered the service with the following characteristics: (1) Intangibility; (2) Simultaneity; (3) Heterogeneity; (4) Perishability. Gronroos (1982) Pointed out that the quality of service can be divided into two categories: technical quality and functional quality; the former refers to the actual customers are obtained from the Services, while the latter refers to the customer to take part in service delivery gained.

Gronroos (1982) pointed out that the quality of service can be divided into two categories: technical quality and functional quality. Technical quality refers to the actual customers that are obtained from the services, while functional quality refers to the customers that take parts in service delivery. Lewis and Booms (1983) proposed that the service quality is the difference between the services to expect and the receiving services. In other words, the service to expect and the receiving services will be similar. Sasser et. al (1978) considers the services level is similar to quality. It includes seven issues, that is, safe, consistency, attitude, integrity, relevance, convenience, and instant. They use the materials, equipments, and personnel to define the three dimensions of service quality. Levitt (1972) considers the quality of service as the results of service to meet the standards.

According to Council of Logistics Management, reverse logistics networks is refers to capacity reduction, recycling, substitution, reuse, disposal, carries on the physical distribution related activity such as product return and re-repair system, the goods re-processing, renewable materials, waste management and clearance of hazardous materials management role. However, product recall has: exchange, repair, re-distribute, bottle recycling and reuse, and product and materials re-use in the reverse logistics activities (Daugherty et al., 2001). Cohen (1988) proposed the main purpose of reverse logistics for enterprises is to achieve economic goals with environmental protection which can save up to 60% cost for re-manufacturing enterprises, production.

Guide and Srivastava (1997) defined seven characteristics that make the product recycling in reverse logistics of recovery with tremendous challenges. Asugman et.al (1997) think business-to provide after-sale service to customers would like to enter the market to other competitors have the potential barriers to entry; Although it is impossible to set up short-term financial performance, but in the long run, managers should not overlook the importance of after-sales service. Anell and Wilson (2002) p ointed out that the after-sales service enterprises have set up a competitive advantage and maintain the importance of profits. For enterprises, the maintenance of competitive advantage is an increasingly difficult job, in the face of the global market, after-sales service can be regarded as maintain key competitive advantages, quality of service related to the ability to obtain customer permanent loyalty to the enterprise. The focus of after-sales service to enhance the overall quality and customer satisfaction, high quality, customer satisfaction will be re-purchase (Slater 1996).

3   METHODOLOGY

In reviewing the literature, three factors, time, price, and inventory level, are usually identified to have significant effect on service quality. To be more specifically, reducing lead time will improve the possibility of on-time delivery that will increase customer satisfaction. Moreover, inventory level is considered within the lead time since different inventory level will result in different waiting time of the component which will have impact on the lead time. Also, the inspection price of mobile communication devices will be different with or without warranty. If the devices are still within the time of the warranty, lower repair price is expected and higher satisfaction will be achieved. Therefore, determining the period of the warranty will also affect the service quality. To conduct the survey of QFD (Quality Function Deployment), the SERVQUAL and five gaps of PZB (Parasuraman et al., 1985) are adapted into reverse logistics of mobile communication devices. Three experts' opinions are considered in mapping the SERVQUAL questionnaire onto service quality factors. Table 1 demonstrates the results of mapping reverse logistics onto PZB SERVQUAL (Parasuraman et al, 1988).

Table 1: Mapping the reverse logistics onto PZB

	Descriptions	PZB	Factors	
			Time	Price
Tangibles	Modern looking equipment.	P1		•
	The physical facilities are visually appealing.	P2		•
	Employees are neat in their appearance.	P3		•
	Materials associated with the service (pamphlets or statements) will be visually appealing.	P4		•
Reliability	When companies promise to do something by a certain time, they do.	P5	•	
	When a customer has a problem, excellent companies will show a sincere interest in solving it.	P6	•	
	Excellent companies will perform the service right the first time.	P7	•	
	Excellent companies will provide the service at the time they promise to do so.	P8	•	
	Excellent companies will insist on error free records.	P9	•	
Responsiveness	Employees of excellent companies will tell customers exactly when services will be performed.	P10	•	
	Employees of excellent companies will give prompt service to customers.	P11	•	•
	Employees of excellent companies will always be willing to help customers.	P12	•	
	Employees of excellent companies will never be too busy to respond to customers' requests.	P13	•	
Assurance	The behavior of employees in excellent companies will install confidence in customers	P14	•	•
	Customers of excellent companies will feel safe in transactions.	P15	•	•

	Employees of excellent companies will be consistently courteous with customers.	P16	●	
	Employees of excellent companies will have the knowledge to answer customers' questions.	P17	●	
	Excellent companies will give customers individual attention.	P18	●	
	Excellent companies will have operating hours convenient to all their customers.	P19	●	
Empathy	Excellent companies will have employees who give customers personal service.	P20	●	●
	Excellent companies will have their customers' best interest at heart.	P21	●	●
	The employees of excellent companies will understand the specific needs of their customers.	P22	●	●

In Table 1, time is more recognized and preferred than price for the effect of post-sale service quality. The reason might be that maintenance price for the mobile communication devices are usually consistent and fair. If a mobile communication device will be repaired with high price, customers might purchase a new device with advanced functions. Also, customers prefer to repair their mobile communication devices within shorter time so that their work or entertainment will not be interrupted. Therefore, time plays an important role in determining post-sale service quality that should be further investigated. To realize the importance of time factor, QFD is adapted and 12 experts' opinions on the relationship (strong relationship (◎), moderate relationship (○), and weak relationship(△)) between factors and SERVQUAL are identified. Table 2 shows the house of quality in QFD analysis. In Table 2, most of strong relationships fall on the repair time while most of weak relationships are on the price (with warranty, without warranty, and inspection). The results show that time is the main factor associated with post-sale service quality. Furthermore, repair time is the main concern of time factor while inspection is the main concern of price factor. The phenomenon might be that customers expect to have low inspection fee and short repair time so that they can make decisions of maintenance or dispose quickly. The price of with or without warranty seems not important for the customers since the difference of the maintenance fee for with or without warranty are fairly acceptable. Moreover, the final score of time outperforms the final score of price. The weighted score or delivery time also outperforms the weighted score of order processing time. The reason might be that order processing time is usually short and would not affect the total time of maintenance.

Table 2: QFD for service quality factors evaluation

			importance	Order processing	Repair	Delivery	With warranty	Without warranty	inspection	summation	Weight summation	rank
				Service quality factors								
				Time			**Price**					
Customer expectation	Tangibles	P1	2	◎	◎	○	△	△	○	18	36	2
		P2	1.5	△						1	1.5	15
		P3	0.5		△					1	0.5	17
		P4	2.5	△	◎	△	△	△	○	12	30	4
	Reliability	P5	2.5		○	◎	△	△	△	11	27.	5
		P6	2			○	△	△	△	6	12	13
		P7	2.5		○	◎	△	△	△	11	27.	5
		P8	3	△	◎	◎				11	33	3
		P9	3	△	◎	△	○	○	◎	18	54	1
	Responsiveness	P1	2.5		◎	○				8	20	10
		P1	2.5		◎					5	12.	12
		P1	2		◎					5	10	14
		P1	3	△	◎				○	9	27	6
	Assurance	P1	2			○	○	○	○	12	24	7
		P1	1			△				1	1	16
		P1	1			△				1	1	16
		P1	1.5	△			○	○	○	10	15	11
	Empathy	P1	1.5		◎		△	△	△	8	12	13
		P1	1.5		◎		○	○	○	14	21	9
		P2	1.5	△	○	○	○	○	△	14	21	9
		P2	2		△	△	○	○	○	11	22	8
		P2	2		◎	△				6	12	13
Summation				12	66	36	24	24	31			
weight				3	3	2	2	2	2.5			
Sum of the weighted score				36	198	72	48	48	77.5			
Average score				102			57.8					
Rank				1			2					

4   CONCLUSION AND FUTURE RESEARCH

This research investigates the primal factors that have significant effect on the service quality. Three factors, time, price, and inventory level, are identified to have significant effect on service quality. To realize the effect of the factors, SERVQUAL and five gaps of PZB are mapped onto the factors and QFD is then adapted to conduct survey of the primal factors. The results show that time is more recognized and preferred than price for the effect of post-sale service quality. The reason might be that maintenance price for the mobile communication devices are usually consistent and fair. Also, repair time is the main concern of time factor while inspection is the main concern of price factor. The phenomenon might be that customers expect to have low inspection fee and short repair time so that they can make decisions of maintenance or dispose quickly. The price of with or without warranty seems not important for the customers since the difference of the maintenance fee for with or without warranty are fairly acceptable. The results also provide an insight in constructing post-sale service for mobile communication devices. Companies should concentrate on shorten the service time for the customers with fair price other than lower the maintenance fee. However, some variables such as inventory level, technology support, and repair process will affect the repair time that can be investigated as further research topics.

REFERENCES

Asugman, G., J. Johnson , and J. McCullough 1997. The role of after-sales service in international marketing *Journal of International Marketing* 5(4): 11-28.

Anell, B.I. and T.L. Wilson 2002. Prescripts: creating competitive advantage in the knowledge economy *Competitiveness Review* 12(1): 26-37.

Cohen, M. 1988. Replace, Rebuild or Remanufacture *Equipment Management* 16(1): 22-26.

Davidow, W.H. and B. Uttal 1989. *Total Customer Service : The Ultimate Weapon* New York: City: Harper Collins Publishers Inc.

Gronroos, C. 1981. Internal Marketing—Theory and Practice. *American Marketing Association Services Marketing Conference Proceedings*: 41-47.

Guide Jr., V. D. R. and R. Srecastava 1997. An Evaluation of Order Release Strategies in a Remanufacturing Environment *Computer Operations Research* 24(1): 37-47.

Juran , J.M. 1974. *Quality Control Handbook. 3rd ed.* New York City: McGraw-Hill Book Company.

LaLinde, B.J. and P.H. Zinszer 1976. *Customer Service: Meaning and Management* Chicago: National Council of Physical Distribution Management.

Lewis. R.C. and B.H. Booms 1983. The Marketing Aspects of Service Quality. In *Emerging Perspectives on Service Marketing.* L. Berry, G. Shostack, and G. Upah (eds.). Chicago, IL: American Marketing: 99-107.

Levitt, T. 1972. *Production-line approach to service* Harvard Business Review 50(4): 41-52.

Parasuraman, A., V.A. Zeithaml, and L.L. Berry 1985 A conceptual model of service quality and itsimplications for future research *Journal of Marketing* 49(1): 41-50.

148

Parasuraman A., V.A. Zeithaml, and L.L. Berry 1988. SERVQUAL: A Multiple-Item Scale for Measuring Consumer Perceptions of Service Quality *Journal of Retailing* 64(1): 12-40.

Slater, S.F and J.C. Narver 1993. Product-Market Strategy and Performance: An Analysis of the Miles and Snow Strategy Types *European Journal of Marketing* 27(10): 33-51.

Sasser, W. E., P.R. Olsen, and D.D. Wyckoff 1978. *Management of Service Operation: Text, Case and Readings*. Boston: Allyn and Bacon.

Slater, S.F and J.C. Narver 1994. Product-Market Strategy and Performance: An Analysis of the Miles and Snow Strategy Types *European Journal of Marketing* 27(10): 33-51.

CHAPTER 15

A Study of Establishing Demand Assessment and Support Network of Elderly Fall-Prevention System

Yen-Jen Chen

Department of Health and Social Work
Yu Da University
Chao-chiao, Miao-li County, ,Taiwan,
Email:bathchen@hotmail.com
yjchen1231@yahoo.com.tw

ABSTRACT

This paper discusses the ways of controlling the risk of disability and national medical costs out of the falls of the elderly via technology and social support networks. By examining the new factors on the elderly falls under the rapid social transformation, the paper explores the possibilities of building a preventive mechanism using the Chinese culture of filial piety to enable the elderly to improve their living quality.

The study took place in Taiwan and collected data via in-depth interviewing and focus group. The corpus includes two focus groups with 22 practitioners and 38 interviews with the elderly who suffer from falls, as well as the caretakers and experts.

The study finds that, although there used to be institutional prevention of the elderly falls, the social transformation has rendered the circumstances unpredictable, such as the falls caused by stumbling on the pets. It also finds that, because of the change in family structure, the problems of caretaking have pressurized the modern families and traditional filial culture, the main family support network for the elderly rehabilitation. The more the elderly are conscious of empowerment and acquire the family and professional support, the better the effects of prevention are.

The elderly are more likely to fall because of physical deterioration, and technology is thus needed to develop auxiliary equipment to minimize the risk of elderly falling. The preventative mechanism should firstly establish the ideas of empowerment and enabling with the elderly so that they can start the process of self-protection. The government's caretaking policies should take into consideration of the elders' physical and mental needs to provide friendly environment and strengthen the social support system. More importantly, the government should encourage the market development by establishing inter-departmental teams to promote elderly welfare technology, so that the fall prevention system can well function.

Keywords: health aging, elderly falls, social support, empower, enable

1   INTRODUCTION

The stumbling of the elderly has become a public health problem for the aging countries. It affects not only the living standard but also life in general for both the individuals and their families. According to the principles of the United Nations' Guidelines for the Action of the Elderly, there are five aims to achieve to make a healthy living for the aged, viz. independence, participation, caretaking, self-achievement, and dignity. Take the principle of independence. The elderly should live in a safe environment where their needs are taken care of and enough varieties are ensured, so as to maintain 'local aging' and 'aging in original domicile'. To judge Taiwan's living conditions on this standard, it is clear that the work to prevent the elderly falls has not been materialized. At present most of the interventional technologies for the elderly falls focus on the responses after the falls and are unable to make pre-event analyses and combine community-based, precautionary technologies (Scamaill et al., 2011). As today's technology can help improve the elderly falls, whether it can be put into effect depends on the elderly's self-consciousness and a well-developed social support system. Hence the paper aims to explore the causes of fall for the elderly in Taiwan, the process of their empowerment, and the establishment of the support system. A qualitative research is conducted so as to construct an effective mechanism of prevention and achieve an active and healthy aging society.

2   LITERATURE REVIEW

One of the most feared accidents in an aged life is to be put into a long-termed care after stumbling. For after stumbling the quality of living is seriously impaired out of the interrelations between the socio-cultural environment and the individuals' health conditions and adaptive mechanism (Sarvimaki, 2000). The question of how to adopt the precautionary mechanism and a sense of self-protection and empowerment thus become important to construct a support system to prevent the elderly falls. Falling means one physically hits the floor and gets hurt out of a loss in balance (Hornbrrok et al., 1994). There are many factors for the elderly to fall. Some study points out that 85% elderly falls occurs at home (Abreu et al., 1998). Even for those who live within communities, one out of three falls at least once a year, and the number triples for those living in the care institutes in America (Sieri & Beretta, 2004). Falling often leads to bone fracture in the hip, which may cause bed confinement and even death. Besides the physical damage, there are also a fear of losing dignity and the danger of depression and anxiety (Howland et al., 1993; Arfken et al., 1994; Howland et al., 1998). Hence many elderly who have the experience of falling are wary of outdoor activities (Huang & Action, 2004; Sieri & Beretta, 2004). Research has suggested that the likelihood of falling decreases if the elderly are properly trained in bodily activity (Wolfe, Jordan & Wolfe, 2004; Bruyere et al., 2005), and if information on medicine is provided and living environment improved (Huang & Acton, 2004). Therefore, to augment the quality of living for the elderly, particularly those who repeatedly fall or falter, it is necessary to examine their individual physical functions and the safety of their living environment, so that an effective prevention system can be constructed. Studies on fall prevention and evaluation find that falls can be best prevented with multi-factored evaluation and intervention (Chang et al., 2004). The prevention and protection against the elderly falls have become a key issue in Taiwan in the spheres of social, medical and economic policies. If it is not properly attended to, more problems on social care and high medical cost are bound to arise (Chen, 2007). Hence we should get hold of the causes and development of falls via an interdisciplinary evaluation team, so that a comprehensive protection system can be established.

As of August 2011, the population of the 65 year-olds and above in Taiwan exceeds 2.5 million, comprising 10.78% of the whole and 14.58% in ratio of old age population dependency. Among the 50,310 reported cases of the elderly accidents, there are 13,630 ones of falling, which comprises 27.1% and the second highest place (Department of Health, 2011). In the previous year, the ratio of the elderly falls is 20.5%, 37% of which belongs to the category of "repeated falling". According to a 2007 longitudinal survey of living condition of the middle to old ages, the ratio of falling for the 65-74 year-olds is 20.5%, with that for 75 year-olds and above being 28.6% (Department of Health, 2007). Besides, the population of the disabled in Taiwan exceeds 3900,000, 40% of which are taken care of by the families and relatives, 18% by foreign or native care-workers, and 10% by institutional care and home services. Generally speaking, even in Taiwan where filial piety is much emphasized, tending the elderly falls by the families and relatives has become problematic. There are less than half of the fallen elderly being taken care of by the family members and relatives, and more and more people are buying home services and paying for care-workers. Although both the elderly and their families are not satisfied with such arrangements, they cannot help but comply with the circumstances. It is on this basis that a preventive mechanism and an empowering sense of self-protection become important.

Following the development of aging countries, the core values of Taiwan's welfare policies for the elderly also evolve around the three pillars of active aging, elderly-friendliness, and generational integration (Ministry of Interior, 2009). Active aging includes health improvement, social participation and safety maintenance. Elderly-friendliness means instituting an obstacle-free environment for the aged people. Generational integration aims to establish a society where there is no discrimination and the aged are dignified. Recently, the ideas of empowerment and enabling have been introduced to health care and intervention in the hope that the mismatch between professional services and the needs can be solved. The perspective of empowering can not only make explicit the resources of the elderly but also provide empathetic views on the preventive mechanism and protective measurements. The empowered individuals are more likely to seek resources and support, interact with the others (Rappaport, 1984 ;Yip,2004.), and acquire positive ideas about themselves (Gibson, 1991; Wallerstein, 1992). The process and results of empowering also help generate a sense of hope, direction and encouragement (Rodwell, 1996), and enhance the sense of control in life including personal attitudes, values and beliefs, particularly the ability to control one's destiny (cited from Chadiha et al., 2004); hence a holistic concept. When the subjects participate in empowerment, their living abilities are often improved, and they are more likely to be in control of the resources and action to achieve the aim of self-protection (Kieffer, 1981; Cox & Parson, 1994). Ellis-Stoll Popkess-Vawter (1998) also suggests a model of the empowerment process including the pre-factors, process and end results to evaluate the family resources, resources of the professional system, and the environment factors of cultivation (cited from Turnbull & Turnbull, 2001). Therefore, to achieve a model of independent and dignified aging, both the elderly and the society should acquire the ideas of empowerment and enabling to construct an effective mechanism protecting the aged from the harm of falling.

3   RESEARCH METHOD

The study adopts the qualitative method. It firstly gathers the causes of the elderly fall and the needs of its prevention via reviewing the literature, and then collects first-hand data from focused group and in-depth interviews. It analyzes the new factors of stumbling and support systems, and explores their strategies for empowerment and models of protection. The professionals in the field of elderly care and the practitioners (e.g. social workers and nurses) of the care institutes, community care spots and elderly associations were targeted and invited for the focused group interviews. Two group interviews were carried out and recruited 22 respondents. The in-depth interviews, on the other hand, focused on the aged, the care-taking family members, and the practicing professionals, who were 25 in total. All together there were 47 people being interviewed in the study.

4   RESEARCH CONCEPT

From the perspectives of the elderly, caretakers and practitioners, the study analyzes the problems and methods of prevention of the elderly fall on the basis of the contextual and experiential factors. It seeks to get hold of the contexts where problems occur and understand the possibilities of using technological facilities. It hopes to construct an effective support system for the benefits of the elderly.

5   RESULTS AND ANALYSIS

Whether at home or in the institute, the elderly are more likely to stumble out of deterioration in bodily functions of responding and balancing. The study finds that social transformation has led to new factors such as pets, elevator, medication, which increase the likelihood of elderly fall. The more damage, the longer time for rehabilitation and the bigger pressure in caretaking. Hence it is important that the inner and outer resource systems be integrated to prevent the aged from falling, the former referring to individual empowerment, the latter a support system. The interviewing data are shown in Table 1.

Table 1 Interviewing responses

New Factors	Problems After Falling
1)Positional low blood pressure	1)Longer time for rehabilitation
2)Over medication	2)Little support from family out of work
3)Mental hurt	3)Applying for alternative caretaking out of
4)Personality	little family resources
5)Pets	4)Buying home service out of incompatibility
6)Environment (such as uneven and	with the foreign caretakers
slippery streets)	5)High pressure for the caretakers
7)Odd reasons	6)Poor public facilities
8)Self-esteem	
9)Previous experiences of falling	

The study differentiates formal and informal support systems for the elderly fall. The informal system comprises of care from family and neighbors who are in close contact with the elderly. While family is the primary system for the prevention of the elderly fall, the pre-fall social network also affects the extent of satisfaction with the care. Hence the two should work in tandem to construct a strong basis of emotional support (Chen, 2009; Chen, 2008). Walsh (1993) points out that family is a basic social unit which provides economic and emotional support and care. Yet in Taiwan, with the abounding of double-income families, children have little time in accompanying their parents and rely on foreign caretakers or institutes to take care of the elderly after fall. It is therefore that most of the elderly take children's accompaniment for demonstrating the traditional value of filial piety, as shown in this study. After the fall, some family members work together, some become alienated, and some opt for foreign caretaking. From these is suggested that family support system is faltering. It is important that family be encouraged and the idea of family empowerment be developed so that the support system can be established.

Besides, rebuilding the support system of neighborhood on the cultural basis of filial piety is also helpful in awakening the consciousness of empowerment in contribution to the wellbeing of the elderly. This study shows that the families mostly consider taking care of the parents as their responsibility. Yet because they have to work at the same time, they cannot make both ends meet. Thus they need formal support system to help provide home services, meal delivery, and community care so that the strategies of fall prevention can be put in effect. Support should also be institutionalized in policies such as "care leave" which encourages the children to accompany parents on subsidies, and formal services like home services, meal delivery and long-distance medical care can cover those elderly who live afar or alone. With the intervention of the wellbeing technologies in the information society, protection mechanism can be more effectively actualized. The analysis of the needs of the social support system is shown in Table 2.

Table 2 Analysis of the needs of the social support system

Formal support system	Informal support system
1) Enhance institutes' on-job education of fall prevention.	1) Formulating family consensus on the treatment of the elderly.
2) Train community seed instructors to promote fall prevention measurement.	2) Refurbish home environment to obstacle-free one.
3) Use slippery-resistant tiles and help bells.	3) Establish community self-help team for a close watch of the elderly.
4) Use furniture appropriate for the elderly.	4) Use hi-tech auxiliaries.
5) Play Tai-Chi to enhance bodily balance.	5) Use night lamps or light of aquarium to prevent falls during the night.
6) Play Wii game to enhance bodily balance.	6) Make step-learning carts and tightening strips to prevent the elderly from falling.
7) Build up fall prevention professional team to reduce the fear of stumbling.	7) Removing the bed close to the toilet.
	8) Teach the elderly to call before getting up from bed.
	9) Help the elderly to build up a self-awareness of getting old.
	10) Combine leisure activities to develop "orange industry".

Generally speaking, the contemporary Taiwan's social support system of fall prevention is fragmentary and limited. The study finds that it is because the departments of the government are not well coordinated and there is a lack of horizontal association between them. For example, there is no sufficient communication between the medical and social welfare systems, and hence the services are either redundant or isolated. It also finds from literature review that the Ministry of Economic Affairs had a fall-prevention project of alarming mattress which made sound when the elderly getting up from bed but failed out of a lack of marketing. It is then clear that there is room to work on between the formal and informal support system. Without policy encouragement, the problem of elderly fall will lead to a big pressure for the family and social systems and a heavy burden on medical cost.

6 STRATEGIES FOR EMPOWERING FALL PREVENTION AND SUPPORT SYSTEM

One of the important effects of the mechanism of fall prevention besides family and community support is a sense of self-empowerment with the elderly. The strategies and methods of fall prevention can be seen in Table 3.

Table 3. Strategies and methods of empowerment

Strategies of Empowerment		
Self-Empowerment	Family Empowerment	Social Empowerment
1) Don't underestimate the speed of physical deterioration and the danger of living environment.	1) Use facilities to prevent falling.	1) Establish a professional support system to boost the confidence of the elderly.
2) Teach the elderly to support themselves with hands in case of falling.	2) Refurbish the bedroom, passageways and bathroom with facilitating equipment.	2) Establish an obstacle-free, elderly only space.
3) Take exercise of massage to keep physical flexibility.	3) Family should be in accompaniment.	3) Set up local watch.
4) Slow down the motion.	4) Family should work more closely after the event of falling.	4) Actualise and distribute services of welfare delivery.
5) Be informed through fall prevention community courses.	5) General need of social welfare resources.	5) Improve public space for the elderly.

Figure 1 shows the framework of empowering social support system and the operating mechanism of fall prevention from the perspective of empowerment. By evaluating resources of family and professional system as well as the environment factors, it is more likely to establish an encompassing protection system. The idea of empowerment depends firstly on the elderly's self-assertion and a sense of controlling the living environment. The more the aged are in control of their life, the better their living quality. On the bases of a sense of self-empowerment and health management as well as the support from family and friendly environment, a better mechanism of healthy aging with policy support and welfare technological facilitation can be set up. For achieving the aim of independent and dignified aging, it is therefore necessary to construct an empowerment-centered framework of fall prevention to analyze the unpredictable and increasing new factors of stumbling.

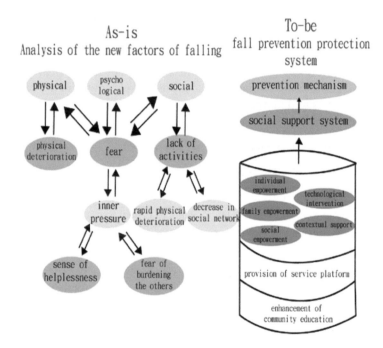

Figure 1 Social Support System of Fall Prevention

7   CONCLUSION AND SUGGESTION

7.1   Conclusion

As accidents are the primary cause of the elderly falls, there are more unpredictable factors hidden within the modern society. Since physical functions deteriorate with aging, once they stumble the outcome can be too heavy for them and their families to bear. Even with a cultural tradition of filial piety, there is a trend to hire professional care services in the place of family care. Hence a multi-layered, well-integrated support system, i.e. community, institute, government policies and industrial resources, is necessary.

On the other hand, the government should establish a trans-professional team of fall prevention to work on customized technological products and service system. In the future Taiwan also has to bring in resources from different fields and encourage the aged to learn using the products to make their life safer, easier, and more active.

7.2 Suggestion

On the basis of the above study, the author makes the following suggestions:

7.2.1 Provide danger indexes and training.

This study finds that the fear incurred from falling affects not only the elderly themselves but also their families. So far the evaluation indexes and training programs have achieved certain effect such as institutes' provision of cautionary warning and on-job education of the caretakers. Besides, it is also found that "nagging" about the importance of self-protection to the elderly is necessary and helps to remind them of the dangers of falling. Certain models of support such as removing the bed closer to the toilet also help prevent stumbling for those who are so coy as not ask any assistance from the others.

7.2.2 Promote a multi-factored preventive system

The elderly need support from family, obstacle-free space and facilities, caretaking teams, volunteers and technology. These are effective measures for the elderly fall prevention. Research on preventive and protection products can contribute to the strengthening of the support, hence a hi-tech industry is necessary for achieving the aim of healthy aging.

7.2.3 Use Tai-Chi or computer games to train physical balance.

It is found from this study that Tai-Chi or games can help train the bodily response and balance. If the aged can learn Tai-Chi or play Wii games, their bodily balance, stability and muscle can be improved and reduce the likelihood of stumbling.

7.2.4 Technological auxiliary system and government policies can solve the problem of a lack of manpower.

Previously the institutes used tightening strips to tie the aged to beds or chairs to prevent them from the risks of falling. Yet such measures are not respectful to the extent of indignity. A lot of respondents suggest that technology can help solve such problems. On the one hand the elderly's activities are under close surveillance, and on the other the lack of manpower can be fixed.

7.2.5 Establish related research within the higher education to develop "Orange Industry".

In face of the aging of the postwar baby boomers, the heath care of the aged should be aided by the power of technology and government policies. There should be related research within the higher education such as leisure and sports, management of aging and health care within the higher education to help develop "Orange Industry" to meet the needs of the elderly.

7.2.6 Expand home services and community service delivery.

In case of the government's social service, there is a need to expand home services, meal delivery, day care center and community care spots to better support the elderly. The policies should integrate resources to form an effective model.

7.2.7 Establish elderly-friendly technology and management system to boost consciousness in self-management of health and empower action.

The home service programs should consider the needs derived from physical deterioration and those from establishing a friendly environment. Research on the products and service systems for surveillance and elderly welfare should be encouraged.

ACKNOWLEDGEMENTS

This research is supported by National Science Council, Taiwan.

REFERENCES

Abreu, N., Hutchins. J., Maston, J., Polizzi, N. and C. 1. Seymour. 1998. Effect of group versus home visit safety education and prevention strategies for falling in comrnunity-dwelling elderly persons. *Home Health Care Management & Practice*,10(4): 57-63.

Arfken, C. L., Lach, H. W., Birge, S. J. and P. Miller. 1994. The prevalence and correlates of fear of falling in elderly persons living in the community. *American Journal of Public Health*, 84(4):565-569.

Bruyere, O., Wuidart, M. A., Di Palma, E., Gourlay, M., Ethgen, O., Richy, F.and J. Y. Reginster. 2005. Controlled whole body vibration to decrease fall risk and improve health-related quality of life of nursing home residents. *Archives of Physical Medicine & Rehabilitation*, 86(2):303-307.

Chadiha, L. A .et al .2004. Empowering African women informal caregivers: A literature synthesis and practice strategies. *Social Work*, 49(1): 97-108.

Chang JT, Morton SC, Rubenstein LZ. et al. 2004. Interventions for the prevention of falls in older adults: systemic review and meta-analysis of randomized clinical trials. *BMJ* 328:680-7.

Chen, Y. J. 2007.*Theories and practices of the elderly welfare: A native perspective*. Taipei: Yeh Yeh Book Gallery.

Chen, Y. J. 2008. Strength perspective: An analysis of ageing in place care model in Taiwan based on traditional filial piety. *Ageing International*, 32:183-204. USA: Publication Springer.

Chen, Y. J. 2009. *Senior service and community care: From the view of multi-services*. Taipei: Wiseman Publishing.

Cox, E. O. and R. J. Parsons.1994. *Empowerment-oriented social work practice with the elderly*. Pacific Grove, C.A.: Brooks/Cole.

Department of Health. 2011. The fourth quarterly brief report of Taiwan patient-safety reporting system, 2011. From   http://www.tpr.org.tw/index03.php, accessed on 8 January 2011.

Department of Health. 2007.The sixth longitudinal survey on the physical, mental and social living of the middle to advanced aged people, 2007. From http://olap.bhp.doh.gov.tw/index.aspx accessed on 10 February 2011.

Gibson, C. H. 1991. A concept analysis of empowerment. *Journal of Advanced Nursing*, 16:345 -361.

Hornbrook, M. C., Stevens, V. J., Wingfield, D. J., Hollis, J. F., Greenlick, M. R. and M. G. Ory.1994. Preventing falls among community-dwelling older person: Results from a randomized trial. *Gerontologist*, 34(1):16-23.

Howland, J., Lachman, M.E., Peterson, E.W., Cole, J., Kasten, L. and A. Jette.1998. Covariates of fear of falling and associated activity curtailment. *The Gerontologist*, 38:549-555.

Howland, J., Peterson, E. W., Levin, W. C., Fried, L., Pordon, D. and S. Bak.1993.Fear of falling among community-dwelling elderly. *Journal of Aging and Health*, 5:229-243.

Huang, T. T. and G. J. Acton.2004. Effectiveness of home visit falls prevention strategy for Taiwanese community-dwelling elders: randomized trial. *Public Health Nursing*, 21(3): 247-256.

Kieffer,C.H.1981.The emergence of empowerment: The development of participatory competence among individuals in citizen organizations: I & II . *The Gerontologist*, 38:549-555.

Ministry of the Interior .2009. *Friendly service program for the elderly: 2009- 2011*. Taipei: Department of Social Affairs.

Rappaport, J. 1984. Studies in empowerment: Introduction to the issues. *Prevention in Human Services*, 3:1-7.

Rodwell, C. 1996. An analysis of the concept of empowerment. *Journal of Advanced Nursing*, 23:305-313.

Sarvimaki, A. 2000. Quality of life in old age described as a sense of well-being, meaning and value. *Journal of Advanced Nursing*, 32(4):1025-1033.

Scanaill et al.2011. Falls prevention in the home: Challenges for new technologies. *Intelligent Technologies for Bridging the Grey Digital Divide* : 46-64.

Sieri, T.and G. Beretta.2004. Fall risk assessment in very old males and females living in nursing homes. *Disability and Rehabilitation*, 26(12): 718-723.

Turnbull, A.P.andH.R.Turnbull.2001. *Amilies, professionals, and exceptionality-collaborating for empowerment* (4ed).U SA: The University of Kansas.

Wallerstein, N. 1992. Powerlessness, empowerment and health: Implication for health promotion programmes. *American Journal of Health Promotion*, 6(3):197-205.

Walsh, P. 1993. The concept of family resilience: Crisis and challenge. Article first published online: 29 JUL 2004. 35(3), 261-281.

Wolfe, R. R., Jordan, D. and M. L. Wolfe.2004. The walk about: A new solution for preventing falls in the elderly and disabled. *Archives of Physical Medicine & Rehabilitation*, 85 (12):2067-2069.

Yip, K.2004. The empowerment model: A critical reflection of empowerment in Chinese culture. *Social Work*, 49(3): 479-487.

CHAPTER 16

Reducing Ergonomic Injuries for Librarians Using a Participatory Approach

Lu Yuan

Southeastern Louisiana University
Hammond, USA
Lu.Yuan@selu.edu

ABSTRACT

This study utilized a participatory ergonomics approach to examine the ergonomic hazards and to reduce musculoskeletal symptoms for librarians in the East Baton Rouge Parish Main Library. A total of 39 employees from 9 different divisions in the Library participated in the study. The results of pre- and post-training ergonomics knowledge tests indicate significant improvement of librarians' understanding of ergonomics principles, whereas the questionnaire responses for both 2-month-post- and 8-month-post- the ergonomics training compared against those before the training have shown positive improvements in the majority of musculoskeletal symptom presence and severity, computer workstation, manual material handling, and perceived control over the work environment. With the identification of ergonomic hazards through RULA (Rapid Upper Limb Assessment) and REBA (Rapid Entire Body Assessment) observations as well as focus group discussions, the study findings accomplish the project's overall objective of assisting librarians to improve ergonomics in the workplace. The results of this study provide foundation for future long-term study of participatory ergonomics to reduce musculoskeletal injuries and disorders for librarians and other service sector workers.

Keywords: participatory ergonomics, librarians, musculoskeletal symptom

1   INTRODUCTION

Technological advancement has shaped the work environment in libraries dramatically since the 1990s. Intensive or long-term use of computers and other electronic tools has become more and more popular in all public service areas and technical operations, particularly cataloging. This has caused librarians to use awkward postures of the head, neck, and upper extremities and to endure increased pressures on the soft tissues against external workstation surfaces. On the other hand, librarians are still involved in extensive and repetitive handling of books, boxes, and other materials, where they usually have to exert excessive strength during different activities and maintain sustained static posture during prolonged holding (Thibodeau and Melamut, 1995).

Both these typical aspects of library work expose librarians to a relatively wider range and higher level of ergonomic hazards than "standard" office-type work does, as they have produced enormous risk and stress on librarians (Chao, 2001). For example, Mansfield and Armstrong (1997) reported that the average yearly numbers of injuries and traumatic musculoskeletal disorders (MSDs) at the Library of Congress whose yearly average was 4917 staff during 1991-1995, are 229 and 47, respectively. These injuries and disorders have caused an average yearly $946,284 workers' compensation cost during that five-year span.

It seems there is a great need to address ergonomic issues within the library environment. Although there is a growing body of literature discussing ergonomics and libraries found in books, journals, and internet sources, there is currently no systematic process to determine needs and evaluate interventions (Tepper, 1996). Rather, the majority of relevant ergonomic research focuses on the examination of hazards in the "standard" office environment. Libraries spend a great amount of time planning the hardware and software implementations of electronic information services, but the human factors and ergonomics are often overlooked (Thibodeau and Melamut, 1995). Thus, it is imperative to explore effective and efficient research methodologies to identify, analyze, and control ergonomic hazards during library work.

One method for introducing and implementing ergonomics is to use the concept of participatory ergonomics, which originated from discussions between Drs. Kageyu Noro and Kazutaka Kogi in Singapore in 1983 (Imada, 1991). As the word "participatory" indicates, this specific concept constitutes the use of participative techniques and various forms of participation in the workplace (Vink and Wilson, 2003). Wilson (1995) defined participatory ergonomics as "the involvement of people in planning and controlling a significant amount of their own work activities, with sufficient knowledge and power to influence both processes and outcomes in order to achieve desirable goals."

The objective of the present study was to utilize a participatory ergonomics approach to reduce musculoskeletal injuries and disorders for librarians in the East Baton Rouge Parish Main Library. Specifically, the study was designed to: provide training of the basic concepts and principles of ergonomics to librarians; identify the ergonomic hazards associated with typical library work; and introduce and then apply the participatory ergonomics approach to mitigate the ergonomic hazards within the library environment.

2   METHODS

Figure 1 illustrates a simplified diagram outlining the participatory ergonomics process used in this study. It should be noted that *evaluation* is one of the most crucial elements involved in the entire research process. Variables that were evaluated are connected to relevant measures within each step using different methods.

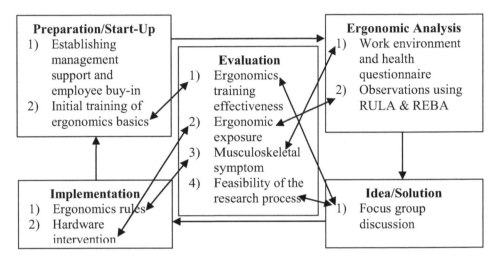

Figure 1 A simplified diagram of the participatory ergonomics process in this study

Details about preparation/start-up, ergonomics training, and work environment and health questionnaire are available in Yuan and Culberson (2011).

2.1   Observations Using RULA & REBA

RULA (Rapid Upper Limb Assessment) was used to assess working postures and required muscle use and force exertion for computer usage of the library work (Lueder and Corlett, 1996; McAtamney and Corlett, 1993). For the manual material handling activities during library work, REBA (Rapid Entire Body Assessment) was used to estimate the risks of entire-body injuries and disorders (Hignett and McAtamney, 2000).

A total of approximately 20 hours of RULA and REBA observations were conducted on representative samples of the library work before the ergonomics training (10 hours in total), 2 months after the training (5 hours), and 8 months after (5 hours). Typical library work tasks/activities that were observed by RULA include: labeling, stamping, and lining books and cataloging books in the Technical

Service Department; answering phone calls in the Reference Division; Checking in/out books in the Circulation Division; and by REBA include: unloading boxes of books and other materials and categorizing in the Shipping/Handling Office of the Circulation Division.

2.2   Focus Group Discussion

Focus group interviews and/or meetings have been employed in many participatory ergonomics research projects (Pehkonen, Takala, and Ketola, et al., 2009). This technique uses a scripted brainstorming method to solicit questions and answers from a group of people representing a wide range of employment. In this study, the employee representatives from different divisions in the Library were invited to participate in the focus group discussion. During the one-hour meeting, the attendees were asked questions about their work activities, safety and health concerns, ergonomic exposures, hazard control, and opinions on the feasibility of the study.

2.3   Implemention of Ergonomics Rules and Hardware

Based on the previous steps of the research process where ergonomic hazards and risk factors have been identified and assessed through ergonomics training, the work environment and health questionnaires, observations, and focus group discussions, the researcher explored some handy posters or brochures illustrating ergonomics rules through consultation with the Barbre Ergonomics Consulting and Training (2011) and Experteyes (2011). Permission has been granted to distribute three ergonomics brochures including Stretches (from Barbre Ergonomics), and Workstation Ergonomics and Manual Handling and Storage (from Experteyes) to the study participants.

It was expected that better ergonomically-designed equipment, typically an ergonomic chair, would be recommended through both observations and focus group discussions. The upper management of the Library under study was supportive regarding expenses for reasonable requests. However, since the Library was in a process of getting a new building at the end of this study, the Director has preferred to make the investment for ergonomic workstation at a later stage.

2.4   Evaluation

Evaluation of the study includes both comparison of intermediate effects before and after the participatory ergonomics intervention and examination of the feasibility of the participatory approach. The effects of the intervention were measured through pre-post-differences in mean scores for ergonomics training test, work environment and health questionnaire, RULA and REBA observations.

The feasibility of the research process was assessed by focus group discussions and satisfaction surveys. At the end of the focus group discussions, the attendees were asked such questions as general opinions about the process, benefits of the project, difficulties with the approach, and barriers in the implementation, etc. to assess feasibility. To measure the success of the intervention, a survey was distributed to the study participants at the end of the intervention phase. Satisfaction with the arrangements of the project, flow of information, implemented changes, support from the management, and support from researcher was evaluated on a five-point scale (1 = very dissatisfied, 2 = fairly dissatisfied, 3 = undecided, 4 = fairly satisfied, 5 = very satisfied).

2.5 Data Analysis

Differences in the average scores of pre- and post- training ergonomics knowledge tests were examined using paired t-tests. Changes in the responses of work environment and health questionnaire, especially the presence and severity of musculoskeletal symptoms, workstation postures, manual material handling experience, and perceived control of the work, were calculated and then summed across librarians to determine the proportion of subjects' responses in each of the three classifications ("improved", "worsened", or "no change"). A McNemar non-parametric test was used to evaluate the statistical significance of the observed changes ("improved" vs. "worsened") for comparisons of two-month-post- VS. pre-training, eight-month-post- VS. pre- training, and eight-month-post- VS. two-month-post- training, respectively. The one-way ANOVA was used to examine the differences in the average RULA and REBA scores at different stages of the research process, including before ergonomics training, 2 months after the training, and 8 months after. The focus group discussion notes were analyzed qualitatively, whereas a descriptive statistics was presented to summarize the satisfaction survey results.

Data were analyzed using PASW (also known as SPSS) Statistics 18.0. In each of the statistical tests described above, the level of significance required to reject the null hypotheses was established at $p < 0.05$.

3 RESULTS

3.1 Demographics of the Study Population

Thirty nine employees representing nine different divisions participated in the study. There were 28 females and 11 males. The average age for the sample population was 43.3 years (range 22 to 72). The subjects have been in their profession for an average of 13.2 years (range 0.4 to 45 years), and they have been working in the East Baton Rouge Parish Main Library for an average of 10.0 years (range 0.3 to 36 years). The majority of subjects are full-time employees (89.7% of the total), and the top two job titles that the subjects hold are Librarian Technician/Assistant (41.0%) and Librarian (30.8%).

Six of the 39 subjects did not return their 8-month-post-training questionnaires, which made the total number of questionnaire responses for this round to be 33.

Among those 6 subjects, one person has retired after completing the 2-month-post-training questionnaire, 2 employees have moved to other branch libraries, and the other 3 people could not be located.

3.2   Pre- and Post- Training Ergonomics Knowledge Tests

The average pre-test score was 37.6 (of 100 points), whereas the average post-test score was 76.3. The mean increase was 38.7. The t value of 11.9 (df = 38) was significant at $p < 0.0001$. Thirty four of the 39 subjects have answered the open-ended question in the post-test, of which the most common answers include adjusting monitor and chair height, and removing clutter from desk, etc.

3.3   Work Environment and Health Questionnaire

Two-Month-Post- VS. Pre-
The numbers (proportions) of changes in subjects' overall health rating for the three categories of "improved", "worsened", or "no change" were 6 (15%), 10 (26%), and 23 (59%). The χ^2 value for the McNemar test was 0.56, which did not show a significant difference between "improved" and "worsened" responses ($p = 0.45$).

There were statistically significantly positive changes in the questionnaire responses to the four specific questions: "break/rest every 2 hours", "hand/wrist positions", "handle more than 50 lbs", and "bend or twist at the waist to handle objects". The changes in other categories of the questionnaire, including the presence and severity of musculoskeletal symptoms and perceived control over the work environment, were not statistically significant; however, there was a trend toward positive improvement.

Eight-Month-Post- VS. Pre-
Subjects responded more "improved" changes 11 (33%) than "worsened" ones 8 (24%) in the overall health rating eight months after the training. The $\chi2$ value for the McNemar test was 0.21, which did not show a significant difference between "improved" and "worsened" responses (p = 0.65).

There are statistically significantly positive changes in the questionnaire responses to three specific questions: "break/rest every 2 hours", "hand/wrist positions", and "supervisor's willingness to listen to work-related problems". The changes in other categories of the questionnaire were not statistically significant; however, there was a trend toward positive improvement.

Eight-Month-Post- VS. Two-Month-Post-
There were 10 (30%) "improved" and 4 (12%) "worsened" changes, respectively, in subjects' overall health rating between eight-month-post- and two-month-post- training questionnaire responses. However, this difference is not statistically significant ($\chi2 = 1.79$, p = 0.18).

The net changes in the ratings of the presence and severity of musculoskeletal symptoms tended to fluctuate, and there were negative net changes in the manual material handling experience. Yet, it has shown positive improvement in the category of computer workstation postures and adjustability. Also, it has been reported that supervisors were more willing to listen to work-related problems and the improvement was statistically significant.

3.4   RULA and REBA

The study did not find any statistically significant differences in the average RULA and REBA scores for the tasks/activities that were observed before and after the ergonomics training. Overall, some typical ergonomic hazards/issues were identified and these include: awkward postures of neck and upper extremities during computer usage and back during material handling, inadequate leg room under desk and work space on desk, improper postures of neck and shoulder and simultaneous computer typing when answering phone calls, and extreme overload on Mondays especially for the Shipping/Handling Office, etc.

3.5   Focus Group Discussion

Twelve subjects from 7 different divisions participated in a total of 3 focus group meetings in February 2011. The important things learned from these meetings include: 1) Heavy lifting, repetition, and sitting at computer for a long period of time are common activities for librarians; 2) There are also health concerns, e.g., customers might be sick and books might also contain viruses and bacteria; 3) Ergonomic furniture should be in place; and 4) Participation in the project was beneficial, and a workstation model during the ergonomics training and ergonomics posters/brochures afterwards would help even more.

3.6   Satisfaction survey

Approximately 85-94% of the subjects felt satisfied or very satisfied with arrangements of project, flow of information, and support from researchers. Only half of the subjects were satisfied with implemented changes and about 27% of the subjects were not so satisfied with support from management; however, there was a general consensus among the majority of those people that they understood the management was waiting to make the investment of ergonomic workstation for the new building that would be broken ground soon.

4   DISCUSSION

The present study utilized a participatory ergonomics approach consisting of ergonomics training, observations, work environment and health questionnaires, focus group discussions, ergonomics brochures to improve ergonomics in the workplace and to reduce musculoskeletal symptoms for librarians in the East Baton Rouge Parish Main Library. The results of ergonomics knowledge tests indicate significant improvement of librarians' understanding of ergonomics principles, whereas the questionnaire responses for both 2-month-post- and 8-month-post- the ergonomics training compared against those before the training have shown positive improvements in the majority of musculoskeletal symptom presence and severity, computer workstation, manual material handling, and perceived control over the work environment. With the identification of ergonomic hazards through RULA and REBA observations as well as focus group discussions, the study findings accomplish the project's overall objective of enhancing ergonomics in the library environment.

The evaluation of library ergonomics training has not been conducted significantly in previous research studies; however, there have been a few publications depicting office and VDT (Video Display Terminal) ergonomics training evaluation (Bohr, 2000; Ketola, Toivonen, and Hakkanen, et al., 2002; Lewis, Fogleman, and Deeb, et al., 2001; Rizzo, Pelletier, and Chikamoto, 1997; Robertson, Amick III, and DeRango, et al. 2009). In particular, Lewis, Fogleman, and Deeb, et al. (2001) evaluated the effectiveness of a VDT training program through comparing the 170 participants' responses to a musculoskeletal symptom questionnaire before and one year after the program. This study did not have a control group not receiving the training, which used the similar design as the present study. However, both two studies shared the same results in demonstrating the effectiveness of ergonomics training to reduce musculoskeletal symptoms.

Since the post-training questionnaires were handed out both two months and eight months after the training, this study only measured the short-term effects on changes in subjects' work behavior and health status. There are some positive net changes in the presence and severity of musculoskeletal symptoms of major body parts when comparing subjects' responses two months and eight months after the training with those before training, respectively. Yet, the comparison of the responses in between two-month-post- and eight-month-post- training did not show one-way pattern at all. It may seem natural to see the decrease of the influence that participation in this project has produced on the subjects' musculoskeletal symptoms when time passes by, even though there have been positive improvements when compared with those ratings before the training. As 13 subjects have reported medical care for existing symptoms, it was not surprised to not see significant changes in the short-term effects that participating in this project could improve the subjects' health conditions. On the other hand, the responses in "other non-work-related activities" indicate that approximately half of the subjects had prolonged use of home computer, did heavy housework such as painting and mowing, and attended fitness program regularly. All these activities could produce confounding effect in the work-relatedness of musculoskeletal symptoms.

The improvement in subjects' computer usage and other work activities and experience indicates the application of ergonomics principles into their daily work life. Although the training tests only examined the subjects' knowledge of ergonomics, it would be surmised that because of the improving knowledge which could be demonstrated by the increasing test scores, subjects tended to apply ergonomics more often during their regular work activities.

The observational data by RULA and REBA, however, could not confirm any statistically significant improvements in subjects' workstation posture and behavior. The researcher felt part of the reasons might be that no hardware intervention, primarily installation of the ergonomic workstation, has been implemented, as the Library is waiting on the new building that it will get in a near future. Also, the distribution of ergonomic brochures have been delayed as the management of the Library would like to purchase the license of these brochures and expand the distribution to all of the employees (approximately 500) working in the Library system. It would be interesting to see if there were any significant differences should the new building be in place, which might unduly indicate the necessity and importance of a follow-up study to continuously helping the librarians improve ergonomics in their new work environment.

The management's decision on not considering any significant changes until the new building is in place might also indirectly explain the less satisfaction ratings on support from management in the subjects' exit survey at the end of the project. Nevertheless, the 8-month-post-training questionnaire results indicate that the research participants felt their supervisors were more willing to listen to their work-related problems. In fact, the management was in the process of selecting and testing a variety of chairs that shall be purchased for different workstations and public service areas for the new building. Based on the conversation with the librarians during the focus group discussions and other casual occasions, it seemed that the majority of librarians understood the management's situation and had been anxiously waiting for the new building.

5 CONCLUSIONS

Overall, the study findings accomplish the project's objective of assisting librarians to improve ergonomics in the workplace and to reduce musculoskeletal symptoms in a short term. The present study provides invaluable baseline information about the ergonomic issues in the library environment, beyond which further research effort is warranted to improve the effectiveness of the library ergonomics program.

ACKNOWLEDGMENTS

The author thanks the East Baton Rouge Parish Main Library employees who participated in the research activities. The support from City of Baton Rouge Risk Management Division is greatly appreciated. This study was funded by the NIOSH Southwest Center for Occupational and Environmental Health (SWCOEH) Pilot Projects Research Training Program (Grant 3T42OH008421-0552). The contents in this article are solely the responsibility of the authors and do not necessarily represent the official views of NIOSH. The author would like to thank Gregory Culberson who contributed to data collection.

REFERENCES

"Barbre Ergonomics Consulting and Training." Accessed March 1, 2011, http://www.barbre-ergonomics.com.

Bohr, P.C. 2000. Efficacy of office ergonomics education. *Journal of Occupational Rehabilitation* 10: 243-255.

Chao, S.J. 2001. Library ergonomics in literature: a selected annotated bibliography. *Collection Building* 20: 165-175.

"Experteyes." Accessed March 1, 2011, http://www.experteyes.com.au/.

Hignett S. and L. McAtamney. 2000 Rapid Entire Body Assessment: REBA. *Applied Ergonomics* 30: 201-205.

Imada, A.S. 1991. The rationale and tools of participatory ergonomics. In: *Participatory Ergonomics,* eds. K. Noro and A.S. Imada. London: Taylor & Francis.

Ketola, R., R. Toivonen, and M. Hakkanen, et al. 2002. Effects of ergonomics intervention in work with video display unit. *Scandinavian Journal of Work, Environment & Health* 28: 18-24.

Lewis, R.J., M. Fogleman, and J. Deeb, et al. 2001. Effectiveness of a VDT ergonomics training program. *International Journal of Industrial Ergonomics* 27: 119-131.

Lueder R. and E.N. Corlett. 1996. A proposed RULA for computer users. Proceedings of the Ergonomics Summer Workshop, UC Berkeley Center for Occupational & Environmental Health Continuing Education Program.

Mansfield, J.A. and T.J. Armstrong. 1997. Library of Congress workplace ergonomics program. *American Industrial Hygiene Association Journal* 58: 138-144.

McAtamney L. and E.N. Corlett. 1993. RULA: a survey method for the investigation of work-related upper limb disorders. *Applied Ergonomics* 24: 45-57.

Pehkonen, I., E-P. Takala, and R. Ketola, et al. 2009. Evaluation of a participatory ergonomic intervention process in kitchen work. *Applied Ergonomics* 40: 115-123.

Rizzo, T.H., K.R. Pelletier, and Y. Chikamoto. 1997. Reducing risk factors for cumulative trauma disorders (CTDs): the impact of preventive ergonomic training on knowledge, intentions, and practices related to computer use. *American Journal of Health Promotion* 11: 250-253.

Robertson, M., B.C. Amick III, and K. DeRango, et al. 2009. The effects of an office ergonomics training and chair intervention on worker knowledge, behavior and musculoskeletal risk. *Applied Ergonomics* 40: 124-135.

Tepper, D. 1996. Participatory ergonomics in a university library. Master's Thesis. Ithaca: Connell University.

Thibodeau, P.L. and S.J. Melamut. 1995. Ergonomics in the electronic library. Bull Med Libr Assoc. 83: 322-329.

Vink, P. and J.R. Wilson. 2003. Participatory ergonomics. Proceedings of the XVth Triennial Congress of the International Ergonomics Association and The 7th Joint conference of the Ergonomics Society of Korea/Japan Ergonomics Society, 'Ergonomics in the Digital Age', Seoul, Korea.

Wilson, J.R. 1995. Ergonomics and participation. In. *Evaluation of Human Work: A Practical Ergonomics Methodology, 2nd ed.,* eds. J.R. Wilson and E.N. Corlett. London: Taylor and Francis.

Yuan, L. and G. Culberson. 2011. Effectiveness of a library ergonomics training program. Proceedings of the 55th Annual Meeting of Human Factors and Ergonomics Society, Las Vegas, NV.

CHAPTER 17

Usability in the Opening of Soft Drinks Packagings: Age Influence in Biomechanical Forces

Danilo Corrêa Silva, Luis Carlos Paschoarelli

UNESP – Univ Estadual Paulista
Bauru, Brazil
danilo@idemdesign.net

ABSTRACT

Packagings have gone through profound changes. New materials and processes allow different designs, which became true consumer appeals. However, these advances do not necessarily lead to improvement in the usability of such products. This paper assesses how the different shapes of soft drinks PET bottles affect the ability of different age groups to open them. The results show that the adults have a greater strength in all situations when compared to youth or elderly subjects. A comparison of collected data and the torque forces necessary to perform the opening reveals a lack of usability and accessibility of these products.

1   INTRODUCTION

The development of new technologies and materials in packaging design provided additional resistance and safety, lower weight and environmental impacts, among others. On the other hand, in the last ten years, packaging has become more difficult to open (Yoxall, Janson, Bradbury, et al. 2006). These advances are not always accompanied by studies of user interaction with packagings, which may harm access to the product, particularly to the elderly users. There are still a few studies involving the accessibility and usability of wide consumption packagings, such as used in soft drinks beverages (PET bottles).

Soft drinks are non-alcoholic and carbonated beverages, which are flavored, acidified, colored and carbonated artificially. Typically, such beverages are made of concentrated syrup, aspartame, caffeine and preservative agents (like benzoic acid and / or its derivatives), which are all mixed together with carbonated water and then bottled (Sádecká and Polomsky 2000). Carbonation is the process of dissolution of CO_2 in the beverage.

Along with technological progress, the packagings have undergone significant changes. Thus, the implementation of new materials, such as polymers, provided mechanical qualities that are sometimes superior to metals or glass, and a freedom of design that made the products more attractive. However, these advances are not always accompanied by improvements in the accessibility or usability of the products.

Most of biomechanical studies involving the manipulation of objects have showed a fluctuation of manual forces related to the individual's age; with peak strength in early adulthood and a gradual decline until the onset of old age (Mathiowetz, Rennells and Donahoc 1985; Voorbij and Steenbekkers 2001). As age advances, people tend to present force equivalent to teenagers or children (Peebles & Norris 2000; 2003). Additionally, such limitations can be increased by a design which does not consider the constraints of elderly people.

A poor packaging design can cause several constraints to consumers / users, including the inability to access the product due the high biomechanical forces needed to open them. This is a common problem in vacuum jars (Yoxall, Janson, Bradbury, et al. 2006; Silva, Paschoarelli, Campos, et al. 2010) and an issue for other types of screw caps packagings. Hence, PET bottles used for bottling soft drinks have different shapes, which may interfere with the user's ability to apply biomechanical forces.

Campos (2010) and Paschoarelli (2009) highlight in their studies about door knobs that differences among genders or age groups were influenced by the product design. In both studies, the spherical shape offered the lowest torque force for groups and leaded to the greatest differences among them. The lever type handles increased significantly the ability to apply forces by those with lower biomechanical capabilities (elderly and women).

The objective of this study is to assess how subject's age and the design of the soft drinks PET bottles affect manual torque forces for opening such packagings. Furthermore, the torque force necessary to open these packagings will be compared with the obtained data, allowing the assessment of the accessibility of these products.

2 METHODS

The evaluation consisted in collecting subject's maximum static torque forces during simulated activities with five soft drinks PET packagings, in which a sensor recorded the peak force for each one of the 180 individuals with each packaging.

2.1 Materials and equipment

For data collection, specific materials and equipment were used (Table 01).

Table 01. Materials and equipment used in the study.

ITEM	VIEW	DESCRIPTION
1		Research protocols: Informed Consent and Identification Protocol.
2		Static torque screwdriver (STS - Mecmesin Ltd., UK) model ST10 – 871 – 101, with 10 N.m capacity;
3		Force Gauge AFG 500N (Mecmesin Ltd., UK), with 500 N capacity and 0.1% of precision. This equipment was used to convert the analog signal of the STS and display the exerted force;
4		Metallic structure with three shelves: the first (1) holds a scale of perception about the activity (not addressed here); the second (2) for equipment of data exhibition and spreadsheets; and the third one (3) to hold the packaging during force exertion. This last one provided a height of 850 mm from the ground;
5		5 different PET bottle packagings commercially available in the city of Bauru – SP – Brazil. These packagings were used "as they are" (cap and body together). At them was coupled the STS.

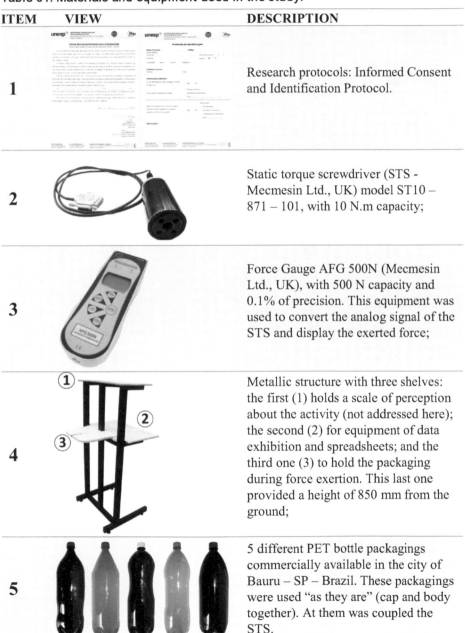

174

The criteria for selection of the packagings were based on their shapes, mainly in the shoulder region (near the cap). Only packagings with the same capacity (2 liters) were selected, so uniformity in the relative proportions of the models was obtained. It was not considered any brand or market criterion in the selection of the packaging models. The first two models (E1 and E2) use a single cap type, which is lower and has thicker slots (Cap 1). The other models (E3, E4 and E5) use a taller cap with thin slots (Cap 2). A detailed illustration of all the five bottles can be found in Figure 1.

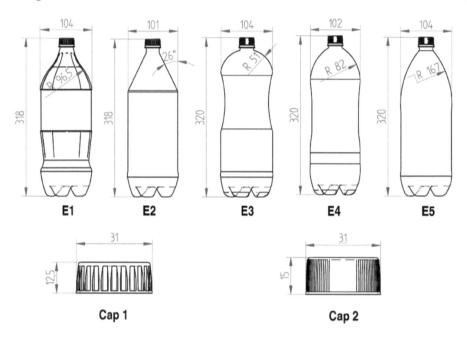

Figure 1. Technical details of the five PET packagings used in this study. Dimensions are in millimeters.

The coupling system developed (Figure 2) consists in two stages of fixation: in the packaging (B) is fixed the first one, through a square hole at the bottom, where a duct of squared section (D) is inserted. This duct is made of a 3 mm thick polystyrene plate, and is fixed and held in position by expandable polyurethane foam (E), which was injected at the bottom of the bottle too. The second stage consists in a duct of square section, made of 1 mm thick stainless steel plate AISI 416 (F), this duct has a smaller size than the previous one, in which the STS (G) is fixed.

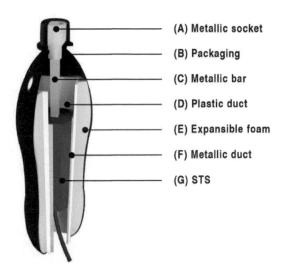

Figure 2. Schema of the STS coupling system.

2.2 Subjects and procedures

A total of 180 subjects of both genders, 60 subjects aged 18-29 years old (youth - mean 23.85, s.d. 3.33 years old), 60 aged 30-55 years old (adults - mean 42.62, s.d. 7 years old) and 60 aged over 55 years old (elderly - average 67.10, s.d. 9.10 years old). The procedures of this study were approved by the Ethics Committee of USC (Protocol: 121/09 – Bauru – SP - Brazil) and all subjects signed an Informed Consent form, following the recommendation of "Norm ERG BR 1002 - Code of deontology of Certified Ergonomists" (Abergo 2003).

The subjects were addressed individually, informed about the aims and the procedures of the study, signed the Informed Consent and fulfilled the Identification Protocol. Then, one subject at time was asked to stand in front of the structure, hold at the shoulder of the bottles with one hand and hold the cap with the other. A grip of opposition between the pulp of the thumb and the side of index finger was used to hold the cap (pulp-lateral between thumb and index fingers). Then the subject was asked to exert his maximum force to open the bottle (counter-clockwise).

There were two attempts for each hand, with a 30 seconds time interval to rest. After all the attempts, the packaging model was switched and the subject was asked to repeat the procedure. The impossibility of opening was not informed to the participants, unless they wondered it. The sequence for applying the models was randomized, as well as the use of right or left hand first. The exerted force for each situation was displayed in the AFG and written down in a spreadsheet.

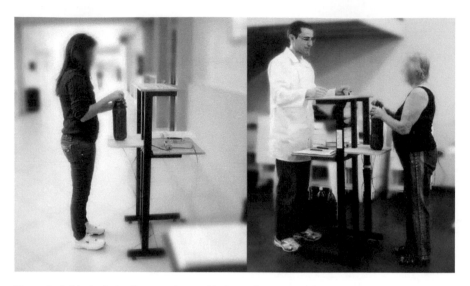

Figure 3. Subjects during the experiment with the packaging models.

2.3   Data analysis

The data were tabulated in electronic spreadsheets for descriptive statistics. To assess differences between the data sets, the *Statsoft Statistica 8®* software was used. The procedures involved the conditions of normality of data sets, according to Shapiro-Wilk's test, and homogeneity, according to Levene's test. The non-normality or non-homogeneity of data has implicated the use of a nonparametric statistical test (Kruskal-Wallis), instead of parametric tests (ANOVA one way).

For comparison of the strength values obtained with the torque forces needed to perform the opening in this type of packaging, the study of Silva, Paschoarelli, and Silva (2012) was used. That study evaluated 48 soft drinks packagings commercially available in the city of Bauru – SP (Brazil), resulting in the average torque required to open up these products (1.37 ± 0.25 N.m). All data groups were considered normal (normal distribution) and the "Probability calculator" of the refereed software was used to estimate the amount of people who would have trouble opening the product.

3   RESULTS AND DISCUSSION

There were no significant differences between the two opening attempts with left or right hand. Since it is an evaluation of maximum torque forces, higher value between the two attempts for each individual was used, since the higher value best corresponds to the biomechanical capacity of the subject. Hence, the results presented below correspond to the values obtained with the maximum torque forces exerted by the subjects.

The results of ANOVA (one way) showed that youth exerted forces slightly higher than elderly in all situations, although there were no significant differences (p = 0.6859). On the other hand, the group of 30-55 years exerted the highest forces in all situations of the experiment, both related to the youth (p = 0.0054) and the elderly (p = 0.0003). Figure 4 shows the graph for torque values exerted by the groups.

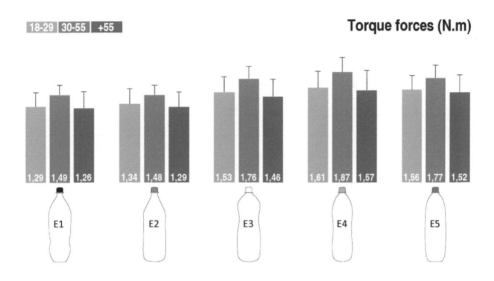

Figure 4. Average torque forces for each age group using the different packagings.

This observation was statistically confirmed for each packaging model. The values of statistical comparisons for each situation can be found on Table 02. The 18-29 group has similar strength to that of the elderly group in all models evaluated.

Table 02 ANOVA one way – pos hoc Tukey / *Kruskal-Wallis p-values for each age group according to packagings model used.

Model	E1*		E2		E3		E4		E5	
Age	18-29	30 -55	18-29	30 -55	18-29	30 -55	18-29	30 -55	18-29	30 -55
30-55	0,0028		0,0489		0,0029		0,0019		0,0054	
+55	1,0000	0,0006	0,6467	0,0042	0,4997	0,0002	0,7874	0,0003	0,8199	0,0008

Hence, comparisons between age groups revealed that the group aged between 30-55 years old exerted the highest forces. These results corroborate that the elderly people have smaller biomechanical forces when compared to adults. However, there were no differences between the torque forces exerted by the youth and elderly.

Paschoarelli (2009) had a similar result only for one of the door knobs analyzed, while all others showed differences between youth and elderly subjects. Campos (2010) has also stated that the occurrence of differences among age groups depends on the model analyzed, highlighting that these differences are, in some cases, increased by the design of the interface.

In this study there were also significant differences among the different models within age groups (Table 03). Apparently there is a division between the first two models, that features close shapes to each other, and use the same type of cap (Cap 1), and the three others, which also have similarities with each other, and use the second type of cap (Cap 2).

Table 03. Comparisons among packaging models within age groups (p-values for ANOVA one way – pos hoc Tukey).

Age	18-29				30-55				+ 55			
Model	E1	E2	E3	E4	E1	E2	E3	E4	E1	E2	E3	E4
E5	0,0004	0,0068	0,9946	0,9218	0,0000	0,0000	1,0000	0,2243	0,0032	0,0150	0,9179	0,9755
E4	0,0000	0,0003	0,7346		0,0000	0,0000	0,1841		0,0003	0,0018	0,6029	
E3	0,0022	0,0248			0,0000	0,0000			0,0491	0,1477		
E2	0,9532				0,9993				0,9920			

These results confirm that design of the bottle is one of the decisive factors in biomechanical capabilities and performance of users. As Peebles and Norris (2003) said, the mere use of slots, one of the attributes of the design of a cap, can influence the forces applied. In this case

In this case, however, the force pattern seems to be the same for each group. The model E4 provided the best biomechanical condition, followed by the models E5, E3, E2, and E1. Thus, it is assumed that an ergonomic design of such packaging can improve the accessibility of the product for all people, despite the age.

Regardless that, the mean values obtained in this evaluation are slightly near the average force required to open soft drinks PET bottles, which can lead users to extreme efforts or even the inability to open these packagings. That finding is corroborated, since the estimative about the amount of people who would have difficulties opening such packagings reveals disturbing conditions of accessibility (Figure 5).

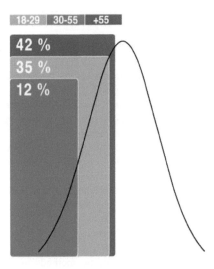

Figure 5. Percentage of users who would have difficulties opening PET bottles.

As shown in Figure 5, for the youth group, 35% of subjects would have difficulties opening these packagings, while for adults this proportion drops to about 12%. The estimative for the elderly are more worrying, since about 42% would have difficulties or even could not open these products.

The closure systems of these bottles have relative uniformity in the opening forces, for both types of caps. Thus the strength requirements appear to be more related to other factors, such the internal pressure of the liquid, instead of the cap type. Anyway, the biomechanical factors related to different user groups correspond to a determinant factor in the design of these systems.

As one of the main activities that determine the characteristics of a product's interface, the design can make them more efficient, comfortable and safe. To do that, the professional must choose between the use of geometric or anthropomorphic shape, texture or smooth surface, colors, materials and many other aspects that should give the subject appropriate conditions of use.

It can be concluded that the ergonomic design of products - especially packaging - depends on biomechanical parameters that improve accessibility for several user's groups. This study used assessment methods that can be, and must be expanded to other types of packaging such as various food, drinks, and medicines, among others. Finally, the results obtained and the estimative made indicate the need to improve the design of PET bottles for soft drinks, making these products more efficient, practical and functional.

ACKNOWLEDGMENTS

This study was supported by FAPESP (Proc. 2009/13477-4 - Proc. 2005/59941-2) and CNPq (Proc. 303138/2010-6).

REFERENCES

Abergo 2003. Norm ERG BR 1002 - Code of deontology of Certified Ergonomists. Brazilian Ergonomics Association, 2003. Available in: <http://www.abergo.org.br/arquivos/Norma% 20ERG%20 BR%201002%20-%20Deontologia.pdf>. Retrieved in august 8, 2005.

Campos, L. F. A. 2010, *Evaluation of manual forces in simulated activities with Brazilian adults individuals of different genders and ages: aspects of ergonomic design.* Master's degree Thesis. Bauru: UNESP. [in portuguese]

Mathiowetz, V, C. Rennells, and L. Donahoc 1985, 'Effect of elbow position on grip and key pinch strength', *The Journal of Hand Surgery*, 1985, pp. 694-697.

Paschoarelli, L. C. 2009. *Ergonomic design: evaluation and analysis of manual instruments in interface user x technology.* Habilitation Thesis. Bauru: UNESP.

Peebles, L. and B. Norris 2000, 'Strength data for designers', *Proceedings of the International Ergonomics Association 14*, IEA, San Diego.

Peebles, L. & B. Norris 2003, 'Filling 'gaps' in strength data for design', *Applied Ergonomics*, 2003, pp. 73-88.

Sádecká, J. and J. Polomsky 2000, 'Eletrophoretic methods in the analysis of beverages', *Journal of Chromatography A*, 2000, p. 266.

Silva, D. C., L. C. Paschoarelli, L. F. A. Campos, J. N. L. Lanutti, F. J. Muniz, and J. C. P. Silva 2010, Evaluation of torque forces used in the packings covers: accessibility and usability, *Proceedings of X International Congress of Ergonomics and Usability of human-technology Interfaces*, LEUI - PUC-Rio, Rio de Janeiro.

Silva, D. C., L. C. Paschoarelli, and J. C. P. Silva 2012, 'Openability of soft drinks PET packagings', *Work: A Journal of Prevention, Assessment and Rehabilitation*, 2012, pp. 1346-1351. DOI 10.3233/WOR-2012-0322-1346.

Voorbij, AIM & STeenbekkers, LPA 2001, 'The composition of a graph on the decline of total strength with age based on pushing, pulling, twisting and gripping force', *Applied Ergonomics*, 2001, pp. 287-292.

Yoxall, A, R. Janson, S. R. Bradbury, J. Langley, J. Wearn, and S. Hayes 2006, 'Openability: Product design limits for consumer packaging', *Packaging Technology and Science*, 2006, pp. 219-225.

CHAPTER 18

Laterality and Usability: Biomechanical Aspects in Prehension Strength

Luis Carlos Paschoarelli 1, Bruno Montanari Razza 2, Cristina do Carmo Lúcio 3, José Alfredo Covolan Ulson 4 and Danilo Corrêa Silva 5

UNESP – Univ Estadual Paulista
Bauru, Brazil
paschoarelli@faac.unesp.br

ABSTRACT

Although technological developments in recent decades have improved the quality of life for many, a number of user interface problems still exist. There are difficulties, for example, manipulating manual instruments by those with specific needs, especially left-handed users. The design of such instruments depends on scientific knowledge of biomechanical forces – especially prehension. The aim of this study was to analyze biomechanical effort during simulated manual activities (compression, traction and torque with 14 different manual interfaces) between dominant and non-dominant hands. Sixty individuals (30 left-handed) participated in the study. Measurements were taken with an advanced force gauge and a static torque transducer. The results indicate that right-handed individuals perform better ($p \leq 0.05$) with the dominant hand (in 12 manuals interfaces), while there were no significant strength differences among the left-handed (except for two manual interfaces). The reasons why left-handed individuals present little difference in strength between the dominant and non-dominant hands are not clear, but could be the result of frequent use of the non-dominant hand to perform many instrumental activities of daily living (IADLs). Moreover, it could be due to greater symmetry in the organization of brain hemispheres compared to the strong lateralization of right-handed individuals. The results provide insight into the dynamics of the manipulation of a number of manual instruments according to right and left-handed groups.

Keywords: Laterality, Biomechanics, Ergonomic Design

1   INTRODUCTION

Although technological developments in recent decades have improved the quality of life for many, these developments have also led to a certain amount of friction regarding user–product interface.

Several such cases are related to the handling of manual instruments in the occupational or everyday activities of users with specific needs, e.g., the left-handed, who favor use of the left side of the body.

It is estimated that 11% of the population is left-handed, with a higher incidence among males (12.6%) than females (9.9%) (Gilbert and Wysocky, 1992). Unfortunately, the inferior market demand they represent is one of the justifications for designing products exclusively for right-handed individuals.

Handedness, or the preferential or partial use of only one of the hands to perform tasks, may be related to the individual's genotype, lateral asymmetry, and/or other cultural and psychological aspects. The first study on left-handedness was carried out in the 17th century, when Thomas Browne investigated mentions of it in the Holy Bible (Perelle and Ehrman, 1994). The subject has since been studied using different approaches such as medical and behavioral conditions, artistic temperament, intelligence quotient, etc. (Annett and Manning, 1989; Coren and Halpern, 1991; and Agtmael, Forrest, and Williamson, 2001) and investigated regarding different cultural or psychological factors, as well as subsequently refuted associations with several diseases, behavioral anomalies and developmental disorders (Coren and Searleman, 1990 and Dellatolas, Bitter, and Curt, et al. 1997).

Besides right- and left-handedness, there are also ambidextrous individuals (Annett, 1970). However, according to Gillies, MacSweeney, and Zangwwill (1960), the ambidextrous are left-handed individuals "adapted" to environments designed for right-handed individuals.

The establishment of handedness in individual development is still not fully understood. Ager, Olivett, and Johnson (1984) and Fullwood (1986) believe that handedness is inconsistent up to 11 years old. Moreover, as age advances, cultural factors can influence handedness (Viggiano, Borelli, and Vannucci, 2001).

Handedness can be influenced by the anthropometry of the human hand. Mohammad (2005) found significant differences between the hands of right and left-handed individuals.

Regarding task performance, Kaya and Orbak (2004) evaluated dentistry students and confirmed that left-handed individuals presented significantly better performance when working on the left side of the patient than right-handed individuals working on the right side. According to Coren and Halpern (1991), the constructed environment and its technological devices have had a negative influence on the functional activities of left-handed individuals and are becoming an increasing risk factor for accidents. Graham and Cleveland (1995) report that manual instruments are usually designed for right-handed individuals, which increases this negative influence. In terms of prevention, Stellman, Wynder, and Derose, et al. (1997) state that left-handed individuals should be considered in the design of manual equipment, including products for everyday use.

The ergonomic design of manual instruments demands knowledge of the biomechanical effort employed while handling them. Since it is not fully known how biomechanical effort varies between the dominant and non-dominant hands of right- and left-handed individuals, the objective of the present study was to analyze the dominant and non-dominant hand grip strength of these two groups during simulated manual activities (compression, traction and torque with different interfaces).

2   MATERIALS AND METHODS

This cross-sectional study was approved by the UNESP Research Ethics Committee (Of. 374/2005-CEP-FMB-UNESP) and followed the Certified Ergonomist Deontology Code (Código de Deontologia do Ergonomista Certificado) - Norm ERG BR 1002 (Abergo, 2003). A free and informed consent form was signed by all participants.

Sixty male university students participated in the study. The Edinburgh Inventory (Oldfield, 1971) was applied to determine handedness; 30 right-handed individuals (CL $\geq$ +38.00 / mean age 22.30 years old – SD 2.45) and 30 left-handed individuals (CL $\leq$ -28.00 / mean age 21.60 years – SD 2.88) were selected.

An advanced force gauge (AFG – Mecmesin, Ltd) was used to analyze the compression strength (thumb) and traction strength (pulpo-lateral, bidigital and tridigital grip). A static torque transducer (STT- Mecmesin, Ltd) was used to evaluate the strength of fixed hand torque (clockwise and counterclockwise), following the recommendations of Caldwell, Chaffin, and Dukes-Bobos, et al. (1974) and Mital and Kumar (1998). Both devices were fixed on mobile vertical stands, which allowed adjustment to the individuals' elbow height, according to the guidelines of Daams (1993) and Smith, Norris, and Peebles (2000), as well as to the procedures adopted by Peebles and Norris (2003). A smooth 40 x 40 mm surface was used in the compression strength task. Fabric-covered 1.20 x 40 mm handles were used for the traction strength task. To evaluate torque strength, a cylinder and smooth-surfaced hexagonal, square and triangular prisms were used.

The subjects were evaluated individually after personal and anthropometric data were recorded. The subjects maintained a comfortable standing posture during the tasks, and standardized instructions were given that emphasized the importance of keeping the trunk erect and static during the activity. The subjects used latex surgical gloves for hygiene control, which did not interfere with the results. Maximal voluntary isometric contractions were performed for all variables with dominant and non-dominant hands. There was a 60 s interval between each activity for each variable.

184

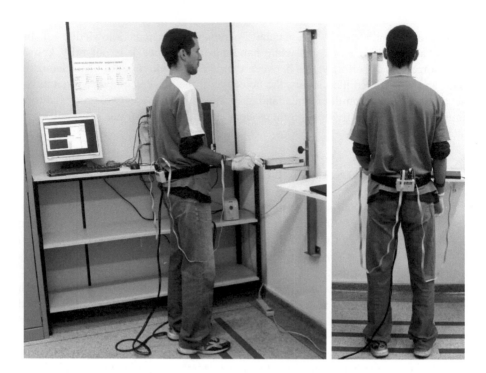

Figure 1 Procedimentos adotados na coleta de dados.

In order to verify significant differences ($p \leq 0.05$) in compression, traction and torque strengths for the dominant and non-dominant hands (36 total interactions), Student's t-test for paired samples was applied, given the normality (Shapiro-Wilk test) of the differences for each pair of measurements (Zar, 1999, p. 161-169).

The Wilcoxon non-parametric test (Zar, 1999, p. 165) was used with the following non-normal variables ($p \leq 0.05$): compress ion/right-handed; traction/1 mm/bidigital/right-handed; traction/40mm/tridigital/left-handed; traction/40mm/ pulpo-lateral/right-handed; torque/prismatictriangular/clockwise/left-handed; torque/cylindrical/clockwise/left-handed; and torque/cylindrical/clockwise and counterclockwise/right-handed.

3   RESULTS

The mean values and respective standard deviations (Kgf) of compression strength (thumb) and traction strength (pulpo-lateral, bidigital and tridigital grip) of the dominant and non-dominant hands of the right- and left-handed individuals are presented in Figure 2. The mean values and respective standard deviations (N.m) of torque strength for cylindrical and hexagonal, square and triangular prismatic interfaces for the dominant and non-dominant hands of left- and right-handed individuals are presented in Figure 3.

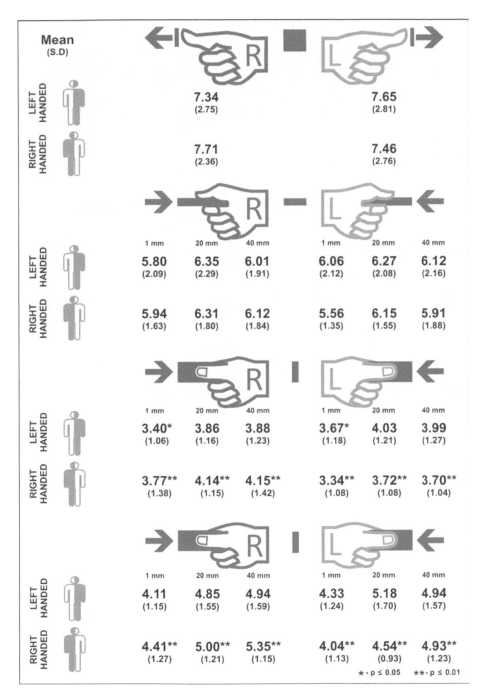

Figure 2 Means and respective standard deviations (Kgf) of the compression (thumb) and traction (with pulpo-lateral, bidigital and tridigital grip) strength in the dominant and non-dominant hands of right- and left-handed individuals.

186

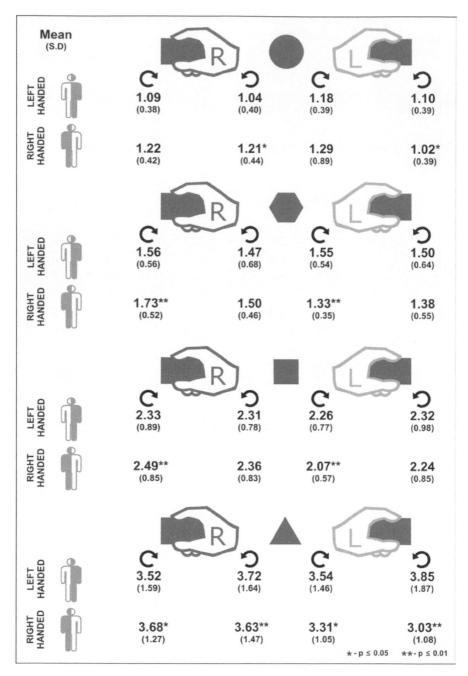

Figure 3 Means and respective standard deviations (N.m) of the grip strength (torque) for cylindrical and hexagonal, square and triangular prismatic interfaces in the dominant and non-dominant hands of left- and right-handed individuals.

4   DISCUSSION AND CONCLUSION

The biomechanical grip strength (compression, traction and torque) values found in the present study indicated that the dominant hand presented better performance than the non-dominant hand in 86.11% of the interactions for both right and left-handed individuals.

Upon comparing the dominant and non-dominant hands of left-handed individuals, the mean values for traction strength in the tridigital grip with the 40 mm interface were identical (4.94 Kgf). Among right-handed individuals, the non-dominant hand presented better performance than the dominant hand only for clockwise torque with the cylindrical handle. Among left-handed individuals, this was observed for traction strength in the pulpo-lateral grip with the 20 mm interface, as well as in the clockwise direction torque strength with the hexagonal and square prismatic handles. No significant difference was observed in any of these interactions (P>0.05).

The performance of left-handed individuals was better than that of right-handed individuals, particularly in the compression task, since for the left-handed the strength of the left hand was 4.2% greater than the right hand, and for the right-handed the strength of the right hand was 3.4% greater than that of the left hand, although it was not significant for either of the two interactions (P>0.05). In all the other interactions where the dominant hand exerted greater strength than the non-dominant hand, the differences were expressively greater between right-handed individuals than left-handed individuals.

Regarding the production of upper-limb biomechanical strength, Schmauder, Eckert, and Schindhelm (1992) point out that left-handed individuals perform better with their non-dominant hand than the right-handed, which has been corroborated by Hoffmann (1997) and Hoffmann, Chang, and Yim (1997). However, Armstrong and Oldham (1999) report no significant differences between the right and left hands in left-handed individuals, in spite of considerable variability.

In the present study, the dominant hand performance of left-handed individuals was significantly better (P≤0.05) only for bidigital grip traction with the 1 mm interface. The dominant hand performance of right-handed individuals was significantly better (P≤0.05 or 0.01) in eleven interactions: bidigital grip traction in 1.20 and 40 mm interfaces; tridigital grip traction in 1.20 and 40 mm interfaces; counterclockwise torque with the cylindrical handle; clockwise torque with hexagonal and square prismatic handles; and clockwise and counterclockwise torque with the triangular prismatic handle.

It is normal among left-handed individuals for the dominant hand to be as strong as or slightly stronger (1-2%) than the non-dominant hand (Crosby, Wehbé, and Mawr, 1994; Schmauder, Eckert, and Schindhelm, 1992). Among right-handed individuals, the dominant hand can be from 6% to 14% stronger than the non-dominant hand (Crosby, Wehbé, and Mawr, 1994; Lindahl, Nyström, and Bjerle, et al. 1994; O'Driscoll, Horii, and Ness, et al. 1992). In the present study, the dominant hand of right-handed individuals was up to 30.07% stronger than the non-dominant hand. According to Hanten, Chen, and Austin, et al. (1999), the reasons for the small strength differences between dominant and non-dominant hands in left-handed individuals are not clear, but could result from either physical conditioning of the non-dominant hand during daily living activities or from greater symmetry in the organization of the brain hemispheres compared to the substantial lateralization of right-handed individuals.

According to Hoffmann (1997), left-handed individuals may develop greater ability in the non-dominant hand due to its frequent use, since it is hard to find products specifically developed for them. Bourassa, McManus, and Bryden (1996) also take this position, reporting that left-handed individuals are forced to develop more ability with their right hand.

Our results support this line of thought, since right-handed individuals presented greater strength differences, most of them strongly significant ($P\leq0.01$), between the dominant and non-dominant hands, whereas among left-handed individuals this difference was minimal, which supports the hypothesis that left-handed individuals, due to being forced to perform a large part of their manual activities with their right hand, have developed balanced capacity between the dominant and the non-dominant hands.

Traction strength, both for right- and left-handed individuals, was influenced by the type of grip. A progression in values was observed between the bidigital grip and the tridigital and pulpo-lateral grips (the strongest), which has also been found in other studies (Dempsey and Ayoub, 1996; Imrhan, 1991; Peebles and Norris, 2003). The pulpo-lateral grip demands a neutral wrist position, especially compared to the tridigital and bidigital postures, for which greater wrist extensions are observed. Dempsey and Ayoub (1996), Imrhan (1991), Shih and Ou (2005) and Lamoreaux and Hoffer (1995) believe that traction grip strength is greater when the wrist in a neutral position than in an extended position.

Regarding torque strength, the results also show a progressive growth in grip strength between the cylindrical handle and the hexagonal, square and triangular prismatic handles. Shih and Wang (1996) observed the same tendency, with the more sharply angled handles allowing greater torques.

The results of the present study underscore the fact that, in most cases, the dominant hand exerts greater strength than the non-dominant hand, which is more significant among right-handed than left-handed individuals. Therefore, left-handed individuals seem to be more apt (also from a biomechanical point of view) to perform activities with both hands.

The apparent advantage for left-handed individuals is in fact due to the lack of appropriate manual instrument designs for them. Questions of accessibility and usability, even though analyzed and debated in scientific studies, are still far from the production sector and deserve the special attention of ergonomic design.

ACKNOWLEDGMENTS

This study was supported by FAPESP (Proc. 05/59941-2) and CNPq (Proc. 303138/2010-6).

REFERENCES

Ager, C. L., B. L. Olivett, and C. L. Johnson. 1984. Grasp and pinch strength in children 5 to 12 years old. *The American Journal of Occupational Therapy* 38 (2): 107-113.

Abergo. 2003. Norm ERG BR 1002 – Code of Deontology of Certified Ergonomists. Accessed February 3, 2007. http://www.abergo.org.br/arquivos/norma_ergbr_1002_deontologia/pdfdeontologia.pdf. [in portuguese]

Agtmael, T. V., S. M. Forrest, and R. Williamson. 2001. Genes for the left-handedness: How to search for the needle in the haystack? *Laterality* 6 (02): 149-164.

Annett, M. 1970. A classification of hand preference by association analysis. *British Journal of Psychology* 61 (03): 303-321.

Annett, M. and M. Manning. 1989. The disadvantages of dextrality for intelligence. *British Journal of Psychology* 80: 213–226.

Armstrong, C. A. and J. A. Oldham. 1999. A comparison of dominant and non-dominant and strengths. *Journal of Hand Surgery - British and European Volume* 24B (04): 421-425.

Bourassa, D. C., I. C. McManus, and M. P. Bryden. 1996. Handedness and eye-dominance: a meta-analysis of their relationship. *Laterality* 01 (01): 5-34.

Caldwell, L. S., D. B. Chaffin, F. N. Dukes-Bobos, K. H. E. Kroemer, L. L. Laubach, S. H. Snook, and D. E. Wasserman. 1974. A proposed standard procedure for static muscle strength testing. *American Industrial Hygiene Association Journal* 35 (04): 201-206.

Coren, S. and D. F. Halpern. 1991 Left-handedness: a marker for decreased survival fitness. *Psychological Bulletin* 109 (01): 90-106.

Coren, S. and A. Searlman. 1990. Birth stress and left-handedness: The rare trait model. In *Left-handedness, behavioral implications and anomalies*, ed. Coren, S. Advances in Psychology (p. 3-32). North-Holland.

Crosby, C. A., M.A. Wehbé, and B. Mawr. 1994. Hand strength: normative values. *The Journal of Hand Surgery* 19A (04): 665-670.

Daams, B. J. 1993. Static force exertion in postures with different degrees of freedom. *Ergonomics* 36 (04): 397-406.

Dellatolas, G., P. T. Bitter, F. Curt, and M. D. E. Agostini. 1997. Evolution of Degree and Direction of Hand Preference in Children: Methodological and Theoretical Issues. *Neuropsichological Rehabilitation* 7 (04): 387–399.

Dempsey, P. G. and M. M. Ayoub. 1996. The influence of gender, grasp type, pinch width and wrist position on sustained pinch strength. *International Journal of Industrial Ergonomics* 17 (03): 259-273.

Fullwood, D. 1986. Australian norms for hand and finger strength of boys and girls aged 5 – 11 years. *Australian Occupational Therapy Journal* 33 (01): 26-36.

Gilbert, A. N. and C. J. Wysocki. 1992 Hand preference and age in the United States. *Neuropsychologia*, 30 (07): 601-608.

Gillies, S. M., D. A. MacSweeney, and O. L. Zangwwill. 1960. A note on some unusual handedness patterns. *Quarterly Journal of Experimental Psychology* 12 (02): 113-116.

Graham, C. J. and E. Cleveland. 1995. Left-handedness as an injury risk factor in adolescents. *Journal of Adolescent Health* 16 (01): 50-52.

Hanten, W. P., W. Chen, A. A. Austin, R. E. Brooks, H. C. Carter, C. A. Law, M. K. Morgan, D. J. Sanders, C. A. Swan, and A. L. Vanderslice. 1999. Maximum grip strength in normal subjects from 20 to 64 years of age. *Journal of Hand Therapy* 12 (03): 193-200.

Hoffmann, E. R. 1997. Movement time of right- and left-handers using their preferred and non-preferred hands. *International Journal of Industrial Ergonomics* 19 (01): 49-57.

Hoffmann, E. R., W. Y. Chang, and K. Y. Yim. 1997. Computer mouse operation: is the left-handed user disadvantaged? *Applied Ergonomics* 28 (04): 245-248.

Imrhan, S. N. 1991. The influence of wrist position on different types of pinch strength. *Applied Ergonomics* 22 (06): 379-384.

Imrhan, S. N. and G. D. Jenkins. 1999. Flexion-extension hand torque strengths: applications in maintenance tasks. *International Journal of Industrial Ergonomics* 23 (04): 359-371.

Imrhan, S. N. and C. H. Loo. 1989. Trends in finger pinch strength in children, adults, and the elderly. *Human Factors* 31 (06): 689-701.

Kaya, M. D. and R. Orbak. 2004. Performance of left-handed dental students is improved when working from the left side of the patient. *International Journal of Industrial Ergonomics* 33 (05), 387-393.

Lamoreaux, L. and M. M. Hoffer. 1995. The effect of wrist deviation on grip and pinch strength. *Clinical Orthopaedics and Related Research* 314: 152-155.

Lindahl, O. A., N. Nyström, P. Bjerle, and A. Boström. 1994. Grip strength of the human hand - measurements on normal subjects with a new hand strength analysis system (Hastras). *Journal of Medical Engineering & Technology* 18 (03): 101-103.

Mital, A. and S. Kumar. 1998. Human muscle strength definitions, measurement, and usage: Part I – Guidelines for the practitioner. *International Journal of Industrial Ergonomics* 22 (01-02): 101-121 [1].

Mital, A. and S. Kumar. 1998. Human muscle strength definitions, measurement, and usage: Part II – The scientific basis (knowledge base) for the guide. *International Journal of Industrial Ergonomics* 22 (01-02): 123-144.

Mohammad, Y. A. A. 2005. Anthropometric characteristics of the hand based on laterality and sex among Jordanian. *International Journal of Industrial Ergonomics* 35 (08): 747-754.

O'Driscoll. S. W., E. Horii, R. Ness, R. R. Richards, and K. An. 1992. The relationship between wrist position, grasp size, and grip strength. *Journal of Hand Surgery* 17A (01): 169-177.

Oldfield, R. C. 1971. The assessment and analysis of handedness: The Edinburgh inventory. *Neuropsychologia*, 09 (01): 97-113.

Peebles, L. and B. Norris. 2003. Filling 'gaps' in strength data for design. *Applied Ergonomics* 34 (01): 73-88.

Perelle, I. B. and L. Ehrman. 1994. An international study of human handedness: The data. *Behaviour Genetics* 24 (03), 217–227.

Schmauder, M., R. Eckert, and R. Schindhelm. 1992. Forces in the hand-arm system: Investigations of the problem of left-handedness. *International Journal of Industrial Ergonomics* 12 (03): 231-237.

Shih, Y. C., and Y. C. Ou. 2005. Influences of span and wrist posture on peak chuck pinch strength and time needed to reach peak strength. *International Journal of Industrial Ergonomics* 35 (06): 527-536.

Smith, S., B. Norris, and L. Peebles. 2000. *Strength data for design safety*. Nottingham: Institute for Occupational Ergonomics / University of Nottingham.

Stellman, S. D., E. L. Wynder, D. J. Derose, and J. E. Muscat. 1997. The Epidemiology of Left-handedness in a hospital population. *Annals of Epidemiology* 07 (03): 167-171.

Viggiano, M. P., P. Borelli, and M. Vannucci. 2001. Hand preference in Italian students. *Laterality* 06 (03), 283–286.

Zar, J. H. 1999. *Biostatistical Analysis*. Prentice Hall: Upper Saddle River.

CHAPTER 19

Innovative Long-distance Interaction between Elderly and Their Beloved Young Children Using a Kinect-based Exercise Game

Tien-Lung Sun, Pin-Ju Liu

Department of Industrial Engineering & Management, Yuan Ze University
Taiwan, R.O.C.
E-mail: tsun@saturn.yzu.edu.tw

ABSTRACT

Most families in developed countries are turning into 'nuclear family' where the elderly does not live together with young generation. The contact between the elderly and their family members are significantly reduced. For older adults to maintain contact or connected with their family members, audio and video based long-distance interactions like phone, Face Time, Skype, etc. are commonly employed. Recent emergence of the Kinect device (http://www.xbox.com/Xbox360) provides a low-cost and convenient full-body motion capture opportunity. Kinect enables innovative long-distance interaction by letting people at distant places drive their avatars to interact with each other in a 3D virtual world. This paper leverages the Kinect-based long-distance interaction techniques to develop a multi-player game to enhance 'connection' between older adults and their beloved young kids. The older adults and their young kids play before Kinect videos. The two game scenes are synchronized through Internet so that both sides see the same game scene. The game scene contains two avatars that are manipulated by the elderly and their lovely kids through two Kinect devices. A posture frame based game playing strategy is developed to force 'body touch' of the avatars in the game scene. The players have to manipulate theirs avatars to avoid colliding with the posture frame while at the same time trying to touch the 'hot points' of the other person's avatar.

The 'body touch' will warm the older adult's heart. An evaluation experiment is planned to examine the feeling of the elderly at different game playing strategies.

Keywords: older adults, long-distance interaction, Kinect, exergame

1   INTRODUCTION

Many developed countries begin to enter the so-called 'aging society' due to rapid growth of aging population. Most families in these countries are turning into 'nuclear family' where the elderly does not live together with young generation. The contact between the elderly and their family members are significantly reduced. The older adults are forced to live lonely by themselves. They have strong desire to be 'connected' to their beloved kids. The lonely feeling not only affects the mental health of the older adults but also their physical health (Rodríguez, et al., 2007).

The most common way for older adults to be 'connected' to their family members is through audio or video based long-distance interaction tools like phone, Face Time, Skype, etc.. Recently, the emergence of Kinect device (http://www.xbox.com/Xbox360) opens a new opportunity for low-cost and convenient full-body motion capture. With Kinect, people can drive their avatars in a 3D virtual world. If the 3D scenes at distant places are synchronized through Internet, then people at distant places could manipulate their avatars to interact in a 3D virtual world. A representative example is the Avatar Kinect from MicroSoft (http://www.youtube.com/watch?v=jb4bzomm9iU), which allows people at different places to chat with full facial and body tracking, as shown in Figure 1. Kinect has created an innovative long-distance interaction method different from traditional voice or video based methods.

Figure 1   Avatar Kinect for long-distance chatting with body tracking

This paper leverages the Kinect-based long-distance interaction techniques to develop a Kinect-based multi-player game to enhance 'connection' between older adults and their beloved young kids. As shown in Figure 2, the older adults and their young kids at a different place play before Kinect videos. The two game scenes are synchronized through Internet so that both sides see the same game scene. The game scene contains two avatars that are manipulated by the elderly and their lovely kids through two Kinect devices. Such avatar-based exercise enables an innovative long-distance interaction that gives the older adults a different 'connection' feeling that the traditional voice or video based interaction can not offer. The detailed design, implementation and usability evaluation experiment of the proposed method are discussed in this paper.

Figure 2 Long-distance interaction using Kinect-based exercise game

2   POSTURE FRAME BASED GAME PLAY

The Kinect-based multi-player game employs a posture frame (PF) based game playing strategy to force 'body touch' of avatars in the game scene. The PF is composed of a set of 3D rectangles that forms the outline of a posture, as shown in Figure 3. The PF is added to the game scene to guide and constrain avatar's actions. The older adults and the young kids have to manipulate theirs avatars to avoid colliding with the posture frame while at the same time trying to touch the 'hot points' of the other person's avatar. The 'body touch' will warm the older adult's heart. If the avatar collides with the PF, special game effects like smoke, exploration etc., are displayed to increase the fun of the game.

Figure 3 Posture frame to guide and constrain avatar's action

The PF-guided game playing scenario is shown in Figure 4. The PF is dynamically generated in the game scene at a distance from the avatar. It then moves toward the avatar until it reaches the avatar and becomes overlapped with it. While the PF moves toward the avatars, the elderly and the young kids have to work together to avoid their avatars colliding with the PF. By doing so, the long-distance greeting process becomes more fun and with more body interactions. After staying overlapped with the avatar for a while, the PF disappears and the next PF is generated, so on and so forth.

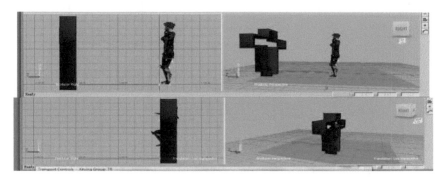

Figure 4 PF-guided game playing scenario

3   KINECT-BASED EXERGAME FOR LONG-DISTANCE INTERACTION

The exercise game scene is developed using a game authoring tool called Unity 3D (http://unity3d.com). The U3D game scene is connected to Kinect using Open NI Kinect SDK(http://75.98.78.94/default.aspx).

Since the older adults may have declined motor ability, chronic illnesses or cognitive disability, it is recommended not to introduce a new, unfamiliar game to the older adults immediately (Pruitt 2003). It is better to let the older adults be familiar with a new game gradually. The Kinect game developed in this paper contains 5 levels of difficulties. As shown in Table 1, the first 4 levels are training scenes to let the older adults gradually become familiar with the game play.

Table 1 Levels of game scenes in our Kinect game

Level		Learning goals
1	Single avatar manipulation	The older adults practice manipulating the avatar in the game scene, as shown in Figure 5.
2	PF-constrained single avatar manipulation	The older adults practice to drive the avatar to stay inside a PF and avoid colliding with the PF.
3	On-line two avatars cooperation	The older adults practice driving his/her own avatar to interact with the young kid's avatar. The older adults have to drive his/her own avatar to touch the 'hot spots'

		of the young kid's avatar.
4	PF-constrained on-line two avatars cooperation	The older adults have to cooperate with the young kids to avoid their avatars colliding with the PF while still touching each other's 'hot spots', as shown in Figure 6.
		Game scene
5	Game playing	The PF starts to move toward the avatars in different behavior parameters.

Figure 5 Single avatar manipulation.

Figure 6 PF-constrained on-line two avatars cooperation.

4   USABILITY EVALUATION

We are planning to conduct usability evaluation experiments to study the effect of different game play strategies on the 'social connection' feeling of the older adults.

4.1 Communication Media

The first experiment will study the effect of communication media on the 'social connection' feeling of the older adults. Traditional long-distance interaction tools like phone, Skype, Face Time, etc. are based on text, voice and videos. The Kinect-based long-distance interaction proposed in this paper is based on avatar's interactions in a virtual world. We shall ask the older adults to compare the Skype based interaction with the Kinect-based interaction.

4.2 Interaction Object

The second experiment considers the effect of interaction objects. Previous researches have suggested that older adults are usually reluctant to accept new technology, especially when they are not familiar with the technology. When playing an on-line social game, the older adults usually feel uncomfortable if the people at the other end of the network are strangers. On the other hand, if the older adults know that people at the other end of the network are family members or friends, they feel more comfortable and are willing to accept the technology (Brox et al. 2011; Gajadhar, etl., 2010). We shall ask the older adults to compare their feelings when playing with stranger, friends and beloved young kids.

4.3 Avatar Body Touch

The third experiment investigates the effect of avatar's body touch during game play. The older adults will compare two game play experiences. The first one is an ordinary treasure collection game in which their avatars do not touch the young kid's avatars. The second one is the PF-constrained game proposed in this paper in which the older adult's avatar touches the kid's avatar. It is expected that game play with body touch will warm the heart of the older adults more.

5 CONCLUSIONS

This paper discusses a Kinect-based multi-player game to enhance long-distance interaction between older adults and their beloved young kids. The older adults and their young kids at two places play before Kinect videos. The two game scenes are synchronized through Internet so that both sides see the same game scene. The game allows the older adults to drive their avatars to interact with the avatar of their beloved young kids in a virtual world. A posture frame based game playing strategy is developed to force 'body touch' of the avatars in the game scene. The players have to manipulate theirs avatars to avoid colliding with the posture frame while at the same time trying to touch the 'hot points' of the other person's avatar.

The new long-distance interaction offers many new advantages over traditional voice or image based communication media. Several usability evaluation experiments will be conducted to study the effect of communication media, interaction objects and body touch on the 'social connection' feeling of the older adults.

ACKNOWLEDGMENTS

This work is partially supported by National Science Council, Taiwan, R.O.C. under contract NSC 99-2221-E-155-021-MY3 and NSC 100-2627-E-155-001.

REFERENCES

Brox, E., Luque, L.F., Evertsen, G.J., and Hernandez, J.E.G., " Exergames for elderly: Social exergames to persuade seniors to increase physical activity", Pervasive Computing Technologies for Healthcare, 2011, Pages 546-549. ISBN: 978-1-61284-767-2

Gajadhar, B.J., H. H. Nap, et al. Out of sight, out of mind: co-player effects on seniors' player experience. Proc. Fun and Games '10.

Pruitt, B. A. 2003. Exercise progressions for seniors. Idea Health & Fitness Source, 21(3), 53-57.

Rodríguez, M.D., Aguirre, A., Moran, A.L., Mayora-Ibarra, O. ," Dealing with Computer Literacy and Age Differences in the Design of a Ubicomp System to Cope with Cognitive Decline in Lonely Elders",Lecture Notes in Computer Science, Volume 4560 LNCS, Issue PART 2, 2007, Pages 451-459

CHAPTER 20

Ergonomics and the Environment for the Elderly

Marie Paiva, Vilma Villarouco, Nicole Ferrer, Mariana Oliveira

Federal University of Pernambuco
Recife, Brazil
mariem.paiva@gmail.com

ABSTRACT

Given longer life expectancy rates and declining fertility rates world-wide, the phenomenon of demographic transition requires the built environment to be adapted and planned for, particularly for services targeted on the elderly. Among these, it is the growth of Long-term Stay Institutions for the Elderly that is the fastest in the world. However, there is a lack of in-depth studies and evaluations of their structures, which shows the need for Ergonomics to look at these spaces in order to understand how they affect the quality of life of their residents. Thus, this study presents a survey of sheltered housing for the elderly in Brazil, an emerging country in which there is starting to be evidence that its population is living longer. This study is based on the Ergonomic Methodology of the Built Environment. Use was made of well-established ergonomic and architectural techniques to make functional and behavioral assessments in order to identify the level of users' satisfaction and the extent to which the spaces are suitable for the activities that are offered.

Keywords: housing for the elderly, ergonomics of the built environment

1   LONG-TERM STAY INSTITUTIONS FOR THE ELDERLY

Worldwide, the proportion of people aged 60 and over can be seen to be growing faster than any other age group in modern society, as a result of both longer life expectancy and declining fertility rates. According to the World Health Organization (WHO), this ageing of the population can be seen as a "success story for public health policies and for socioeconomic development, but it also challenges society to adapt, in order to maximize the health and functional capacity of older people as well as their social participation and security" (WHO, 2011).

Neri (2007) states that this demographic reality should be considered when planning and adapting several services that cater to this age-group because the new profile is that of an aging population that has psychosocial and biological peculiarities that should be considered.

The World Health Organization confirms that the physical environment, as evidenced by architectural designs, focuses on functionality, which means the positive interaction between the individual whatever his/her state of health and personal (age, gender, etc.) and environmental (physical, cultural, etc.) factors. (WHO, 2003).

In less developed countries, it is expected that the number of older people will have risen from 400 million in 2000 to 1.7 billion by 2050 (WHO, 2011). Born and Boechat (2006) and Camarano and Pasinato (2004) claim that of the specialized services for the elderly, it is Long-term Stay Institutions for the Elderly that are the fastest growing in Brazil and the world as a whole.

Long-term Stay Institutions for the Elderly (LSIE), according to the Brazilian National Agency for Safeguarding Health, ANVISA, consist of "governmental or non-governmental institutions, residential in character, set up as collective residences for people aged 60 and over, with or without family support, which provide for their freedom and dignity and citizenship" (2005, p.1).

Given that the ideal home environment is one that provides a sense of independence, this space needs to be adapted to the reduced capacities of the elderly. To Perracini and Prado (2007), adapting the environment to the elderly provides a sense of normalcy or invariance vis-à-vis the personal losses associated with disorders common to aging.

Thus, institutional spaces should promote the autonomy, independence and privacy of the elderly, while considering the ambience and identity. It should be stressed how important the relationship of the elderly with the physical environment is as the latter contributes directly to determining the quality of aging and what degree of autonomy the elderly can have.

Ergonomics of the Built Environment, since its focus of study is on human beings as users of space such as to make this environment more appropriate for them, seeks to promote their safety and an individual's quality of life while they carry out their activities. Therefore, Villarouco (2009) states that the environment is fundamental if functional needs are to be met, and these besides physical and cognitive aspects have formal ones, with reference to psychological aspects of the users.

Therefore, by making use of different social approaches in the way that sheltered housing for the elderly is conceived, this article aims to analyze three long-term stay institutions in Recife, a city in the northeast region of Brazil, with a view to investigating positive and negative aspects of the built environment for the quality of life of the resident users.

2   METHODS

Taking into account the situation of the aging of Brazilian society and the prospects for housing for the elderly in Brazil, this research is characterized as an applied analysis that addresses the issue in a qualitative way, and takes a multiple-case study approach. Therefore, it works with a purposive sample, selected from the viability of access in accordance with the research.

The sample comprises three institutions in Recife, Brazil, one of which is private and profit-making, one a public institution, maintained from government funds and one managed from mixed funds that come from government maintenance funds and philanthropic and private contributions.

Thus, with the objective of extending out to the different configurations of Long-term Stay Institutions for the Elderly by the evaluation of the built environment as to its suitability and ambience, besides its impact on functionality, the Ergonomic Methodology of the Built Environment (VILLAROUCO, 2009) was used, which undertakes an ergonomic approach so as to understand, evaluate and propose recommendations for environments which are in continuous interaction with the user.

This methodology consists of an ergonomic evaluation conducted in two general stages: the first is of a physical order and the second one of perception. During the physical assessment phase, this methodology is divided into three distinct steps. Initially, there is the Global Analysis of the Environment, which is the first contact with the space in which one seeks an understanding of the environment and the activities carried out in it, besides characterizing the main problems and demands, which point to the need for intervention.

Later, during the phase of Identifying the Environmental Configuration, all the physical and environmental conditioning factors are identified. It is to this phase that gathering all the data from environmental is attributed such as dimensioning, lighting, ventilation, noise, temperature, layout, displacements, cladding materials and accessibility conditions, which give rise to the initial hypotheses on the question of the influences of the space on the conduct of work activities.

In the third and last stage of a physical order – the Evaluation of the Environment in Use –, care is taken to observe the environment in action, with a view to identifying its suitability, viewing the extent to which it facilitates or hinders the conduct of the activities which it hosts.

The phase of a perceptive order should be inserted into studies of environmental psychology, or environmental perception in which there is the need to adopt tools that aid identifying variables of a more cognitive and perceptual character. The use of one of these tools helps the researcher to understand employees' perceptions in relation to work spaces. From these data, a check is made on what factors are most strongly linked to the motivational aspects.

After collecting these data, all the variables should be joined together so that, after a complete analysis, the Ergonomic Diagnosis can be drawn up. Having done this, one moves on to the Ergonomic Recommendations, in an attempt to mitigate and even solve the spatial barriers in the environment which make it difficult for the elderly to undertake activities in it.

3   RESULTS

Collective dwellings for the elderly in Brazil are subject to national laws, such as the Resolution of the Collegiate Directorate - ANVISA-RDC No. 283/2005 which lays down norms for operating Long-term Stay Institutions for the Elderly. Norms of the Brazilian Association for Technical Norms - ABNT-NBR No 9050/2004 for accessibility are also applied, the purpose of which are to see to it that users can use buildings, urban furniture and equipment autonomously and safely.

By means of collecting data on Institutions 1, 2 and 3, some instance of non-compliance were identified. Table 01 below identifies the degree to which their operation and accessibility to them comply with the laws. This is shown as: full compliance (represented by OK); partial compliance (represented by P highlighted in light gray); or non-compliance (highlighted in dark gray.) As to the items in the Table that do not reflect the representations given above, the legislation does not apply.

On analysing Table 01, it is seen that the instances of non-compliance in institutions are mostly issues related to accessibility. In some cases, compliance is partial and in others there may even be non-compliance which adversely affects the quality of life of the elderly and their safety.

Another important aspect that stands out is the absence of a fire prevention and detection system in the institutions investigated. Also the unsuitable floor coverings assume importance since they are elements that have the potential to cause accidents because people may slip and fall down.

Incompatible dimensioning of specific facilities for employees in Institutions 1 and 2 was detected, while this area was not covered in the third area. There being no bathrooms and changing rooms for staff leads them to share the facilities with the elderly, which the law does not permit.

Table 1 Compliance with legislation for the functioning of sheltered housing in Brazil for the elderly. The table identifies the degree to which their operation and accessibility comply with the laws: full compliance (represented by OK); partial compliance (represented by P highlighted in light gray); or non-compliance (highlighted in dark gray.)

LEGISLATION	LSIE 1		LSIE 2		LSIE 3	
	RDC 283	NBR 9050	RDC 283	NBR 9050	RDC 283	NBR 9050
FUNCTIONAL ASPECTS						
External access	OK	-	OK	-	OK	-
Ramps	P	P	OK	OK	P	P
Circulation	P	P	OK	OK	OK	P
Hand rail	P				P	OK

Support bar	P		P		OK	OK
Guard rails	-	OK	-	P	P	
Dormitory	P	-		-	OK	-
Private dormitory bathroom		-	P	-		-
Wash-hand basin	-	P	-		-	
Toilet bowl	-	P	-	P	-	P
Emergency alarm	-		-		-	OK
Collective bathroom	-	OK	-	P	OK	P
Activities Room	OK	-	OK	-	P	-
Kitchen	OK	-	OK	-	OK	-
Laundry	P	-		-	OK	-
Store-room		-		-	OK	-
Staff facilities	P		P			
Consultation Room		-	P	-	OK	-
Cladding/ Flooring		-	P	-	P	-
Fire detection		-		-		-

3.1   Characterization of LSIE 1

The first institution is an organization with a private source of funding, with occasional support from volunteers, but is mostly run with private finance.This Long-term Stay Institutions has thirty six residents, of whom thirty are female and six male. 40% of them are totally dependent (as per RDC No. 283/2005, degree of dependence III). According to the management of the institution, the exact number of residents with a diagnosis of cognitive impairment or behavioral problems is not known. However, management stated that about 60% of those who are dependent have Alzheimer's disease at a moderate or advanced stage.

The building that houses the institution was not designed for this purpose. It is characterized as three houses that are adjacent to each other, adapted from their original functions of residential use without observing the requirements of Brazilian norms for collective residential buildings for the elderly.

In general, it is seen that there is a lack of open and green areas set aside for the well-being of the elderly, and the few existing spaces have poor access. Thus, the entrance hall, which has one roofed area and another without a roof is widely used by the independent elderly where they can sit, read newspapers and watch television, and do physical activities. It is the most frequented area but insufficient in size to accommodate the total number of residents comfortably. As to the living rooms of the three houses, they are places where mainly the dependent elderly, who are monitored by caregivers during the day, stay.

The bedrooms are separated by gender and have dimensions that can hold from one to four beds. Individual bedrooms with private bathrooms are registered. The rooms have dimensions that are suitable for the resident's furniture and objects (Figure 1). According to the administration, as to accommodating the elderly, the affinity among residents, the level of their functional capacity and the interest of the client in the choice of bedroom are all respected.

Under a general context, there are accessibility issues in the home related to sizing environments, unsuitable handrails, and there only being stairs to access the residents' bedrooms. Inadequacies were also observed as to there not being non-slip floor coverings, the sizing of the leisure and social area of living, and the shortage of green areas.

Figure 1 Illustration of a bedroom in the Long-term Stay Institution for the Elderly 1 (From the archive of the authors, 2010).

3.1   Characterization of LSIE 2

The second institution is as an organization funded with public money, the main source being a government subsidy from the Institute of Social Welfare and Citizenship – IASC (in Portuguese), an independent agency linked to the Department of Social Welfare of the City Hall of Recife, and created to consolidate the social assistance policy of the municipality. It may also receive donations.

Although its maximum occupancy is for 24 residents, during the field research, twenty-one residents were identified, of whom seventeen were male and four female. Of the total number of elderly, nine showed dependence with regard to walking on the march and to undertaking everyday activities (EDAs) like taking a bath, feeding themselves and getting dressed. However, despite the profile of the residents of this institution being of independent elderly, it receives on a temporary basis, the dependent elderly, while they await transfer to an appropriate institution for the care of the elderly whose profile is that of dependency.

The residents of the Long-term Stay Institution for the Elderly 2 are elderly people taken from the streets, who have been abandoned by their family or have no affective ties, and who are taken there by the military police of Recife, sent on to hospitals or, more commonly, guided by IASC (Institute for Social Assistance and Citizenship).

With a built area of approximately 395 m², the organic structure of the LSIE is distributed in the adaptation of two residential buildings (one for the common and administrative area and the other for the bedrooms), which were refurbished and extended in 2003 for the purposes of housing the elderly.

The Institution has four multi-bed rooms, two for males, with no bathroom, for the elderly with greater independence, and one male and one female, both with en suite bathrooms. All rooms have more than four beds, which exceeds the standard recommended by the norm (RDC No. 283/2005). The rooms, which have unsuitable furniture, allow for neither privacy nor territoriality (Figure 2).

Even though maintenance is barely adequate, the institution is endowed with a substantial green area. The entrance hall area is in contact with the green area, and has benches, which gives the environment a homely and welcoming air.

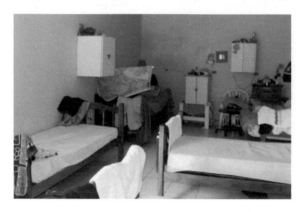

Figure 2 Illustration of a multi-bed room in Long-term Stay Institution for the Elderly 2 (From: the archives of the authors, 2011).

3.1   Characterization of LSIE 3

The third institution is an organization with a mixed source of funds, which consist of a government subsidy, ongoing support from volunteers and private participation. With a maximum capacity of 175 residents, there are currently one hundred and twenty elderly residents in the Institution, of whom sixty are male and sixty female. Of the total, between the men and women, the functional capacity of 70% of residents is preserved and 30% are considered dependent. According to the directorate, there is no record of there having been a bedridden elderly person.

The structure of the building has a pavilion morphology which at one time was a feature of the social exclusion models for the care of psychiatric patients, those suffering from infectious diseases and the homeless (VERAS, 1997).

The structure of the institution consists of nine buildings and a church, laid out on uneven terrain, and they are connected by covered sidewalks, ramps and passageways.

The spaces are large with high ceilings, typical of the architectural design, which gives the environment a gloomy atmosphere. This feeling is exacerbated by the deficit in natural light despite several openings.

The bedroom pavilions are separated by gender, male and female, and connected directly to the blocks of common use for dining and leisure, and those for kitchen and laundry services.

The bedrooms are separated by a plaster wall partition, with a capacity for two, three or four occupants, which makes the space more private and gives the elderly their identity back. The infirmary units have a capacity for four or six elderly people (Figure 3) and are dimensioned to have a large open area between beds.

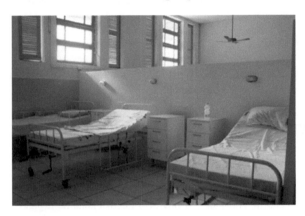

Figure 3   Illustration of infirmary with six beds in the Long-term Stay Institution for the Elderly 3 (From the archive of the authors, 2011).

4   CONCLUSIONS

Brazil, just like other developing countries, faces a rapidly aging population, and society is not yet prepared to supply the new demand in their socio-spatial needs. As they have to accelerate the process of accommodating the elderly in a new area of the built environment, most institutions in Brazil work in spaces that have long since been fitted out, and adaptations are not always adequate for the demands of the new tasks.

However, it is noticeable that for the purposes of financial investment, the option is made for basic adaptations that allow the institution to operate immediately. These do not always match the built environment to the real needs of the elderly, who effectively are the agents of this space, and end up having to put up with the physical barriers present in the environments. Another important aspect of the presence of these obstacles is that accidents occur when activities are undertaken.

It is also seen that the current legislation to guide these adaptations is far from rigorous and is subject to misinterpretation, which ends up weakening even more the situation of these institutions. Adequate control or inspection of collective housing is also difficult.

Thus, one can understand that recommendations for suitable environments for the elderly and those that meet the limitations of aging go beyond parameters established in legislation.

Therefore, the environmental perception that the user has of the environment because he/she has experienced it justifiably is of fundamental importance, according to Bins-Ely (2003) and Villarouco (2008), due to the interaction of the human-environment system, and because the formal (psychological) and functional (physical and cognitive) needs of elderly users are met.

Therefore, it is concluded that it is essential to make the environments suitable, since the healthy aging comes about through active engagement with life, being a function of autonomy that generates improvement in the quality of life.

ACKNOWLEDGMENTS

The authors are grateful to FACEPE for their support by granting a Scientific Initiation Scholarship for this research study.

REFERENCES

ASSOCIAÇÃO BRASILEIRA DE NORMAS TÉCNICAS. NBR 9050/04 - *Acessibilidade a edificações, mobiliários, espaços e equipamentos urbanos*. Rio de Janeiro, 2004.

BINS ELY, Vera H. M. Ergonomia + Arquitetura: buscando um melhor desempenho do ambiente físico. In: Anais do 3º Congresso Internacional de Ergonomia e Usabilidade de Interfaces Humano-Tecnologia: Produtos, Programas, Informação, Ambiente Construído. Rio de Janeiro: LEUI/PUC-Rio, 2003.

BORN, T.; BOECHAT, N. S. A qualidade dos cuidados ao idoso institucionalizado. In: FREITAS, E.V. et al. *Tratado de Geriatria e Gerontologia*. 2. ed. Rio de Janeiro: Guanabara Koogan, 2006. p. 1131- 1141.

BRASIL. Agência Nacional de Vigilância Sanitária - ANVISA. Resolução da Diretoria Colegiada - *RDC Nº 283, de 26 de Setembro de 2005- Regulamento Técnico que define normas de funcionamento para as Instituições de Longa Permanência para Idosos*. Brasília, 2005. Available at: <http://portal2.saude.gov.br/saudelegis/leg_norma_pesq_consulta.cfm>. Accessed in Dezembro 2009.

CAMARANO, A. A.; PASINATO, M. T. M. *Os novos idosos brasileiros: muito além dos 60?*.1 ed. Rio de Janeiro: IPEA, 2004, 22p.

NERI, A. L. Qualidade de vida na velhice e subjetividade. In:_____. *Qualidade de vida na velhice: enfoque multidisciplinar*. São Paulo: Ed. Alínea, 2007. cap. 1. p. 13-59.

ORGANIZAÇÃO MUNDIAL DE SAÚDE (OMS), CIF: *Classificação Internacional de Funcionalidade, Incapacidade e Saúde*. São Paulo: Universidade de São Paulo, 2003.

PRADO, A. R. de A.; PERRACINI, M. R. *A construção de ambientes favoráveis aos idosos*. In: NERI, A. L. Qualidade de vida na velhice: enfoque multidisciplinar. São Paulo: Ed. Alínea, 2007. cap. 9. p. 221-229.

VERAS, R A Reestruturação do Abrigo Cristo Redentor: o macroasilo transformado em uma minicidade. PHYSIS: Revista Saúde Coletiva, Rio de Janeiro, 7(2): 85-104, 1997.

VILLAROUCO, Vilma. An ergonomic look at the work environment WORLD CONGRESS ON ERGONOMICS, XVII, 2009, Beijing-China. *Anais...* Beijing-China: IEA, 2009.

VILLAROUCO, Vilma. Construindo uma metodologia de avaliação ergonômica do ambiente. *In:* Anais do XV Congresso Brasileiro de Ergonomia – ABERGO. Porto Seguro - Bahia, 2008.

WHO, World Health Organization. Available at: <http://www.who.int/topics/ageing/en/>. Accessed in November 2011.

CHAPTER 21

(Less than) Perfect Aging

Ng Yuwen Stella, John Brian Peacock, Tan Kay Chuan

National University of Singapore
Singapore
yuwen@nus.edu.sg

ABSTRACT

Clear articulation of human characteristics, capabilities, behavior, and performance limitations due to aging is confounded by individual differences, which may be very large. In this paper, the authors chose to study athletics records in their research approach to aging as it is not confounded by disease, disuse and disinterest. These age-based data from standardized tests are obtained from the World Masters Athletics organization website.

The records of each 5 year age group are represented as a percentage, with the 35 – 39 year old age group as the baseline. These percentages are then modeled with a parametric function and analyzed for each individual event as well as the event categories (race walks, sprints, distance, marathons and jumps), and with the men's and women's records analyzed separately. Overall, it was observed that the records for jump events declined faster than marathons and distance events, followed by walk and sprint events. Also, for all event categories, women's records declined faster than men's. One-way analysis of variance (ANOVA) is used to study the differences in the events.

This study reports an analysis of age-based athletic records and demonstrates the boundaries of human physical performance as affected by age, without the confounding effects of disease, disuse and disinterest.

Keywords: *Aging, sports performance, masters athletics*

1   INTRODUCTION

With rapidly advancing medical technology and declining birth rates, most industrialized countries are experiencing a demographic shift towards an aging population (Lee 1999; Anderson and Hussey 2000). As such, there is increasing attention in the research of the various aspects of aging. In this study, we focused on the analysis of age-based athletics records to model the boundaries of human physical performance with age.

Human factors and ergonomics specialists constantly face challenges due to variability of individuals, tasks and contexts (Wilson 2000; Salvendy 2006; Pikaar, Koningsveld et al. 2007). As these athletics records are constrained by the design of standardized tests of performance, they reflect the true boundaries of perfect aging. World records are also more empirically stable than sample means and variances as they are, by definition, based on very large sample sizes.

2   LITERATURE REVIEW

Researchers all over the world have been actively studying the effects of age on human physiological functions for many years and the method of using athletics records to infer physical capabilities is not new (Hill 1925; Riegel 1981; Young 1997; Grubb 1998; Baker, Tang et al. 2001; Tanaka and Seals 2003; Baker and Tang 2010). Tanaka and Seals (2003) used running and swimming performances to assess physiological function capacity (PFC) and noticed that PFC "decreased only modestly until age 60–70 years but declined exponentially thereafter". This is consistent with Sehl and Yates (2001) whose study revealed a linear decline in most organ systems of healthy people between the ages 30 – 70 years. However, it is insufficient to study the effects of aging until age 70 years. Compared with the 1980s, the population of people aged 65 and older have increased significantly and it is projected to continue increasing rapidly in all countries (Anderson and Hussey 2000). Thus, there is an increasing need to study the effects of aging beyond age 70 years.

As records for explosive activities like jumping and sprinting peak at an earlier age (Schulz and Curnow 1988), it is expected that they decline more quickly than endurance events such as long distance running. In 2001, Baker and colleagues used an exponential function to model the track running events and a linear function to model the walking and field records of masters athletes published in September 1999. This study revealed that "walking events declined more slowly than the running events, which declined more slowly than the jumping events".

Due to the change in weights of throwing events across the different age groups, there was no standardized way to compare the records across the age groups. Thus, the throwing events were left out of the study in 2001. However, in 2010, Baker and Tang expanded their study to compare the analysis of athletics events done in 2001 with records in swimming, rowing, cycling, triathlon and weightlifting. Thus, they further concluded that "weightlifting showed the fastest and greatest decline" and "rowing showing the least deterioration".

3   HYPOTHESIS

The objective of this research is to develop a model that represents the boundaries of perfect aging. It is hypothesized that there are different rates of decline for different types of events – sprints, distance, marathon, walks and jumps

– and the use of percentage performances is an effective way of demonstrating this age-related decline.

4   METHODS

The age-based athletics records used in this study were obtained from the World Masters Athletics records online (WMA 2011) and the age ranges used are 35 – 90 for all men's events and 35 – 85 for women's outdoor events and 35 – 80 for women's indoor events. Performances of each 5-year age group for all events are represented as a percentage of the 35 – 39 year old age group.

$$\text{Performance percentage} = 100\% \times \frac{\text{Performance at each age group}}{\text{Performance of } 35 - 39 \text{ age group}}$$

For timed events, the reciprocal is used (i.e. the performance of the 35 – 39 year old age group will be used in the numerator to calculate the performance percentage).

$$\text{Performance percentage} = 100\% \times \frac{\text{Performance at } 35 - 39 \text{ age group}}{\text{Performance of each age group}}$$

Tables 4.1, 4.2, 4.3 and 4.4 below show the performance percentages calculated for each event over their respective age ranges. These performance percentages are then used for further analysis.

Table 4.1: Performance percentages for men's outdoor events

	M40	M45	M50	M55	M60	M65	M70	M75	M80	M85	M90
100m	97%	93%	92%	87%	85%	81%	78%	74%	69%	62%	57%
200m	97%	92%	90%	86%	84%	80%	76%	72%	65%	59%	52%
400m	96%	91%	89%	88%	85%	81%	77%	70%	65%	57%	48%
800m	94%	91%	87%	84%	80%	77%	74%	67%	61%	51%	42%
1500m	95%	91%	87%	84%	80%	76%	71%	66%	59%	52%	44%
1 mile	96%	91%	87%	85%	79%	78%	72%	68%	54%	48%	37%
3000m	93%	88%	86%	84%	79%	76%	70%	67%	57%	53%	40%
5000m	94%	90%	87%	83%	80%	78%	70%	67%	59%	52%	41%
10000m	94%	89%	87%	83%	78%	77%	71%	68%	60%	51%	42%
Marathon	94%	89%	87%	83%	78%	77%	71%	68%	60%	51%	42%
High jump	96%	91%	89%	85%	79%	77%	71%	67%	60%	45%	36%
Pole vault	93%	88%	86%	80%	78%	72%	69%	62%	58%	53%	48%
Long jump	97%	87%	81%	78%	69%	65%	56%	51%	47%	38%	24%
Triple jump	90%	86%	80%	75%	71%	64%	61%	57%	51%	44%	38%
3km walk	93%	84%	81%	77%	71%	67%	60%	56%	50%	46%	37%
5km walk	98%	93%	92%	89%	91%	74%	74%	69%	64%	58%	53%
10km walk	94%	89%	88%	83%	82%	74%	71%	67%	63%	56%	53%
20km walk	96%	91%	91%	83%	83%	76%	72%	70%	63%	49%	50%
30km walk	96%	93%	91%	85%	83%	76%	70%	68%	63%	55%	51%
50km walk	100%	91%	91%	85%	85%	74%	73%	63%	62%		

Table 4.2: Performance percentages for women's outdoor events

	W40	W45	W50	W55	W60	W65	W70	W75	W80	W85
100m	97%	95%	92%	81%	78%	76%	73%	68%	58%	54%
200m	97%	92%	85%	80%	77%	75%	70%	64%	54%	48%
400m	94%	90%	88%	82%	76%	74%	66%	59%	50%	38%
800m	98%	95%	86%	82%	75%	72%	65%	60%	52%	39%
1500m	99%	97%	85%	80%	76%	69%	65%	60%	53%	38%
1 mile	98%	89%	86%	79%	74%	68%	63%	54%	48%	39%
3000m	92%	91%	86%	83%	76%	70%	63%	59%	54%	35%
5000m	95%	91%	86%	81%	77%	71%	66%	59%	54%	39%
10000m	98%	96%	87%	84%	80%	74%	66%	62%	53%	36%
Marathon	98%	96%	87%	84%	80%	74%	66%	62%	53%	36%
High jump	96%	95%	84%	82%	72%	68%	62%	56%	49%	34%
Pole vault	89%	88%	80%	77%	73%	68%	65%	58%	53%	47%
Long jump	85%	77%	77%	77%	75%	67%	69%	70%	52%	45%
Triple jump	94%	80%	77%	72%	68%	66%	60%	54%	44%	31%
3km walk	90%	83%	80%	76%	76%	66%	60%	51%	45%	38%
5km walk	90%	86%	87%	79%	78%	74%	67%	58%	57%	44%
10km walk	89%	85%	86%	79%	76%	74%	69%	62%	58%	50%
20km walk	89%	89%	86%	78%	77%	75%	69%	62%	56%	53%
30km walk	91%	90%	81%	75%	74%	72%	64%	53%	41%	

Table 4.3: Performance percentages for men's indoor events

	M40	M45	M50	M55	M60	M65	M70	M75	M80	M85	M90
100m	96%	93%	91%	89%	85%	83%	79%	76%	73%	67%	57%
200m	95%	91%	90%	88%	85%	81%	77%	71%	66%	60%	51%
400m	97%	95%	90%	89%	85%	84%	80%	72%	67%	56%	44%
800m	100%	94%	91%	86%	81%	80%	78%	71%	63%	52%	42%
1500m	96%	91%	86%	82%	78%	75%	71%	67%	62%	49%	37%
1 mile	99%	91%	89%	84%	80%	76%	71%	69%	58%	46%	40%
3000m	95%	89%	87%	85%	78%	75%	70%	68%	64%	44%	38%
High jump	91%	89%	87%	80%	75%	70%	69%	61%	59%	52%	47%
Pole vault	91%	87%	80%	77%	69%	63%	55%	50%	47%	41%	31%
Long jump	94%	87%	81%	79%	74%	67%	65%	59%	51%	45%	39%
Triple jump	92%	85%	79%	77%	72%	67%	61%	55%	50%	44%	37%

Table 4.4: Performance percentages for women's indoor events

	W40	W45	W50	W55	W60	W65	W70	W75	W80
100m	98%	96%	87%	83%	81%	78%	76%	66%	62%
200m	96%	90%	87%	83%	80%	75%	71%	62%	56%
400m	93%	91%	85%	81%	76%	72%	66%	61%	42%
800m	97%	88%	82%	78%	74%	71%	65%	57%	52%
1500m	99%	91%	85%	81%	75%	69%	66%	60%	44%
1 mile	100%	98%	93%	90%	80%	77%	68%	59%	44%
3000m	96%	94%	86%	82%	75%	72%	67%	59%	54%
High jump	90%	83%	79%	75%	72%	66%	65%	59%	55%
Pole vault	79%	74%	72%	64%	65%	63%	46%	42%	31%
Long jump	97%	84%	79%	70%	67%	62%	57%	52%	43%
Triple jump	88%	80%	74%	72%	65%	63%	55%	49%	50%

Thereafter, a regression analysis was done to fit the parametric models to each event. In this analysis, events with less than 7 data points (i.e. age range of less than 35 years) are omitted. With these parametric equations, the events are compared individually as well as grouped into the categories – sprint (Outdoor 100m, 200m, 400m, Indoor 60m, 200m, 400m), distance (Outdoor 800m, 1500m, 1 mile, 3000m, 5000m, 10000m, Indoor 800m, 1500m, 1 mile, 3000m), walk (Race walk 3000m, 5000m, 10K, 20K, 30K, 50K), jump (Outdoor and indoor pole vault, high jump, long jump, triple jump).

5   RESULTS

All masters athletics records for all events declined with age and can be adequately modeled with a quadratic function as shown in Figure 5.1 below. Thus, Figure 5.1 illustrates the overall boundary of aging for both men and women combined.

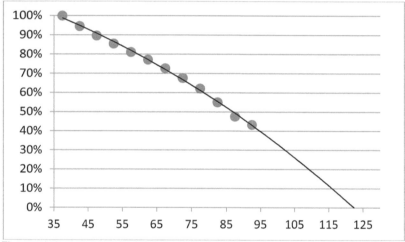

Figure 5.1: Average percentage performance for all men and women events combined

However, there are different rates of decline for different events. Therefore, in our analysis of the effects of aging on athletic performance, we have split the athletics events into 5 categories – jumps, walks, sprints, distance and marathons. Despite splitting the running events (sprint: 400m and less; distance: above 400m, excluding marathon) with reference to the anaerobic and aerobic energy systems exchange at approximately 90 seconds (Maud and Foster 2006), a one-way ANOVA performed on the individual events within each category showed that there are significant differences within all categories except walks for men, but insignificant differences within all categories except jumps for women.

Table 5.1 below shows the projected age at which the performance percentage goes to zero for each group of events. It is observed that in general, women deteriorate faster than men in all categories and the rate of decline is the fastest for women marathon and the slowest for men marathon.

Table 5.1: Projected intercept for each group of events using parametric model

	Projected intercept (Age)
Women marathon	105.6
Women distance	107.3
Women walk	113.2
Women sprint	114.1
Women jump	117.8
Men sprint	117.9
Men jump	119.8
Men distance	123.5
Men walk	127.2
Men marathon	141.3

Also, with the parametric functions, the rate of decline for each event at all ages can be obtained by solving for the gradient of the respective functions at each required age. In this study, we looked at the rate of decline at the end of each decade within the studied age range. For men's events, pole vault shows the greatest rate of decline at ages 40 and 50 years, followed by the 1 mile run at ages 60, 70 and 80 years. For women's events, pole vault also shows the greatest decline rate at age 40 years; followed by long jump at age 50 years, 50km race walk at ages 60 and 70 years, and 80m hurdles at age 80 years. Table 5.2 below shows the rates of decline at each decade from 40 to 80 years.

Table 5.2: Gradient of respective function at various ages from 40 – 80 years

	40	50	60	70	80
Women marathon	-0.0076	-0.0099	-0.0121	-0.0143	-0.0165
Women distance	-0.0063	-0.0088	-0.0113	-0.0138	-0.0163
Women walk	-0.0059	-0.0081	-0.0104	-0.0126	-0.0148
Women sprint	-0.0055	-0.0077	-0.0099	-0.0121	-0.0143
Women jump	-0.0102	-0.0109	-0.0116	-0.0123	-0.0130
Men sprint	-0.0004	-0.0036	-0.0068	-0.0100	-0.0132
Men jump	-0.0092	-0.0100	-0.0107	-0.0114	-0.0122
Men distance	-0.0026	-0.0050	-0.0074	-0.0098	-0.0121
Men walk	-0.0062	-0.0074	-0.0086	-0.0098	-0.0110
Men marathon	-0.0001	-0.0021	-0.0041	-0.0061	-0.0081

The parametric functions for each individual event are shown in Tables 4.5 and 4.6 below. They are ranked from highest to lowest rates of decline at age 40 years (i.e. gradient of each function at $x = 40$).

Table 4.5: Parametric functions of men's events

Event	Parametric Equation	R^2	Age range	Data points
Indoor Pole Vault	$y = 0.0000011x^2 - 0.0121402x + 1.4437133$	0.995	35-90	12
Outdoor Pole Vault	$y = -0.000049x^2 - 0.006662x + 1.317250$	0.991	35-90	12
Outdoor Long Jump	$y = -0.000044x^2 - 0.005303x + 1.220476$	0.984	35-95	13
Outdoor Triple Jump	$y = -0.000048x^2 - 0.004946x + 1.219851$	0.987	35-95	13
Outdoor High Jump	$y = -0.00002x^2 - 0.0071x + 1.2703$	0.997	35-95	13
Indoor Triple Jump	$y = -0.0000517x^2 - 0.0045164x + 1.2050897$	0.986	35-95	13
Indoor High Jump	$y = -0.0000186x^2 - 0.0068358x + 1.2619869$	0.994	35-95	13
Indoor Long Jump	$y = -0.0000635x^2 - 0.0028570x + 1.1706293$	0.988	35-95	13
Race Walk 5000m	$y = -0.0000279x^2 - 0.0047182x + 1.1980052$	0.991	35-90	12
Race Walk 20K	$y = -0.0000413x^2 - 0.0035152x + 1.1919196$	0.995	35-95	13
Race Walk 50K	$y = -0.0000690x^2 - 0.0011812x + 1.1221935$	0.950	35-80	10
Outdoor 1 Mile	$y = -0.0002x^2 + 0.0095x + 0.8303$	0.989	35-95	13
Race Walk 30K	$y = -0.0000647x^2 - 0.0012782x + 1.1471056$	0.967	35-80	10
Race Walk 10K	$y = -0.0000800x^2 + 0.0011247x + 1.0594212$	0.978	35-90	12
Race Walk 3000m	$y = -0.0000755x^2 + 0.0009823x + 1.0726995$	0.971	35-90	12
Indoor 1 Mile	$y = -0.0001143x^2 + 0.0042886x + 0.9893431$	0.983	35-90	12
Indoor 1500m	$y = -0.0001090x^2 + 0.0040524x + 0.9750233$	0.976	35-90	12
Indoor 3000m	$y = -0.0001260x^2 + 0.0061553x + 0.9135185$	0.967	35-90	12
Outdoor 10000m	$y = -0.0001x^2 + 0.0049x + 0.9402$	0.980	35-90	12
Indoor 800m	$y = -0.0001360x^2 + 0.0078026x + 0.8885250$	0.980	35-90	12
Indoor 60m Hurdles	$y = -0.0001152x^2 + 0.0065742x + 0.8768598$	0.974	35-90	12
Outdoor 3000m	$y = -0.0001x^2 + 0.0069x + 0.8803$	0.986	35-95	13
Indoor 400m	$y = -0.0001612x^2 + 0.0118463x + 0.7569069$	0.985	35-90	13
Outdoor 400m	$y = -0.0002x^2 + 0.0152x + 0.6478$	0.985	35-100	14
Indoor 200m	$y = -0.0001624x^2 + 0.0123330x + 0.7216505$	0.973	35-95	13
Outdoor 1500m	$y = -0.0001x^2 + 0.0074x + 0.8804$	0.992	35-100	14
Outdoor 100m	$y = -0.0001x^2 + 0.0075x + 0.8583$	0.988	35-100	14
Outdoor Marathon	$y = -0.0001x^2 + 0.0079x + 0.8812$	0.985	35-90	12
Outdoor 5000m	$y = -0.0001x^2 + 0.0085x + 0.84$	0.983	35-95	13
Outdoor 200m	$y = -0.0001x^2 + 0.0088x + 0.83$	0.987	35-100	14
Outdoor 800m	$y = -0.0001x^2 + 0.0089x + 0.8269$	0.986	35-95	13
Indoor 60m	$y = -0.0001767x^2 + 0.0151579x + 0.6290564$	0.960	35-100	15

Table 4.6: Parametric functions of women's events

Event	Parametric Equation	R2	Age range	Data points
Indoor Pole Vault	$y = -0.0000168x^2 - 0.0106383x + 1.3378882$	0.925	35-80	10
Indoor Long Jump	$y = -0.0000347x^2 - 0.0089229x + 1.3725022$	0.983	35-90	12
Indoor Triple Jump	$y = -0.0000138x^2 - 0.0102027x + 1.3525566$	0.966	35-90	12
Indoor High Jump	$y = 0.0000241x^2 - 0.0125235x + 1.3999769$	0.976	35-85	11

Outdoor Long Jump	$y = -0.0000596x^2 - 0.0047878x + 1.2231040$	0.975	35-90	12
Outdoor High Jump	$y = -0.0000133x^2 - 0.0082077x + 1.2912830$	0.989	35-90	12
Outdoor 1 Mile	$y = -0.0000663x^2 - 0.0038284x + 1.2389545$	0.997	35-85	11
Outdoor Pole Vault	$y = -0.0000949x^2 - 0.0012254x + 1.1039898$	0.917	35-80	10
Outdoor Triple Jump	$y = -0.0000766x^2 - 0.0021318x + 1.1449678$	0.986	35-90	12
Outdoor 1500m	$y = -0.0000847x^2 - 0.0013179x + 1.1819422$	0.979	35-85	11
Indoor 1500m	$y = -0.0000954x^2 - 0.0001724x + 1.1430135$	0.981	35-80	10
Outdoor 2000m Steeplechase	$y = -0.0001021x^2 + 0.0005386x + 1.1157328$	0.988	35-75	9
Outdoor Marathon	$y = -0.0001114x^2 + 0.0012838x + 1.1074182$	0.992	35-90	12
Indoor 800m	$y = -0.0000999x^2 + 0.0006635x + 1.0985060$	0.971	35-85	11
Race Walk 20K	$y = -0.0001054x^2 + 0.0011200x + 1.0761853$	0.965	35-80	10
Outdoor 100m	$y = -0.0000587x^2 - 0.0020066x + 1.1605793$	0.986	35-90	12
Outdoor 200m	$y = -0.0000798x^2 - 0.0002551x + 1.1110441$	0.991	35-90	12
Race Walk 10K	$y = -0.0000562x^2 - 0.0020157x + 1.1158256$	0.978	35-90	12
Outdoor 5000m	$y = -0.0001057x^2 + 0.0020447x + 1.0579862$	0.992	35-85	11
Race Walk 3000m	$y = -0.0000780x^2 - 0.0001047x + 1.0758910$	0.977	35-85	11
Outdoor 800m	$y = -0.0001308x^2 + 0.0043630x + 1.0171253$	0.992	35-90	12
Indoor 60m Hurdles	$y = -0.0000897x^2 + 0.0012105x + 1.0619091$	0.973	35-5	9
Race Walk 5000m	$y = -0.0000702x^2 - 0.0002249x + 1.0562635$	0.978	35-90	12
Indoor 60m	$y = -0.0000717x^2 - 0.0000120x + 1.0988357$	0.985	35-90	12
Outdoor 3000m	$y = -0.0001293x^2 + 0.0046588x + 0.9823924$	0.980	35-85	11
Indoor 3000m	$y = -0.0001406x^2 + 0.0056529x + 0.9736177$	0.977	35-85	11
Indoor 200m	$y = -0.0001183x^2 + 0.0044896x + 0.9793282$	0.993	35-90	12
Outdoor 10000m	$y = -0.0001602x^2 + 0.0081437x + 0.9159909$	0.985	35-85	11
Outdoor 400m	$y = -0.0001592x^2 + 0.0082510x + 0.8868137$	0.993	35-90	12
Indoor 400m	$y = -0.0001743x^2 + 0.0096627x + 0.8495953$	0.979	35-85	11
Race Walk 50K	$y = -0.0002479x^2 + 0.0162879x + 0.7680835$	0.887	35-65	7
Indoor 1 Mile	$y = -0.0002343x^2 + 0.0161915x + 0.7316731$	0.994	35-85	11
Outdoor 80m Hurdles	$y = -0.0003089x^2 + 0.0257765x + 0.4504762$	0.995	40-80	9

6   DISCUSSION

In this study, a quadratic function was used to model the decline in performance percentage of the athletics records as it is more intuitive to the layperson. Although Stones and Kozma's study (as cited in Baker, Tang et al. (2001)) in 1980 found that an exponential fit was better than power models for running events, our study showed that the quadratic fit also achieves a high coefficient of determination for all of the events. This could be due to the different age range used as Stones and Kozma studied a period of 30 years from ages 40 to 69, while we studied up to the age of 90 in this study. In addition, a quadratic decline of athletics performance fits the quadratic relationship of decrease in muscle area and the number of muscle fibers with increasing age as found by Lexell, Taylor et al. (1988).

On average, the ages at which performance percentage goes to zero for men and women are significantly different ($p < 0.05$) at ages 123.5 years and 111.9 years respectively. This implies that women's athletics performance declines faster than men's. Furthermore, there is also a significant difference between the men's and women's rate of decline for all the event categories except race walks ($p = 0.08$) based on the gradient of the quadratic functions over the 40 – 80 years age range. This observation supports the study done by Netz and Argov (1996), which indicated that "women deteriorate more than men in the aging process".

An analysis of the rates of decline of the 5 categories at each decade from 40 – 80 years showed that there is a significant difference in the rates of decline in the earlier decades (40 – 60 years) but no significant difference at ages 70 and 80 years. This may be due to lower participation rates for athletes aged 70 years and above. The WMA Outdoor Track and Field Championships 2011 showed that the average number of participants in the M80 (W80) category is about 40.2% (21.7%) of the number of participants in the M40 (W40) category for all athletics events (WMA 2011), therefore, leading to a relatively high rate of decline across all events from ages 70 – 85 years.

7 CONCLUSION

In summary, this study has shown that a quadratic function is a good fit in modeling the boundary of perfect aging using athletics records. With these quadratic functions, it was found that there are significant differences between the rates of decline for men and women in sprints, distance, marathons and jumps, but not for walks. There are also significant differences in the rates of decline for the 5 categories from age 40 – 60 years, but the rates of decline among these categories are similar from age 70 – 80 years. As this study has produced the boundary of the human performance using the world records, future research may consider encumbrance factors in the model, so as to develop a more realistic model.

REFERENCES

Anderson, G. F. and P. S. Hussey (2000). "Population aging: a comparison among industrialized countries." Health Affairs 19(3): 13.

Baker, A. B. and Y. Q. Tang (2010). "Aging Performance for Masters Records in Athletics, Swimming, Rowing, Cycling, Triathlon, and Weightlifting." Experimental Aging Research 36(4): 25.

Baker, A. B., Y. Q. Tang, et al. (2001). "Percentage Decline in Masters Superathlete Track and Field Performance With Aging." Experimental Aging Research 29(1): 19.

Grubb, H. J. (1998). "Models for Comparing Athletic Performances." Journal of the Royal Statistical Soci 47(3): 13.

Hill, A. V. (1925). "The Physiological Basis of Athletic Records." The Scientific Monthly 21(4): 20.

Lee, W. K. M. (1999). "Economic and Social Implications of Aging in Singapore." Journal of Aging & Social Policy 10(4): 20.

Lexell, J., C. C. Taylor, et al. (1988). "What is the cause of the ageing atrophy?: Total number, size and proportion of different fiber types studied in whole vastus lateralis muscle from 15- to 83-year-old men." Journal of the Neurological Sciences 84(2-3): 20.

Maud, P. J. and C. Foster (2006). Physiological Assessment of Human Fitness. United States of America.

Netz, Y. and E. Argov (1996). "Gender Differences In Physical Fitness In Old Age." Perceptual and Motor Skills 83(1).

Pikaar, R. N., E. A. P. Koningsveld, et al. (2007). Meeting Diversity in Ergonomics, Elsevier.

Riegel, P. S. (1981). "Athletic Records and Human Endurance: A time-vs.-distance equation describing world-record performances may be used to compare the relative endurance capabilities of various groups of people." American Scientist 69(3): 6.

Salvendy, G. (2006). Handbook of Human Factors and Ergonomics. New Jersey, John Wiley & Sons.

Schulz, R. and C. Curnow (1988). "Peak Performance and Age Among Superathletes: Track and Field, Swimming, Baseball, Tennis, and Golf." Journal of Gerontology 43(5): 8.

Sehl, M. E. and F. E. Yates (2001). "Kinetics of Human Aging: I. Rates of Senescence Between Ages 30 and 70 Years in Healthy People." Journal of Gerontology 56A(5): 11.

Tanaka, H. and D. R. Seals (2003). "Invited Review: Dynamic exercise performance in Masters athletes: insight into the effects of primary human aging on physiological functional capacity." J Appl Physiol 95: 11.

Wilson, J. R. (2000). "Fundamentals of ergonomics in theory and practice." Applied Ergonomics 31(6): 11.

WMA (2011). "WMA Outdoor Track and Field Championships 2011 Results." Retrieved January, 2012, from http://www.world-masters-athletics.org/results/2011.

WMA (2011). "World Masters Athletics Records." Retrieved July, 2011, from http://www.world-masters-athletics.org/records.

Young, A. (1997). "Ageing and physiological functions." Philosophical Transactions: Biological Sciences 352(1363): 7.

CHAPTER 22

A Comparison Study of the Effects on Road Crossing Behavior between Normal and Parkinson Disease

Yang-Kun Ou,1 * Chin-Hsien Lin,2 Yung-Ching Liu,3*

1,3Department of Industrial Engineering and Management, National Yunlin University of Science and Technology, Yunlin, Taiwan
2Department of Neurology, National Taiwan University Hospital, Taipei, Taiwan
Email address: g9621802@yuntech.edu.tw1

ABSTRACT

As society ages, a large amount of human factors related research has been carried out into the subject of the safety of the elderly in their daily lives. However, most research focuses on the general elderly population and there is a serious lack of research into elderly sufferers of Parkinson's disease, who receive a substantial amount of attention in medical circles. This research proposed pedestrian road-crossing experiment to compare of two elder groups (the ordinary vs. with Parkinson's) to examine the relationship between medical measuring scale test results and the pedestrian road crossing behavior of the different groups. Four factors were involved in this mixed factorial experiment study, including the age (normal older vs. Parkinson's patient; between -subjects), vehicle speed (40 km/hr vs. 60 km/hr vs. 80 km/hr; within-subjects), time gap (5 seconds vs. 7 seconds vs. 9 seconds; within-subjects) and time of day (midday vs. dusk; within-subjects). The safety margin, subjective confidence and crossing pace were collected as the dependent variables. Results showed that the Parkinson patient's safety margin was negative than those of the normal elder and thus caused the Parkinson patient to take higher risk then crossing the road. The pedestrian made their decisions majorly based on the factor of distance. The faster the vehicle speeds, the higher risk for driver and pedestrian. In addition, drivers in the time of the dusk had longer response time to detect the pedestrian presence. The results of this research will clarify the correlation between medical scale test indicators and allowing the results of medical testing to be provided for further safety management and control and following related aid equipment design concepts to be provided, helping patients and improving the lives of the elderly.

Keywords: Parkinson's disease, age, road crossing, safety margin

1   INTRODUCTION

Approximately 1.3 million people die from road traffic injuries worldwide each year, accounting for 2.2% of global deaths and ranked as the ninth leading cause of death in 2004 [1]. Almost half of the global traffic injury deaths are vulnerable road users, such as pedestrians and motorcycle riders. In the United States, data show that 4378 pedestrians were killed and 69,000 pedestrians were injured in road traffic crashes in 2008 [2]. According to the statistic data of Taiwan Ministry of Transportation in 2009, the most common causes of traffic fatalities are motorcycle accidents (44.3%) and then followed by pedestrians (31.5%).

Given the crossing behavior of pedestrians and road environment both contribute to pedestrians' exposure to risks on the road accidents, safe road crossing requires performance of multiple competing tasks, including visual sensory function, switching the focus of attention between disparate spatial locations, and motor tasks [3]. Specifically, when crossing a road, pedestrians have to determine whether the remaining time before a vehicle reaches them is long enough for crossing, and adapt their behaviors to the continuous perception of oncoming vehicles. Therefore, pedestrians need to determine the remaining time (depending on time gaps between the coming vehicle and the pedestrian) and relate it to the time needed to cross the road (depending on environmental factors such as road width, and individual factors such as walking speed). Crossing is possible if the remaining time is greater than the crossing time, which allows a margin of safety for pedestrians to cross the road [4]. A number of studies have shown that the characteristics of pedestrians, such as old age; and traffic environment, such as vehicle speed and time of the day, are risk factors contributing to the crash events while crossing road [4-7]. One previous study showed that much greater risk-taking behavior was taken among old age group (age older than 75 years old) than young age pedestrians, resulting in increased risk of road accidents, although some conflict results existed [6, 8]. As populations age, patients with neurodegeneration disorders increase exponentially and the road safety is an important issue to be addressed [9].

Parkinson's disease (PD) is a progressive neurological disorder that is primarily caused by dopaminergic neuron degeneration in the substantial nigra and predominantly affects motor function of patients [10]. In light of the pathological changes of PD and the close connections between frontal cortex and striatum, PD patients are expected to have deficits in cognitive function [11]. Patients with PD can have executive and visuo-spatial dysfunction [12], poor attention, and deterioration of visual perception [13], even in early stages or drugs naïve PD [12, 14]. These motor and non-motor features both could hamper the safety in road-users with PD while crossing the street and results in vehicle collision events.

Using a simulated simple street-crossing situation, we conducted a case-control study to test the hypothesis that patients with PD have a greater risk in crossing street than age-matched controls. We further identified the clinical and environmental risk predictors of unsafe crossing behaviors in PD patients.

2 Materials and Methods

2.1 Participants

A total of 81 participants, including 31 PD patients and 50 control subjects without evidence of PD, were included in this study. Informed consent was taken from the study participants and the study was approved by the institutional ethics board committee. The subjects were recruited from the neurology outpatient clinics at National Taiwan University Hospital Yun-Lin branch and all patients fulfilled the diagnostic criteria for PD.[19] Each participant underwent a standard neurological and neuropsychological examination, including Mini-Mental Status Evaluation (MMSE).[20] Patients with PD also received the evaluation of Unified Parkinson's Disease Rating Scale (UPDRS). The inclusion criteria for patients were idiopathic PD with mild to moderate level of disease severity (Hoehn and Yahr Stages, H&Y stage I-III).[21] Exclusion criteria were history of brain surgery, other neurologic or psychiatric disorders, score of MMSE less than 24 as well as impairments of visual acuity or hearing ability.

2.2 Visual and Cognitive Testing Battery

All of the participants were tested a battery of cognitive, and visual tasks ，Visuoconstructional ability was tested using Rey-Osterrieth Complex Figure Test, CFT) [13]; Clock Drawing Test (CDT) was tested Visuospatial ability [23]; Trail Making Test (TMT) parts A and B was tested executive function [25]; Visuoperceptual Function was evaluated using the Visual Form Discrimination (VFD) [24]; The useful field of view (UFOV)(Visual Resources, Inc) task was used to measured visual attention. UFOV task has been used successfully with AD and PD patient's drivers to predict traffic accident [22]. The UFOV task was divided into three subtasks to determine a driver's risk of accident involvement: 1) information processing speed, 2) divided attention and 3) selective attention. Sum of 3 subtests of the UFOV task (UFOVTOT) was used in our analyses.

2.3 Experiment design

Four factors were involved in this mixed factorial experiment study, including the elder groups(normal older vs. Parkinson's patient; between-subjects), vehicle speed (40 km/hr vs. 60 km/hr vs. 80 km/hr; within-subjects), time gap (5 seconds vs. 7 seconds vs. 9 seconds; within-subjects) and time of day (midday: 11:00-13:00 vs. dusk:17:50-18:10; within-subjects).The midday was adapted from the central weather bureau, Taiwan. To avoid vehicle lamp effects, the vehicle was open the lamp for the two conditions (midday and dusk). A total 18 traffic scenes were randomly assigned to the participant and it was 2 trials for each road crossing condition. Thus, they viewed a total of 36 traffic scenes.

The street-crossing simulation scene, which was pre-recorded from a real street, was projected to the 17-inch LCD (Figure 1A). The visual scenes represented a two-way street with 3.5 m width with vehicles moving from left to right (in reference to the pedestrian's standing at the sidewalk) (Figure 1B). The images refresh rate was 30Hz. A press button used to record the participants' responses while they decided to cross the road was connected to a personal computer. Traffic sound effects were broadcast from 2-channel amplifiers.

Dependent variables including: (1) 10 meters walk speed: normal walking pace and fast walking pace. (2) Remain time (the time period remains for pedestrian safely crossing road) which can be obtained by (time gap – the time of participant can safely crossing-road with his walking pace). (3) Safety margin which can be obtained by (the time remaining – the time of the pedestrian walked normally in 3.5 meters distance), and the normal walking speed was obtained in this study by asking the participants walked normally in 10 meters distance.

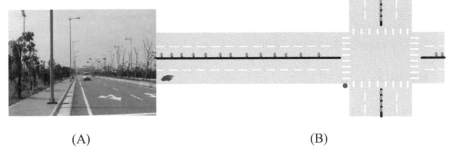

(A) (B)

Figure 1. Experimental setup: A: a vertical view of roadway (rural road with straight 2 lanes); B: Scenery example from the participant's point of view at the starting position.

2.4 Procedure

Participants were first given information about the purpose of this experiment and the task they were instructed to perform. First, the subjects were tested on a battery of cognitive and visual tasks (see section 2.2). The cognitive test took approximately 1 hour, with a ten-minute break for participants, if required.

The second part was the street-crossing test. The participants gave their consent for this study and then started to practice 4-10 training trials to get familiar with this experiment. The participants then walked 10 meters with two different speed paces twice (normal walking pace and fast walking pace) and their walking times were collected. In the real trial, each participant was seated at a desk approximately 60cm in front of a 17-inch LCD monitor located at eye height with a computer keyboard placed in front of them. They were instructed to place their index finger on the "space" key on the keyboard. The participants subjectively assessed the time of walking 3.5 meters. The task was then based on their time estimation to make the judgment by pressing the space button at the point in which they had decided it was safe to cross the road. At this moment, the participants should respond as quickly as possible by pressing the space key and say "pass" loudly. The experimenter then collected the times from the start to the participants pressing the space key. A total of 18 experimental trials were randomly assigned to the participants and there were 2 trials for each road crossing condition. The street-crossing experiments required approximately 1 hour to complete. All participants were compensated for their time and effort.

2.5 Data analysis

The variance of the results was analyzed using SPSS v.12.0 statistical software (SPSS Inc., Chicago, IL), and post hoc analyses were conducted using the least significant differences (LSD) test. The level of significance used for all analyses was α <.05. The odds ratios (OR) was analyzed using logistic regression and describes the strength of association between two binary data values. The OR analysis was developed to predict relative risks of street crossing. The ratio of exposure in diseased subjects was then compared to the ratio of exposure in non-diseased subjects. For the binary outcome, takes a value 1 with a probability of failure cross road and a value 0 with probability of success cross road.

3. Results

3.1 Demographics

Table 1 shows the results of demographics. The groups did not differ in age, education and MMSE. Similarly, no differences emerged on the 3 stages of Parkinson disease (PD). The pedestrians with PD performed significantly worse on both the cognitive and visual battery (see Table 1). Under the cognitive battery, a control group showed significantly higher scores on CFT-copy and CFT-recall than the PD groups. The groups did not differ in CDT and TMT-A. However, we observed significantly longer reaction time in TMT-B during PD than in the control group. Under the visual battery, the control group showed significantly faster reaction time on UFOV-1, UFOV-2 and UFOVTOT than the PD groups. In addition, no significant difference was found between the control groups and PD in UFOV-3 and VFD. Three stages of Parkinson's disease are discussed separately. Table 2 shows the neuropsychological test results and the results showed that stage I for PD in CFT-copy scores was significantly better than stage II and stage III. No differences emerged on other neuropsychological tests.

Table 1. Clinical demographics of patients with PD and control subjects.

	Controls(n=50)	PD patient(n=31)	P-values
Demographics			
Sex(M, F)	27,23	16,15	0.835
Age, y	67.74 (5.44)	65.32 (8.07)	0.071
Education, y	10.64(5.49)	9.61 (4.51)	0.233
MMSE	27.52 (2.25)	26.61 (2.17)	0.070
Cognitive tests			
CFT-Copy	31.39 (5.69)	26.52 (6.00)	<0.001*
CFT-Recall	12.58 (7.29)	7.50 (7.56)	0.003*
TMT-A	57.23s(52.39)	58.86s(33.26)	0.180
TMT-B	84.82s(43.27)	107.05s(54.13)	0.011*
CDT	15.58 (0.95)	14.94 (2.11)	0.134
Visual tests			
UFOV-1	46.40(43.71)	103.35(90.85)	0.002*
UFOV-2	203.08(160.40)	312.32(184.33)	0.010*
UFOV-3	343.84(111.39)	384.00.73(139.19)	0.122
UFOVTOT	593.32(265.97)	799.68(380.62)	0.009*
VFD	26. 47(4.48)	26.16(3.83)	0.287

MMSE, mini-mental state examination; CDR, clinical dementia rating; CFT-copy, complex figure test-copy; CFT-recall, complex figure test-copy; TMT-A, trail making test subtest A, TMT-B, trail making test subtest B; CDT, clock drawing test; UFOV, useful field of view task; VFD, visual form discrimination; y, years.

Table 2. Clinical data of PD patients with different disease severity.

	PD stage I (n=13)	PD stage II (n=10)	PD stage III (n=8)	p-values
Demographics				
Sex(M, F)	8,5	3,7	5,3	0.263
Age, y	64.69(9.70)	63.60 (7.61)	68.50(5.18)	0.295
Education, y	10.15(3.41)	9.00(5.75)	9.50 (4.87)	0.802
MMSE	26.92(1.89)	26.90(1.13)	25.75(1.98)	0.394
UPDRS part I	2.23(1.59)	1.60(1.43)	3.63(1.41)	0.036*
UPDRS part II	4.38(4.03)	7.20(3.23)	13.75(3.88)	0.001*
UPDRS part III	10.15(4.08)	15.70(7.39)	21.88(4.12)	<0.001*
Cognitive tests				
CFT-Copy	29.26(4.77)	22.42(6.52)	27.20(4.60)	0.043*
CFT-Recall	10.26(7.06)	6.63(9.10)	4.09(4.93)	0.076
TMT-A	56.51(27.98)	60.74(39.80)	60.31(36.66)	0.967
TMT-B	98.68(27.42)	90.75(41.01)	138.97(85.40)	0.432
CDT	15.31(1.38)	14.40(3.06)	15.00(1.77)	0.573
Visual tests				
UFOV-1	103.23(108.80)	99.80(81.29)	108.00(80.85)	0.837
UFOV-2	265.54(202.23)	325.70(176.72)	371.63(164.37)	0.474
UFOV-3	369.54(148.20)	375.90(155.11)	417.63(112.58)	0.867
UFOVTOT	738.31(422.51)	801.40(382.06)	897.25(330.58)	0.681
VFD	25.92(4.54)	26.30(3.65)	26.38(3.20)	0.993

MMSE, mini-mental state examination; CDR, clinical dementia rating; CFT-copy, complex figure test-copy; CFT-recall, complex figure test-copy; TMT-A, trail making test subtest A, TMT-B, trail making test subtest B; CDT, clock drawing test; UFOV, useful field of view task; VFD, visual form discrimination; y, years.

3.2 Crossing road behavior

The baseline crossing behavior assay showed that the crossing time, either under normal or fast pace conditions, of patients with PD was significantly longer than control subjects. Under normal condition, the mean crossing time for 10 meter distance was 3.5 ± 0.6 seconds in control group and 5.0 ± 1.4 seconds in PD patient group (P<0.01). Under the fast pace condition, the mean crossing time for the same distance was 2.5 ± 0.4 seconds in control group and 3.9 ± 1.2 seconds in PD patient group (P<0.01). Among patients with PD, the mean crossing time of patients with H-Y stage III was significantly longer than patients with H-Y stage I and II in either situation (normal condition: stage I: 4.6 ± 1.3 seconds, stage II: 4.6 ± 0.9 seconds, stage III: 6.1 ± 1.3 seconds; fast pace condition: stage I: 3.6 ± 1.2 seconds, stage II: 3.6 ± 0.9 seconds, stage III: 4.8 ± 1.3 seconds, P<0.01).

In this study, the time remaining for the pedestrian was defined as [the time gap – the time in total for the pedestrians to decide to cross the road], and the results showed that the time remaining for the pedestrian at dusk was significantly longer than that at midday [midday: 3.48s vs. dusk: 4.25s, p<0.001]. In addition, the approaching vehicle speed will affect the pedestrian's walking remaining time, the faster the vehicle speed, the longer the time remaining the pedestrian selected (80km/hr (3.46s) < 60km/hr (3.87s) < 40km/hr (4.28s), p<0.001]. The time remaining increased with the similar trend as the time gap increased [time gap 5s (2.81 s) < time gap 7s (3.87s) < time gap 9s (4.91s), P<0.001]. However, for the two different age groups, no difference was found in their remaining time decision [the control group (3.70s), the PD group (4.03s), P=0.21]. No significant difference was found on 3 stages of PD in remaining time.

The safety margin was obtained by (the time remaining - the average fast and normal crossing time of the pedestrian in 3.5 meters distance), and the average crossing time was obtained in this study by asking the participants walked in 10 meters distance. Based on the formula, if the safety margin value was greater than zero, then the pedestrian walked crossing the road was "safe", otherwise, if the safety margin value was less than zero, the pedestrian walking cross the road was "in danger". The ANOVA's results revealed that the four main factors were significant differences [the age group, p=0.003; the time of the day, p<0.001; the vehicle speed, p<0.001; the time gap, p<0.001]. The control group performed safer than the PD group (0.78s vs.-0.22s); the day in the dusk time was safer than the time in the midday (0.62s vs. -0.05). The approaching vehicle increased its coming speed will negatively impact upon the pedestrian's safety (40km/h (0.63s) > 60km/h (0.28s) > 80km/h (-0.06s)), while increasing the time gap will increase the safety margin (9s(1.20s)>7s(0.21s)>5s(-0.55s)). The ANOVA's results revealed that the PD stage I (0.42s)was significant safer than the PD stage III (-1.25s)(p=0.02).

Table 3 shows the regression results of the control group and PD groups, and the results showed that the PD pedestrians have approximately 1.329 times greater risk crossing a road than do the control groups. Moreover, crossing at midday is approximately 2.394 times riskier than crossing at dusk. When the speed is 80km/hr and 60km/hr, the risk of crossing the road is 40km/hr of 2.389 and 1.496 times.

When it takes longer to cross the road (such as 9s), it is safer to cross the road. The time gaps for 5 and 7 seconds to cross the road is 4.050 and 1.5026 times that of 9 seconds. Overall, dusk, lower speed, longer time gap, and control groups are factors that improve the safety of crossing a road. From the battery we found that CFT-Copy, CFT-Recall, TMT-B and VFD are significantly associated with safe road crossing. When CFT-Copy and CFT-Recall are reduced by 1 score, the risk of safely crossing the road increases 1.036 and 1.038 times, respectively. When TMT-B is increased by 30s, the risk of harm while crossing the road is increased by 1.145 times. When VFD is reduced by 1 score, the risk of harm while crossing the road is increased by 1.078 times.

4   CONCLUSIONS

The findings in this study support our hypothesis that pedestrians with early stage PD make more errors of decision making in crossing road (indexed by negative value of safety margin) than age, gender, education level-matched neurologically normal pedestrians. We demonstrated that early stage of PD per se is an independent risk factor for unsafe crossing road behaviors, even considering the potential confounding effects of general cognitive functions (MMSE) and other traffic environmental factors. Additionally, we found that traffic environmental factors also affect the road safety in both PD and control subjects. Specifically, fast coming motor vehicle speed, short time gap and midday of the day would increase the risk of pedestrian's crossing road behaviors.

Road-crossing is a dynamic activity, which requires synchronized actions of pedestrians in facing different traffic scenes, such as visual perception and discrimination of coming vehicles, cortical flexibility and executive function to judge and decide the crossing time, and activation of the relevant motor controls to cross the road. Previous studies have shown that old age is the major risk factor of unsafe crossing road for pedestrians [7]. However, the information regarding the role of cognitive functions in crossing road in this aged population is scarce. In our study, we found that visual-spatial dysfunction (CFT-copy and CFT-recall) and decreased executive flexibility (TMT B-A) have an impact on the correct decision making in crossing road behaviors in both PD and control groups. In line with the information procession demands of the crossing road, measures of visual-constructional abilities (CFT-copy), visual perception (VFD), visual memory (CFT-recall), and executive function (TMT B-A) correlated significantly with the outcome of safe crossing road. In our studied subjects, the mean age of enrollment is 66.81 years. As age-related cognitive decline is not uniform across cognitive domains, previous studies have suggested "frontal aging hypothesis" that cognitive abilities mediated by the frontal lobe are highly vulnerable to the aging effects [26]. Therefore, in support of our findings, the age related frontal executive dysfunction partially contributed to poor crossing road behaviors in our study populations.

Given that factors of traffic environment also contribute to the road safety, we confirmed previous studies that fast coming motor vehicle speed (80 km/hr>

60km/hr> 40 km/hr), decreased time gap between vehicle and pedestrians (5sec > 7sec> 9sec) and mid-day increased risk of unsafe crossing road behaviors in both PD and control subjects [4]. Previous studies have shown that pedestrians tend to select a shorter time gap at 60 km/h than at 40 km/h [4]. In supportive of our observations, these findings confirmed the effect of coming vehicle speed on time gaps selected in street crossing situations. The observed decrease in the time gap with the vehicle speed increase resulted in a smaller safety margin and a higher percentage of unsafe decisions regardless of PD or control subjects. The risk association was significantly increased in the PD patients. These results clearly show that vehicle speed is an important risk factor that affects crossing decisions for all pedestrian populations, especially in PD patients. Notably, we also observed that time of the day also impacted on the risk of crossing behaviors. We found crossing streets in the mid-day (11AM-13PM) is more dangerous than in the dusky day (18PM). Our findings opposed to the previous observation that pedestrian/motor vehicle collisions occurred most frequently in the late afternoon/early evening when pedestrian and vehicle volumes are higher than other time period [5]. One of the possible reasons to explain our findings maybe the participants were more cautious to decide the crossing time in the dusky time period than in the mid-day, which dusky environment made the pedestrians to be vigilant.

In conclusion, our study indicates that PD patients with early-stage disease are at risk for motor vehicle collisions when crossing street, owing to both motor and non-motor cognitive dysfunction. Future large sample studies are warranted to confirm our findings. Training programs, portable stimulator device and modification of traffic markers to compensate the visual-spatial disabilities of PD patients are required to improve road safety in PD patients.

ACKNOWLEDGMENTS

We would like to thank all of the subjects who participated in this study. The grant is supported from the National Taiwan University Hospital Yun-Lin Branch (NTUHYL100.N001).

REFERENCES

1. Peden M, S.R., Sleet D, Mohan D, Hyden AA, Jarawan E et al., World Report on Road Traffic Injury Prevention. Geneva: WHO; 2004. 2004.
2. Administration., N.H.T.S., Traffic Safety Facts 2008 Data: Pedestrians. Washington, DC: National Center for Statistics and Analysis. 2008.
3. Owsley, C., et al., Visual processing impairment and risk of motor vehicle crash among older adults. JAMA, 1998. 279(14): p. 1083-8.
4. Lobjois, R. and V. Cavallo, Age-related differences in street-crossing decisions: the effects of vehicle speed and time constraints on gap selection in an estimation task. Accid Anal Prev, 2007. 39(5): p. 934-43.
5. National Highway Traffic Safety Administration (NHTSA) notes. Ann Emerg Med, 2008. 52(4): p. 453-4.

6. Oxley, J.A., et al., Crossing roads safely: an experimental study of age differences in gap selection by pedestrians. Accid Anal Prev, 2005. 37(5): p. 962-71.

7. Koepsell, T., et al., Crosswalk markings and the risk of pedestrian-motor vehicle collisions in older pedestrians. JAMA, 2002. 288(17): p. 2136-43.

8. Holland, C. and R. Hill, The effect of age, gender and driver status on pedestrians' intentions to cross the road in risky situations. Accid Anal Prev, 2007. 39(2): p. 224-37.

9. de Lau, L.M. and M.M. Breteler, Epidemiology of Parkinson's disease. Lancet Neurol, 2006. 5(6): p. 525-35.

10. Dickson, D.W., et al., Neuropathological assessment of Parkinson's disease: refining the diagnostic criteria. Lancet Neurol, 2009. 8(12): p. 1150-7.

11. Foltynie, T., et al., The cognitive ability of an incident cohort of Parkinson's patients in the UK. The CamPaIGN study. Brain, 2004. 127(Pt 3): p. 550-60.

12. Cooper, J.A., et al., Cognitive impairment in early, untreated Parkinson's disease and its relationship to motor disability. Brain, 1991. 114 (Pt 5): p. 2095-122.

13. Uc, E.Y., et al., Visual dysfunction in Parkinson disease without dementia. Neurology, 2005. 65(12): p. 1907-13.

14. Yu, R.L., et al., Advanced Theory of Mind in patients at early stage of Parkinson's disease. Parkinsonism Relat Disord, 2012. 18(1): p. 21-4.

15. Gelb, D.J., E. Oliver, and S. Gilman, Diagnostic criteria for Parkinson disease. Arch Neurol, 1999. 56(1): p. 33-9.

16. Folstein, M.F., S.E. Folstein, and P.R. McHugh, "Mini-mental state". A practical method for grading the cognitive state of patients for the clinician. J Psychiatr Res, 1975. 12(3): p. 189-98.

17. Hoehn, M.M. and M.D. Yahr, Parkinsonism: onset, progression and mortality. Neurology, 1967. 17(5): p. 427-42.

18. Edwards, J.D., et al., Reliability and validity of useful field of view test scores as administered by personal computer. J Clin Exp Neuropsychol, 2005. 27(5): p. 529-43.

19. Ehreke, L., et al., Is the Clock Drawing Test a screening tool for the diagnosis of mild cognitive impairment? A systematic review. Int Psychogeriatr, 2010. 22(1): p. 56-63.

20. Eslinger, P.J. and A.L. Benton, Visuoperceptual performances in aging and dementia: clinical and theoretical implications. J Clin Neuropsychol, 1983. 5(3): p. 213-20.

21. Reitan, R.M., The relation of the trail making test to organic brain damage. J Consult Psychol, 1955. 19(5): p. 393-4.

22. Gunstad, J., et al., Patterns of cognitive performance in middle-aged and older adults: A cluster analytic examination. J Geriatr Psychiatry Neurol, 2006. 19(2): p. 59-64.

CHAPTER 23

Glare Sensitivity of Pilots with Different Ages and Its Effects on Visual Performance of Nighttime Flying

Yihong LIU, Mengdi Dong, Yaojie SUN, Yandan LIN

Institute for Electric Light Sources, Fudan University; Engineering Research Center of Advanced Lighting Technology, Ministry of Education
Shanghai, P.R.China
ydlin@fudan.edu.cn

ABSTRACT

The glare in the visual field during nighttime flying is a potentially insecure problem for the pilots as it reduces the contrast of the visual target and causes psychological discomfort. However, with the increase of the age of the pilots, the glare sensitivity will be changed for the same lighting condition because of the different physiological condition of the eyes. This may result in the variance of the visual performance in nighttime flying especially for the reaction speed to the obstacle out of the cockpit.

In this article several lighting conditions with difference glare source luminance, background luminance and glare source size will be set up to simulate different degrees of glare. Participants will be divided into two groups according to the age, namely the young group and the old group. The reaction time was selected as the evaluation criteria of the visual performance. Subjective evaluation for each lighting condition is also necessary as it is the direct performance about the glare sensitivity physiologically and psychologically. With rigorous data analysis, the glare sensitivity of different age groups was figure out according to the change tendency of the visual performance based on the experiment data.

Keywords: nighttime flying, aging, glare sensitivity, visual performance, reaction time, discrimination threshold size

1 INTRODUCTION

Glare is a lighting condition which can result in visual discomfort and reduction of the object visibility because of the unreasonable luminance distribution as well as the extreme luminance contrast temporally and spatially in the visual field. As the glare may pose the risk of visual discomfort and the safety of the people, the prevention of it is one of the most significant evaluation criteria of the lighting quality in all the field of lighting industry.

Driving the aircraft is a task which depends on vision at a high level and the pilots receive 80%~90% of the information via the visual scene. A comfort and steady lighting condition is necessary for the pilots to make out the information and signal on the display screen, instrument panel, light guide plate as well as the information outside the cockpit and to make quick and reasonable reaction to avoid accident and make correct route. Even a slight mistake would precipitate flight accidents. Majority of these mistakes result from the abrupt change of the lighting condition at the pilot's position especially when in nighttime flying as the eyes have adapted to the dim ambient environment. If the eyes of the pilot were exposed at the strong lights suddenly, the adaptation status and the visual function would be changed and the availability of the information would be influenced which bring in unsafe factor to flight driving.

The study on the influence of glare for the pilots mostly focus on the influence of the existence of the glare on the pilot's reaction and performance compared with when there is no glare. How the form, position, extend or some other capability of the glare lead to the performance changing is rarely researched and the lighting design of the cockpit for preventing the glare disturbance just followed some exist standards. In addition, with the increasing of the age, a number of changes of visual system occur in its structure and capabilities, resulting in reduced absolute sensitivity to light, reduced visual acuity, reduced contrast sensitivity, and greater sensitivity to glare, and related to youth, the elderly also need more time to recover from glare exposure [1].

In this article, a deep discussion focus on the glare sensitivity of different age groups and how the different glare sensitivity affect the reaction time of the pilots in nighttime flying which is based on the characteristic of pilots' visual task is made to provide theoretical foundation for prevention of the glare and development of the comfort lighting environment design in the cockpit.

2   METHOD

2.1   Experiment design

Among the most researches [2-6] which explore the effect of glare on the visual performance or task performance, three kinds of evaluation indicators have been widely used by these researchers:

I. The indicators which reflect the reaction speed of the observer, for example, the reaction time to detect the stimulus.

II. The indicators which reflect the detection extent of the observer for the stimulus. There are two indicators can be used usually, the one is the distance between the stimulus and the observer   when the observer can detect the stimulus, the size of the stimulus is fixed as constant, the other is the size of the stimulus when the observer can detect or discriminate the stimulus, the distance between the stimulus and the observer is fixed as constant.

III. The indicators which reflect the luminance contrast sensitivity of the observer. The most widely used is the luminance contrast threshold which is the luminance contrast between the stimulus and the background around it when the observer can just detect the stimulus target.

Based on the study method of the research domestic and international on the effect of glare on visual performance, in this paper, a scene of encountering the glare when in nighttime flying is simulated in the laboratory. The levels of glare can be changed by adjusting the parameters related to the glare rating, such as the surface luminance of the light source, the solid angle of the light source and the luminance of background environment.

Simple visual tasks are arranged to explore the mechanism of the effect of glare on the pilot's visual performance in nighttime flying.

2.1.1   Extent of the glare

There are several factors which can influence the experience of glare extent including the luminance, size, situation of the glare source and the luminance contrast between the glare source and the background environment.

In this paper, taking account of the operability of the experiment, the rationality the experiment time and improving the efficiency of the experiment, three key parameters are selected as variables to produce different glare extent, they are the luminance of the glare source, the luminance of the background and the solid angle of the glare source, the other parameters related are fixed as constant. The values assigned to the parameter are according to the real situation or make equal ratio miniature.

2.1.2 Subjective evaluation

Here a nine point scale is adopted to evaluate the discomfort glare degree of different lighting conditions, which is deBoer rating investigated by deBoer and Schreuder and was widely used as a criteria in evaluation of the glare outdoor.

The description of deBoer rating is given in the following table.

Table1. DeBoer rating

Point	Implication
1	Unbearable
3	Disturbing
5	Just admissible
7	Satisfactory
9	Unnoticeable

The deBoer rating method is a Multi-label scale. The scales allow the experimenter to pinpoint the actual discomfort of a particular glare stimulus.

In this paper, for each lighting condition, the observer should give a subjective evaluation for it based on deBoer rating.

2.1.3 Visual tasks

The whole experiment focus on the effect of glare on the reaction time of the pilot in nighttime flying, mimicking the scene that the pilot encountering the glare from the lighting installation on the runway when landing, The visual task is a black dot with the size of 12.9 minute of arc and eccentric angle of $2°$ presenting on the projection screen by self-develop software. The observers press the button on the wireless keyboard after perceiving the target appear and the time from the target formation to observers taking action can be recorded by the software, namely the reaction time. The start point is controlled by the observers by pressing the button "enter" on the wireless keyboard. Before the target appearing, there is a sound signal which to draw the observers' attention to look at the position where the sight fixation is carefully and to eliminate random error, the target presents or not after the sound signal is random. If the target presents, the observer should press the button "0" and the target will disappear immediately. If the target doesn't present, the observer should do nothing but waiting for the next sound signal 3 seconds later. For each test under a certain lighting condition, the judgment and reaction as above may be taken for about 15 times and each test will be repeated for 3 times to get enough data. The average reaction time for each observer under a certain lighting condition can be calculated by the software.

2.2 Apparatus

The experiment apparatus used for investigating the effect of glare on reaction time is shown in figure 1. The whole apparatus is put in a dark room which can avoid interference of other light outside. As shown in figure 1, the cockpit model is made according to the real condition with the proportion of 1:1, 2 indicates the operation panel and 1 indicates the display screen on it.

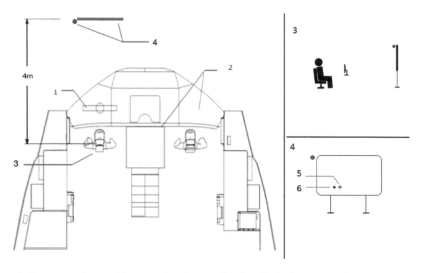

Figure1. The apparatus used to measuring the reaction time (1.display screen, 2.operation panel, 3.subject, 4.outside background and glare source, 5.sight fixing point, 6.task target)

The platform of the experiment is setup by simulating the scene that he pilot encountering the glare from the lighting installation at the airport when landing. There is a big projection screen mimicking the background environment of the cockpit in the presence of the observer with a distance of 4m as 4 indicates. The glare source used in the experiment which is a LED road light's module with the nominal power of 23W, it is hung in the upper left corner of the screen and the observing angle is 10° which is in accordance with the real situation. The red cross mark at the middle bottom of the screen is used for fixing the sight of the observer to avoid looking at the glare source straightly which may result in directly influence on visual function, 6 indicates the target projected on the screen by the projector mimicking the obstacle outside the cockpit, the eccentric angle of the target is 2°. As mentioned above, the target can present on the background screen randomly in a permission time period by programming. The observer press the button on the wireless keyboard after seeing the target appear on the screen, then the reaction time can be recorded by a feedback control system.

2.3   Parameter setup

The parameter setup in detail is shown in table 2. All of the three parameters related to the glare rating mentioned in 2.1.1is selected as variables.

Table2. Parameters setup in the research on the effect of glare on reaction time

Parameters		Values
Constants	Adaptation time(min)	5
	Surrounding luminance(Ls: cd/m2)	0.1
	CCT of glare source(K)	6500
	Tilt angel of the glare source(°)	12
	Observing angle(°)	10
	Observing distance(m)	4
	Observing way	Binocular
	Eccentric angle(°)	2
	Size of the target(min of arc)	12.9
Variables	Glare source luminance(Lg: cd/m2)	800/8000/80000
	Background luminance(Lb: cd/m2)	1/8
	The solid angle of the glare source(ω: sr)	0.56x10-5/4.4x10-5

With different combinations of the three variables, there are 12 lighting conditions for test in total.

2.3   Participants

There are 21 observers participating in the whole experiment procedure. And according to different ages, they are divided into two groups: the young group aged from 23-25 and the old group aged from 50-69.All the participants have the normal corrected visual acuity (1.2 and above), no eye disease or color blindness based on Pseudoisochromatic Plates Test. They all participate in the experiment voluntarily.

2.4   Test procedure

Both within-subject and between-subject design are involved in the experiment. For each observer, he must be informed the whole procedure of the experiment and take some practice before the formal experiments to avoid the practice effect which may increase the random error.

After several tests of practice, 21 participants will take the experiment under 12 lighting conditions in turn. Before each test, 5-10 minutes is taken for dark adaptation. Then the lighting condition is adjusted according to Table 2, the order of the lighting condition present to each observer is different to eliminate random error. After the observer taking the test of reaction time under each lighting condition, they should give the subjective evaluation for the lighting condition. Between different lighting conditions there should be enough time for resting to avoid visual fatigue.

The whole experiment is made up of adaptation stage, measurement stage and resting taking about 60 minutes in total and there is only one participants for each time.

3 Results and discussion

3.1 Subjective evaluation

The average ratings of different lighting conditions given by the observers of the two age groups are as shown and compared in figure 2.

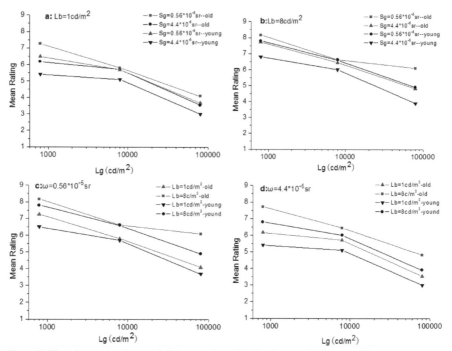

Figure2. The change tendency of deBoer rating with the three variables for different age groups (Lg- glare source luminance, Lb-background luminance, ω-solid angle)

It can be seen from the figure 2 that the deBoer rating of both the young group and the old group increase with decreasing light source luminance, solid angle, and increasing background luminance. What's more, the tendency of the relationship between deBoer rating and the logarithm of glare source luminance is approximate linearity.

In order to confirm the effect extent of each variable related to glare on the mean reaction time for target detection further, variance analysis is made which can calculate the effect extent of each variable on the dependent variable. Useful information of variance analysis is shown in table3. It should be noticed that the luminance of glare source is expressed by its logarithmic form.

Table3. Within-subject variance analysis results

Variable	The Young		The Old	
	F	Sig.	F	sig
Lg	35.686	0.000	39.633	0.000
Lb	16.107	0.000	23.756	0.000
ω	9.244	0.000	5.939	0.000

It can be read from table that the effect extent of the three parameters mentioned are different, the most significant effect is the glare source luminance, followed by the background luminance, the solid angel. It can also be read that the change of luminance of the glare source and background caused more change of discomfort glare feeling by the elderly compared to the younger group.

3.2   Reaction time

According to the data get in the experiment, the results are shown in figure 2. The horizontal coordinate is the glare source luminance and the vertical coordinate is the mean reaction time of the 10 young subjects or 11 old subjects. To make a comparison between young group and old group, the two parts results are shown in the same figure. It can be seen there is a change tendency that mean reaction time increases with the glare source luminance, and it's obviously that the mean reaction time of older group is much longer than young group.

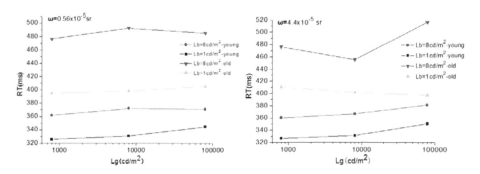

Figure3. The change tendency of reaction time with the three variables for different age groups (RT-mean reaction time)

In order to confirm the effect extent of each variable related to glare on the mean reaction time for target detection further, variance analysis is also made to confirm the influence of the three variables on the reaction time.

According to the results of the variance analysis, the weight of the effect of the three variables on the mean reaction time is different. For both the young group and older group, the change of the background luminance has the most significant effect on reaction time($F=404.257$, $P<0.001$), and the glare source luminance has a less significant effect($F=4.769$, $P<0.05$)while the glare source size doesn't have a significant effect on reaction time($P>0.05$).When it goes to the effect of age, the variance analysis result shows that the difference of mean reaction time between young group and old group is significant($P<0.001$), the old subjects' mean reaction time is longer than the young subjects dramatically (at least 60 ms).

3.3   Associate the subjective evaluation with reaction time

According to the experiment data, the change of the subjective evaluation (psychological) and the change of reaction time (physiological) can be associated and shown in figure 5. It can be read that with the decreasing of the deBoer rating, which means more intense of the feeling of discomfort glare, the reaction time prolongs significantly, especially for the elder people.

The results mentioned above can provide several guidance of the lighting design of the airport:

Firstly, it should pay more attention to the increase of the luminance of the background. The increase of the average luminance of the runway of the airport can quicken the reaction speed to the obstacle outside the cockpit significantly. And it should make a good weight between the glare source luminance and the background luminance. As the decrease of the glare source luminance can result in the decrease of the background luminance directly, it's better to decrease the light emitted directly to the observer's eyes and to ensure enough background luminance on the land, the luminaires of the airport requires good optical design and high output efficiency.

What's more, with the increasing age, the capability of fast reaction to the obstacle outside the cockpits shrivels significantly, for the elder pilot, it should draw support from some facilities to prevent the effect of direct glare such as the eye shield or safety goggles.

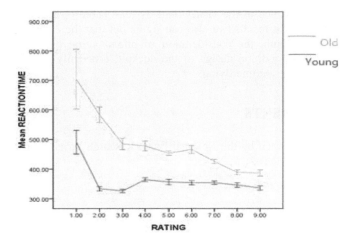

Figure5. Relationship between the subjective rating and the reaction time for the two age groups

4   Conclusion

This paper is focus on the glare sensitivity of pilots with different ages and its effects on visual performance of nighttime flying by synthesizing the existing study method of this field. Subjective evaluation based on deBoer rating and reaction time to the target outside the cockpit are selected as the dependant variables to evaluate the visual performance. Here are several conclusion achieved in this research:

I.   For the research on the effect of glare on the reaction time, three parameters are selected as the variables to adjust the extent of the glare, they are the background luminance, glare source luminance and solid angle. Results show that the background luminance has the most significant influence on the reaction time, followed by glare source luminance, while the solid angle has no obvious influence on the reaction time. Reaction time decreases with increasing background luminance and decreasing glare source luminance. And under the same lighting condition, the reaction time of the old group is always longer than the young one (at least 60ms).

II.   The subjective evaluation of discomfort glare based on deBoer rating increases with increasing background luminance and decreasing glare source luminance and solid angle for both the two age groups. And with the increase of the deBoer rating, the reaction time presents the decline tendency, especially for the elderly people.

The conclusion above can provide some guidance to the lighting design on prevention the effect of glare for the airport and the cockpit as well as the factor of aging. The increase of the glare source luminance can result in the increasing of the reaction time for detecting the target outside the cockpit. However, the influence extent of the glare source luminance on the visual performance can be offset by some measures. For the reaction time, as increasing the background luminance can decreasing the reaction time significantly, the influence can be offset by increasing the background luminance in the visual field by reasonable optical design for the luminaire. From the test results, we also can figure out that the elderly have longer reaction time than youth, the consideration of pilots' aging problem should be involved in lighting condition design of the cockpit, especially in preventing the effect of glare during nighttime flying.

ACKNOWLEDGMENTS

This work is supported by the National Basic Research Program of China (973 Program No.2010CB734102).

REFERENCES

Frank, S., Age and glare recovery time for low-contrast stimuli. *Proceedings of the Human Factors and Ergonomics Society Annual Meeting.*

Gibbons, R.B., et al, A Review of Disability and Discomfort Glare Research and Future Direction, TRB Biennial Visibility Symposium.

Theeuwes, J., Alferdinck, J.W.A.M. and Perel, M., Relation between Glare and Driving Performance, Human Factors, 2002.

Xicheng Liu, The Effect of Glare on Visual Task in Nighttime Driving, Chong Qing University, China, 2007.

Fenzhou Shi, Yu Wang and Yijun Lu, The Effect of Glare on Visual Performance in Mesopic Vision of the Navy, 28(1), 2007.

Aguirre, R.C., E.M. Colombo and J.F. Barraza, Effect of glare on simple reaction time. Optical Society of America, 2008, 25(7): p. 1790-1798.

CHAPTER 24

Training Standing Balance in Older Adults Using Physical Activity Controlled Games: A Comparison Study

Jun-Lin Lin, Hung-Chjh Chuan

Yuan Ze University
Chung-Li, Taiwan
jun@saturn.yzu.edu.tw, chuanchc@gmail.com

ABSTRACT

Nintendo Wii Balance Board (WBB) is a valid tool for accessing standing balance (Clark et al., 2010, Pagnacco et al., 2011), and has been widely used for balance rehabilitation (Young et al., 2011, Shih et al., 2010, Gokey and Odland, 2010, Guderian et al., 2010, Loureiro et al., 2010, Brown et al., 2009, Rogers et al., 2010, Padala et al., 2009). However, standing on a force platform often poses fear of falling for the elderly with balance problem, and this is especially true when the force platform is taller than the floor, or when the standing area of the force platform is not large enough, as in the case of WBB. On the other hand, Microsoft Kinect provides a natural user interface using posture and avoids the need and fear of standing on a platform. The objective of this work is to study the feasibility of replacing WBB with Kinect for accessing standing balance. Consequently, WBB-controlled video games that are used for balance rehabilitation can be ported to Kinect to improve the safety of the game-playing activities for the elderly.

Since the main function of WBB is to capture the movement of the center of pressure (COP), the first part of this work studies the correlation between the COP data from WBB and the joints' data from Kinect. A system had been built to simultaneously capture COP data and joints' data from WBB and Kinect, respectively. Then, the effectiveness of Kinect was measured by comparing the movement of the joints against the movement of the COP.

Thirty healthy adults (22 males and 8 females aged 22 to 46 years old) participated in this part of the study. Each participant was asked to stand on both feet on a WBB for 30 seconds. Head joint and Shoulder_Center joint showed excellent test-retest reliability on the range of movement along the left-right direction, and on the path length (ICC= 0.737~0.934). Furthermore, Head joint and Shoulder_Center joint also showed excellent inter-device reliability against COP on the range of movement along the left-right direction (ICC= 0.802, 0.821). This finding suggests that Head joint data and Shoulder_Center joint data are more correlated to the COP than other joints' data are, but further work is needed to verify whether this finding remains true under different test conditions, e.g., standing on one foot.

The second part of this work compared the controllability and the fear of falling when using either WBB or Kinect to play games. Two video games were developed such that both can be configured to use either WBB or Kinect to play the games. Three male elderly (age 65, 68 and 93) participated in this part of the study. Each participant was asked to play both games using Kinect or WBB, and then a short questionnaire was taken. The results showed that two out of three participants felt that the Kinect has better controllability than WBB, and all participants agreed that Kinect incurs less fear of falling than WBB.

Keywords: Wii Balance Board, Kinect, center of pressure

1   INTRODUCTION

Physical activity controlled video game platforms, such as Nintendo Wii or Microsoft Kinect, have gained much popularity for the past few years. These game platforms have also been used in rehabilitation to improve patients' motivation over traditional repetitive exercises (Chang et al., 2011, Brown et al., 2009). However, most commercial games on these platforms were not originally designed for rehabilitation purpose, and thus they might be too difficult or dangerous for rehabilitation patients without additional safety precautions.

For examples, Nintendo Wii Fit is a video game that had been adopted in many rehabilitation studies (Gokey and Odland, 2010, Guderian et al., 2010, Loureiro et al., 2010, Odland et al., 2010, Rogers et al., 2010, Padala et al., 2009). Wii Fit requires the players to stand on or step on/off a force platform, called Wii Balance Board (WBB), whose size is about $50 \times 32 \times 6$ cm^3. Standing on a WBB often poses fear of falling for patients with balance problem. Some safety precautions for WBB include providing additional support from auxiliary machinery or caretakers, or placing the WBB such that the top surface of the WBB is at the same height as the floor.

Compared to WBB, Kinect sensor uses cameras to capture players' movement, and does not require players to stand on a platform, and thus reduces the fear of falling. This motivates the study of comparing Kinect against WBB for rehabilitation exercises.

One key function of WBB is to capture the center of pressure (COP). Recent study has shown that WBB is valid tool for assessing standing balance (Clark et al., 2010). It remained unclear whether Kinect can provide reliable measurements. The first objective of this study is to evaluate the reliability of Kinect. The skeletal tracking function from Microsoft Kinect SDK is used to track the positions of 20 joints of the player in 3D space. The movements of four joints (i.e., Head, Shoulder_Center, Spine, and Hip_Center) are used to assess the reliability of Kinect and their correlation to COP.

Another key function of WBB is to control objects in video games. The second objective of this study is to evaluate the possible of replacing WBB with Kinect to play the same video games. Two video games were developed in this study, and both games can be played using either WBB or Kinect. Three elderly were asked to play both games and then to take a questionnaire. The results give a comparison on the controllability and the fear of falling between Kinect and WBB for video games.

2   METHODS

2.1   Reliability Test

Thirty young (age = 25.4 ± 7.6 years), injury free individuals (gender = 22 male, 8 female; body mass = 63.68 ± 12.42 kg) were participated in the reliability test. Each participant was tested on two occasions, completed within two days and at least 24 apart.

On each of the two test occasions, participants were asked to stand on both feet on a WBB with eyes closed and feet together for 30 seconds while facing a Kinect sensor. A software platform has been built to simultaneously capture both the COP data from WBB and the joints' data from Kinect. Data from both WBB and Kinect were sampled at 30 Hz. This frequency is the highest stable sample rate that Kinect can attain, but it is lower than the 40 Hz sample rate used in (Clark et al., 2010).

Four joints' data (i.e., Head, Shoulder_Center, Hip_Center and Spine) from Kinect and the COP data from WBB were used in this study. Five outcome measures were calculated for each of these joints and COP. They are the ranges of movement along the left-right direction (denoted as r_x), along the front-rear direction (denoted as r_y), and the path lengths along the left-right direction (denoted as p_x), along the front-rear direction (denoted as p_y), and on the horizontal plate (denoted as p).

To access the within-device test-retest reliability of these outcome measures, a two-way, random-effect, single measure intraclass correlation coefficient (ICC(2,1)) model was used for each of these joints and COP. Points estimates of the ICCs were interpreted as excellent (0.75~1), modest (0.4~0.74), or poor (0~0.39) (Fleiss, 1999).

To evaluate the possibility of using the joints' data to assess COP data, a two-way, random-effect, single measure intraclass correlation coefficient (ICC(2,1)) model for each outcome measure was also built for each of these joints against COP.

2.2 Acceptance Test

Three male elderly (age 65, 68 and 93) were participated in the acceptance test. All participants had no prior experience playing somatosensory video games, and only one (age 93) had prior history of falling.

A 2D shooter game and a 3D car racing game had been developed for this study. Both games can be configured to use either WBB or Kinect to play the games. Each participant was asked to play the 2D shooter game using WBB, and then play both games using Kinect, and finally play the 3D car racing game using WBB. Afterward, a short questionnaire was taken to collect the user experience on controllability and fear of falling on using WBB and Kinect.

3 RESULTS

Both the COP data from WBB (Table 1) and the Head joint data from Kinect (Table 2) showed excellent test-retest reliability on outcome measures r_x, p_x, p_y and p, but only modest reliability on outcome measure r_y. The Shoulder_Center joint from Kinect (Table 3) showed excellent test-retest reliability on all three path length measurements (i.e., p_x, p_y and p), but only modest reliability on both range measurements (i.e., r_x, and r_y). Both Spine joint and Hip_Center joint showed poor test-retest reliability on most outcome measures (Tables 4 and 5).

Table 1. Test-retest reliability for COP from WEB

measure	Day 1	Day 2	ICC (95% CI)
r_x	2.718 (0.997)	2.6 (0.839)	0.737 (0.517, 0.865)
r_y	3.350 (1.112)	2.766 (1.02878)	0.579 (0.281, 0.775)
p_x	41.041 (13.184)	37.410 (8.943)	0.753 (0.543, 0.874)
p_y	42.144 (15.260)	38.870 (12.828)	0.779 (0.585, 0.888)
P	65.744 (19.918)	60.099 (15.794)	0.780 (0.587, 0.889)

Table 2. Test-retest reliability for Head joint from Kinect

measure	Day 1	Day 2	ICC (95% CI)
r_x	3.007 (1.246)	2.871 (1.164)	0.753 (0.542, 0.874)
r_y	4.528 (2.591)	4.001 (3.1667)	0.680 (0.428, 0.834)
p_x	33.001 (53.378)	33.066 (43.305)	0.912 (0.823, 0.957)
p_y	57.552 (191.926)	63.305 (177.013)	0.933 (0.865, 0.968)
P	73.223 (203.4)	78.773 (186.016)	0.934 (0.866, 0.968)

Table 3. Test-retest reliability for Shoulder_Center joint from Kinect

measure	Day 1	Day 2	ICC (95% CI)
r_x	2.666 (1.193)	2.477 (1.071)	0.658 (0.395, 0.821)
r_y	3.764 (1.878)	3.151 (2.171)	0.730 (0.506, 0.861)
p_x	27.321 (41.265)	24.847 (30.684)	0.860 (0.727, 0.931)
p_y	47.645 (160.629)	44.828 (130.044)	0.959 (0.916, 0.98)
p	61.505 (171.203)	57.571 (135.741)	0.955 (0.907, 0.978)

Table 4. Test-retest reliability for Spine joint from Kinect

measure	Day 1	Day 2	ICC (95% CI)
r_x	2.382 (1.613)	2.48 (2.254)	0.236 (-0.13, 0.545)
r_y	2.768 (1.682)	2.344(1.746)	0.5 (0.176, 0.726)
p_x	26.260 (39.984)	42.869 (133.027)	0.252 (-0.113, 0.557)
p_y	19.370 (39.358)	36.213 (113.857)	0.211 (-0.155, 0.527)
p	36.232 (60.024)	60.644 (185.223)	0.248 (-0.117, 0.554)

Table 5. Test-retest reliability for Hip_Center joint from Kinect

measure	Day 1	Day 2	ICC (95% CI)
r_x	2.417 (1.766)	2.509 (2.507)	0.214 (-0.152, 0.359)
r_y	2.593 (1.807)	2.134(1.633)	0.362 (0.007, 0.635)
p_x	28.576 (45.693)	52.903 (187.199)	0.152 (-0.215, 0.481)
p_y	18.939 (36.9)	43.974 (163.888)	0.124 (-0.241, 0.459)
p	37.762 (62.445)	73.768 (261.615)	0.151 (-0.216, 0.48)

The ICCs for output measurements r_y, p_x, p_y and p between the COP data and each of the four joints' data showed poor reliability, and thus is omitted here. Table 6 showed the ICCs for output measurement r_x between the COP data and each of the four joints' data, using the data collected on the first day. Only Head and Shoulder joints showed excellent reliability on assessing r_x of COP.

Table 6. Reliability test for r_x between the COP and joints

	COP	Joint	ICC (95% CI)
Head	2.7175 (0.997)	3.007 (1.246)	0.802 (0.625, 0.901)
Shoulder_Center	2.7175 (0.997)	2.666 (1.193)	0.821 (0.657, 0.91)
Spine	2.7175 (0.997)	2.381 (1.613)	0.493 (0.167, 0.722)
Hip_Center	2.7175 (0.997)	2.417 (1.766)	0.418 (0.074, 0.673)

For the acceptance test, two participants felt that Kinect is better than WBB in terms of controllability, but one participant felt no difference. In terms of fear of falling, all three participants felt that Kinect is safer than WBB.

244

4   DISCUSSION

Previous study showed that WBB can provide reliable measurement of the moving length of COP (Clark et al., 2010). This study conducted a similar experiment using the positions of four joints. Among the positions of the 20 joints collected by Kinect, only four joins (i.e., Head, Shoulder_Center, Hip_Center and Spine) are located on the vertical central line of the human skeleton. Since our experiment requested the participants to stand on both feet and hold still, these four joints are likely to be more correlated to the COP than the other 16 joints are. Our reliability test showed that only Head and Shoulder_Center joints can provide excellent test-retest reliability. Since these two joints are higher than Hip_Center and Spine joints, this result indicates that the measurements for the joints on the upper body are more reliable than that on the middle body. Further experiments are needed to verify whether this result remains true for other postures, e.g., standing on one foot.

COP and joints are two different things, so are their paths of movement. Our experimental result showed low correlation between the moving lengths of the COP and each of the four joints. One exception is the measurement for the range of movement along the left-right direction. Both Head and Shoulder_Center joints have excellent correlation to the COP on this measurement. Since the center of mass of a human body is about at the 55% of his/her height, one would expect Hip_Center and Spine joints to have better correlation to the COP than Head and Shoulder_Center joints have. However, the poor test-retest reliability of Hip_Center and Spine joints might be the cause of their poor correlation to the COP. Another possible cause is inside the black box of Kinect SDK about the accuracy of these joints' positions for various body builds.

Compared to WBB, Kinect provides a natural user interface using posture and avoids the need and fear of standing on a platform. However, a common safety precaution for using physical activity controlled games for rehabilitation is to have a caretaker standing near the player to provide instantaneous help whenever needed. This could pose a problem for Kinect because it is not trivial to distinguish one person from another using Kinect SDK when the two persons overlap from the Kinect camera's perspective. One simple remedy is to require the caretaker to stand on the side of the player but not in front of or behind the player. However, this arrangement might not be feasible for patients who need continuous support from the caretakers to stand up and play the game. Further work is needed to use Kinect video games for rehabilitation exercises.

ACKNOWLEDGMENTS

The authors would like to acknowledge the support of NSC Grant 99-2221-E-155-048-MY3.

REFERENCES

Brown, R., Sugarman, H. & Burstin, A. 2009. Use of the Nintendo Wii Fit for the Treatment of Balance Problems in an Elderly Patient with Stroke: A Case Report. International *Journal of Rehabilitation Research*, 32, S109-S110.

Chang, Y. J., Chen, S. F. & Huang, J. D. 2011. A Kinect-based system for physical rehabilitation: A pilot study for young adults with motor disabilities. *Research in Developmental Disabilities*.

Clark, R. A., Bryant, A. L., Pua, Y. H., Mccrory, P., Bennell, K. & Hunt, M. 2010. Validity and reliability of the Nintendo Wii Balance Board for assessment of standing balance. *Gait & Posture*, 31, 307-310.

Fleiss, J. L. 1999. *The design and analysis of clinical experiments*, New York, Wiley.

Gokey, H. & Odland, L. M. 2010. Balance Indices in Elderly Women Following 4 Weeks of Training with Nintendo's (R) "Wii Fit". *Medicine and Science in Sports and Exercise*, 42, 594-594.

Guderian, B., Borreson, L. A., Sletten, L. E., Cable, K., Stecker, T. P., Probst, M. A. & Dalleck, L. C. 2010. The cardiovascular and metabolic responses to Wii Fit video game playing in middle-aged and older adults. *The Journal of sports medicine and physical fitness*, 50, 436-42.

Loureiro, A. P. C., Stori, F. R., Chen, R., Ribas, C. G. & Loureiro, C. C. 2010. Comparative Study of Conventional Physiotherapy and Virtual Rehabilitation Using WII Fit for Improvement of Balance in Parkinson's Disease Patients. *Movement Disorders*, 25, S714-S714.

Odland, L. M., Adams, L., Woods, R. & Sears, L. 2010. Health Outcomes In Middle-aged Women Following 10 Weeks Of Training With Nintendo's (R) "Wii Fit (TM)". *Medicine and Science in Sports and Exercise*, 42, 779-779.

Padala, K. P., Padala, P., Stergiou, N., Bissell, M. A., Davis, S., Malloy, T., Potter, J. & Burke, W. J. 2009. Wii-Fit for Balance and Gait in Skilled Nursing Facility: A Retrospective Study. *Gerontologist*, 49, 50-51.

Pagnacco, G., Oggero, E. & Wright, C. H. 2011. Biomedical instruments versus toys:a preliminary comparison of force platforms and the Nintendo Wii balance board - biomed 2011. *Biomedical sciences instrumentation*, 47, 12-7.

Rogers, N. L., Slimmer, M. L., Amini, S. B. & Park, E. Y. 2010. Wii Fit for Olders Adults: Comparison to a Traditional Program. *Medicine and Science in Sports and Exercise*, 42, 593-594.

Shih, C. H., Shih, C. T. & Chu, C. L. 2010. Assisting people with multiple disabilities actively correct abnormal standing posture with a Nintendo Wii Balance Board through controlling environmental stimulation. *Research in Developmental Disabilities*, 31, 936-942.

Young, W., Ferguson, S., Brault, S. & Craig, C. 2011. Assessing and training standing balance in older adults: A novel approach using the 'Nintendo Wii' Balance Board. *Gait & Posture*, 33, 303-305.

CHAPTER 25

Cognitive Performance and Age

Petra Alfred and Valerie J. Rice

Army Research Laboratory-Army Medical Department Field Element
Ft. Sam Houston, San Antonio, TX
petra.alfred@amedd.army.mil

ABSTRACT

Does age slow you down or does experience make you sharper? The purpose of this study was to investigate age differences in cognitive performance among a group of healthy service members and veterans. The Automated Neuropsychological Assessment Metrics (ANAM, 2007) was administered to groups of younger (< 39 years, n = 19) and older (≥ 39 years, n = 21) adults. Independent samples t-tests were employed to compare the groups' mean scores on the tests within ANAM of code substitution-learning, mathematical processing, matching to sample, and code-substitution delayed. Results indicate a significant difference between younger and older participants for the tests of code substitution-learning, mathematical processing, and code substitution-delayed (p<.10). Younger participants had faster mean reaction times and throughput on the code substitution tests, while older participants had faster reaction times on mathematical processing. There were no age differences on the matching to sample test (p>.10). These findings support taking age into consideration when using the ANAM to screen for cognitive impairments such as those that occur with TBI and other injuries or illnesses. In addition, they support the notion that some cognitive functions sharpen with age while others slow down.

Keywords: cognitive performance, age, ANAM, code substitution, mathematical processing, memory, reaction time, spatial processing, sustained attention

1   INTRODUCTION

There are many myths as well as facts associated with cognitive changes that result from the normal aging process. In cases of illness such as dementia and Alzheimer's disease, the effects on memory and cognition are drastic and well documented in the scientific literature.   However, among healthy adults, the cognitive changes resulting from normal aging are less concrete, more subtle, and often debated as not existing, or existing only as myth. As an example, for many years cognitive decline was not thought to begin until after age 60 (Hedden and Gabrieli, 2004), but recent research has revealed declines between ages 45 and 55 in reasoning and memory (Singh-Manoux et al., 2012). Visuospatial performance has also been found to suffer the effects of aging (Libon et al., 1994).

The majority of active duty service members are young, with the target recruitment age being 17 to 28 years of age. However, senior service members may be in their 50's or even early 60's (Army Policy Message 06-06), thus awareness of age-related cognitive (and physical) functioning are important considerations for the military services.

The U.S. Army has begun administering the Automated Neuropsychological Assessment Metric (ANAM, 2007) to soldiers prior to their deployment to establish a baseline of cognitive functioning, as part of their cognitive protocol in accordance with the National Defense Authorization Act HR 4986, 2008, section 1673. Comparisons can then be made between the baseline and subsequent performance following deployment, including exposure to Improvised Explosive Devices or suffering from injury or insult to the brain. Although the preference is to have baseline data for every soldier, this is not always possible, and age and gender related normative data may be necessary.

The purpose of this study is to examine age-related differences in cognitive performance on selected neurocognitive tests from the ANAM4. The tests include code substitution-learning, mathematical processing, and code substitution-delayed.

2   METHODS

2.1   Procedures

The forty participants in this study were recruited from a population of healthy military service members at Ft Sam Houston and the San Antonio Area Medical Center (SAAMC) in San Antonio, TX. First, they were briefed on the study and provided their informed consent.   Then, they completed a brief demographic questionnaire, including questions about their age, gender, ethnicity, and military service. Then, they completed the ANAM4 test battery on a laptop computer with an external mouse. The ANAM was administered to each participant individually, in a small, quiet office, with the door closed to reduce distractions. One researcher was present during each test administration. The first ANAM administration served as a practice session, to account for learning effects. After performing other tasks for approximately one hour, participants took the ANAM again. Data from the second administration were analyzed for this paper.

2.2   Instrumentation

The Automated Neuropsychological Assessment Metric (ANAM4) is a computerized assessment tool, comprised of a battery of neurocognitive tests to measure various cognitive domains/functions. For this study, the researchers chose to examine the cognitive domains of attention/ concentration (measured by the code substitution-learning and mathematical processing tests), memory (working memory measured by mathematical processing and matching to sample tests and delayed memory measured by code substitution-delayed test), and visuospatial ability (measured by code substitution-learning and matching to sample tests) (Table 1) (C-SHOP, 2007). This particular version of the ANAM was developed by the Department of Defense (DoD) to be used in screening for concussion (Bleiberg, Kane, Reeves, Garmoe, & Halpern, 2000). The battery takes approximately 20 minutes to complete.

Table 1 Selected ANAM4 Tests

Test List	Domain / Function
Code Substitution- Learning	Visual search, sustained attention, and encoding
Mathematical Processing	Basic computational skills, concentration, and working memory
Matching to Sample	Spatial processing and visuospatial working memory
Code substitution – delayed	Delayed Memory

Code Substitution-Learning. During Code Substitution-Learning, a series of nine digit-symbol pairs are provided in a key on the top of the screen (Figure 1). The user is presented with a series of digit-symbol pairs on the bottom of the screen. The user has to decide if the pair matches the digit-symbol pair shown above. If yes, they hit one key; if no, they hit a different key. Mean reaction time for selecting the correct response (measured in milliseconds), as well as 'throughput' (the number of correct responses made in a one minute interval) are computed by the ANAM system.

Figure 1. ANAM Code substitution-learning

Matching to Sample. In Matching to Sample, the user is shown a pattern created in a 4 x 4 grid, where each cell is either colored in one color or another, forming a visual pattern (Figure 2). The pattern is removed and a few seconds later two grids appear on the screen. The user must pick the grid pattern that matches the pattern shown initially. Mean response time for correct responses and throughput are automatically recorded.

Figure 2. ANAM Matching to Sample

Mathematical Processing. During Mathematical Processing, users are presented with a math problem which includes both adding and subtracting a series of three single-digit numbers (Figure3). If the resulting number is greater than five, one button is pressed, if it is less than five, a different button is pressed. Mean reaction time for correct responses and throughput are computed by system.

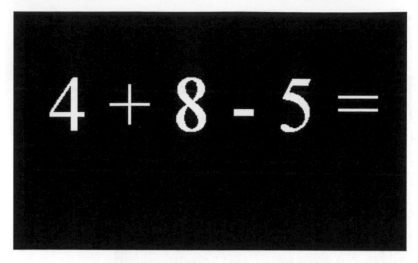

Figure 3. ANAM Mathematical Processing

Code Substitution-/delayed. During the Code Substitution-Delayed task, users are shown a single digit-symbol pair, and need to decide if the pair is correct or incorrect, using their memory of the pairs from the previous Code substitution-learning test (Figure 4). This is a memory task since a period of time has elapsed since the initial code substitution-learning task, and their attention has been shifted to performing other tasks. If the digit-symbol pair presented matches what they recall from the earlier task, they are to press one button. If it does not match, they are to press another button. Mean response time for correct responses and throughput will be collected.

Figure 4. ANAM Code-substitution delayed

2.3   Analysis

Key ANAM variables were examined descriptively, first with histograms to check for normality, and then with side-by-side boxplots to visually inspect for differences between younger and older participants and to identify potential outliers. All variables met the assumption of normality to allow the researchers us to proceed with the t-test analysis. However, extreme values were seen in the boxplots for: 1) code substitution-delayed (mean reaction time); 2) code substitution-delayed (throughput); and 3) matching to sample (mean reaction time). Independent t-tests were used to compare the mean ANAM performance of younger versus older participants on mean reaction time (for correct responses) and throughput (the number of correct responses per minute of available response time) for the four ANAM tests. The t-tests were first run with all participants included, and a second time with the extreme values removed. Since the removal of outliers did not change the significance of the t-tests, all values (including extreme values) are included in the reporting of the results.

3   RESULTS

Forty volunteers participated in this study, and were divided into two groups: a "younger" (< 39, n=19) and an "older" group (≥ 39, n = 21). The median age of participants was 39, with a mean age of 40 (Table 2). The median was used as the division point, as it more equally divided the groups.   Participants had an average of 17 years of military service, between 1 and 2 deployments, and their most recent deployment occurred within the last 3 to 4 years.   The majority were male (70%), Caucasian (77.5%), and in the Active duty Army (87.5%). None of the participants reported ever having a concussive event, or a diagnosis of post-traumatic stress disorder.

Table 2. Demographics – Age, Military Service & Deployments

	$M(SD)$	Minimum	Maximum
Age, years	40.83 (10.01)	22.17	58.83
Years current military service	17.47 (8.65)	2.91	35.33
How many times deployed	1.58 (1.41)	0	6

Means and standard deviations, as well as the t-statistic and associated p-values are provided in Table 3. Due to the importance of annotating differences due to age and the small sample size, a relaxed alpha value of .10 was used to test for significance.   The researchers were willing to accept the increased risk of false positives that occurs with a relaxed alpha due to the importance of these findings for establishing normative data for the Army and for soldiers.

Independent samples t-test results indicate a significant difference between younger and older participants for Code substitution-learning, Mathematical processing, and Code substitution-delayed. Specifically, younger participants had faster reaction times (t(38) = -1.76, p < .10) and greater throughput (t(38) = 2.08, p < .05) on the code substitution-learning task. Younger participants were also faster at the code substitution-delayed (memory) task (t(38) = =1.94, p < .10) and had greater throughput (t(38) = 1.73, p < .10). On the mathematical processing task, older participants had significantly faster mean reaction times for correct responses (t(38) = 2.928, p < .01), but did not differ from their younger counterparts in terms of throughput (p > .10). No differences were found between younger and older participants on the Matching to Sample test (p > .10).

4   DISCUSSION

Overall, younger participants outperformed older participants in the cognitive domains of visual search, sustained attention, and encoding, and delayed memory as measured by both code substitution tests on the ANAM4. Their mean reaction time for correct responses was faster than the older participants. This speed also enabled them to have greater throughput, meaning more correct responses in a given period of time. These findings are consistent with a study finding that older subjects were slower than younger subjects on reaction time tasks, although that study's task was a simple time detection task and their older volunteers were older than those included in this research (Racliff, Thapar & McKoon's, 2001). Our findings are also consistent with Jenkins, Myerson, Joerding & Hale's (2000) findings that older adults were generally slower than younger adults on all speeded tasks, and specifically on tasks requiring learning visual information such as the ANAM code substitution test.

Table 3. Independent Samples T-Test Results

ANAM Test & Measure	YOUNGER (N = 19)	OLDER (N = 21)		
	$M(SD)$	$M(SD)$	t	p
Code Substitution – Learning				
Mean reaction time (*msec*)	1202.26 (247.76)	1334.82 (228.59)	-1.760	.086
Throughput	50.76 (10.01)	45.04 (7.32)	2.079	.044
Mathematical Processing				
Mean reaction time (*msec*)	2739.61 (951.07)	2119.49 (310.55)	2.928	.006
Throughput	23.04 (9.71)	25.68 (5.13)	-1.087	.284

Matching to Sample				
Mean reaction time (*msec*)	1660.87 (490.33)	1716.10 (364.44)	-.407	.686
Throughput	37.12 (11.17)	33.08 (6.48)	1.416	.165
Code substitution – delayed				
Mean reaction time (*msec*)	1206.61 (277.03)	1355.96 (207.67)	-1.941	.060
Throughput	46.72 (15.23)	39.25 (12.09)	1.727	.092

No difference was found between younger and older participants on the Matching to Sample test which tests spatial processing and visuospatial working memory. These findings are inconsistent with Libon and colleagues (1994) who found visuospatial test performance declines with age, due to age-related decline in executive functions. Again, the age of their population was considerably older than those participating in this study ($\geq$ 75 yrs of age) (Ibid).

The researchers did find that older participants did better than younger participants in their basic computational skills, concentration, and working memory as measured by their reaction time for correct responses on the mathematical processing test. There was, however, no difference in their throughput on the math test. This means the older group was able to more quickly respond for correct responses, but they did not necessarily answer more problems correctly in a given period of time. This may be due to the older participants having more experience with math problems during their lifetime, making them faster at identifying correct answers. Additionally, it could mean that older participants have better concentration or working memory as these domains are also measured by the ANAM mathematical processing test.

For the military, these findings mean that when cognitive ability is being used either to potentially screen applicants for traumatic brain injury or other illnesses or for jobs or tasks that require attention/concentration, memory, and visuospatial ability (e.g., scanning maps for critical information such as friendly or enemy locations), age should be considered. In the absence of individual baseline data or performance data , age-related normative values may need to be developed for the tests of code substitution and mathematical processing with a larger sample of military service members.

5   LIMITATIONS

While this study presents interesting findings regarding the relationship between age and cognitive performance on four ANAM4 tests, additional research should be conducted with a larger sample size to examine age-related effects on the other

ANAM4 tests and before generalizing the differential effects of age on cognitive performance and to other military and civilian populations. In addition, with a larger sample size, a preferred statistical approach would be to use a regression analysis.

A potential confounding variable affecting the results on the mathematical processing test is individual math ability. An individual's math experiences such as whether they completed high school-level geometry or college-level calculus, and their grades in math courses, were not recorded or taken into consideration in the analysis, and may be another confound, one that is not taken into account by the ANAM developers when they devised this test. Other limitations include the sample size and lack of sufficient numbers of women.

6 CONCLUSIONS

Age appears to be related to ANAM performance among military service members and veterans, specifically in the areas of visual search, sustained attention, and encoding, and delayed memory, with younger adults achieving faster and superior scores compared with older adults. There was no difference between younger and older adults in spatial processing and visuospatial working memory. Older adults were quicker at performing mathematical computations, which embody computational skills, concentration, and working memory. These results warrant additional research on the effects of age on different cognitive dimensions measured using the ANAM, with larger numbers of volunteers, and with a mixed gender population.

DISCLAIMER

The views expressed in this article are those of the authors and do not reflect the official policy or position of the Department of the Army, Department of Defense, or the US Government.

REFERENCES

Army Policy Message 06-06. Change to Maximum Age Criteria. 5 April 2006. Retrieved 30 January 2012 from: http://www. armyreenlistment.com/Messages/Policy/PM_06_06_ age.pdf

Automated Neuropscyhological Assessment Metrics (Version 4) [Computer Software]. (2007). Norman, OK: Center for the Study of Human Operator Performance (C-SHOP), University of Oklahoma.

Bleiberg, J., Kane, R. L., Reeves, D. L., Garmoe, W. S., & Halpern, E. (2000). Factor analysis of computerized and traditional tests used in mild brain injury research. The Clinical Neuropsychologist, 14(3), 287–294.

C-SHOP (2007). ANAM4 TBI-MIL: User Manual. Center for the Study of Human Operator Performance, University of Oklahoma, Norman, OK.

Hedden, T. and Gabrieli, J.D. (2004). Insights into the ageing mind: A review from cognitive neuroscience. Nat Rev Neurosci, 5: 87-96.

Jenkins, L., Myerson, J., Joerding, J.A. and Hale, S. (2000). Converging evidence that visuospatial cognition is more age-sensitive than verbal cognition. Psychology and Aging, 15(1), 157-175.

Libon, D.J., Glosser, G., Malamut, B. L., Kaplan, E., Goldberg, E., Swenson, R. and Prouty Sands, L. (1994). Age, executive functions, and visuospatial functioning in healthy older adults. Neuropsychology, 8(1), 38-43.

National Defense Authorization Act for FY 2008, HR 4986, section 1673. 28 Jan 2008. Improvement of medical tracking system for members of the Armed Forces deployed overseas. Retrieved 30 January 2012 from: www.dod.gov/dodgc/olc/docs/pl110-181.pdf.

Ratcliff, R., Thapar, A. and McKoon, G. (2001). The effects of aging on reaction time in a signal detection task. Psychology and Aging, 16(2), 323-341.

Singh-Manoux, A., Kivimaki, M., Glymour, M.M., Elbaz, A., Berr, C., Ebmeier, K.P., Ferrie, J.E., and Dugravot, A. (2012). Timing of onset of cognitive decline: Results from Whitehall II prospective cohort study. BMJ, 344: 1-8, retrieved 30 Jan 2012 from

CHAPTER 26

An Attempt to Model Three-dimensional Arm Movement Time
Effects of Movement Distance, Approach Angle to Target and Movement Direction

Atsuo MURATA, Takanori AKIYAMA and Takehito HAYAMI

Graduate School of Natural Science and Technology, Okayama University
Okayama, Japan
murata@iims.sys.okayama-u.ac.jp

ABSTRACT

Technologies on three-dimensional human interfaces are paid more and more attention in recent years. However, there are few studies that clarified the condition of angle and distance under which we feel the three-dimensional movement to an object easy to point. The aim of this study was to explore how the movement distance and the approach angle to the object affected the pointing movement and to model the three-dimensional movement. In the experiment, five targets were installed on the surface of a rectangular solid. The approach angle and the movement distance to the surface were controlled as experimental parameters, and thus the movement time to the target was measured. We examined how the movement time changed as a function of the approach angle and the movement distance. The movement time tended to increase when the movement distance was short. Moreover, with the increase of approach angle, in particular, when the movement was carried out on the opposite side of a preferred hand, the movement time tended to increase. When the movement was conducted on the same side with a preferred hand, the approach angle did not affect the movement time. In other words, the relationship between the approach angle and the movement time was different between clockwise and counterclockwise movements. On the basis of the finding, an attempt was made to model the arm movement separately according to the clockwise and counterclockwise movements.

Keywords: Fitts' law, three-dimensional modeling, movement trajectory, dispersion of trajectory, movement time.

1   INTRODUCTION

Recently, technologies on three-dimensional human interface are paid more and more attention, and a few studies related to three-dimensional remote manipulation technologies are reported. On the other hand, there are few studies that clarified the condition of movement distance and approach angle to targets under which one feels the three-dimensional movement to an object easy to point. Thus, the aim of this study was to explore how the movement distance and the approach angle to targets affected the arm movement and to model the movement time.

Although Murata and Iwase (2001) made an attempt to extend Fitts' law to a three-dimensional arm movement, the approach angle to targets was not taken into account. Murata and Iwase (2001) carried out an experiment in which participants were required to perform three-dimensional pointing movements under the manipulation of target size, distance to target, and direction to target. As expected, the duration of the three-dimensional movements was rather variable and affected markedly by direction to target. The variance in the movement times was not statistically explained by the conventional Fitts' model. The conventional model was extended by incorporating a directional parameter into the Fitts' model. This led to fit better than the conventional Fitts' model.

In real-world situations, however, the case where arm movement is carried out for a fixed approach angle to targets is rarely observed. In order to understand arm movement properly and apply it to a three-dimensional human interface, not only the direction to a target but also the approach angle to a target must be considered in modeling three-dimensional arm movement time. The aim of this study was to investigate the effect of the approach angle to a target on the movement time and to model the movement time taking into account the effect of approach angle to a target. Accot and Zhai (Accot and Zhai, 1997; Accot and Zhai, 2001; Accot and Zhai 2003) investigated the movement trajectory during a two-dimensional movement in a HCI (Human- Computer Interaction) task.

This paper aimed at modeling three-dimensional movement by considering the approach angle to a target as a main factor in the modeling. In the modeling, the method proposed by Accot and Zhai (Accot and Zhai, 1997; Accot and Zhai, 2001; Accot and Zhai 2003) and Murata and Iwase (2001) was used and extended to a three-dimensional arm movement by taking the variation of movement trajectory into account. The clockwise and the counterclockwise arm movements were modeled separately on the basis of the movement characteristics obtained in the experiment.

2   METHOD

2.1   Participants

Five male graduate or undergraduates (from 21 to 26 years old) participated in the experiment. They were all healthy and had no orthopedic or neurological diseases.

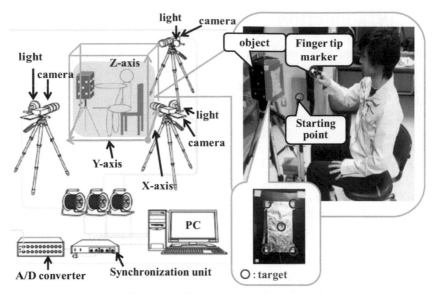

Figure 1 Outline of experimental setup.

2.2 Apparatus

Using the A/D conversion system of TRIAS (AD converter & video measurement system, DKH) and utilizing the change of voltage due to finger tip contact with the target plane, the arm movement time was measured. The measurement of arm movement time was synchronized with the motion capture system.

The motion capture system consisted of the following equipments.

· PH-1461A BCAM SYNC GENERATOR(DKH)
· PH-770 USB 16CH A/D conversion unit(DKH)
· High speed 1394 camera(DKH) (three sets)
· ProLite E2008HDS(Iiyama)
· i-Works(Applied)
· Trigger switch(DKH)

The connection of PH-770 USB 16CH A/D converter with i-Works via an USB cable corresponded to the basic unit of TRIAS. Using the voltage change in this basic system, the arm movement time was measured. Connecting PH-1461A BCAM SYNC GENERATOR with three high speed 1394 cameras via a coaxial cable enabled us to synchronize the arm movement time with the motion capture system of a finger tip during the arm movement. The outline of experimental setup is summarized in Figure 1. The approach angle is explained in Figure 2.

2.3 Design and Procedure

The experimental factors were movement distance, (10cm and 20cm from the

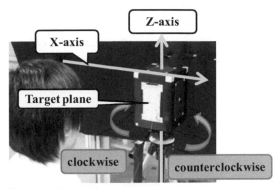

(Approach angle)
0 deg: Target plane is parallel to X-axis.
Coronal plane of participants is parallel to X-axis.
0, 10, 20, 30, 40, 50, 60, and 70 deg for both
clockwise and counterclockwise.

Figure 2 Explanation of approach angle.

start point of arm movement), and the approach angle to a target (0, 10, 20, 30, 40, 50, 60 and 70 degrees for both counterclockwise and clockwise). The starting point of arm movement was fixed throughout the experiment. The approach angle represents the relationship between the center of the target plane where targets are arranged and the coronal plane. The approach angle of 0 degree means that the target plane is parallel to the coronal plane (see Figure 2).

In the experiment, five targets were installed on the surface of a rectangular solid. The movement distance and the approach angle to a target were controlled as experimental factors. The arm movement time to a target was measured for each combination of the movement distance and the approach angle in order to investigate how these parameters affected the movement time. During the arm movement, the motion of a finger tip was captured using the motion capture system above mentioned. The movement trajectory of finger tip was used to define the index of difficulty in the arm movement and model the arm movement time.

3   RESULTS

3.1   Effects of movement direction to target

The arm movement time tended to increase when the movement distance was short. In particular, when the movement was carried out on the opposite side of a preferred hand, the movement time tended to increase with the increase of the approach angle. When the movement was conducted on the same side with a preferred hand, the approach angle did not affect the movement time.

As for the clockwise approach, the arm movement time increased linearly with the increase of approach angle. On the other hand, concerning the counterclockwise approach, the arm movement time showed a sinusoidal change as a function of approach angle (see Figure 3).

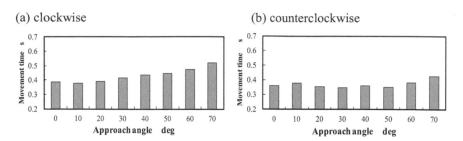

Figure 3   Arm movement time as a function of approach angle ((a) clockwise, (b) counterclockwise).

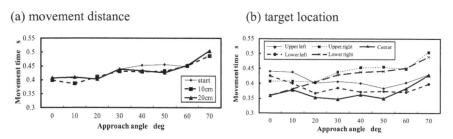

Figure 4 (a) Movement time as a function of approach angle and movement distance. (b) Movement time as a function of approach angle and location of target.

In Figure 4(a), the movement time is plotted as a function of approach angle and movement distance. In Figure 4(b) the movement time is plotted as a function of approach angle and location of a target. No statistically significant difference of movement time was detected among the conditions of movement distance. Contrary to this, there was a statistically significant difference of movement time among the conditions of target locations.

3.2   Effect of approach angle to target

It was also examined how the arm movement trajectory affected the movement time by measuring the movement trajectory of a finger tip with the three-dimensional motion capture system. The trajectories were compared between longer and shorter movement times. Examples of trajectory of a finger tip during the arm movement are shown in Figure 5. In Figure 5, the trajectories for both clockwise and counterclockwise movements are depicted.

4   DISCUSSION

4.1   Effects of movement direction to a target

It has been clarified that the movement distance had little effect on the

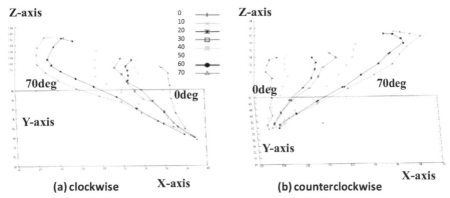

Figure 5   Movement trajectory as a function of approach angle. (a) clockwise, (b) counterclockwise.

movement time. Rather, the location of a target to be pointed affected the movement time to a further extent. According to the location of a target, the approach angle to a target differs. This might suggest that the movement time in the three-dimensional space is more remarkably affected by the difference of movement trajectory than that performed in the two-dimensional space such as HCI tasks. This makes us notice that the accurate modeling of movement trajectory leads to a reliable modeling of arm movement time.

4.2   Effect of approach angle to a target

As shown in Figure 5, the movement trajectory of the clockwise direction is symmetrical to that of the counterclockwise direction. With the increase of approach angle to target, the participant must rotate his or her movement to a larger extent in order to point a target. This is verified by the prolonged movement time with the increase of approach angle as shown in Figure 4(a). It must also be noted that the relationship between the approach angle and the movement time in the clockwise direction is different from that in the counterclockwise direction (cf. Figure 3(a) and (b)).

The reason can be inferred on the basis of the result by Boritz, Booth and Cowan (1991). They showed that the relationship between the direction of movement and the movement time depends on the dominance of hand (hand preference). Therefore, we hypothesized as follows: The movement time of the clockwise direction is different from that of the counterclockwise direction. The movement time in the counterclockwise direction is shorter than that in the clockwise direction. As shown in Figure 5, this hypothesis was verified in the range of this experiment. It must also be noted that all participants in this study were right-handed. Figure 3 is also indicative of the different modeling of arm movement trajectory between clockwise and counterclockwise directions or between preferred and non-preferred hands. Future research should model the three-dimensional arm movement by taking this property into account.

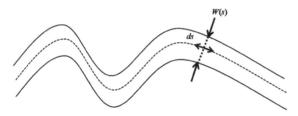

Figure 6 Explanation of Acott's model in two-dimensional movement.

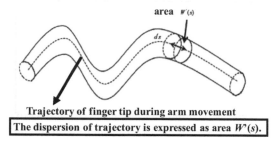

Figure 7 Explanation of the dispersion of trajectory area W'(s).

4.3   Modeling of movement time

On the basis of such data processing, the index of difficulty of arm movement was proposed by extending the result of Accot and Zhai (Accot and Zhai 1997; Accot and Zhai 2001; Accot and Zhai 2003). [2]-[4]. Accot and Zhai proposed the following index of difficulty.

$$ID_c = \int_C \frac{ds}{W(s)} \qquad (1)$$

In Accot and Zhai (Accot and Zhai 1997; Accot and Zhai 2001; Accot and Zhai 2003), they proposed a new index of difficulty in pointing movement on the basis of a concept that the inverse of the path width along the movement trajectory should be integrated. The modeling of movement trajectory by ds and $W(s)$ is depicted in Figure 6. They introduced a curvilinear abscissa as the integration variable. If C in Eq.(1) is a curved path, the index of difficulty can be defined through this path. The hypothesis in this model is that the movement time is linearly related to ID_c.

This model was extended to the three-dimensional arm movement as shown in Figure 7. It was assumed that the dispersion of trajectory can be expressed as an area $W'(s)$ shown in Figure 7. As a result, the following equation for defining the index of difficulty was found to be proper for the modeling of arm movement considered in this study.

$$ID_c = \int_C \frac{ds}{W'(s)} \qquad (2)$$

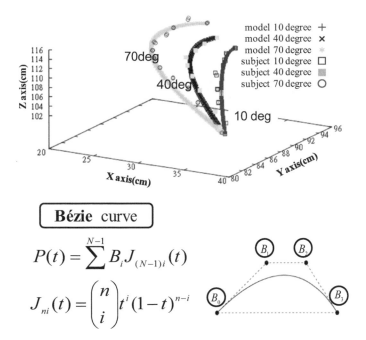

Figure 8 Approximation of trajectory of finger tip by Bézier curve during the arm movement.

where $W'(s)$ represents the area of hypothesized circle in the trajectory of a finger tip during the arm movement. The arm movement time MT was modeled by

$$MT = a + bID_c \qquad (3)$$

where a and b are empirically determined parameters. The $W'(s)$ was derived using the movement trajectory of a finger tip which was approximated by Bézier curve.

In Figure 8, examples of modeling arm movement trajectory using Bézier curve are depicted. Bézier curve can be derived as follows.

$$P(t) = \sum_{i}^{N-1} B_i J_{(N-1)i}(t) \qquad (4)$$

$$J_{ni}(t) = \binom{n}{i} t^i (1-t)^{n-i} \qquad (5)$$

On the basis of such a modeling of three-dimensional arm movement, the MT was formulated as a function of ID_c. The relationship between ID_c and MT is shown in Figure 9 (The contribution R^2 was $0.97^2 = 0.94$). The proposed model was found to explain 94% of the variation of arm movement time and provide a satisfactory modeling of arm movement time.

The counterclockwise arm movement is modeled as follows. The effects of the approach angle to a target differ between clockwise and counterclockwise movements. In case of the clockwise movement, both the movement time and the movement trajectory increased linearly with the increase in the approach angle to a

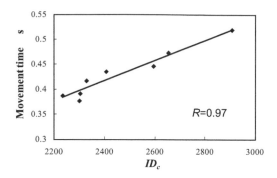

Figure 9 Relationship between ID_c and movement time (clockwise).

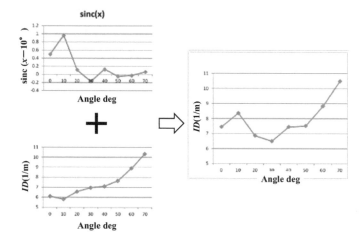

Figure 10 Modeling of ID (Index of Difficulty) for counterclockwise movements using sinc function.

target. On the other hand, in case of the counterclockwise movement, the relationship between the approach angle and the movement time did not correspond well with the relationship between the approach angle and the movement trajectory. Therefore, as shown in Figure 10, the difference between the movement time and the movement trajectory for reach approach angle was modeled using the sinc function in Eq.(6).

$$\begin{cases} \mathrm{sinc}(x) = \dfrac{\sin x}{x} \\ \mathrm{sinc}(x) = \dfrac{\sin \pi x}{\pi x} \end{cases} \quad (6)$$

Different from the clockwise movement, the counterclockwise movement was modeled by means of the following index of difficulty IDcc (Eq.(7)). The parameters in Eq.(7) were estimated as c=1, d=2.65, e=13.5, and f=0.14. The variable x corresponds to the approach angle to a target.

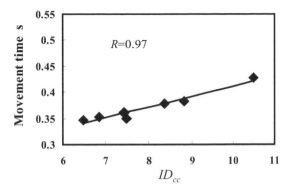

Figure 11 Relationship between ID_{cc} and movement time (counterclockwise arm movement).

$$ID_{cc} = c\int_c \frac{d's}{W'(s)} + d\frac{\sin(e(x-f))}{e(x-f)} \quad (7)$$

The relationship between ID_{cc} and MT is shown in Figure 11 (The contribution R^2 was $0.97^2 = 0.94$). The modeling of counterclockwise arm movement also provided a satisfactory result.

The disadvantage of this model is that the ID_c cannot be obtained easily and conveniently, because the movement trajectory of each participant must be obtained before modeling. In future research, an easy estimation and determination method of movement trajectory should be proposed.

The proposed equation for defining the index of difficulty was found to be proper for the modeling of arm movement considered in this study. It provided a correlation coefficient R of 0.97 to model the relationship between the index of difficulty ID_c or ID_{cc} and the arm movement time in the arm movement. Future work should propose a model which can treat a variety of three-dimensional objects.

REFERENCES

Accot, J. and S. Zhai 1997. Beyond Fitts' law: Model for trajectory-based HCI tasks. In. eds. S. Pemberton. M. Beaudouin-Lafon, R. J. K. Jacob. CHI97 Conference Proceedings: Human Factors in Computing Systems. Atlanta, Georgia, USA, 295- 302.

Accot, J. and S. Zhai 2001. Scale effects in steering law tasks. In. eds. M. Beaudouin-Lafon, R. J. K. Jacob. Proceedings of the ACM CHI 2001 Human Factors in Computing Systems Conference. Seattle, Washington, USA, 1-8.

Accot, J. and S. Zhai 2003. Refining fitts' law models for bivariate pointing. In. eds. G. Cockton and P. Korhonen. Proceedings of the ACM CHI 2003 Human Factors in Computing Systems Conference, Ft. Lauderdale, Florida, USA, 193-200.

Boritz, J., K. S. Booth and W. B. Cowan 1991. Fitts' law studies of directional mouse movement. In. Graphics Interface '91, Calgary, Alberta, Canada. 216-223.

Murata, A. and H. Iwase 2001. Extending Fitts' law to a three-dimensional pointing task. *Human Movement Science* 20: 791-805.

CHAPTER 27

Workplace Solutions for Elderly Workers

V. Melcher, H. Widlroither, V. Brückner

Fraunhofer-Institute for Industrial Engineering
Stuttgart, Germany
vivien.melcher@iao.fraunhofer.de

ABSTRACT

The older generation is willing to use new technologies and innovations to improve their living and working situation. Therefore it gets more and more important to remove obstacles that prevent older people from using new technologies. Innovative workplace solutions can help elderly workers to be flexible in their working environment and therefore to handle their livelihood independently.

The following paper describes the conception of a workplace solution for the elderly at the Fraunhofer Institute for Industrial Engineering in Stuttgart, Germany. Currently available workstations have been conceptually revised to fulfill the requirements of elderly workers. A prototype for an all-in-on solution has been developed including latest technology and best practice solutions for integrating a PC, keyboard, and an ergonomically revised mouse. The prototype was evaluated in a comparative test with users. As comparison standard input devices and elderly related solutions available on the current market were used.

Keywords: workplace, elderly workers, OASIS

1   INTRODUCTION

The societies of the European countries will grow older significantly in the next few decades. The baby boomers of the 50ies and 60ies are retiring in the next years. In Germany, for example, the number of working inhabitants will decrease for more than 7 percent until 2020, according to the German Federal Agency for Statistics (2009). This trend comes along with an increasing percentage of elderly workers. The ratio of young workers to elderly workers requires the appropriate inclusion of elderly people into work life.

Furthermore the working situation itself is changing. With increasing age the employees are going to carry out less physical workings. Instead there will be an increasing amount of knowledge workers working full time on a computer. Additionally, working time and free time are merging more and more. Flexible workers are working in their company and at home on the personal computer or professionally mobile on the go. This fact also refers to elderly workers. To incorporate older workers, flexible workplace solutions are needed (Federal Institute for Occupational Safety and Health, 2008).

Standard workplaces have not been adapted to the needs of older workers yet. Increasing restrictions in the musculoskeletal system, a decreasing visual function, a decreased ability to perceive visual contrasts and concomitant glare sensitivity (Hawthorn, 2000) make it more difficult for elderly workers to do their tasks in a proper way with standard workplace solutions.

As a second trend personal computer become more and more important for the elderly as an instrument for communication (Thayer, and Ray, 2006). and for keeping the contact to the outerworld (Mann, Belchior, Tomita, and Kemp, 2005). More and more senior citizens are living in sheltered homes with no fortune or space for personal computers. Affordable smart home solutions can replace the standard personal computer and help the elderly to use new technologies in a proper way.

Within the EU funded project OASIS (www.oasis-project.eu) integrated services have been developed to enhance the quality of life of elderly people. They are classified into three main categories considered vital for the quality of life enhancement of the elderly: Independent Living Applications, Autonomous Mobility, and Smart Workplaces Applications. The main scope of the web-based smart workplace application is to facilitate the multimodal user access to work environment, infrastructures and innovative IT solutions, covering home workers, mobile workers and other types of flexible workers and hiding at the same time the complexity of use (OASIS Consortium, 2007). To follow a holistic approach next to the web-based service a hardware solution for a flexible workplace was conceptualized and realized in a first prototype.

2   THE WORKPLACE CONCEPT

The main aim of the workplace concept was to provide an all-in-one solution that fits to the smallest available space in the home of the elderly and provides high comfort. Furthermore the requirements of elderly workers and PC users should be taken into account for a revision of existing input devices. A literature research of existing ergonomic and elderly friendly solutions available on the current market was conducted to extract the essential key parameters and problems of standard solutions. Furthermore an age simulation suite was used to evaluate standard workplaces and their components.

As a result a portable all-in-one concept was developed consisting of a hard drive component a projector for projecting the desktop screen on a wall, an integrated keyboard with wrist support and an interface for smartphone integration (figure 1). Additional USB and monitor ports allow the users a personalised completion of the workplace with their favourite input and output devices.

The projector can be used to project the desktop to any wall within the elderly home. The idea beneath is that the screen size can be adjusted to fit the requirements of the users without being dependable on a hardware monitor size. The projector is horizontally and vertically adjustable to change the projection angle on the wall and to provide the user with more flexibility with the keyboard position. This is important to avoid forced postures while interacting with the device/ keyboard. Since the system as it whole is portable the user can, for instances, project pictures or videos to the living room wall or recipes for cooking to the kitchen wall. For work related text entry tasks one or more additional hardware monitors can be connected.

Figure 1 Fraunhofer IAO workplace concept with all-in-one solution, desktop projection and mouse (Fraunhofer IAO).

The keyboard is equipped with enlarged letters to ensure high readability. In front of the keyboard a gel wrist rest is integrated into to the computer case. The wrist rest supports wrist and hand of the user, avoids an unhealthy snapping of the wrist angle and provides a comfortable placing of the user's arms on the desk.

A depression next to the keyboard can be used to visually integrate a touchscreen smartphone into the workplace and to connect it technically with the system. For one thing the smartphone can be used to synchronize data for easy access from home and on the go. For another thing the touchscreen surface can be used as a configurable input device. Either the touchscreen can be used as a touch pad for navigation, or additional shortcuts can be configured and placed on the surface to provide access to often used functions like "e-mail" or "web browser". Additionally the workplace keyboard grants easy text input to the small display device as a substitute of the software keyboard of the smartphone.

Beside the compact computer element a revised mouse concept was created. The mouse embeds the hand and the wrist and avoids unnecessary rowing motions of the arm. The concept is based on 3D navigation by tilting, pushing and pulling the mouse. Extended arm and hand movements for navigation are not required. The mouse is supplemented with a wrist rest to provide comfort and wrist support.

Based on the concept a first hardware prototype was developed using a 3D plotter for the case. Existing standard hardware solutions were integrated to form the overall structure (figure 2). The all-in-one PC system was composed of a Netbook, a digital projector for desktop projection, and a keyboard. A standard compact keyboard was used and equipped with enlarged letters and a wrist support for optimal comfort (figure 3).

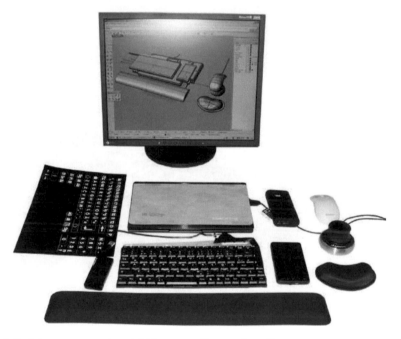

Figure 2 Prototype components and CAD concept (Fraunhofer IAO).

Figure 3 All-in-one computer with keyboard and integrated smartphone (Fraunhofer IAO).

The mouse consists of a 3D mouse system with an ergonomically formed superstructure fitted to the requirements of left-hander and right-hander, and supplemented with an integrated wrist rest. The mouse buttons were integrated in the superstructure (figure 4).

Figure 4 3D mouse system with ergonomically formed superstructure and wrist rest (Fraunhofer IAO).

3.   WORKPLACE EVALUATION

To evaluate the workplace concept and get first indications for improvement, a user test with elderly workers and pensioners was conducted at the Fraunhofer laboratories in Stuttgart, Germany. Within the user test the workplace prototype with mouse was compared to a standard keyboard and mouse (Logitech®Access™ Keyboard and Microsoft® Optical Wheel Mouse) and a keyboard with enlarged keys and labeling for vision impaired elderly users, available on the market (GETT Gerätetechnik GmbH, BIGKEY keyboard).

The main aim of the evaluation was to get a first impression on how the users rate the new concept and its components. Therefore the users were asked to write a short text with each keyboard. The think-aloud technique was used to get to know the impressions of the users. Afterwards the users had to assess each keyboard and mouse in a questionnaire with focus on handling issues. The comparison of the workplace to standard keyboard and mouse and to the special keyboard was used to classify the usefulness and usability of the new concept. Due to restrictions in prototype construction the workplace prototype did not fulfill the ergonomic standard in its entirety (e.g. the design height was too high). That is why the focus of the evaluation was based on the concept and not entirely on the realization of the prototype.

3.1   Results

Ten users took part in the user test (m= 6; f= 4), two of them were already retired. The mean age was 57.3 years (min= 47; max= 72). All of the users had wide experience with computer and keyboard usage. Half of the users were skilled in touch-typing.

Logitech keyboard: The standard keyboard is an established tool which is most suitable for touch-type writing (e.g. writing with all ten fingers). It is the common input device of the users in office and private places. For a majority of the participants the keys are hard to read, even with glasses, because of the low contrast and the small labeling. Having regard to vision impaired people this is a considerably disadvantage of the standard keyboard. This was especially a problem for users without touch-typing skills. They had to search for every single letter on the keyboard. Without glasses it would have been very difficult for them to write the text in an appropriate time. Most of the users reported painful physical strains while typing on standard keyboards. This is mainly due to the missing wrist support and the twisting of the hands while typing. The key depressions were rated as very comfortable to get hold of the keys.

GETT BIGKEY keyboard: The BIGKEY keyboard was designed by using very big keys with clear distances between the keys. Even though the key size was rated as too big, the main criticism of the participants referred to the distance between the keys. Especially users with touch typing skills had difficulties with typing. They made more typing errors than users without touch-typing skills. All users announced that they won't use the keyboard for a prolonged time. The big keys and the distance between the keys led to wide hand and finger movements and concomitant to reported strain and pain. The missing key depressions made it difficult to press the keys without side-slipping.. The keys were rated as relatively hard to press which goes along with an additional exertion of force. The visual rating was beneficial because of the good readable labeling with big black letters

Fraunhofer IAO workplace: The visual appearance of the overall workplace was rated as appealing. It´s integrated wrist rest was described as very comfortable and relieving, even though some participants noted that it is too elevated. Figure 6 shows the result of the questionnaire item "wrist support keyboard". Here the workplace prototype was rated best, followed by the standard keyboard and the special keyboard.

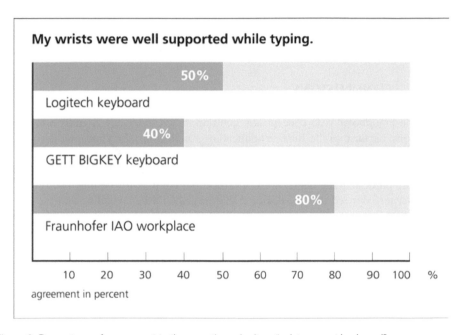

Figure 6 Percentage of agreement to the questionnaire item "wrist support keyboard".

Contrast and size of the labeling were also positive observed aspects. All participants declared that the key labeling was easy to read.

The all-in-one concept was appreciated well by the users. Especially the desktop projection was described as useful for the private environment to project videos and pictures. The main scope of application is attributed to the business area. The smartphone integration was assessed as useful by smartphone users. Particularly the possibility to use the keyboard as input device for the smartphone was described as a benefit. Easy data synchronization was rated as a good support in handling of everyday business tasks.

Microsoft optical wheel mouse: The optical wheel mouse is the common navigation device used by the participants. While some noted that the size could be slightly bigger all of the interviewed persons handled the device very well. According to the opinion of the users it was easy to navigate precisely with the optical wheel mouse. In fact all participants felt very confident with the device.

Fraunhofer workplace mouse: The visual appearance of the device was pleased by most of the interviewed persons. The construction and surface was rated as comfortable. Even though the developed mouse prototype is based on an unfamiliar handling concept all participants identified the way of use independently and in short time. Nevertheless all users had difficulties to navigate with the device. For some users it was stressful to target a specific point on the screen. In addition they said that the buttons are hard to reach. The difficulty in use is mainly based on the integrated, unfamiliar 3D-mouse linked with the superstructure, which make the device slightly shakily. Minor movements of the hand lead to extended cursor movements on the screen. The transmission of the hand movement to the cursor movement on the screen made it difficult to aim at a target on the screen. For that reason the participants would prefer the familiar standard mouse device.

4   CONCLUSIONS

The Fraunhofer IAO concept is a good approach to elderly friendly workplace solutions, because of the flexible application possibilities, the appealing design, the compact construction and the ergonomic wrist support. Enlarged key labeling improves the readability and increases user satisfaction. Furthermore a good pressure point and key depressions offer optimal comfort for typing. The results of the user test showed that the device fashioning should be appealing and the design should avoid stigmatism to provide a product that is accepted by the users. Nevertheless, the results also showed that familiar solutions are still preferred by the users. It seems difficult for users to adopt to new devices. Particularly the unfamiliar navigation concept of the mouse caused the main difficulties. The challenge is to transfer the advantages of standard solutions to new devices. Thus, the concept has to be revised. In a next step the prototype will be improved by using new materials to reduce the weight of the all-in-one computer. Moreover the overall construction has to be revised to fulfill ergonomic standards. Further research is needed to investigate if the 3D navigation concept can be accepted by the users by improving the transfer mechanism of hand movements to cursor movements. An idea would be to equip the mouse prototype with a buffer material at the connection between 3D mouse and superstructure to cushion the device movement. In consequence only major movements of the mouse would be transferred as curser movements. As an advantage a trembling of hands would be neglected.

The number of test participants was sufficient for a first evaluation, to find major ergonomic issues and to get a first impression on how the concept was perceived. For further tests with an improved prototype a larger number of test users are intended.

REFERENCES

BAuA. 2008. Alles grau in grau? Älter werdende Belegschaften und Büroarbeit. Accessed December 16, 2011 http://www.baua.de/de/Publikationen/Broschueren/A46.pdf?_blob=publicationFile&v=7

Mann, W. C., Belchior, P., Tomita, M. R., & Kemp, B. J. (2005). Computer use by middle-aged and older adults with disabilities. *Technology and Disability*, 17(1), 1–9.

Hawthorn, D. 2000. Possible implications of aging for interface designers. *Interacting with Computers*, 12(5), 507–528.

OASIS Consortium 2007. *(OASIS) Grant Agreement no 215754 – Annex I - Description of Work*. European Commission, Brussels, Belgium.

Statistische Ämter des Bundes und der Länder. 2009: Demografischer Wandel in Deutschland- Auswirkungen auf die Zahl der Erwerbspersonen. Heft 4. Stuttgart

Thayer, S. E., & Ray, S. (2006). Online communication preferences across age, gender, and duration of internet use. *Cyberpsychology & Behavior*, 9(4), 432–440.

CHAPTER 28

Survey Report of Recent Researches on Usability Evaluation for Elderly People in Japan

Tomokazu Shimada, Michiko Ohkura

Shibaura Institute of Technology
Tokyo, JAPAN
nb11104@shibaura-it.ac.jp

ABSTRACT

Since Japanese society continues to age rapidly, we must help elderly people live safe, peaceful, and comfortable lives by considering the usability of such industrial products and environments as IT devices, home electric appliances, public devices, and public facilities. However, usability evaluation methods remain unestablished because little research has addressed usability for elderly people. In this paper, we surveyed recent research on the usability of industrial products for elderly people in Japan and classified it and the usability evaluations. We identified future necessary research directions.

Keywords: Elderly people, Usability, Usability evaluation methods, Industrial products

1   INTRODUCTION

According to the National Institute of Population and Social Security Research, by 2050 in Japan, one person in three is expected to be 65 years old or over (Shiizuka, 2011). In such an aging society, products, services, and environments for the elderly and the handicapped must be designed based on various concepts, such as universal design, accessible design, inclusive designs, design for all, and barrier free (Sagawa, 2011). In the development process of products, we must consider various characteristics of users, including age and abilities. Since information technology has been changing so rapidly, IT devices and home appliances have made remarkable progress in functionality and performance. As information technology progresses, the positive effects of using electronic devices for elderly people would be strengthened more than for younger generations because such equipment was assumed to compensate for the decline in the functional ability of seniors (H. Inagaki. and Y. Gondo. 2005). However, usability evaluation methods for elderly people have not been established because little research has focused on usability for elderly people. In this paper, we introduce our survey of recent research on the usability evaluations for elderly people in Japan.

2   DEFINITION OF USABILITY

In this section, we introduce the main definitions of usability used in this survey.

2.1   Defined by ISO9421-11

According to Kurosu (2003), standard ISO9421-11 defines usability as "The extent to which a product can be used by specified users to achieve specified goals with effectiveness, efficiency, and satisfaction in a specified context of use."

2.2   Defined by Neilsen

According to (Nielsen, J. 1994), usability has multiple components and is traditionally associated with following five usability attributes:

Learnability: The system should be easy to learn so that the user can rapidly start getting some work done with the system.

Efficiency: The system should be efficient to use, so that once the user has learned the system, a high level of productivity is possible.

Memorability: The system should be easy to remember, so that the casual user is able to return to the system after some period of not having used it, without having to learn everything all over again.

Errors: The system should have a low error rate, so that users make a few errors during the use of the system, and so that if they do make errors they can easily recover from them. Further, catastrophic errors must no occur.

Satisfaction: The system should be pleasant to use, so that users are subjectively satisfied when using; they like it.

2.3   Defined by user engineering

M. Kurosu et al. (1999) defined user engineering as usability that includes such concepts as operability, cognition, and comfort. We used these three elements of usability in our survey.

3 CLASSIFICATION OF USABILITY

In this section, we classify the industrial products, the subject characteristics, the evaluation items, and the usability evaluation methods from our survey results.

3.1 Classification of industrial products

We classified industrial products into the following categories: IT devices and home appliances, public devices, public equipment and facilities, and medical-related products. Table 1 shows the categories and a list of the industrial products and environments that belong to each. Next we describe each category.

IT devices and home appliances: This category includes mobile phones whose functionality and performance progress has been remarkable, web browsers that elderly users use more often as internet services continue to grow, high-functionality TV remote controls, and such home appliances as rice cookers.

Public devices: Experimental evaluations of usability for ATM (Automatic Teller Machine) have been conducted to ameliorate the difficulties for elderly people in using IT devices. Even though ATMs could be classified as IT devices or home appliances, we classified them as public devices based on the public situations and environments where elder people use them.

Public equipment and facilities: Experimental evaluations of usability have been conducted for stairways where elderly people may be involved in accidents and guide signs at railway stations. We classified such industrial products as stairways and railway stations that need to be considered in daily environments as public equipments and facilities in contrast to the environments for IT devices and home appliances.

Medical-related products: Evaluation experiment of pharmaceutical package design has been conducted for elderly people to prevent from human errors when they take medicine and classified medicine as medical-related products.

3.2 Characteristics of subjects

Age and sex: Definition of elderly or old person is a person who is older than 65 years old. Many researches on usability for elderly people have been conducted with subjects older than 65 years old, both males and females. In addition to elderly

Table 1 Categories of industrial products

Categories	Industrial products
IT devices and home appliances	- Mobile phones (touch screens, touch buttons) - Web browsers and automatic devices - Home appliances and displays
Public devices	ATMs
Public equipments and facilities	Stations and stairways
Medical-related products	Medicine: displays of medical information

Table 2 Usability evaluation items classified by industrial products

Categories		Evaluation items
IT devices and Home appliances	Mobile phones	- Accessible properties of mobile phone buttons - Total manipulation image
	Web browsers	- Relation between mouse movements and tiredness - Operating and searching times - Ratio of mistakes per operation - Average speed and movement efficiency of mouse
	Automotive devices	- Relation between character size and contrast - Area of useful field of view (UFOV) - The focusing of attention for elderly
	Home appliances	- AIST cognitive-ability-assessment test - Behavior analysis, behavioral observation, eye tracking - Most preferred remote control - Learning ratio of each remote control
	Displays	-Visual characteristics of elderly
Public devices		- Utterances and behavior during operations - Confirmation times - Error analysis: number of operation errors and their recurrence - Auditory and visual guidance
Public equipments and facilities		- Utterances and behavior - Visibility of color coordination hue for stairways - Impression of spaces
Medical-related products		- Visibility and Recognition of drug label

people, experiments for younger people, mainly in 20's, have been conducted to compare with characteristics of elderly people.

Living area and living environment: According to A. Hasizume (2011), the usage ratio of new communication media, such as mobile phones and PCs, in rural areas is lower than in urban areas. Therefore, we must consider the living areas of elderly people for usability evaluations.

Past experiences: The past experiences of individual users influence the usability of industrial products. Elderly people tend to construct mental models of them based on past experiences, including the number of years they have used PCs or mobile phones, the number of mobile phones and devices they have owned, and so on. Therefore, the past experience of users is one characteristic that we must consider for usability evaluations.

Table 3 Usability evaluation items classified by research areas

Categories		Evaluation items
Ergonomics engineering	Movement	- Accessible properties of mobile phone buttons - Most preferred remote control in subjective evaluation - Motor abilities on web navigation - Relation between mouse movements and tiredness
	Visual	- Gazing time - Relation between character size and contrast - Area of useful field of view (UFOV) - The focusing of attention for elderly
	Hearing	Evaluations of auditory guidance
	Touch	Effect of button properties
Cognitive engineering		- Accessible properties of mobile phone buttons - Utterances and behavior during operations - Evaluations of auditory guidance - Visibility and recognition of drug labels - Error analysis: number of operation errors and their recurrence of the same errors
Kansei engineering		- Physiological measurement - Evaluation of impression of spaces - Visibility and recognition drug labels - Distinction of confused state

Cognitive characteristics: To consider the aged-related decline of the cognitive function for usability evaluations, AIST-cognitive aging test is used to screen senior citizens as subjects. From these test results, all subjects were sorted into four groups: no decline in cognitive function, decline in planning function, decline in attention, and decline in working memory.

3.3   Classification of evaluation items

We also classified the usability evaluation items based on three viewpoints: category of industrial products, research areas related to usability, and three elements of usability defined by user engineering. We also created two cross-classification tables to find distinctions among research areas related to usability. Each table shows examples of the evaluation items of usability.

Category of industrial products: We classified the usability evaluation items based on the category of industrial products shown in Table 1. The classification results are shown in Table 2.

Table 4 Usability evaluation items classified by three usability elements

Categories	Evaluation items
Operability	- Relation between amount of mouse movement and tiredness - Accessible properties of mobile phone buttons - Ratio of mistakes per operation - Relationship between display information, depth of display layer and mean navigation, thinking and mouse operation time - Average speed and movement efficiency of mouse - Achievement time of task - Evaluations of auditory and visual guidance
Cognition	- Relationship between index of difficulty and operation - AIST cognitive-ability-assessment test (cognitive functions) - Response time - Auditory guidance - Visibility and recognition of drug labels - Error analysis: number of operation errors and their recurrence of the same errors - Relationship between display information, depth of display layer and mean navigation, thinking and mouse operation time - Ratio of mistakes per operation
Comfort	- Total manipulation image - Physiological measurement - Impression of spaces - Visibility and recognition of drug labels - Distinction of confused state and evaluations of auditory guidance

Research areas related to usability: Ergonomics, cognitive engineering, and kansei engineering are research areas that are related to usability in user engineering. We classified the usability evaluation items based on these research areas (Table 3).

Three elements of usability defined by user engineering: As we described in Section 2.3, user engineering defines three elements of usability: operability, cognition, and comfort. We classified the evaluation items of usability based on these three elements (Table 4).

Cross-classification: Based on the results of the three above classifications, we created two cross-classification tables to find the usability evaluation features that depend on industrial products, research areas, and the three usability elements. We made the following cross-classifications:
- between industrial products and research area related to usability (Table 5)
- between industrial products and three elements of usability (Table 6)

Table 5 Cross classification: industrial products and research areas

		Research areas		
		Ergonomics	Cognitive	Kansei
Industrial products	IT devices and home appliances	- Accessible properties of mobile phone buttons - Most preferred remote control - Motor abilities of web navigation - Average speed and movement efficiency of mouse - Achievement time of task - Gazing time	- Relationship between index of difficulty and operation time - AIST cognitive-ability-assessment test - Learning ratio of each remote control - Auditory and visual guidance - Eye tracking - Area of useful field of view	- Effect of button properties - Physiological measurement - Eye tracking
	Public devices	- Response time - Utterances and behavior - Auditory guidance - Confirmation times of operation - Confirmation times of operation	- Response time - Utterances and behavior - Auditory guidance - Error analysis - Confirmation Times of operation	- Auditory and visual guidance
	Public equipment and facilities	- Visibility of color coordination hue for stairways	- AIST cognitive-ability-assessment test - Utterances and behavior	- Impression of spaces
	Medical-related products	- Visibility of drug labels		- Visibility and recognition of drug labels

Table 6 Cross classification: industrial products and three elements of usability

		Three elements of usability		
		Operability	Cognition	Comfort
Industrial products	IT devices and home appliances	- Relation between mouse movement of and tiredness of elderly people and younger people - Accessible properties of mobile phone buttons - Ratio of mistakes per operation - Operation and search times	- Relationship between index of difficulty and operation - Ratio of mistakes per operation - AIST cognitive-ability-assessment test - Operation, search, and response times	- Effect of button properties - Total manipulation image - Physiological measurement
	Public devices	- Response time - Operation and search times	- Auditory guidance response time - Analyze errors - Operation and search times	
	Public equipment and facilities	- Visibility of color coordination hue for stairways	- AIST cognitive-ability-assessment test - Visibility of color coordination for stairways	- Impression of spaces
	Medical-related products	-Visibility of drug label	- Visibility of drug labels	

284

4   CONCLUSION AND FUTURE DISCUSSION

In this report, we survey recent research on the industrial products of usability for elderly people. From our classification and analysis results of the usability methods for elderly people, we identified future research on usability evaluations from different viewpoints. From the analysis of our cross-classifications, the number of research related to kansei engineering is less than other areas, including ergonomics and cognitive engineering. Also, comfort related to kansei, one of the three elements of usability, is also less than the other two usability elements. Even though negative feelings for elderly people have been investigated in kansei engineering, we found little research on such positive feelings as "wakuwaku."

Since many elderly people abandon using IT devices, improving the usability of such industrial products is one approach to help elderly people benefit from IT devices. Because research on the positive feelings experienced by elderly people has not been investigated, we must focus on them to support elderly people. As future work, we believe that research on the positive feelings of elderly people will continue to grow and contribute to Japanese aging society.

REFERENCES

H. Shiizuka. 2011. Aging Society and Its Related Problems-Toward Extending Healthy Longevity. *Journal of Japan Society of Kansei Engineering,* 9 3: 131–139.

K. Sagawa. 2011. Accessible Design –Design to address the needs of older persons and persons with disabilities and its international standardization. *Journal of Japan Society of Kansei Engineering,* 9 3: 140-146.

H. Inagaki and Y. Gondo. 2005. Actual conditions of use of information technology, electronic devices in the oldest old. *IEICE Technical Report. Welfare Information Technology*, 105 186: 67-72.

M. Kurosu. 2003. An Introduction to Usability Engineering, *IPSJ Magazine, 44 2:*122-127.

Nielsen, J. 1994. Usability Engineering. Morgan Kaufmann Publishers.

M. Kurosu. et al. 1999. Introduction to User Engineering. Kyoritsu. Pub.

A. Hasizume. 2011. Use of Communication Media among Seniors. *Journal of Japan Society of Kansei Engineering,* 9 3: 153–159.

K. Kajita. et al. 2005. Web browser's interface that considers characteristic of elderly people. *IEICE Technical Report. Welfare Information Technology*, 105 186: 62-67. (in Japanese)

Y.Suzuki, et al. 2004. Web Interface Design based on Elderly User's Cognitive Characteristics (2)-Experimental Reports of Influence of Elderly User's Cognitive Performance on Web Usability. *Proceedings of the Symposium on Human Interface*, 983-988. (in Japanese)

R. Takahashi, et al. 2005. Basic Study on Web Design that is Friendly for Older Adults-Effects of Perceptual, Cognitive and Motor Functions and Display Information on Web Navigation Time. *The Japanese Journal of Ergonomics*, 44 1: 1-13. (in Japanese)

A. Taketomi. H. Takada., and Y. Matsuura. 2008. *JSME annual meeting,* 2008 7:195-196.

H. Akatsu, et al. 2004. Usability Research for the Elderly People. *Oki Technical Review 55,* 71:3.

H. Akamatsu, et al. 2011. Dynamic guidance: synchronized auditory and visual guidance for elderly people: case study of automatic teller machines. *The Japanese Journal of Ergonomics*. 47 3: 96-102. 21

H. Sawashima, et al. 2002. A Massive Usability Test of IT devices for Elderly People Use: A Case Study of Errors for using Automatic Teller Machine (ATM). *The Japanese Journal of Ergonomics*, 38: 244-245. (in Japanese)

M. Kitajima, et al. 2008. Usability of Guide Signs at Railway Stations for Elderly Passengers: Focusing on Planning, Attention, and Working Memory. *The Japanese Journal of Ergonomics*, 44 3: 131-143. (in Japanese)

M. Kwon, et al. 2009. Color Coordination for Stairways with High Visibility for the Elderly People: For Prevention of Falling Accidents. *Bulletin of Japanese Society for Science of Design*, 56 3: 99-108. (in Japanese)

T. Aoto, et al. 2006. Assessment of relief level of living space using immersive space simulator. *IEICE technical report. Welfare Information technology*, 105 684: 31-34. (in Japanese)

A. Izumiya, et al. 2009. The evaluation of pharmaceutical package design for the elderly people. *Correspondences on human interface*, 11 2: 59-62. (in Japanese)

A. Izumiya, et al. 2007. Examination of method of displaying medicine to prevent human error (VII): The evaluation experiment of pharmaceutical package design to administer once a week intended for female senior citizens. *Proceeding of Japan Ergonomics Society 48th Conference*, 366-367. (in Japanese)

K. Ueda, et al. 2009. Usability evaluation of mobile phone for the elderly: The effect of properties of the button. *Proceedings of the Symposium on Human Interface*, 673-676. (in Japanese)

A. Chikawa, et al. 2009. Usability Evaluation of Touch Screen Mobile Phone for the Elderly. *Proceedings of the Symposium on Human Interface*. 677-680.

N. Hara and T. Shida. 2005. Characteristics of Cognitive Aging in Appliance Operation. *Matsushita Technical Journal*. 51 4: 369-373.

Y. Suzuki, et al. 2008. A Study on Human Interface Design of IT-based Equipments Adapted for Elderly Persons' Cognitive Character (1). *Proceedings of the Symposium on Human Interface*. 689-692. (in Japanese)

K. Komine, et al. 2001. Evaluation of Usability of TV Remote Controls for Elderly People in a GUI Environment. *The Journal of the Institute of Image Information and Television Engineers*. 55 10: 1345-1352. (in Japanese)

R. Fukuda. 2009. Difficulties in Use of "high-tech" Home Appliances for Older Users Experimental consideration based on behavior analysis, eye tracking, and physiological measurement during use of rice cookers. *Keio SFC journal*. 9 2: 51-62. (in Japanese)

K. Yamaba, M. Nagata. 2002. Experiments on Visual Characteristics of Elderly. *A publication of Fundamentals and Materials Society*. 122 8: 736-741.

T. Miura, et al. 2007. Aged-related on the mechanism to control visual attention. *The Fundamental Research On Communicational Function Of The Disabled And The Aged Persons*, 155-162.

M. Shino, R. Takatani, M. Kamata. 2011. Proposal of Distinction Method of Confused State Based on Cognition and Manipulation of Older Person in Using Information Devices. *The Transaction of Human Interface Society*. 13 2: 147-15. (in Japanese)

Y. Sakaguchi. 2000. Decision of Character Size and Color for on-Board Displays Based on Human Vision. *R&D Review of Toyota CRDL*, 35 2: 11-18.

CHAPTER 29

Upper Extremity Interaction Modeling for Persons with Advanced Osteoarthritis

David J. Feathers, Ph.D.

Cornell University
Ithaca, New York
djf222@cornell.edu

ABSTRACT

Millions of adults worldwide live with hand osteoarthritis (HOA), and great increases in arthritis-attributable activity limitations (AAAL) are expected in the upcoming decades. Advances in anti-inflammatory and analgesic medications give persons with HOA the opportunity to be more active and the capacity to reduce AAALs by remaining active. OA-appropriate ergonomic design of products and interactive environments can promote activity thereby creating a positive step to address AAALs. This chapter outlines the basic framework of 3D physical interaction modeling and analysis (PIMA) of hand-intensive activities, and centers on issues of data consistency for 3D anatomical landmarks used to dimension HOA for ergonomic applications. 3D measurements for the active OA hand require new considerations for anatomical landmark data collection quality, such as joint inflammation, tenderness and pain, and visual and haptic issues that influence 3D landmark detection and registration. New ways to capture HOA interactive hand strategies for analysis of grasping, pinching, and other hand movement strategies can shape new ergonomic designs. Capturing diversity in PIMA for the active OA hand represents initial steps to establish an evidence-base supporting the development of additional ergonomic design considerations, intervention strategies, diverse digital human models, and new ways to analyze interactive movement for this large and growing population.

Keywords: arthritis, 3D modeling, upper extremity, physical interaction modeling.

INTRODUCTION

Osteoarthritis (OA) is considered to be the most common age-related joint disorder in the world (WHO, 2002; Zhang et al., 2002) and a leading cause of pain and associated disability in many countries in the world (Fransen et al., 2011). Hundreds of millions of older adults worldwide live with hand osteoarthritis (HOA), a progressive disease that does not remit. In the United States, prevalence of HOA is described by clinical and radiographic definitions outlined by the American College of Rheumatology (ACR) classification guidelines (Altman et al., 1990; CDC, 2011). Lawrence et al., 2008 report that 27.2/100 US residents aged 26 years and older have mild, moderate, or severe HOA based on radiographic evidence collected in the Framingham OA study (Zhang et al., 2002). Osteological and rheumatologic changes from HOA correlate poorly with clinical signs (Bagge et al., 1991), which makes symptomatic prevalence rates difficult to estimate (Lawrence et al., 2008). Symptomatic prevalence rates from HOA are generally second only to knee OA (Dillon et al., 2007).

Arthritis-attributable activity limitations (AAAL) are defined by Hootman and Helmick (2006) as, "an answer of "yes" by a patient with doctor-diagnosed arthritis, to the question 'Are you now limited in any way in any of your usual activities because of arthritis or joint symptoms?'" (p. 227). Hootman and Helmick (2006) report from the National Health Interview Survey (NHIS, 2003) that 16.9 million adults in the United States report AAALs, accounting for 7.9% of the total US population in 2003, and is expected to rise to 25 million (9.3% US population) by 2030. The prevalence of individuals reporting AAALs within the growing population of doctor-diagnosed arthritis group is expected to remain at approximately 37% over the next three decades, reflecting a growing US populous reporting limitations of activities.

OA-appropriate ergonomic design of products and interactive environments can invite activity thereby creating a positive step to address AAALs. Ergonomic design intervention strategies require an evidence base that includes both standardized static and functional measurements, and new measures of interaction to examine, analyze, and improve design interventions. This chapter outlines two basic themes of research concerns for the OA population, focusing exclusively on data for individuals with HOA. The first theme considers physical interaction modeling and analysis, which uses 3D data collection of the hand, the artifact, and the hand-artifact interaction. The second theme discusses new concerns for registering 3D surface landmarks of the HOA population which is needed to accurately and consistently measure hand-intensive physical interaction.

Physical Interaction Modeling and Analysis (PIMA) of Hand-Intensive Activities for the Active Arthritic Hand

When asked to perform hand-intensive activities, all individuals will have adaptive strategies that best match their capacities to task requirements through the lens of product or environmental features. The capacity to capture these adaptive

288

strategies in 3D space is now facilitated by advanced measurement systems (e.g. light-based scanners, electromechanical probes, and stereophotogrammetry equipment). Combining traditional human body measurement descriptions and dimensions with new anthropometric technologies has garnered attention from researchers in ergonomic design, clothing and product design, clinical settings, military applications, and environmental design (Bougourd *et al.*, 2000; Feathers, 2004; Kouchi and Mochimaru, 2011; Paquet and Feathers, 2004; Paquette et al., 2000; Sims et al., *in press*; Weinberg *et al.*, 2004; Wong *et al.*, 2008).

Physical Interaction Modeling and Analysis (PIMA) is an effort by researchers at the Digital Anthropometry and Biomechanics Laboratory (DAB Lab) at Cornell University to combine traditional anatomical landmark identification and registration techniques used in standard anthropometry with the multidimensional capacities of new measurement technologies (Feathers, 2009; Feathers, 2011). PIMA has four stages:

<u>Stage 1</u>: *Structural Modeling*: The 3D structure of an individual's anatomy is measured through at least two types of 3D measurement technologies and a traditional method to capitalize on the strengths of each technology. The DAB Lab uses 3D light-based scanning to capture 3D surfaces of both the human participant and the artifacts to which they will be interacting and a 3D electromechanical digitizer to collect anatomical landmark data (using traditional techniques of palpation and visual inspection), and feature data of the artifact. Figure 1 shows the right hand of a HOA participant, a 72-year-old female in standard and non-standard postures. 3D structural modeling efforts in both standard and non-standard postures can directly translate into ergonomic design. Current research in capturing and

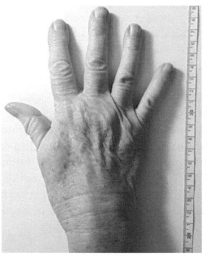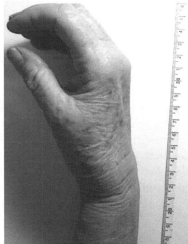

'Flat' Hand, Dorsal View   'Relaxed' Hand, Lateral View

Figure 1   Osteoarthritis of the Hand, dorsal and lateral views. Participant: Female, 72 years old.

creating solid 3D models of comfortable, perhaps more 'natural' hand postures, will benefit healthy populations (Rogers *et al.*, 2008), arthritic populations (Feathers, 2011), and other populations in upcoming studies.

Stage 2: *Functional Modeling*: Traditional measures of hand strength and dexterity are measured and 3D scans are taken during this data collection. This combination allows the researchers to understand normative strength and dexterity tests to be compared to adaptive techniques with an associated quantitative outcome. For example, a power grip dynamometer measurement is taken in a standard manner (Crosby and Wehbe, 1994) within the 3D scanner. Studies with individuals that have HOA report muscle weakness and lower grip strength values than that of non-HOA subjects (Dhara *et al.*, 2009; O'Reilly *et al.*, 1997; Dominick *et al.*, 2005). Alternative methods of generating power grip force, such as pronating or supinating the forearm at the wrist may increase this grip force for some individuals, and compensate for limitations in functional range of motion in the hand (Hume *et al.*, 1990). These alternative or adaptive strategies are collected, scanned, and analyzed as per a specific research protocol.

Stage 3: *Physical Interaction Modeling*: This stage has several steps, starting from physical interaction with basic objects such as a grip cone to complex objects such as multi-touch screens with gesture recognition. This 4D (3D + time) interaction of the individual with a product or environment is captured with either serial 3D scanning images and/or a video-based motion analysis system. Supplemental data, such as forces generated within this interaction (captured with transducers), offer a more complete picture of multi-dimensional interaction. Figure 2 outlines two forms of evidence, photographic and 3D scans of an HOA participant. The task of opening

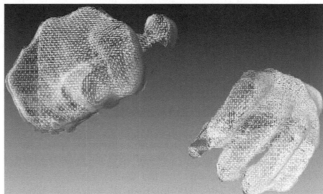

Figure 2 Adaptive strategies balance force requirements, user strength, and task demands. Illustrated here: PIMA of opening a screw-top bottle. **Left**: Photograph of participant with HOA executing this task by altering the product orientation to assist grip strength. **Right**: 3D scan of participant hand positions at the start of this task. Screw-cap is colored green and is loacted in between the right thumb and index finger. The body of the bottle has been removed from view to emphasize hand positioning.

a screw-top bottle can place demands that can bring about alternative strategies. In this case, greater torque can be applied to the bottle top by tilting the body of the bottle so that the right hand (that is manipulating the cap) and upper extremity is experiencing a reduction in ulnar deviation and shoulder abduction. 3D data from standardized and non-standardized hand positions are valuable for preliminary analysis of hand-held and hand-manipulations for fit, functional use, and physical perceived affordances of product interaction. This analysis combines active hand movement strategies that balance user strength, product features, and performance requirements.

The Framingham study notes that writing, handing or fingering small objects are issues present in 54.3% of the HOA sample (Zhang *et al.*, 2002). This self-reported functional limitation data underscores the need for further research in 3D anthropometry and responsive physical interaction data. Rogers *et al.*, 2008 offer a detailed set of 3D landmark-based measurements for healthy hands intended to capture more functional grips and pinches typically adopted by active hands. These detailed landmarks can be captured on the active arthritic hand to assess fit, use, and perceived physical affordances.

Stage 4: *Multi-Dimensional Analysis*:   Matching individual capacities, personal adaptive strategies, and task requirements can offer new ergonomic design solutions. Stage 4 represents the combination and analysis of static, dynamic, and functional use for participants for each research protocol. This combination can be analyzed using existing and novel methods of understanding 3D movement strategies (Armstrong *et al.*, 2009; Chaffin, 2002; Jacquier-Bret *et al.*, 2012) and creating new digital human models/simulations of the active arthritic hand.

A 3D anthropometry dataset for the HOA population in the United States has been assembled (Feathers, 2011) and additional data will be collected in China in late spring 2012. This research includes techniques to consistently measure the HOA hand, and collect 3D and 4D physical interaction data with products and interactive environments. The next section investigates issues of 3D data quality in the theme of interactive research.

Data Quality Concerns for Capturing 3D Anatomical Landmarks of the Active Arthritic Hand

Three-dimensional anatomical landmarks have the capacity to characterize structural and functional aspects of the hand and hand-object interaction by capturing specific points on the skin surface, most typically associated with a synovial joint. Recent 3D anthropometric studies have investigated landmarking accuracy for healthy populations (Feathers, 2009; Kouchi and Mochimaru, 2011). Research on errors in 3D landmarking and measurements for arthritic populations have begun (Feathers, 2011), and data quality assurance for anatomical landmarks for individuals with HOA will provide responsive measurements for appropriate ergonomic design of products and environments.

Consistent and reliable anatomical landmark identification, digitization, and resultant measurement of a 3D landmark/delimitation (3DLD) relies on information gathered from bottom-up and top-down processing simultaneously (Feathers, 2004). Visual cues of ink marks, stickers or bony protrusions and haptic palpation assist the anthropometrist in digitizing an anatomical landmark consistently (Kouchi and Mochimaru, 2011). However, with a progressive disease such as HOA, haptic and visual cues change within an individual, and are variable in the presentation across individuals. Figure 3 illustrates the impact visual and haptic cues have on the registration of anatomical landmarks of the thumb (metacarpophalangeal joint I and interphalangeal joint I).

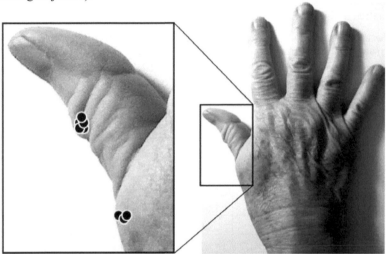

Dorsal View of an HOA Participant Hand

Figure 3 Repeated digitizations of the active arthritic hand, dorsal view of female participant, 72 years old. Anatomical skin-surface digitizations of the base of the first metacarpophalangeal joint and the interphalangeal joint are represented here as black dots with a light colored outline, symbolizing 3D variation present in digitizations.

The structural changes in musculoskeletal anatomy across the stages of HOA have been well-documented (e.g. Goldin *et al.*, 1976; Haugen *et al.*, 2011; Kwok, *et al.*, 2011; Lawrence *et al.*, 2008; Seltzer, 1943; Zhang *et al.*, 2002). The relative rates of degeneration for specific hand joints are known (Hochberg *et al.*, 1994; Kwok *et al.*, 2011; Sowers, *et al.*, 1991) as is the influence of OA degeneration on functional activities (e.g. Liang and Jette, 1981; Sowers *et al.*, 1991). These structural and functional datasets can provide the basis of predictions for digitization consistency in 3D space relative to the skin surface under repeated measurement conditions. Table 1 relates landmark-specific concerns for symptomatic prevalence, radiographic evidence of joint space narrowing, and the potential impact on 3D digitization of anatomical landmarks of the hand.

Graded 3D digitization consistency is outlined in the right-most column (see Table 1), but there are other anatomically-related issues that may add to the complexity of this prediction algorithm- including gross muscle atrophy and variable diarthodial presentations (i.e. swelling of a particular joint) across diurnal endogenous and exogenous changes related to activity of the OA hand. For example, the exogenous stresses placed on the hand during a hand-intensive activity such as knitting may reduce landmark digitization consistency for specific joints of the hand, such as the metacarpophalangeal joint of the thumb, due to an inflammation response from the intense activity. A history of hand use should accompany a participant profile, such as the hand activity level (i.e. ACGIH Worldwide; 2001; Franzblau *et al.*, 2005), in addition to clinical HOA indices such as the AUSCAN Osteoarthritis Hand Index (Bellamy *et al.*, 2002), and other clinical evaluation tools.

Table 1 Selected Hand Anatomical Landmarks, HOA Symptomatic Prevalence, Progressive Joint Space Narrowing, and Predicted 3D Digitization Consistency

Anatomical Joint and Landmark (see NOTES)	Prevalence[1] of Symptomatic HOA	JSN[2] of Radiographic HOA				Graded Impact of HOA on 3D Digitization[3]
		0	1	2	3	
MCP I	1.7	1.47	1.12	0.54	-.-	0-1
DIP I	7.5	0.95	0.72	0.39	0.16	3-5
MCP II Lateral	1.1	1.47	1.12	0.54	-.-	0-2
MCP II	1.1	1.47	1.12	0.54	-.-	3-5
PIP II	6.5	1.05	0.79	0.47	0.18	2-4
DIP II	11.9	0.95	0.72	0.39	0.16	3-5
MCP III	1.1	1.47	1.12	0.54	-.-	0-1
PIP III	8.9	1.05	0.79	0.47	0.18	2-5
DIP III	11.9	0.95	0.72	0.39	0.16	4-5
MCP IV	0.3	1.47	1.12	0.54	-.-	0-1
PIP IV	8.0	1.05	0.79	0.47	0.18	2-5
DIP IV	8.5	0.95	0.72	0.39	0.16	3-5
MCP V Lateral	0.5	1.47	1.12	0.54	-.-	0-1
MCP V	0.5	1.47	1.12	0.54	-.-	0-1
PIP V	6.0	1.05	0.79	0.47	0.18	2-4
DIP V	9.6	0.95	0.72	0.39	0.16	3-5

NOTES: Landmark list represents: dorsal aspect of the respective landmark, unless otherwise noted. MCP=metacarpophalangeal joint; PIP= proximal interphalangeal joint; DIP= distal interphalangeal joint.
[1] From: Zhang et al., 2002: Figure 1, p. 1024: reporting in (%) of symptomatic hand osteoarthritis among the participants in the Framingham Study. *Sample*: women, right hand, n=663. *(Self-report of participants- mark the joints that have 'pain, aching, or stiffness').* Prevalence rates assumed lateral landmarks for MCP II and MCP V.
[2] From: Kwok et al., 2011: JSN= Joint space narrowing (mm), scored by OARSI atlas. No digit-specific data available, averages displayed across each digit.
[3] Hypothesized impact in 3D anatomical landmark digitization consistency as compared to a non-HOA control group: Grades (0-5): 0= minimal change in consistency (< 1mm); 1= 1mm-1.9mm; 2= 2.0mm-2.9mm; 3= 3.0mm-3.9mm; 4= 4.0mm-4.9mm; 5=5.0+mm).

CONCLUSIONS

OA-appropriate ergonomic designs can serve the overarching goal of reducing AAALs. To create OA-appropriate ergonomic designs, consistent and reliable 3D measurement and 4D interactive modeling of the OA population must be collected and analyzed. Future design improvements will be shaped by collecting these unique grasping, pinching, and opposition techniques for persons with HOA brought about by changes due to inflammation, bone remodeling, muscular weakness, and atrophy. Further, the progression of HOA should be accounted for in future designs, including designing in response to research evidence that DIP degeneration occurs at a faster rate than in PIP joints (Hochberg *et al.*, 1993; Sowers, *et al.*, 1991). Ergonomic design can accommodate and perhaps mitigate this progressive disease trajectory. Effective ergonomic design should be evaluated along with advances in anti-inflammatory and analgesic medications which offer persons with OA the opportunity to be more active and offer the capacity to reduce AAALs by remaining active.

ACKNOWLEDGMENTS

Funding for the phase one physical interaction modeling study on the Active Arthritic Hand was provided by the Department of Design and Environmental Analysis, College of Human Ecology, Cornell University.

REFERENCES

ACGIH Worldwide. 2001. Hand Activity Level TLV.
Altman, R. 1991. Classification of disease: osteoarthritis. *Seminars in Arthritis and Rheumatism*, 20(6): 40-47.
Altman, R., Alarcon, G., Appelrouth, D., Bloch, D., Borenstein, D., Brandt, K., Brown, C., Cooke, T., Daniel, W., Gray, R., Greewald, R., Hochberg, M., Howell, D., Ike, R., Kapila, P., Kaplan, D., Koopman, W., Longley, S., McShane, D., Medsger, T., Michel, B., Murphy, W., Osial, T., Ramsey-Goldman, R., Rothschild, B., Stark, K., Wolfe, F. 1990. The American College of Rheumatology criteria for the classification and reporting of osteoarthritis of the hand. Reprinted for: *Arthritis and Rheumatism*, 33(11): 1601-1610.
Armstrong, T., Best, C., Bae, S., Choi, J., Grieshaber, D., Park, D., Woolley, C., Zhou, W. 2009. Development of a kinematic hand model for study and design of hose installation. In: Duffy, V. (ed.) Digital Human Modeling. HCII 2009, LNCS 5620. pp. 85-94. Berlin: Springer-Verlag.
Bagge, E., Bjelle, A., Eden, S., Svanborg, A. 1991. Osteoarthritis in the elderly: clinical and radiological findings in 79 and 85 year olds. *Annals of Rheumatic Diseases*, 50: 535-539.
Bellamy, N., Campbell, J., Haraoui, B., Gerecz-Simon, E., Buchbinder, R., Hobby, K., MacDermid, J. 2002. Clinimetric properties of the AUSCAN Osteoarthritis Hand

Index: an evaluation of reliability, validity and responsiveness. *Osteoarthritis and Cartilage*, 10: 863-869.

Bougourd, J., Dekker, L., Ross, P., Ward, J. 2000. A comparison of women's sizing by 3D electronic scanning and traditional anthropometry. *The Journal of Textile Institute: Part 2*, 91(2): 163-173.

Center for Disease Control (CDC) 2011. Osteoarthritis: Prevalence. Accessed on October 2, 2011. http://www.cdc.gov/arthritis/basics/osteoarthritis.htm

Chaffin, D. 2002. On simulating human reach motions for ergonomics analysis. *Human Factors of Ergonomics Manufacturing*, 12(3): 235-247.

Crosby, C., Wehbe, M. 1994. Hand strength: normative values. *Journal of Hand Surgery*, 19(4): 665-670.

Dhara, P., De, S., Pal, A., Sengupta, P., Roy, S. 2009. Assessment of hand grip strength of orthopedically challenged persons affected with upper extremity. *Journal of Life Science*, 1(2): 121-127.

Dillon, C., Hirsch, R., Rasch, E., Gu, Q. 2007. Symptomatic hand osteoarthritis in the United States: prevalence and functional impairment estimates from the third U.S. National Health and Nutrition Examination Survey, 1991-1994. *American Journal of Physical Medicine and Rehabilitation*, 86(1): 12-21.

Fransen, M., Bridgett, L., March, L., Hoy, D., Penserga, E., Brooks, P. 2011. The epidemiology of osteoarthritis in Asia. *International Journal of Rheumatic Diseases*, 14: 113-121.

Feathers, D. 2004. Digital human modeling and measurement considerations for wheeled mobility device users. *SAE Transactions*, 113(1): 70-77.

Feathers, D. 2009. Mitigating anthropometric measurement variation: Applied tolerances for static anthropometric landmark digitizations in three-dimensional space. In: *The 17th World Congress on Ergonomics: An International Ergonomics Association Triennial Congress*.

Feathers, D. 2011. 3D Anthropometry and physical interaction modeling for persons with arthritis. In: *Proceedings of the 2nd International Conference on 3D Body Scanning Technologies*.

Franzblau, A., Armstrong, T., Werner, R., Ulin, S. 2005. A cross-sectional assessment of the ACGIH TLV for hand activity level. *Journal of Occupational Rehabilitation*, 15(1): 57-67.

Goldin, R., McAdam, L., Louie, J., Gold, R., Bluestone, R. 1976. Clinical and radiological survey of the incidence of osteoarthrosis among obese patients. *Annals of Rheumatic Diseases*, 35: 349-353.

Haugen, I., Bøyesen, P., Slatkowsky-Christensen, B., Sesseng, S., van der Heijde, D., Kvien, T. 2011. Associations between MRI-defined synovitis, bone marrow lesions and structural features and measures of pain and physical function in hand osteoarthritis. *Annals of Rheumatic Diseases*, doi:10.1136/annrheumdis-2011-200341.

Hochberg, M., Lethbridge-Cejku, M., Scott, W., Plato, C., Tobin, J. 1993. Obesity and osteoarthritis of the hands in women. *Osteoarthritis and Cartilage*, 1: 129-135.

Hootman, J., Helmick, C. 2006. Projections of US prevalence of arthritis and associated activity limitations. *Arthritis and Rheumatism*, 54(1): 226-229.

Hume, M., Gellman, H., McKellop, H., Brumfield, R. 1990. Functional range of motion of the joints of the hand. *The Journal of Hand Surgery*, 15A(2): 240-243.

Jacquier-Bret, J., Gorce, P., Rezzoug, N. 2012. The manipulability: a new index for quantifying movement capacities for upper extremity. *Ergonomics*, 55(1): 69-77.

Kouchi, M., Mochimaru, M. (2011) Errors in landmarking and the evaluation of the accuracy of traditional and 3D anthropometry. *Applied Ergonomics*. Vol. **42**: 518-527.

Kwok, W., Kloppenburg, M., Rosendaal, F., van Meurs, J., Hofman, A., Bierma-Zeinstra, S. 2011. Erosive hand osteoarthritis: Its prevalence and clinical impact in the general population and symptomatic hand osteoarthritis. *Annals of Rheumatic Diseases*, 70: 1238-1242.

Lawrence, R., Felson, D., Helmick, C., Arnold, L., Choi, H., Deyo, R., Gabriel, S., Hirsch, R., Hochberg, M., Hunder, G., Jordan, J., Katz, J., Kremers, H., Wolfe, F. for the National Arthritis Data Workgroup. 2008. Estimates of the prevalence of arthritis and other rheumatic conditions in the United States. Part II. *Arthritis Rheum* 2008;58(1):26–35.

Liang, M., Jette, A. 1981. Measuring functional ability in chronic arthritis. *Arthritis and Rheumatism*, 24(1): 80-86.

National Health Interview Survey (NHIS) 2003. Public Access data documentation. URL: http://www.cdc.gov/nchs/data/series/sr_10/sr10_235.pdf

O'Reilly, S., Jones, A., Doherty, M. 1997. Muscle weakness in osteoarthritis. *Current Opinions in Rheumatology*, 9(3): 259-262.

Paquet, V., Feathers, D. 2004. An anthropometric study of manual and powered wheelchair users. *International Journal of Industrial Ergonomics*, 33(3): 191 - 204.

Paquette, S., Brantley, J., Corner, B., Li, P., Oliver, T. 2000. Automated extraction of anthropometric data from 3D images. *Proceedings of the IEA 2000/HFES 200 Congress*, 6: 727-730.

Dominick, K., Jordan, J., Renner, J., Kraus, V. 2005. Relationship of radiographic and clinical variables to pinch and grip strength among individuals with osteoarthritis. *Arthritis and Rheumatism*, 52: 1424-1430.

Rodgers, M., Barr, A., Kasemsontitum, B., Rempel, D. 2008. A three-dimensional anthropometric solid model of the hand based on landmark measurements. *Ergonomics*, 51(4): 511-526.

Seltzer, C. 1943. Anthropometry and arthritis I: Differences between rheumatoid and degenerative joint disease: Males. *Medicine*, 22(2): 163-188.

Sims, R., Marshall, R., Gyi, D., Summerskill, S., Case, K. 2011. Collection of anthropometry from older and physically impaired persons: Traditional methods versus TC^2 body scanner. *International Journal of Industrial Ergonomics*. (in press) 1-8.

Sowers, M., Zobel, D., Weissfeld, L., Hawthorne, V., Carman, W. 1991. Progression of osteoarthritis of the hand and metacarpal bone loss: A twenty-year follow-up of incident cases. *Arthritis and Rheumatism*, 34: 36-42.

Weinberg, S., Scott, N., Neiswanger, K., Brandon, C., Marazita, M. 2004. Digital three-dimensional photogrammetry: Evaluation of anthropometric precision and accuracy using a Genex 3D camera system. *The Cleft Palate-Craniofacial Journal*, 41: 507-518.

Wong, J., Oh, A., Ohta, E., Hunt, A., Rogers, G., Mulliken, J., Deutsch, C. 2008. Validity and reliability of craniofacial anthropometric measurement of 3D digital photogrammetric images. *The Cleft Palate-Craniofacial Journal*, 45: 232-239.

World Health Organization (WHO) 2002. *World Health Report 2002. Reducing Risks, Promoting Healthy Life*. Geneva, World Health Organization.

Zhang, Y., Niu, J., Kelly-Hayes, M., Chaisson, C., Aliabadi, P., Felson, D. 2002. Prevalence of symptomatic hand osteoarthritis and its impact on functional status among the elderly. *American Journal of Epidemiology*, 156: 1021-1027.

CHAPTER 30

Ergonomic Analysis to Support Surgery-Averse Individuals that Live with Chronic Pain

David J. Feathers, Ph.D.

Cornell University
Ithaca, New York
djf222@cornell.edu

ABSTRACT

Chronic pain is costly and frequently disabling, especially in later life. When an individual reports pain while performing a daily activity, the medical condition is often the primary target for an intervention. If the patient is surgery-averse, the palliative benefits of pain medication may temporarily facilitate performance but do not address the potentially hazardous pain avoidance behavior that can injure musculoskeletal structures involved in compensation, nor does it simplify performance expectations relative to medication type and dosage. This chapter outlines a task analytic approach to activities of daily living (ADL) performance in the home environments of older adults who are surgery-averse and live with chronic pain from osteoarthritis (OA). Two areas of focus are considered: kinematic analysis of compensatory upper extremity behavior arising from hip and knee pain, and the impact perceived physical affordances of the built environment have on the compensatory behaviors that assist in ADL performance. New physical interaction considerations for supportive designs interventions intended to mitigate pain and enhance function when performing everyday activities are outlined. Analyzing the task-specific, pain-relevant outcomes concerning the match between the physical environment and user capabilities addresses the interactive demands of ADLs for persons living with chronic pain and creates new opportunities for ergonomic intervention strategies to support the growing OA population.

Keywords: chronic pain, osteoarthritis, built environment, ergonomic design.

INTRODUCTION

In February 2010, the CDC and Arthritis Foundation launched, "The National Public Health Agenda for Osteoarthritis," which outlined an action plan to take the steps necessary to reduce the burden of osteoarthritis (OA), which places severe limits on daily activities and quality of life for over 27 million Americans (CDC, 2010). This agenda listed recommendations centering on disease self-management, exercise and weight management, evidence-based interventions (including workplace environment improvements to reduce OA onset and progression), and enhanced research and evaluation tools. Traditional tools of ergonomic assessment and analysis such as REBA (Hignett and McAtamney, 2000), The Strain Index (Moore and Garg, 1995), and RULA (McAtamney and Corlett, 1993); in addition to ergonomic analysis tools for inclusive design (e.g. Inclusive Design Toolkit: Clarkson, et al., 2007) can offer a reliable evidence-base for the OA population in response to this public health agenda. Additionally, longitudinal datasets such as the Baltimore Longitudinal Study on Aging (Verbrugge, et al., 1996) provide needed data on life-span function changes to be accommodated by responsive ergonomic design intervention strategies.

Gathering and analyzing evidence for functional assessment of elderly and OA populations has been subject of many review articles (e.g. Collier, 1988; Liang and Jette, 1981). Similarly, OA outcomes measures range from joint-specific, self-administered questionnaires, such as the "Knee Injury and Osteoarthritis Outcome Score" (KOOS) (Roos, et al., 1998), to the WOMAC instrument which has been studied in relationship with drug therapy and OA pain assessment (Bellamy, et al., 1988; Bellamy, 1989). These outcomes measurement tools allow for reliable metrics of clinical performance and reliable ways to assess functional capacities for this population, and can be used with non-disease specific outcomes measurement tools such as the Barthel ADL Index (Mahoney and Barthel, 1965) to expand the evidence base for ergonomic interventions.

Osteoarthritic Pain of the Hip and Knee During Daily Activities

Debilitating consequences of chronic pain include progressive disability, self-imposed limitations of physical activity, and chronic pain during activities of daily living (Buer and Linton, 2002). Individuals with chronic pain self-regulate their motor functional response to the iterative and instantaneous matching of capabilities across changing task demands while concurrently self-monitoring and adjusting to changes in pain level (Flordyce, G., 1976; Flordyce et al., 1982; Leeuw, et al., 2007; Lethem et al., 1983; Kerns, et al., 1991; Turk and Okifuji, 2002; Vlaeyen et al., 1995). Chronic pain can influence performance of daily activities such as ambulation (Al-Zahrani and Bakheit, 2002; Rejeski et al., 1996) and self-feeding (Voog et al., 2003), as well as basic movements like lifting tasks (Hodges and Moseley, 2003; Hodges and Richardson, 1996).

Chronic pain has the capacity to influence the calibrated response of changing task demands of basic activity of daily living (ADL) and instrumental activities of daily living (IADLs). When an individual reports pain during an ADL/IADL, the medical condition is often the primary target for an intervention. For example, an individual with OA of the hip reports pain during bathing activities, specifically when they are attempting to get out the tub in their home environment. They will likely receive pain medication to reduce inflammation and joint replacement surgery may be indicated. If the patient is averse to surgery, the palliative benefits of the medication may temporarily facilitate ADL/IADL task completion but do not address two (of many) outstanding issues:

1. The potentially hazardous musculoskeletal movements in response to pain that can injure alternative/compensatory support structures, such as the upper extremity, as it assists in the transfer out of the tub; and,
2. Calibrating personal ADL/IADL performance expectations relative to medication type and dosage. Pain medication in this cohort can have serious and debilitating side effects (Bjarnason, et al., 1993; Laine, 2003).

A deeper understanding of the interaction between the older user's capabilities (e.g. range of motion at the hip and knee) and the construction of the living environment (e.g. height of bathtub wall) through evidence-based studies will benefit individuals with functional limitations, including individuals with OA. In the case of the bathing example, a bathtub wall height greater than *n inches* will cause external rotation of the hip greater than *n degrees* that elevates pain levels for *n percentage* of this OA cohort. Understanding and reporting pain-relevant outcomes concerning the match between the tasks performed in a specific physical environment and user capabilities, addresses the interactive demands of ADL/IADL for persons living with chronic pain and creates new opportunities for ergonomic practitioners to support their clients.

TASK ANALYSIS OF ADL/IADLS OF OLDER ADULTS WHO ARE SURGERY-AVERSE AND LIVE WITH CHRONIC PAIN FROM OA

Functional limitations for older adults have been well documented (e.g. Clark *et al.*, 1990; Collier, 1988; Czaja, *et al.*, 1993; Fisk and Rogers, 1997; Rogers, *et al.*, 1998; Seidel *et al.*, 2009). These limitations have the capacity to enhance personal or compensatory strategies when an individual is performing ADL/IADL. These personal strategies involve a match between their physical capabilities and specific physical task demands of daily activities in the home environment (Clark *et al.*, 1990; Czaja, *et al.*, 1993; Grandjean, 1973; Lawton, 1990; Rogers, *et al.*, 1998;).

Comprehensive task analytic studies have been performed for older adults which combined observations, video analysis, and an ADL survey to understand the physical demands in a multitude of environments (e.g. Clark *et al.*, 1990; Czaja, *et al.*, 1993). Environmental barriers and ADL performance have been investigated using human factors approaches (e.g. Faletti, 1984; Lawton, 1990); full scale modeling of the home environment (Feathers and Steinfeld, 2008); computer-aided simulation and modeling such as HADRIAN (Marshal, et al., 2010), architecture and environmental psychology (Steinfeld and Danford, 1999).

Recent studies have begun to bridge together precise kinematic data with task-specific ADL/IADL performance (Alexander, *et al.*, 2001; Aizawa *et al.*, 2010; Henmi, *et al.*, 2006; Magermans, *et al.,* 2005). For example, in a study of twenty healthy adults, Aizawa *et al.*, 2010 investigated three-dimensional movement patterns of the upper extremity (shoulder, elbow and wrist) for a series of movements related to ADL/IADL tasks (e.g. touching opposite axilla- for bathing) and performing an ADL activity (combing hair; pouring water into a glass). The majority of these studies use able-bodied populations, but can be compared with kinematic studies of the OA population for basic activities (Hicks-Little, *et al.,* 2009) and help shape future research steps for diverse studies of ADL/IADL performance.

Surgery-averse individuals reporting chronic pain of the hip and knee from OA require environmental accommodations as a strategy to address pain levels and reduce dependence on palliative and anti-inflammatory medications. To address environmental accommodations for persons living with chronic pain from OA of the hip and/or knee, a case study from an on-going three-stage research study is outlined. Stage 1 offers a detailed survey of home environments for persons living with chronic pain living with OA of the hip and knee. This survey is given by occupational therapists that care for the individuals on a regular basis. Stage 2 explores targeted kinematic and biomechanical studies in a laboratory environment based on task analysis of home activities. These investigate individual differences across age-matched, abled-bodied adults and the OA chronic pain cohort. Design interventions are placed *in situ* for stage 3. Post-intervention follow-ups are performed with the occupational therapists in set durations.

The following case study example highlights two research considerations that are primarily associated with stage 1 observation and will be a part of stage 2 analysis and interpretation:
1) Kinematic analysis of compensatory upper extremity behavior arising from hip and knee pain during a specific daily activity: preparing a meal;
2) The impact perceived physical affordances of the built environment have on the compensatory behaviors that assist in ADL performance.

Kinematics of Preparing a Meal

Preparing a meal is a physically-demanding task, requiring the individual to ambulate with hand-held objects, assume awkward postures and perform one-handed grasping of heavy, liquid-filled objects. Precision is needed for food preparation and strength is requested under time pressure. Capability limitations in strength, dexterity, balance for older adults performing home-related tasks have been studied from micro or with-in task activities, such as the 180-degree standing turn (Meinhart-Shibata, *et al.*, 2005), which is performed when retrieving objects out of an oven or refrigerator, to longitudinal studies of functional loss for older adults performing the IADL of cooking (Seidel, *et al.,* 2009).

Complicating the translation from assessment studies to real-world user trial studies are issues regarding compensatory behavior during the task. For example, a 72 year-old women with chronic pain from OA of the left hip was observed in her home environment as she prepared and cooked canned soup in her galley-style apartment kitchen in New York City. Task analysis showed three actions that exhibited pain-related compensatory behavior during this task.

Compensatory sequence one: retrieval of sauce pan from bottom drawer (figure 1A). Due to space constraints, the sauce pan is located on the bottom drawer of the lower cabinet with the handle located 14 inches off the floor. The participant could not bend her left hip due to OA pain, rather, she aligned herself perpendicular to the lower handle and performed a right lateral bend at the torso with an extended and slightly abducted right arm to open the drawer. Her pot and saucer handles are always placed facing up, which facilitated retrieval.

Figure 1A: Kitchen Drawers   Figure 1B: Kitchen Stove-Top

Compensatory sequence two: After pouring the liquid contents in the sauce pan, the participant maneuvered to place the sauce pan on her stove-top (see figure 1B). The kitchen countertop is in-line with the stove-top, which required the participant to transfer the sauce pan laterally to the stove-top. This can be performed with either abduction of the arm at the shoulder or a right pivot of the body to be placed in front of the stove. Since she cannot place too much weight on her left lower extremity, and preferred not to abduct her arm as the combined weight of the soup and pan was too heavy, she preferred to pivot with her right leg to be at an angle to the stove. She achieved this pivot with a combined push with her right hand placed on the edge of the countertop (internal rotation of the arm) and dorsiflexion her right foot to 'spin' on her heel, excluding her left leg from any weight-bearing activity.

Compensatory sequence three: Cleaning the sauce pan required a lateral movement approximately 5.5 feet to the left of the stove. This transfer was achieved by several forward strides with the right leg, accompanied by dragging of the sauce pan across the countertop with the left hand. The right hand acted as a support during the transfer process, with the palm flat on the surface of the counter, wrist in hyperextension. When at the sink, the participant reached with her right hand for the handle 22" from the countertop edge.

Figure 2: Kitchen Sink

The Impact Perceived Physical Affordances Have on ADL Compensatory Behaviors

Functional assessment in terms of affordances and perceived affordances within the built environment have been studied in a variety of areas such as stair climbing/ambulation in adults (Bertucco and Cesari, 2009; Konczak, *et al.*, 1992) and motor functional development in children (Gabbard, *et al.*, 2008). To date however, there have been no studies that have coded the built environment and associated perceived affordances in terms of ADL physical interaction for persons living with chronic arthrogenic and/or osteogenic pain from OA. This level of coding can be critical for assessing ergonomic design interventions.

Perceived physical affordances of the built environment played an important role in the development of compensatory behaviors that assisted in ADL performance. The three compensatory strategies outlined above were examples of environment-disease-function/ability-task actions that allowed the participant to perform a musculoskeletally demanding task. For sequence one, it was observed and reported by the participant that the handle shape played a critical role in opening the lower drawer. It was observed that she used the tip of her finger to reach and begin to open the drawer. She later reported that, at times, she will lean her arm and body on the side of the countertop to rest during the reaching task. Sequence two demonstrated that the thickness of the countertop allowed the participant to push with her hand to give her body a turn. The flatness of the countertop in sequence three supplied hand support in compensation for her lower extremity issues. She was observed using other horizontal surfaces to support her ambulation throughout her apartment.

CONCLUSIONS

The case study outlined new physical interaction considerations which can lead to supportive designs interventions intended to mitigate pain and enhance function when performing everyday activities. Coding physical environmental interactions during ADL/IADL activity to include perceived physical affordances of environmental features has many benefits, such as increased detail of physical interaction descriptions for better design interventions, and understanding where potentially hazardous compensatory mechanisms could be originating from and designed out. Designing unambiguously safe use can be informed by this detailed coding and allow for reduced pain and enhanced function (e.g. positioning and curvature of a grab bar to reduce radial deviation and afford improved hand placement).

Activity for persons with OA has been shown to positively influence independence and reduce pain for persons with arthritis (Alexander, *et al.*, 2001; Miller, *et al.,* 2003; Murphy, *et al.*, 2008; Suomi and Collier, 2003). Understanding compensatory behaviors can assist in supporting movement strategies, which can influence activity for persons with chronic pain. Designing to support safe strategies of movement can be supported with this course of research.

ACKNOWLEDGMENTS

Pilot research data was collected by researchers at Weill Cornell Medical Center (L.S., R.B., T.D.). This research was partially funded through an NIH pilot grant from the Cornell Roybal Center for Translational Research on Aging and the Translational Research Institute on Pain in Later Life at Cornell University.

REFERENCES

Aizawa, J., Masuda, T., Koyama, T., Nakamaru, K., Isozaki, K., Okawa, A., Morita, S. 2010. Three-dimensional motion of the upper extremity joints during various activities of daily living. *Journal of Biomechanics*, 43: 2915-2922.

Alexander, N., Galecki, A., Grenier, M., Nyquist, L., Hofmeyer, M., Grunawalt, J., Medell, J., Fry-Welch, D. 2001. Task-specific resistance training to improve the ability of activities of daily living-impaired older adults to rise from a bed or a chair. *Journal of the American Geriatrics Society*, 49: 1418-1427.

Bellamy N, Buchanan WW, Goldsmith CH, Campbell J and Stitt LW. Validation study of WOMAC: A health status instrument for measuring clinically important patient relevant outcomes to antirheumatic drug therapy in patients with osteoarthritis of the hip or knee. The Journal of Rheumatology. 1988;15:1833-1840.

Bellamy N. Pain assessment in osteoarthritis: experience with the WOMAC osteoarthritis index. Seminars in Arthritis and Rheumatism. 1989;18(4 Suppl 2):14-17.

Bertucco, M., Cesari, P. 2009. Dimensional analysis and ground reaction forces for stair climbing: Effects of age and task difficulty. *Gait & Posture*, 29: 326-331.

Bjarnason, I., Hayllar, J., MacPherson, A., Russell, A. 1993. Side effects of non-steroidal anti-inflammatory drugs on the small and large intestine in humans. *Gastroenterology*, 104(6): 1832-1847.

Brooks, R., Callahan, L., Pincus, T. 1988. Use of self-report activities of daily living questionnaires in osteoarthritis. *Arthritis Care and Research*, 1(1): 23-32.

Buer, N., Linton, S. 2002. Fear-avoidance beliefs and catastrophizing: occurrence and risk factor in back pain and ADL in the general population. *Pain*, 99: 485-491.

Clark, M., Czaja, S., Weber, R. 1990. Older adults and daily living task performance. *Human Factors*, 32: 537-549.

Clarkson, P. J., Coleman, R., Hosking, I., Waller, S. 2007. *Inclusive design toolkit*. Cambridge, UK: University of Cambridge, Engineering Design Centre.

Clarkson, P. J., Coleman, R., Keates, S., Lebbon, C. 2003. Inclusive Design: Design for the Whole Population. London: Springer-Verlag.

Collier, I. 1988. Assessing functional status of the elderly. *Arthritis Care and Research*, 1(1): 45-52.

Czaja, S., Weber, R., Nair, S. 1993. A human factors analysis of ADL activities: A capability-demand approach. *Journals of Gerontology*, 48: 44-48.

Faletti, M. 1984. Human factors research and functional environments for the aged. In: Altman, I., Lawton, M., Wohlwill, J. eds. Elderly People and the Environment. New York: Plenum Press. pp. 191-237.

Feathers, D., Steinfeld, E. 2008. Subjective ratings of accessibility using full-scale bathroom environments. In: *Proceedings of the Human Factors and Ergonomics Society Annual Meeting*, 723-727.

Fisk, A., Rogers, W. 1997. Handbook of Human Factors and the Older Adult. San Diego, CA: Academic.

Fordyce, G. 1976. Behavioral concepts in chronic pain illness. Mosby, St. Louis.

Fordyce, G., Shelton, J., Dundore, D. 1982. The modification of avoidance learning pain behaviors. *Journal of Behavioral Medicine*, 5: 405–414.

Gabbard, C., Caçola, P., Rodrigues, L. 2008. A new inventory for assessing affordances in the home environment for motor development. *Early Childhood Education Journal*, 36: 5-9.

Grandjean, E. 1973. Ergonomics in the Home. New York: John Wiley.

Guccione, A., Jette, A. 1988. Assessing limitations in physical function in patients with arthritis. *Arthritis Care and Research*, 1(3): 170-176.

Henmi, S., Yonenobu, K., Masatomi, T., Oda, K., 2006. A biomechanical study of activities of daily living using neck and upper limbs with an optical three-dimensional motion analysis system. *Modern Rheumatology*, 16(5), 289–293.

Hicks-Little, C., Peindl, R., Hubbard, T., Scannell, B., Springer, B., Odum, S., Fehring, T., Cordova, M. 2011. Lower extremity joint kinematics during stair climbing in knee osteoarthritis. *Medicine & Science in Sports & Exercise*, 43(3): 516-524.

Hignett, S., McAtamney, L. 2000. Rapid entire body assessment (REBA). *Ergonomics*, 31: 201-205.

Hlatky, M., Boineau, R., Higginbotham, M., Lee, K., Mark, D., Califf, R., Cobb, F., Pryor, D. 1989. A brief self-administered questionnaire to determine functional capacity (The Duke Activity Status Index). *The American Journal of Cardiology*, 64(10): 651-654.

Jette, A. 1980. Functional status index: Reliability of a chronic disease evaluation instrument. *Archives of Physical Medicine and Rehabilitation*, 61(9): 395-401.

Kerns, R., Haythornwaite, J., Rosenberg, R., Soutwick, S., Giller, E., Jacob, M. 1991. The pain behavior checklist (PBCL): Factor structure and psychometric properties. *Journal of Behavioral Medicine*, 14(2): 155-167.

Konczak, J., Meeuwsen, H., Cress, M. 1992. Changing affordances in stair climbing: The perception of maximum climbability in young and older adults. *Journal of Experimental Psychology: Human Perception and Performance*, 18(3): 691-697.

Laine, L. 2003. Gastrointestinal effects of NSAIDs and Coxibs. *Journal of Pain and Symptom Management*, 25(2): 32-40.

Lawton, M. 1985. The impact of the environment on aging and behavior. In: Birren, J., Schaie, W. eds. Handbook of the Psychology of Aging. (2nd ed.) New York: Van Nostrand Reinhold. pp. 276-301.

Leeuw, M., Goossens, M., Linton, S., Crombez, G., Boersma, K., Vlaeyen, J. 2007. The Fear-Avoidance Model of Musculoskeletal Pain: Current state of scientific evidence. *Journal of Behavioral Medicine*, 30(1): 77-94.

Lethem J, Slade PD, Troup JD, Bentley G (1983) Outline of a fear-avoidance model of exaggerated pain perception. Behav Res Ther 21:401–408

Liang, M., Jette A. 1981. Measuring functional ability in chronic arthritis: a critical review. *Arthritis and Rheumatism*, 24: 80-86.

Mahoney, F., Barthel, D. 1965. Functional evaluation: the Barthel Index. *Maryland State Medical Journal*, 14:61-65.

McAtamney, L., Corlett, N. 1993. RULA: A survey method for the investigation of work-related upper limb disorders. *Applied Ergonomics*, 24: 91-9.

Meinhart-Shibata, P., Kramer, M., Ashton-Miller, J., Persad, C. 2005. Kinematic analysis of the 180^0 standing turn: effects of age on strategies adopted by healthy young and older women. *Gait & Posture,* 22(2): 119-125.

Miller, G., Rejecski, W., Williamson, J., Morgan, T., Sevick, M., Loeser, R., Ettinger, W., Messier, S. 2003. The Arthritis, Diet, and Activity Promotional Trial (ADAPT): design, rationale, and baseline results. *Controlled Clinical Trials*, 24: 462-480.

Moore, J., Garg, A. 1995. The Strain Index: a proposed method to analyze jobs for risk of distal upper extremity disorders. *American Industrial Hygiene Association Journal*, 56: 443-458.

Murphy, S., Strasburg, D., Lyden, A., Smith, D., Koliba, J., Dadabhoy, D., Wallis, S. 2008. Effects of activity strategy training on pain and physical activity in older adults with knee or hip osteoarthritis: A pilot study. *Arthritis & Rheumatism*, 59(10): 1480-1487.

Roos, E., Roos, H., Ekdahl, C., Lohmander, L. 1998. Knee injury and osteoarthritis outcome score (KOOS) – validation of a Swedish version. *Scandinavian Journal of Medicine & Science in Sports*, 8: 439-448.

Roos, E., Roos, P., Lohmander, L., Ekdahl, C., Beynnon, B. 1998. Knee injury and Osteoarthritis Outcome Score (KOOS): Development of a self-administered outcome measure. *The Journal of Orthopaedic Sports Physical Therapy*, 28(2):88-96.

Seidel, D., Hjalmarson, J., Freitag, S., Larsson, T., Brayne, C., Clarkson, P. J. 2011. Measurement of stressful postures during daily activities: An observational study with older people. *Gait & Posture*, 34: 397-401.

Suomi, R. and Collier, D. 2003. Effects of arthritis exercise programs on functional fitness and perceived activities of daily living measures in older adults with arthritis. *Archives of Physical Medicine and Rehabilitation,* 884, 1589-1594.

Steinfeld, E., Danford, G.S. 1999. Enabling Environments: Measuring the Impact of Environment on Disability and Rehabilitation. New York: Plenum Publishers.

Turk, D., Okifuji, A. (2002) Psychological factors in chronic pain: evolution and revolution. *Journal of Consulting and Clinical Psychology*, 70(3): 678-690.

Verbrugge, L., Gruber-Baldini, A., Fozard, J. 1996. Age differences and age changes in activities: Baltimore Longitudinal Study of Aging. *Journal of Gerontology: Social Sciences*, 51B(1): S30-S41.

Vlaeyen, J., Kole-Snijders, A., Rotteveel, A., Ruesink, R., Heuts, P. 1995. The role of fear of movement/(re)injury in pain disability. *Journal of Occupational Rehabilitation*, 5: 235-252.

CHAPTER 31

Digital Tablets as a Tool for Digital Inclusion of Elderly People

Rosana Gonçales Oliveira Rocha 1, Luis Carlos Paschoarelli 2,
João Roberto Gomes Faria 3

UNESP - Univ Estadual Paulista
Bauru, Brazil
rosanagoncales@gmail.com

ABSTRACT

The technological advance, the changes in lifestyle, the growth of internet and the more democratic access to broadband have led more and more people to access the internet, by a variety of reasons ranging from a search for fun to shopping and relationships. In the case of elderly users, another factor is included as critical to the growing desire to join the big network – socialization and independence. Establishing communication with friends and children, surfing with no help, searching the internet for issues that may assist you daily and even make ATM transactions easier are some of the reasons to prove that, opposite to what was supposed, the elderly are able and even feel eager to learn about computers. Thus, the aim of this study is to bring a reflection on the use of digital tablets for elderly users, based on available literature.

Keywords: Ergonomics, Accessibility, Elderly Users, Digital Tablets

1   DIGITAL INCLUSION OF ELDERLY PEOPLE: MAKING TECHNOLOGY POSSIBLE FOR EVERYONE

Much has been studied with respect to digital inclusion and how to transform technology into a democratic tool and accessible for everyone - from the child, who is in a growing and continuous learning process, acquiring and developing cognitions, until the audiences that due to the loss of such cognitions are at a disadvantage in learning and even in social life. In this last scenario, people with disabilities and elderly people are a highlight, who different from a child are in a process of decline of their abilities and cognitions, which hinders learning, especially with regard to new technologies, new equipment and digital world in general.

One of the main factors which make these studies necessary and urgent is the constant growth of the elderly population worldwide. The decrease in the number of children per family, social planning, the increase in life expectancy as a result of medical advances and improved quality of life, have led countries to see their population aging. According to a study done by Joel E. Cohen, et al. (2010), life expectancy from 2045-50 in rich countries shall increase from 76 to 82 years old and in poor families from 63 to 73 years old. Worldwide, the average life expectancy shall be around 74 years old.

Another factor to be considered is the growing advances in technology, **hard to keep up with** even for the ones with total access to technology, and even more difficult for those outside such progress as well as the constant **emergence** of new electronic devices.

In an increasingly digital world, being excluded from this universe means having limited access to basic things, like managing a simple elevator control or making withdrawals and payments from an ATM, which should be part of anyone's everyday life.

In Brazil, according to data from PNAD (National Household Sample Survey) published by IBGE (Brazilian Institute of Geography and Statistics), 2010, advances in Internet access among people on their 50's were the highest among age groups: 40.4%. In contrast to what might be expected, the elderly, in general, are not closed for learning nor to computer learning and they are fully aware of the advantages that access to digital might bring to their lives.

Alexy (2000), Opalinski (2001), White, MacConnell et al (2002), Burdick, Kwon (2004), Czaja, Hiltz (2005), Dickinson, Gregor (2006), Czaja, Lee (2007), among other researchers described the benefits that computer use bring to elderly people. Benefits which go from socialization, guaranteeing ongoing employability and finding the way out of the condition of social isolation until the improvements of conditions such as depression, which has its consequences leading to loneliness and feelings of lack of ability. So, among other benefits, technology may be able to provide the elderly with a sense of importance and ability - including them in a world previously possible only for younger people or for the ones in full operating activity, who are not retired or removed from work.

However, some barriers are put forward, even when there is the desire for learning. In case of elderly users, the loss of cognition is an important blockade for the access to computers. Decrease in visual acuity, motor coordination and reasoning are factors that can raise difficulties on learning. Some studies are already concerned with the development of friendly interfaces for older users. As quoted by Kobayashi et al. (2011), "(...) application developers have little correct understanding of how senior users interact with senior-friendly interfaces." Nevertheless, the first objective is to check how the user's interaction with new equipments will be with regard to the change of the device of data entry. This might happen because the traditional mouse is replaced by a simple touch of the fingertip, and the three-dimensional keyboard, which required a certain pressure of touch varying according to the brand or model, is replaced by a flat keyboard, which exists only in digital screen, activated with a smooth touch, with no pressure.

Riviere, Thakor (1996) carried out a study with subjects of varying ages, including older people with an **average age** of 72 years old, pointing out that the motor disability resulting from the use of the mouse became inaccurate and nonlinear. With training, this difficulty can be reduced, however, such difficulty can become a discouragement for the person in the process of digital inclusion.

We also consider this discussion, the fact that mobile devices such as digital tablets, are becoming popular and are no longer exclusively in the business world to reach the average consumer. Governments of developing countries like Brazil and India, have spared no efforts to create opportunities to provide devices for all social classes, and especially to schools, given their mobility and practicality.

It is for research to verify all the issues that are relevant to the functionality and try to provide funds for developers of applications and hardware to solve the possible problems of access for each group of people. For Ramakrishna (2012), "in a few years, a video call will be as easy to do as a phone call." Having minimum knowledge, enabling basic access to new devices that have emerged is making information accessible for everyone, regardless their social status, age, ability or any other factor that is presented today as unique.

Being outside all this technology is being unable to take advantage of what motivates technology - being democratic, accessible and unconditionally making people's lives easier, shortening distances and bringing generations closer to each other.

2   WHY DO ELDERLY PEOPLE WANT TO ACCESS INTERNET?

As previously mentioned, there are several reasons why an older person desires to access a great computer network, beyond the sense of inclusion to a group that once seemed the selected and privileged, provided with skills considered beyond reach. A survey conducted by Rocha et al. (2010) with a group of nineteen individuals over 60 years old, from which 6 men and 13 women recruited for being common elderly internet users (without any technical knowledge) who use the internet at least once a day for more than three months and who also attend computer courses, and none of the participants presented any physical limitations (disability) or disease that made it difficult or prevented the development of the activity, all literate and in the same condition of reading and interpreting texts, with an average age of 69, 74 (± 6.95) years old, with time spent in the computer course between 3 and 12 months (8.79 ± 2.7 months). The results obtained are synthesized on Table 1.

Table 1. Reasons for accessing Internet reported by subjects of study and percentage of report.

Reason for accessing internet	%
newsreaders	73,68
search for general information	68,42
search for information on health issues	57,89
search for informações on benefits	47,37
access to e-mail	42,11
contact with friends	21,05
access to orkut	15,79
access to blogs	10,53
access to facebook	5,26
others	5,26

Although access to social media and blogs is not the main attraction of the subjects studied, there were spontaneous reports of users who had their own blogs, which were often updated, at least once a week. Others with **accounts on** facebook and orkut, posting mainly photos.

3   RELATED STUDIES

So far, there are few studies that tested the use of digital tablets as a tool for digital inclusion specifically of elderly users. Kobayashi et al. (2011), presented the test results conducted with a sample of 20 elderly Japanese subjects, aged between 60 and 70 years old, 14 women and 6 men. 12 of which had two years of prior experience with mobile phones. There was no report if the subjects had previous experience with digital tablets with 9-inch screen or bigger. Some participants reported that they had difficulties in daily life due to problems arising from age, such as the decline in hearing and vision. The study aimed to evaluate the interaction of elderly users with mobile devices, more specifically smartphones, for which was used an 9.7-inch iPad, 147.8 × 197.1 mm multi-touch screen with 768 ×

1024 pixel format and a 3.5-inch touch iPod,, 49.3 × 74.0 mm multi-touch screen with 640 × 960 pixel format, both running iOS 4.2. The results showed that error rates when touching the iPod device with 30, 50 and 70 pixel format (smaller size) were, respectively, 39%, 6.5% and 5.6%, far above the acceptable 4% in the literature about Human Computer Interaction; while on the larger iPad device, error rates ranged from 13.6%, 1.4% and 1.7%. These results, especially referring to touches on larger targets, point to a possible suitability of the digital tablets of 9 inches or larger, as a tool for digital inclusion of elderly users.

In contrast to science, virtual and printed media has been emphatic in stating that digital tablets are tools perfectly suited for use by elderly people (Shapiro, 2011) and indicate intuitiveness as its most attractive point. This assumed quality of the equipment, was also explored by the manufacturers to disclose the news. The TV commercial, which can be accessed on YouTube (http://www.youtube.com/watch?v=ndkIP7ec3O8), like many others in the same case found in such portal, shows a group of older people exploring the equipment and surprised to see how a fingertip touch could take them to a world of infinite possibilities.

Regarding the use by the elderly of traditional desktop computers, equipped with monitor, central data processing (CPU) and keyboard, all as separate objects, or even laptops, where CPU, keyboard and mouse are part of a whole, although not working as a touchscreen technology, the literature is more extensive, particularly with regard to computer use for Internet access.

Czaja (2007) talks about the changes on cognition: decrease of coordination, concentration and memory ability, vision, hearing, information processing, comprehension, learning, memory, Burdick and Kwon (2004), describe the elderly user's right to have access to information that can positively change their daily life, such as access to information about health and rights, job opportunities, taking them out of social isolation, promoting contact with distant family and friends, making financial transactions easier without getting away from home, increasing the sense of usefulness, improving self-esteem and welfare; Sales et al (2009), describe in a book the experience of teaching computer use for elderly people; Kachar (2002) reports a study where he identified that elderly people have special needs and difficulties, which can be overcome with appropriate strategies for learning and respecting the learning pace of each one. Also according to the researcher, the contact with the Internet brings significant benefits to the lives of these subjects, Freese et al (2006), the use of internet helps in the prevention of brain aging; in contrast to researchers who talk about the benefits of the Internet; Nie (2001), reports that the elderly who have a wide network of social contacts, Internet use does not change and even affects socialization, because the time spent on virtual world prevents them from meeting with friends and relatives; White et al (1999), White et al. (2002), Karavidas et al (2005), Dickinson and Gregor (2006), analyzing several studies, state the scientific evidence that confirm the statement that computer use improves the welfare of elderly people.

4   BARRIERS TO BE TRANSPOSED

After considering the cognitive barriers faced by the elderly due to aging, such as loss of motor coordination, decline in visual acuity, lack of familiarity with the technology and equipment, feeling of inability and inadequacy of the challenge of learning (Czaja and Lee, 2003; Hendrix, 2000; Mann, Belchior et al, 2005), another major difficulty that arises from the contact with the old traditional desktop computers is with the data entry devices, even the essential ones, such as the mouse and the keyboard. Cognitive changes associated with age, as the reduction of coordination, memory, sight, learning among others (Ownby et al, 2011), transform these devices in barriers to be transposed, in particular in regards to motor coordination. Studies show that, if considered separately, the mouse is a device that requires greater effort from the elderly public to enable the initial access to computer (Riviere and Thakor, 1996; Berkov, 2007, Arch, 2008; Kressig and Echt, 2001). Controlling the mouse, positioning accurately, clicking, dragging, are difficult tasks for the elderly due to the decline and loss of motor coordination. For those presenting disabling problems, for example osteoarthritis, the task becomes even more complicated. Smith et al (1999) reported results of research conducted with young and old people, whereas the older participants presented more difficulty than the younger ones while performing tasks with the mouse. This difference in performance, attributed to age was observed in complex tasks such as click and double click, which require greater coordination.

Any difficulty which may be a barrier for elderly users to access the Internet can be considered forfeiture of rights for information access that can positively change their day to day and even take them out of social isolation, promoting sense of usefulness, self-esteem and welfare (Burdick and Kwon, 2004). Since the mouse is the most commonly used device for data entry, we restrict ourselves to its observation, not extending to the use of joysticks, touch pad on laptops, trackball, among others.

5   TOUCHSCREEN

The studies found with elderly users using the touchscreen are for non accurate monitors as the ones used by banks, and smartphones, mobiles with many functions, including the internet connection. Stone (2008), Stossel (2010), presented studies showing the interaction of elderly people with touchscreen interface that indicate the feasibility of using this technology for digital inclusion of these users.

The touch screen technology, born in the 60's is an electronic visual display that can detect the presence and location of a touch within the display area (Wikipedia, 2010). The increase in touch accuracy and the improvement of methods of interaction, resulted in capacitive touchscreen, instead of previously used in user interface technology, the resistive ones. The resistive screen technology was developed and patented in 1977 by Samuel C. Hurst. It is composed of several layers of flexible material (usually polyester on the glass), which does not allow multi-touch input, it is fragile and presents low visibility to light (Jetmais, 2011). A capacitive touch screen is common in existing equipment, such as digital tablets, that work with the concept of storing energy between two layers of glass which continuously conduct current. The interaction is possible only through the fingers, the visibility is good in the presence of light, it has good long-lasting quality and bears multi-touch. Another characteristic of this type of screen is touch accuracy to small targets, such as smartphones. However, the high accuracy of these small devices, due to the size of the target can be considered a problem for elderly users, especially in relation to text input through the digital keyboard of devices, as shown by the study of Kobayashi (2011). On the other hand, when analyzing the digital tablets, whose formats vary from 8.9 to 10.1 inches, keyboards come in pretty good sizes, allowing enough accuracy to touch. Yet, this data lacks ergonomic study to check if the size of the digital device keys is really suitable for various groups of people able to access the equipment, if they have the minimum possible size to allow data entry with comfort and without problems, mainly for the subjects of this discussion – the elderly users.

6   CONSIDERATIONS AND FUTURE STUDY

Studies with the use of digital tablets are just the beginning. Many aspects must still be tested and checked for several hypothesis that point to touch screen interface of this equipment as a possible solution for problems of accessibility by the elderly in digital media, specifically the Internet. However, it is necessary to show from scientific studies, if these devices are in fact prepared to serve these users safely and efficiently, promoting digital inclusion with no harm to their welfare and health.

Future studies must then test the use of digital tablets for elderly users. Another aspect to be studied is to compare the performance of elderly users on computer learning and Internet access from traditional desktop computers and digital tablets. An important aspect to be considered in these studies is the operating system of digital tablets, since it constitutes one of the factors that might enable user's access.

SPECIAL THANKS

Special thanks for Professor Marcia Barros de Sales, a researcher at Federal University of Santa Catarina (UFSC, Santa Catarina, Brazil), who dedicated her studies to digital inclusion of elderly and who spared no efforts to contribute to the studies initiated at São Paulo State University. Even away from digital means, she has contributed when possible, guiding us and providing information on the studies and results achieved along the years dedicated to her research.

REFERENCES

Alexy, E. M. 2011. Computers and caregiving: Reaching out and redesigning interventions for homebound older adults and caregivers. *Holistic Nursing Practice* 14(4): 60-66, 2000.

Arch, A. 2018. *Web Accessibility for Older Users: A Literature Review.* Accessed November 12, 2011. www.w3.org/TR/wai-age-literature/#arfl.

Burdick, D. C. and W. O. Kwon. 2004. *Gerotechnology:* Research and Practice in Technology and Aging. New York: Springer Publishing Company.

Cohen, J. E. 1995. *How Many People Can the Earth Support?* New York: W. W. Norton.

Czaja, S. J. and S. R. Hiltz. 2005. Digital Ids for an Aging Society. *Communications of the ACM* 48(10): 43-44.

Czaja, S. J. and C. C. Lee, 2003. Designing computer systems for older adults. In J. A. Jacko & A. Sears (Eds.), *The human–computer interaction handbook:* Fundamentals, evolving technologies and emerging applications (pp. 413–427). Mahwah, NJ: Erlbaum, 2003.

Czaja, S. J. and C. C. Lee, 2007. The impact of aging on access to technology. *Universal Access in the Information Society* 5: 341–349.

Dicknson, A. and P. Gregor. 2006. Computer use has no demonstrated impact on the well-being of older adults. *International Journal of Human-Computer Studies* 64(8): 744-753.

Freese, J., S. Rivas, and E. Hargittai. 2006. Cognitive ability and internet use among older adults. *Poetics* 34 (04): 236-249.

Hendrix, C. C. 2008. Computer use among elderly people. *Computers in nursing* 18(12): 62–68.

IBGE. Brazilian Institute of Geography and Statistics. *Pnad (National Research for samples of households).* Rio de Janeiro: 2010. [in Portuguese].

Jetmais. Accessed October 29, 2011. http://jetmais.com/2010/como-funcionam-as-telas-touch-screen-resistivas-e-capacitivas.

Kachar, V. The Seniors and digital inclusion. *The World Health Magazine,* 26(3), p. 376-381, 2002 [in portugues].

Karavidas, M., N. K. Lim, and S. L. Katsikas. 2005. The effects of computers on older adult users. *Computer Human Behavior* 21(05):697-711.

Kobayashi, M. et al. 2011. Elderly User Evaluation of Mobile Touchscreen Interactions. *Computer Science* 6946: 83-99.

Kressig, R. W. and K. Echt. 2001. Exercise Prescribing: Computer Application in Older Adults. *The Gerontologist* 42(02): 273–277.

Mann, W. C., P. Belchior, M. R. Tomita, and B. J. Kemp. 2005. Computer use by middleaged and older adults with disabilities, *Technology and Disability* 17: 1–9.

Nie, N.H. 2001. Sociability, interpersonal relations, and the internet. *American Behavioral Scientist* 45(03): 420-435.

Opalinski, L. 2001. Older adults and the digital divide: assessing results of a web-based survey. *Journal of Technology in Human Services* 18: 203-221.

Ownby, R. L. et all. 2011. Cognitive Abilities That Predict Success in a Computer-Based Training Program. *The Gerontologist* 48(02): 170–180.

Ramakrishna, S. Easy as calling. *The Sao Paulo State Journal,* notebook Link. São Paulo, p. L3, 20/02/2012 [in portugues].

Riviere, C. N. and N. Thakor. 1996. Effects of age and disability on tracking tasks with a computer mouse: Accuracy and linearity. *Journal of Rehabilitation Research and Development* 33: 6-15.

Rocha, R. G. O.; Paschoarelli, L. C.; Silva, J. C. Accessibility of websites for elderly users: ergonomics and visual comfort. *11th International Congress of Ergonomics and Usability of Human-Technology Interfaces:* Product, Information, Built Environment and Transport. Proceedings. Manaus, 2011 [in portugues].

Sales, M. B. ; Mariani, A. C. ; Alvarez, A. M. *Computing for the third. age.* 1. ed. Rio de Janeiro: Ciência Moderna, 2009. 2000. 122 p. [in portugues].

Shapiro, D. 2011. *The iPad* - Especially Good for Old People. Accessed November 12, 2011. http://www.associatedcontent.com/article/2645291/the_ipad_especially_good_for_old_people. html?cat=15.

Smith, M. W., J. Sharit, and S. J. Czaja. 1999. Aging, Motor Control and the Performance of Computer Mouse Tasks. *Human Factors* 41(03): 389-396.

Stone, R.G. 2008. Mobile Touch Interfaces for the Elderly. In: *Proceedings of ICT, Society and Human Beings*, July 22-24.

Stôßel, C. and L. Blessing. 2010. Mobile Device Interaction Gestures for Older Users. In: *Proceedingsof the 6th Nordic Conference on Human-Computer Interaction* (NordiCHI 2010): Extending Boundaries, pp. 793–796.

White, H. and E. McConnell, et al. 1999. Surfing the net in later life: a review of the literature and pilot study of computer use and quality of life. *Journal of Applied Gerontology* 18(03): 358-378.

White, H. and E. McConnell, et al. 2002. A randomized controlled trial of the psychosocial impact of providing internet training and access to older adults. *Aging & Mental Health* 6(3): 213-221.

CHAPTER 32

Comparing Muscle Activity and Younger and Older Users' Perceptions of Comfort When Using Sheet Switches for Electrical Appliances

Yasuhiro Tanaka[1], Yuka Yamazaki[2], Masahiko Sakata[3], Miwa Nakanishi[1]

[1] Keio University, Fac. of Science & Technology, Dept. of Administration Engineering
Hiyoshi 3-14-1, Kohoku, Yokohama 223-8522, Japan
[1] tanakayasu63-425@a8.keio.jp, miwa_nakanishi@ae.keio.ac.jp
[2] Mitsubishi Electric Corp.,
5-1-1, Ofuna, Kamakura, 247-8501, Japan
[2] Yamazaki.Yuka@cs.MitsubishiElectric.co.jp
[3] Mitsubishi Electric Group, Central Melco Corp.,
3-26-33, Takanawa, Minato, Tokyo, 108-0074, Japan
3kiai106@ct-melco.co.jp

ABSTRACT

This study focuses on sheet switches, which are widely used as user interfaces for electrical appliances, and clarifies the comfort level required by users. Because the number of elderly users of electrical appliances is increasing in Japan, we focus on these users and compare their characteristics with those of younger users. Our goal is to design a user interface for electrical appliances that considers the physical, cognitive, and psychological characteristics of users. Thus, in this study, we particularly examine the psychological characteristics of comfort and the physical aspects of muscle activity to determine the parameters required to enable easy usage of electrical appliances by all users. Moreover, we explore the relationship between comfort and muscle activity.

Keywords: Comfort, Muscle activity, Sheet switches, Electrical appliances, Elderly users

1 INTRODUCTION

Electrical appliances are indispensable in our daily lives. Although such appliances have already been developed with high levels of functionality and performance, modern usage demands product lines that offer further diversity and comfort. Therefore, our goal is to create a novel design that considers the physical, cognitive, and psychological characteristics that constitute the essence of user comfort.

This study focuses on sheet switches, which constitute a large majority of user interfaces for electrical appliances, and examines the relationship between operator comfort and muscle activities.

In Japan, the senior citizen population continues to increase [1]. Moreover, one of the most prominent research institutes in Japan, the Mitsubishi Research Institute (MRI), has predicted that in 2030, the Japanese market of electrical appliances for the elderly will have expanded to one and a half times of that in 2010 [2]. Many studies have focused on the sociological aspects of machine operation by the elderly [3]-[6] relative to visual [7]-[11] and audio [12]-[17] parameters. However, there is not much research on user comfort while performing simple actions such as touching an interface of an electrical appliance; in particular, to the best of our knowledge, research on sheet switches operated by the elderly has not yet been conducted. Thus, the need for such a study is apparent. From this viewpoint, we attempt to examine the relationship between comfort and muscle activity and clarify the specific characteristics of sheet switches required for comfortable operation by all users.

2 METHODS

2.1 Participants

In this study, two groups of 15 healthy men and women with average ages of 76.6 and 22.0 years representing older and younger users, respectively, participated in the experiment. During the course of the study, both groups operated commonly used electrical appliances such as rice cookers and washing machines. We comprehensively explained the test to all participants, each of whom tendered informed consent.

2.2 Experimental Parameters

In this study, we used model switches that imitated sheet switches of commonly used electrical appliances. Each model switch was installed on an aluminum base, as shown in Fig. 1.

Each model switch was set with the following two parameters related to comfort when touching the sheet switch, which was determined on the basis of our previous experiment:

(1) Depth: the length from the top of the switch to the reaction point (unit: mm) (Fig. 2).

(2) Reaction force: the force needed to complete a switch by touch without a plastic overlay (unit: N).

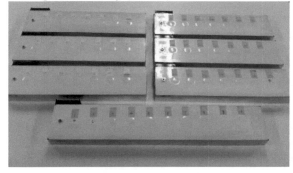

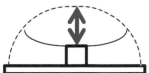

Figure 1 Model switches.   Figure 2 Switch depth

We assigned seven levels of depth and six levels of reaction force to the model switches, which were designed to the same specifications as those presently installed in the electrical appliances or those expected to be designed. Therefore, 42 types of sheet switches were designed, as shown in Table 1. The size of each model switch was adjusted to 18.0 mm in diameter, which is the industry standard for common electrical appliances.

An LED lamp was included on the model switches, which indicated participant reaction.

Table 1. Levels assigned to each model switch parameter.

	[1]	[2]	[3]	[4]	[5]	[6]	[7]	[8]	[9]	[10]	[11]	[12]	[13]	[14]
Depth[mm]	0.25	0.5	0.8	1.1	1.4	1.7	2.0	0.25	0.5	0.8	1.1	1.4	1.7	2.0
Reaction force[N]	0.5	0.5	0.5	0.5	0.5	0.5	0.5	1.0	1.0	1.0	1.0	1.0	1.0	1.0

	[15]	[16]	[17]	[18]	[19]	[20]	[21]	[22]	[23]	[24]	[25]	[26]	[27]	[28]
Depth[mm]	0.25	0.5	0.8	1.1	1.4	1.7	2.0	0.25	0.5	0.8	1.1	1.4	1.7	2.0
Reaction force[N]	1.3	1.3	1.3	1.3	1.3	1.3	1.3	1.6	1.6	1.6	1.6	1.6	1.6	1.6

	[29]	[30]	[31]	[32]	[33]	[34]	[35]	[36]	[37]	[38]	[39]	[40]	[41]	[42]
Depth[mm]	0.25	0.5	0.8	1.1	1.4	1.7	2.0	0.25	0.5	0.8	1.1	1.4	1.7	2.0
Reaction force[N]	2.6	2.6	2.6	2.6	2.6	2.6	2.6	3.5	3.5	3.5	3.5	3.5	3.5	3.5

2.3 Experimental Task

The participants operated each of the 42 model switches in a vertical manner, as shown in Fig. 3. To examine all model switches, users were instructed to use the same approach as they normally would when using a rice cooker or washing machine.

The order of testing trials was randomized to eliminate order effects.

Figure 3 Vertical touch.

2.4 Experimental Environment

In normal practice, electrical appliances such as rice cookers and washing machines are operated from a standing position. However, considering that the participants included the elderly who required more than 1 h to complete the testing trials, we designed the experimental environment such that the tasks could be performed in a sitting position. According to some electrical appliance manufacturers, rice cookers and washing machines are designed under the assumption that the appliance is placed at a height of approximately 100 cm from the floor, which is indeed a common practice in many homes. Thus, considering that the location of the users' arms changes by approximately 30 cm between standing and sitting positions, the model switches were placed approximately 70 cm above the floor.

2.5 Data

The participants evaluated the comfort experienced upon touching each model switch by assigning scores ranging from−2 to +2. Answers such as "between +1 and +2" were allowed and recorded as a score of +1.5. Furthermore, when the LED lamp did not react upon touching the model switch, its score was recorded as −2.

In addition, electromyography (EMG) was performed on the muscles in the participants' forearms and hands to determine the physical characteristics of touch comfort and discomfort. On the basis of our previous study [18], the lumbrical muscles of the hand (Fig. 4), the flexor carpi ulnaris (FCU) muscles (Fig. 5), and the flexor carpi radialis (FCR) muscles (Fig. 6) were selected as the measurement points because these muscles reacted significantly when the sheet switch was touched. To eliminate individual errors and environmental effects, reference data were set prior to the testing trials in the following manner: In vertical touch testing, a digital force gauge that indicated the level of force in real time was set at the same height as the model switches, and the participants pressed it with a force of 10 N for 2 s. Then, the EMG data measured at the three points were used as reference data. Moreover, the instant at which the participants' fingers touched a model switch was recorded using a digital video camera.

2.6 Experimental Systems

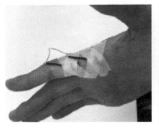 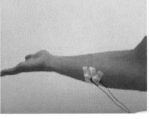 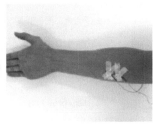

Figure 4 Lumbrical muscles.   Figure 5 FCU muscles.   Figure 6 FCR muscles.

In this experiment, we used an EMG recording system (MP150/EMG100C, Biopac Systems, Inc.), a digital force gauge (FPG-5, Nidec-Shimpo Corporation), and a digital video camera (EX-FC100, Casio).

3 ANALYSIS

3.1 Subjective Scores

The actual scores assigned to each model switch by the participants were averaged and standardized with every generation and used as subjective scores.

3.2 EMG Data

The interval for analysis was set as the time duration from the instant a finger touched the top of a switch to the instant it was withdrawn. Figure 7 shows an example of the EMG data of each muscle. We calculated the integrated electromyogram (IEMG) of the EMG data and used the IEMG ratio as a measure of muscle activity:

IEMG ratio = (IEMG of trial data)/(IEMG of reference data).

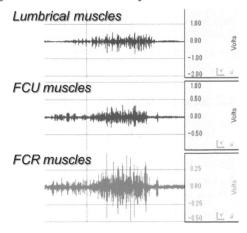

Figure 7 Example of EMG data.

4 RESULTS

4.1 Results of Comfort

To examine the relationship between the degree of user comfort experienced upon touching the model switches and the parameters of the switches, we focused on the relationship between the subjective scores of the model switches and one

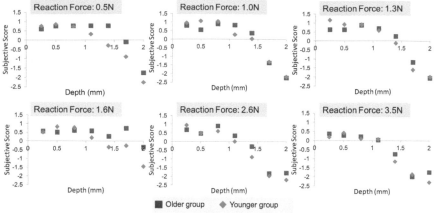

Figure 8 Relationships between subject scores and depth.

switch parameter, i.e., depth.

Figure 8 compares the relationship between the subjective scores and depth for the older and younger groups. The charts show the subjective score for each combination of depth and the common reaction force. We focused on the case of minimal reaction force (not exceeding 1.6 N). Comparison of the results for the older and younger groups revealed that the older group assigned high scores to the model switches with a greater depth and the younger group favored the model switches with a shallow depth. That is, the older group experienced comfort when they touched switches that felt soft to the touch, whereas the younger group experienced comfort when they touched switches that generated a small "clicking" sensation.

4.2 Results of Muscle Activity

To examine the characteristics of muscle activity when the participants touched each switch, we focused on the relationship between the subjective scores and the activity level of each muscle. Figures 9, 10, and 11 compare the activity levels of the lumbrical, FCU, and FCR muscles for both groups at each depth. The significant differences of each group are highlighted in the figures.

For the older group, each muscle activity with respect to depth slightly changed. In contrast, for the younger group, muscle activity increased with depth. In addition, when depth was minimal, some results showed that muscle activity was high, regardless of depth.

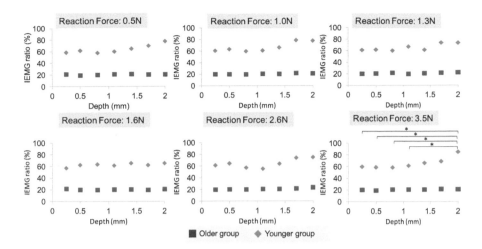

Figure 9 Relationship between muscle activity and depth of lumbrical muscles.

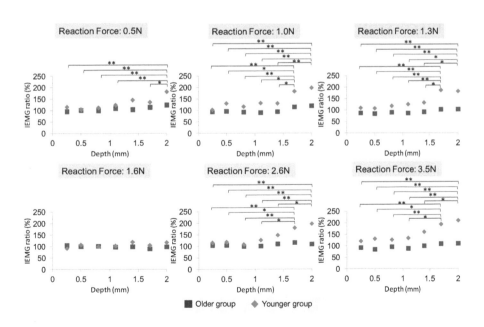

Figure 10 Relationship between muscle activity and depth of FCU muscles.

322

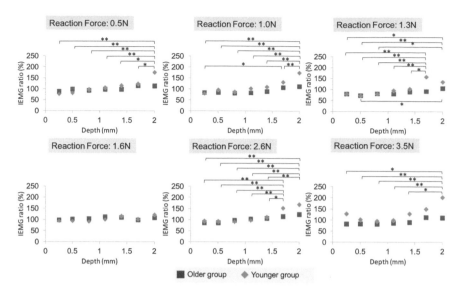

Figure 11 Relationship between muscle activity and depth of FCR muscles.

5 DISCUSSION

In this study, depth was chosen as the main parameter. Greater depth equated to greater force required to activate a switch through a plastic overlay. On the basis of this finding, we compared the results for the older and younger groups and discussed the characteristics of comfort required by both groups while touching the sheet switches.

In our previous study, the Research Institute of Human Engineering for Quality Life (HQL) showed that older users tend to touch sheet switches on electrical appliance more forcefully than younger users in order to achieve a feeling of security [19]. The present study revealed that regardless of switch characteristics, older users tend to touch sheet switches forcefully in order to achieve a feeling of security. Moreover, the responses from the switches reveal their comfort levels. The results indicate that the older group experienced greater comfort using switches that felt soft to the touch because this type of switch cushioned the impact upon touch. In contrast, the younger group altered their muscle activities according to the depths of the sheet switches to determine the amount of pressure necessary for comfortable operation. As a result, the younger group required lesser muscle activity to experience comfort while activating the sheet switches.

Furthermore, we examined the results of muscle activity at various switch depths. When users touched the sheet switches with shallow depths, higher muscle activity was required than that needed for deeper switches. This was because higher muscle activity was required to restrain the pressing force. Thus, we concluded that, in general, muscle activity was high. As a result, the muscle activity was not proportional to the sheet switch parameters. That is, designs that appear to require more relative force for operation actually need lower muscle activity, which relates to a higher level of user comfort.

6 CONCLUSION

In this study, we clarified the characteristics required for comfortable sheet switch operation for older and younger users. To determine muscle usage, we examined the characteristics of the muscle activities for each age group upon touching switches. Our results indicated that muscle usage and comfort levels differ according to operator age. Older users experienced comfort from operating switches that felt soft to the touch, whereas younger users experienced comfort when the switch generated a clicking sensation. Moreover, this study indicates that surface parameters are not reliable for determining the degree of muscle activity and comfort levels while designing sheet switches.

The Japanese market for electrical appliances for the elderly continues to expand; therefore, we propose including switches that feel soft to the touch while designing electrical appliances, in order to provide a higher comfort level for elderly users. Although the effects are small, these products could contribute to increased enjoyment of daily activities such as cooking and cleaning.

REFERENCES

1. Population Projection for Japan, National Institute of Population and Social Security Research, (2008)
2. Nikkei Electronics, 1044, pp. 30–49, Nikkei Business Publications, (2010)
3. Fujikake, K., M. Mukai, H. Kansaku, M. Miyoshi, M. Omori, and M. Miyao. 2004. Readability for PDAs and LCD monitors among elderly people. *Ergonomics* 40(4): 218–227.
4. Mori M. and S. Horino. 2004. Usability enhancement of cellular phone considering elderly people's use. *Ergonomics* 40: 254–255.
5. Hirano K., T. Daimon, and K. Toyoda. 2004. Preliminary Study on Operational Characteristics of Aged People in Predictive Methods for Japanese Text Input on Mobile Phone. *Ergonomics* 40: 272–273.
6. Yamagiwa T. and I. Yoshimura. 2003. A Study of Haptic Perception when using a Mobile Device: Argument Focusing on the Influence of Rotation. *Japan Industrial Management Association* 54(2): 83–94.
7. Akutsu M., H. Hayashi, Y. Ikezawa, A. Ouyama, Atsushi S., and H. Takeuchi. 2007. An Experimental Study on Requirements of Computer for the Aged. *Ergonomics* 43(2): 272–273.

8. Ishihara K., M. Nagamuchi, H. Osaki, S. Ishihara, and A. Tsuji. 1998. Affects of age-related yellowing crystalline lenses on daily life. *Ergonomics* 43(1): 9–16.

9. Sakagami M., Y. Nakashima, M. Takamatsu, S. Nakajima, Mima K. 2007. Fundamental Research of Visual Barrier-free for Elderly people–On the LED Information Board–. *Ergonomics* 43(2): 388–389.

10. Fujikake K., T. Sakurai, K. Miura, S. Hasegawa, H. Takada, M. Omori, Y. Matsumura, R. Honda, and M. Miyao. 2006. Evaluation of the display quality of LCD for elderly people–Comparative study using Stabilometers–. *Ergonomics* 42: 420–421.

11. Moriwaka M., A. Murata, and M. Kawamura. 2006. Display design in the light of age, task difficulty, character size and background color. *Ergonomics* 42(2): 132–138.

12. Kawata A. and I. Fukumoto: A study of the universal alarms for both young adults and the elderly. 2000. *Ergonomics* 36(5): 261–272.

13. Kurakata K., Y. Kubo, T. Kizuka, and Y. Kuchinomachi. 1998. Relationship between hearing levels of the elderly and preferred volume levels of television. Ergonomics 34: 222–223.

14. Kuchinomachi Y. 1999. The characteristics of the elderly people and universal design. *Ergonomics* 35(2): 66–69.

15. Kurakata K., K. Matsushita, Y. Kubo, and Y. Kuchinomachi. 2000. Auditory signals of electric home appliances (Third report): Temporal ringing pattern. *Ergonomics* 36(3): 147–153.

16. Kurakata K., K. Matsushita, Y. Kuba, Y. Kuchinomachi: Audio signals in electric home appliances evaluated in terms of the hearing ability of older adults (Second report). 1999. *Ergonomics* 35(4): 277–285.

17. Kurakata K, Y. Kuba, Y. Kuchinokahi, K. Matsushita. 1998. Audio signals in electric home appliances evaluated in terms of the hearing ability of older adults. *Ergonomics* 34(4): 215–222.

18. Tanaka Y., Y. Yamazaki, M. Sakata, and M. Nakanishi. 2011. Characteristics of Comfortable Sheet Switches on Control Panels of Electrical Appliances–Comparison using Older and Younger Users– , Proceedings of the 14th HCI International 2011, USA, on CD-ROM, (2011)

19. Research Institute of Human Engineering for Quality Life (HQL), Research study on Construction of a Database for Physical Functions, (2000)

CHAPTER 33

Anthropometric Lot Sizing of Garments

Denis A. Coelho, Isabel L. Nunes

Universidade da Beira Interior
Covilhã, Portugal
Faculdade de Ciências e Tecnologia, Universidade Nova de Lisboa
Caparica, Portugal
denis@ubi.pt, imn@fct.unl.pt

ABSTRACT

European Norm 13402-3 sets standards for clothes sizing designation. Depending on the kind of garment, between one to up to four anthropometric dimensions need to be considered in defining the clothing size system. According to the distribution of the anthropometric dimensions of a specific population, the quantity needed or proportion for the various clothes and garment size categories varies greatly. In order to develop a process that enables clothing distribution managers to foresee the relative proportion of each category size that is expected to be consumed, assuming a homogeneous appeal of the garments irrespective of their size, across the population, a set of considerations supported by mathematical calculations, need to be established. To this end, it must be noted that the combination of anthropometric dimensions, described by their Gaussian statistical parameters (mean and standard deviation) is mathematically feasible, as long as correlation factors between anthropometric dimensions are known for the population at hand (these may only be extracted from original data sets). The chapter proposes deploying correlation factors between anthropometric dimensions involved in the European clothes sizing standard, and presents a method for garments lot sizing for point of sale application, informed by the correlation factors, which may be retrieved from literature, and by the statistical parameters of the population (actual or inferred). The chapter demonstrates the approach presented for various clothing item types, including men shorts (sizing based on one anthropometric dimension) and women brassieres (sizing based on two anthropometric dimensions). The mathematical formulation presented and demonstrated in the chapter is systematized in examples solved by spreadsheet calculations and is intended to support the management of orders of garments by size at their point of sale considering the reduced cycle of fashion (about 3 months). The chapter also emphasizes that reporting of anthropometric data should consider not only the mean and standard deviation of individual dimensions as has been the common practice in the field of human factors and ergonomics, but should also be accompanied by correlation charts between the anthropometric dimensions, given reduced availability of this kind of data.

Keywords: sizing system, statistical parameters, variance, correlation

1   INTRODUCTION

The approach of garment lot sizing to consistently assure supply of size categories according to the target population's distribution of anthropometric dimensions may be beneficial as society transitions to a more sustainability led economy with smart use of resources (Coelho, 2012). Authors acknowledge that other factors such as aesthetic appeal, or comfort and pleasure (Coelho and Dahlman, 2002) may influence the consumer's choice more substantially than availability of size, while the garment is on sale. The problem of defining size categories for garments has been the object of extensive inquiry for several population groups (for references see for example Mpampa, Azariadis and Sapidis, 2010). In Europe, EN 13402-3 (2004) sets standards for clothes sizing based on body dimensions, measured in centimeters. Rather than questioning the adequacy of the size categories set forth in the standard, this chapter looks at the clothes sizing problem from an order quantity perspective at the point of sale. Assuming that the point of sale serves a specific geographical area, for which anthropometric dimensions of the population served by the retail outlet can be sampled (e.g. looking at sales records of garment sizes for a specified time interval) or is known or inferred, for any order quantity the store makes to its suppliers, the quantity for each garment size can be estimated, so that the store runs out of the item in approximately the same time for all item sizes. Pheasant's (1998) statistical treatment of anthropometric dimensions, based on the Gaussian formulation (anthropometric dimensions described by their estimated mean values – m, and standard deviation parameters - sd) is a basis on which the mathematical formulation presented is developed.

2   STATISTICALLY BASED METHOD FOR LOT SIZING OF GARMENTS – FITTING TO ONE ANTHROPOMETRIC DIMENSION

Considering a simple example, men's shorts, these are described in EN 13402-3 as a case where only one anthropometric dimension is fitted and used as a basis to develop their sizing system. This dimension is the waist girth. Hence, considering an order quantity of N, for a sizing system developed in step intervals of I, with the intervals centered around the clothing size (for men's waist girth, the standard defines steps of 4 cm), for a population with a known or estimated mean and standard deviation of waist girth, the following steps should be followed.

1 – List the category sizes in centimeters and the transition values between categories. This should yield a list of m category sizes and $m+1$ transition

points, with the category sizes apart from each other of the value I, and the transition points apart from the immediate category values by a distance equal to $I/2$.

2 – Using the anthropometric dimension's statistical parameters, m and sd, compute the cumulative percentile of the Gaussian distribution for each of the $m+1$ transition points. An intermediate calculation is necessary, in order to transform the transition point value (x_p) into the equivalent dimension (z_p) for a normal distribution with mean equal to 1 and standard deviation equal to 0, such that $z_p = \dfrac{x_p - m}{sd}$. The cumulative normal distribution or percentile (p) is hence obtained from z_p.

3 – For each two consecutive transition points, compute the difference between their two respective percentiles, starting from the biggest value down to the lowest one. This will yield m percentile differences, which will be the basis for computing the order quantity for each size category, which is found in the middle point between the consecutive transition pairs.

4 – Multiply the m percentile differences by the total order quantity, N, to obtain the order quantity per size category, n_i.

5 – Check the total sum of $\sum_{i=1}^{m} n_i$, in case this is much smaller than N, then more size categories above the biggest size category must be considered, and, or, below the smallest size category, and reiterate from step 1, until this total sum is fairly close to N.

2.1 Application example – fitting to one anthropometric dimension

Consider a population of adult men, with mean of 84.3cm and standard deviation of 6.7cm, with respect to waist girth. EN 13402-3 (2004) stipulates size categories for men's waist fitting garments from 72 cm to 132cm with size categories at 4 cm intervals (I), and considering a total order quantity (N) of 25.

Table 1 shows the results of the computations described, yielding a total of 24 for the sum of the order sizes for the size categories considered initially. As the results suggest, for this total order quantity, it would not be feasible to consider size categories above size 96, or below size 72, as these would yield order quantities smaller than 1. However, if the total order quantity were one order of magnitude greater, then it would make sense to consider more size categories, for the population parameters used for the computations.

Table 1 Numerical results for the example of application of the statistically based method for lot sizing of garments fitting one anthropometric dimension

Size categories	Transition points	Percentile	Percentile difference	Order quantity
-	70	1,64	-	-
72	-		4,57	1
-	74	6,21	-	-
76	-	-	11,14	3
-	78	17,35	-	-
80	-		19,22	5
-	82	36,57	-	-
84	-	-	23,45	6
-	86	60,01	-	-
88	-	-	20,24	5
-	90	80,25	-	-
92	-	-	12,36	3
-	94	92,62	-	-
96	-	-	5,34	1
-	98	97,96	-	-

3   STATISTICALLY BASED METHOD FOR LOT SIZING OF GARMENTS – FITTING TWO ANTHROPOMETRIC DIMENSIONS

A garment where at least two anthropometric dimensions need to be fit, is the brassiere, disregarding shoulder strap length, if focusing only on bust girth and underbust girth. EN 13402-3 considers underbust size categories from 60cm to 125cm at 5cm intervals, and a cup size range from 11 to 31 cm, at 2cm intervals. Cup size categories are represented by a letter code (AA, A, B, ... H, J, K). The appropriate cup size dimension can be obtained by subtracting the underbust girth from the bust girth for each individual woman. Let the underbust girth be considered dimension 1 ($d1$), and the difference between the bust girth and the underbust girth, leading to the cup designation, be considered dimension 2 ($d2$). Once the correlation coefficient (r) between $d1$ and $d2$ is known or estimated, the following approach to lot sizing of the garment size categories with fitting to two anthropometric dimensions can proceed as follows.

1 – Consider only dimension 1, to begin with, and proceed as recommended for sizing of garments fitting one anthropometric dimension, considering the actual population mean ($m1$) and standard deviation ($sd1$) for the underbust girth. Once the size categories' order quantities have been obtained, dimension 2 shall be considered.

2 – Consider the correlation coefficient (r) between $d1$ and $d2$. Correct the standard deviation of dimension $d2$. Say, if 20% of the variance in variable $d2$

is explained by the variance of variable d1, then only 80% of the variance of variable *d2* is following the normal distribution, for each array of *d2* size categories, within one each fixed *d1* size category. Since the standard deviation is the square root of the variance, the corrected standard deviation for dimension 2 becomes $sd2' = \sqrt{(1 - r^2)}.sd2$

3 – For each size category considered for dimension 1 (*cli*), calculate the local value for the corrected mean of dimension 2 (*m2|cli*) using the following equation.

$$m2 \mid cli = r\frac{sd2}{sd1}cli + m2 - r\frac{sd2}{sd1}m1$$ This will yield local corrected

values for the mean of dimension 2, to be used in conjunction with the corrected standard deviation for dimension 2 (non-dependent on dimension 1size categories).

4 – List the category sizes in centimeters and the transition values between categories for dimension 2 (cup size). This should yield a list of *m* category sizes and *m+1* transition points, with the category sizes apart from each other of the value *I* (for this garment 2cm), and the transition points apart from the immediate categories by a value equal to *I/2* (in this case, 1cm).

5 – Using the order quantities obtained for each category size of dimension 1, parse these into order quantities for the category sizes of dimension 2 using the approach presented for a single anthropometric dimension fitting garment.

3.1   Application example – garment fitting within two anthropometric dimensions

Considering a population of adult women with underbust girth with a mean (*m1*) of 83.5cm and a standard deviation (*sd1*) of 7.2cm, and with cup size dimension with a mean (*m2*) of 15.8cm and a standard deviation (*sd2*) of 2.1cm, the order size quantities for dimension 1 are presented in Table 2, for a total order (*N*) of 250 brassieres.

According to the calculation results presented in Table 2, total order quantity at this stage sums a total of 249 units, which is very close to the starting value of 250 units. Let the correlation coefficient between the two dimensions be 0.33, and the corrected standard deviation for dimension 2 (*sd2'*) becomes 1.98. The results of calculation of the local value for the corrected mean of dimension 2 for each size category considered for dimension 1 (*m2|cli*), are presented in Table 3.

Finally, the order quantities for the combined sizing categories are obtained, and presented in Table 4, yielding a total of 245 units. Interestingly, given the correlation coefficient between the two anthropometric dimensions, the values are skewed across the diagonal, and there is some dependency visible between underbust girth and cup size, which was created by the corrected local means for cup size, depending on underbust girth. This skew in local mean values for dimension 2 is compensated overall across the matrix by the corrected standard deviation which is smaller than the standard deviation of the anthropometric dimension viewed by itself. The results obtained are presented in graphical form in Figures 1 and 2. The charts show the aforementioned skew.

Table 2 Order size quantities for dimension 1 of the application example
for two anthropometric dimensions fitting garments

Size categories	Transition points	Percentile	Percentile difference	Order quantity
-	62,5	0,18	-	-
65	-	-	1,14	3
-	67,5	1,31	-	-
70	-	-	5,01	13
-	72,5	6,33	-	-
75	-	-	13,90	35
-	77,5	20,23	-	-
80	-	-	24,24	61
-	82,5	44,48	-	-
85	-	-	26,60	66
-	87,5	71,07	-	-
90	-	-	18,36	46
-	92,5	89,44	-	-
95	-	-	7,97	20
-	97,5	97,41	-	-
100	-	-	2,18	5
-	102,5	99,58	-	-

Table 3 Results for the local value for the corrected mean of dimension 2
for each size category considered for dimension 1

Size categories for d1	65	70	75	80	85	90	95	100
Corrected local mean for d2	14,0	14,5	15,0	15,5	15,9	16,4	16,9	17,4

As a verification of the mathematical formulation, the mean, standard deviation and correlation coefficients of the results are calculated and compared with the original assumptions, as shown in Table 5. Slight variations occur, which may be explained in part by rounding to the unit of the order sizes, which was done twice (at first for the order quantities considering only underbust girth size, and later in the process, for the final order quantities matching two dimensions). Another source of non convergence between results may be attributable to the somewhat forced comparison of a continuous variable with a discrete one and the computation of statistical parameters intended for continuous variables, applied to step variables.

Table 4 Order quantities for the combined dimensions of the application example (lot sizing of brassiere size categories for a total order size of 250 units; d1 – underbust girth; d2 – cup size)

d2 / d1	65	70	75	80	85	90	95	100
AA	0	1	2	2	1	1	0	0
A	1	4	9	12	9	4	1	0
B	1	5	14	23	23	14	5	1
C	0	2	8	18	22	17	8	2
D	0	0	2	5	9	8	5	1
E	0	0	0	1	1	2	1	0
F	0	0	0	0	0	0	0	0

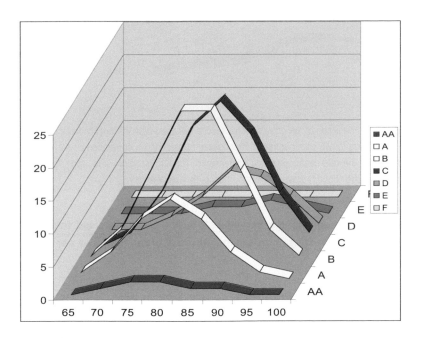

Fig.1 Graphical depiction of the order quantity results obtained from the application example for garment fitting to two anthropometric dimensions, considering a correlation coefficient of 0.33 between the two anthropometric dimensions (underbust girth and cup size for brassieres) – each ribbon represents a cup size category

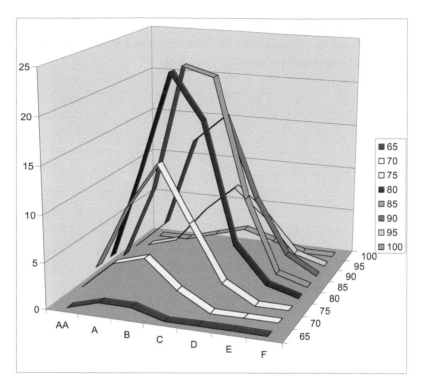

Fig.2 Graphical depiction of the order quantity results obtained from the application example for garment fitting to two anthropometric dimensions, considering a correlation coefficient of 0.33 between the two anthropometric dimensions (underbust girth and cup size for brassieres) – each ribbon represents an underbust girth size category

Table 5 Comparison among assumed distribution parameters and values obtained from the results of the computations of order quantities per size categories

	Mean of d1	Standard deviation of d1	Correlation coefficient (Pearson)	Mean of d2	Standard deviation of d2
Assumption	83.50	7.20	0.330	15.80	2.10
Result	83.43	7.00	0.322	16.09	2.63
Difference (%)	-0.08%	-2.9%	-2.4%	+1,8%	+25%

The results also lead to challenge assumptions not only about clothes lot sizing approaches in practice, but also on common practices of computer model mannequins such as those commonly considered of 5th percentile female or 95th percentile male, which are not representative of the actual proportions between body segment dimensions. Moreover, there is growing evidence that there are several body types where different proportions are typical, irrespective of size, as these proportions define the type and maybe more identifiable if specifically sought in future studies.

4   DISCUSSION

The lot sizing methods presented depend on the knowledge of correlation coefficients between anthropometric dimensions that are relevant for each particular garment. While literature on anthropometric dimensions does report on these coefficients (e.g. Barroso et al., 2005), specific correlations for application within the formulations presented are not readily available. Hence, future studies are needed to unveil these in a more consistent and widespread manner, across the globe.

Transition points were set in the mathematical formulation, and the examples of application exactly at the same distance of two consecutive size category central values. Transition points may be easily moved closer to the size category above or below them, according to the assumptions about snugness or slack that people may prefer, according to the size of the garment, and the fashion trends. The fashion cycle is typically very short (about 3 months), hence tools that may be used by garment distribution managers to optimize merchandise availability at points of sale need to be accurate. Naturally, it is acknowledged that other factors besides availability of size are important to make the sale.

5   FUTURE STUDIES

This study of garment lot sizing raises several important issues to the garment industry that should be considered in future work.

Since nowadays customers are more demanding and have higher expectations the methodology presented here is intended to evolve towards what we can call a 'customer-centric production order'. For that purpose the methodology requires further developments. One of the issues to take in consideration is that companies often make the mistake of treating costumers all the same, which may lead to products that are designed for everyone and that satisfy no one. These days' customer preferences never stand still. Therefore it is important to define and prioritize customer segments, which means that a pure statistically based lot sizing approach may result inadequate from a 'customer-centric' standpoint.

This way we envisage a fuzzy logic (Zadeh, 1965) based methodology where objective attributes (e.g. anthropometric dimensions) are combined with subjective attributes (like aesthetic appeal, garment style, functionality or comfort) in a model designed to support the decision-making process regarding customer segmentation and, departing from there, the lot sizing. Customer segmentation can be seen in the context of 'collaborative customization' (this concept was developed from the concept of 'mass customization', which is the customization and personalization of products and services for individual customers at a mass production price) where companies talk to individual customers to determine the products that best meet the customers' preferences (Pine, 1992). The use of fuzzy logic has several advantages namely it allows the representation of knowledge, the formulation of complex inference processes, and the aggregation, in a coherent mathematical form, of both objective and subjective data. Thus fuzzy logic allows capturing and treating imprecise and vague concepts, supporting also the use of natural language (Nunes, 2009).

This 'customer-centric production order' approach can also contribute to a sustainable supply chain (Figueira et al, 2012) i.e. to a responsible management of resources use, namely of raw materials like fabrics or sewing thread and zero waste. In fact, customers that are fair trade, environment and social issues conscious are more likely to buy from companies' that adopt sustainable management practices. Therefore the fuzzy based methodology can also consider these types of customer concerns as attributes to use in the decision-making process.

REFERENCES

Barroso, M.P., Arezes, P.M., da Costa, L.G. and Miguel, A.S. (2005) Anthropometric study of Portuguese workers, *International Journal of Industrial Ergonomics*, Vol. 35, pp. 401-410.

Coelho, D. A. (2012) Inaugural Editorial: A new human factors and ergonomics journal for the international community is launched, *International Journal of Human Factors and Ergonomics* 1 (1), 1-2.

Coelho, D. A., Dahlman, S. (2002) Comfort and Pleasure, in *Pleasure with Products: Beyond Usability* (edited by William S. Green and Patrick W. Jordan), London: Taylor & Francis, 322-331.

Figueira, S., Cruz-Machado V., Nunes, I. L. (2012) Integration of human factors principles in LARG organizations – a conceptual model, *Work: A Journal of Prevention, Assessment and Rehabilitation*, Vol. 41, pp.1712-1719.

Mpampa, M. L., Azariadis, P. N., Sapidis, N. S. (2010) A new methodology for the development of sizing systems for the mass customization of garments, *International Journal of Clothing Science and Technology*, Vol. 22 Iss: 1, pp.49 – 68.

Nunes, I. L. (2009) FAST ERGO_X – a tool for ergonomic auditing and work-related musculoskeletal disorders prevention, *Work: A Journal of Prevention, Assessment, & Rehabilitation* , Vol. 34 Iss. 2, pp. 133-148.

Pheasant, S. (1998) *Bodyspace: Anthropometry, Ergonomics and the Design of Work*, second ed. Taylor & Francis, London.

Pine II, J. (1992). *Mass Customization: The New Frontier in Business Competition*. Boston, Mass.: Harvard Business School. ISBN 0-87584-946-6.

EN 13402-3 (2004) Size designation of clothes – Part 3: measurements and intervals, Final Draft.

Zadeh, L. A. (1965). *Fuzzy sets, Information and Control*, Vol. 8 Iss: 3, pp. 338-353.

CHAPTER 34

A Study on Development of the Korean Stress Model for Industrial Accident Prevention -Focused on the Married Workers-

Youngsig Kang[1], Sunghwan Yang[2], Taegu Kim[3]*

[1]Dept. of Occupational Health & Safety Engineering, SEMYUNG University, Jecheon, Chungbuk, 390-711, South KOREA
kys@semyung.ac.kr

[2]Dept. of Prosthetics & Orthotics, Korea National College of Rehabilitation and Welfare, Pyeongtaek, Gyeonggi, 459-070, South KOREA
shying@hanrw.ac.kr

[3*]Dep. of Occupational Health & Safety Engineering, College of Biomedical Science & Engineering, INJE University & AEI, Gimhae, Gyeongnam, 621-749, South KOREA
tgkim@inje.ac.kr

ABSTRACT

The causative factors of industrial accidents are an unsafe act and an unsafe condition. Stress, a fundamental cause of industrial accidents, causes unsafe condition as well as unsafe action. In this paper, focused on married workers, we will develop a Korean stress model (KSM) which considers life stress, job stress, and workplace stress in order to prevent the fundamental industrial accident. The results of this paper provide decisive strategies and methods to eliminate the fundamental cause of industrial accidents.

Keywords: Industrial accident, Unsafe action and Unsafe condition, Life stress, Job stress, Workplace stress, KSM (Korean stress model)

1   INTRODUCTION

Korea has established solution strategies by fundamental causes of occupational accidents with a change of complex industrial structure. However Korea has reached to establish preventive measures for industrial accident by fundamental cause analysis of occupational diseases.

The unique factor that appears as a common denominator among all personal factors related to the cause of accidents is high level stress from the point of view that the accident occurs (Kang, et al., 2008). Also, the decision on the priority of these stresses is very important for prevention of industrial accidents.

These results of industrial stress have appeared as industrial accidents, occupational diseases, cancer, and depression. In addition Korean industrial accident rate has been continuously stagnant at 0.7 for 13 years (KOSHA, 2008-2011, Kim, et al., 13). Also, industrial accident rate in 2010 was estimated to be 0.69 (KOSHA, 2008-2011). Accordingly, it is required to acutely look at the fundamental prevention measures of industrial accidents. The existing research domestically and overseas provides the following opinions. In the case of cooperative work evaluated on response numbers by cognition, a condition which is not stressful allowed a higher performance of work and decreased the error in high level work with minimal cooperation more than low level work with minimal cooperation. In contrast, condition which is not the stress performed a higher performance work and decreased the error in minimal cooperation work of low level than minimal cooperation work of high level.

These results were verified by experiments (Larson, 1973). Solutions for the workplace were analyzed and evaluated for risk assessment economic stimulus and support of small to medium sized industries (Cooper and Cartwright, 1997).

The human behavior influenced by physical, psychological, and environmental factors was analyzed relative to the importance by effectively applying hierarchical analysis techniques (Jung, 2000).The life stress model fitted to Korean circumstances for small to medium sized industries was developed for the prevention of life stress. This model was put to practical use for safety management education in real industrial fields. The prevention of stress for unmarried workers integrated the stress of life, job and workplace. These stress factors were developed in the stress model for Korea through questionnaires and were proposed for the strategies of industrial accident prevention (Kang et al., 2011).

Recently, the first priority policy in the advanced safety & health countries is the stress factors. So the prevention of industrial accidents recognizing the stress factors is actively pursued. At present, advanced safety & health countries have actively pursued a preventive policy based on the stress factors for industrial accidents above all other factors (Kang, et al., 2011).

However, a paper on industrial stress which is a combination of job stress, life stress, and workplace stress has not appeared yet for prevention of industrial accidents.

Therefore, this paper is to develop a Korean stress model for the prevention of industrial accidents from induced stress (life, job, and workplace). Accordingly, in the national strategy, this model will establish a solution that dramatically minimizes industrial accidents as well as fatal accidents. Also this model is to be put to practical use for safety management education in the real world workplace.

2 DEVELOPMENT OF KOREAN STRESS MODEL FOR MARRIED WORKERS

To prevent fundamental factors of unsafe acts and conditions which induced occupational accidents and diseases, the stress factors fitting Korean circumstance 33 factors extracted from a paper on stress and an existing Life Change Unit (LCU) model for Korea (Kang, et al., Kim, 2005, Yang, et al., 2007, Cho, et al., 2005, Kang, et al., 2008). The LCU model Korean life stress factors were proved to effect accidents. These factors are the death of parents, death of spouse, death of close friends, changes in family member's health, unemployment, jail term, personal injury or illness, sex difficulties, mortgage over $10,000, change in religious activities, change in living conditions, change in social activities, legal trouble, divorce and so on (Kang, et al., 2011, Yang, et al., 2007, Kang, et al., 2008).

2.1 Methodology

First, we analyze foreign stress models and the domestic stress model, Form this we identify the stress factors. Then we decide the life, job, and workplace stress factors to minimize industrial accidents.

Secondly, we decide priority and sample response rate by stress type based on proposed questionnaire of public institutions, manufacturing companies, construction companies, and service companies for the development of a systematic Korean stress model fitting the real conditions in Korea.

Third, we statistically test each stress factor for evaluation of stress intensity.

The intensity evaluation of stress factors develops a Korean stress model based on the priority from results of normal testing with each stress factor at a significance level of 0.05 (α).

From the above, we suggest a prevention method fitting the Korean real conditions in the national strategy.

3   CASE STUDY

3.1   Scope

The decision on stress factors that effect industrial accidents is extracted from existing stress models and job stress models. From this, the questionnaire is composed of 33 stress factors. We evaluate the priority of stress factors on workers and managers in the fields. The areas for the sample survey are selected across the whole country including Seoul, Gyeonggi Province, Incheon City, Chungcheong Province, Gyeongsang province, and Busan.

The business categories for the survey are mainly manufacturing, construction, and service companies where many industrial accidents occur.

3.2   Result

In Table 1, data was collected from field workers and managers of each company with less than 300 workers by simple random sampling. There were 1,500 questionnaires sent and 1,258, that is,- 83.9%, were returned.

Table 1. Data of the participating enterprises

No.	Selection Items	Frequency (No. of data)	Percentage (%)
	Residence area		
1	City of Seoul	175	13.9
2	Gyeonggi Province	613	48.7
3	Incheon	204	16.2
4	Chungcheong Province	30	4.4
5	Gyeongsang Province	55	4.4
6	City of Busan	42	3.3
7	Other	139	11.1
	No. of workers		
8	Above 300	354	28.1
9	100-300	221	17.6
10	50-100	284	22.6
11	Under 50	399	31.7
	Industrial Classification	353	28.1
12	Manufacturing company	755	60.0
13	Construction company	150	11.9
14	Other		
	Length of service		
15	Above 10	606	48.2
16	5-10	291	23.1
17	1-5	235	18.7
18	Under 1	126	10.0
	Sex		
19	Man	1,190	94.6
20	Women	68	15.4

	Age		
21	Over 50	279	22.2
22	40	533	42.4
23	30	420	33.4
24	20	26	2.1
Total		1,258	100.0

Table 2 shows a Korean stress model that evaluated the strategy and prevention factors by a priority criterion matrix in order to dramatically minimize industrial accidents for married safety & health workers and managers all over the country.

Table 2. Korean Stress Model by the Married Workers

Rank	Category	Frequency	Sample response rate
1	Death of spouse	235	0.187
2	Death of parents	216	0.172
3	Death of close family members	152	0.121
4	Changes in family member's health	74	0.059
5	Death of close friends	70	0.057
6	Unemployment	69	0.055
7	Heavy business	63	0.050
8	Personal injury or illness	63	0.050
9	Troubles with boss	62	0.049
10	Work condition and time pressure	61	0.048
11	Employment instability	60	0.048
12	Unpleasant workplace environment	56	0.045
13	Change in living conditions	55	0.044
14	Irregular work	51	0.041
15	Serious sound and vibration	49	0.039
16	Responsibility between organization and role conflict	41	0.033
17	Corporate culture of Korea	39	0.031
18	Change in social activities	38	0.030
19	Inappropriate resting hours	37	0.029
20	Shift to different line of work	35	0.028
21	Simple repetition work	34	0.027
22	Unnatural work posture	33	0.026
23	Divorce	28	0.022
24	Work with inappropriate posture	27	0.021
25	Work with excessive force	22	0.017
26	Sex difficulties	20	0.016
27	Excessive drinking and smoking	17	0.014
28	Jail term	16	0.013

29	Environment of low temperature	14	0.011
30	Mortgage over $10,000	13	0.010
31	Low lighting (under 300 lux)	12	0.010
32	Legal trouble	9	0.007
33	Change in religious activities	3	0.002

The following conspicuous stress factors were found to have the same priority in Table 2

(1) Death of spouse
(2) Death of parents
(3) Death of close family member
(4) Changes in family member's health
(5) Death of close friend
(6) Unemployment
(7) Heavy business
(8) Personal injury or illness
(9) Troubles with boss
(10) Work condition and time pressure
(11) Employment instability
(12) Unpleasant working environment

Therefore, among those stress factors, the death of spouse and parents causes more serious stress than any other stress factors in this paper.
In Table 3, items with a significance level of 0.05 (α) are:

(1) Death of spouse
(2) Death of close family members
(3) Changes in family member's health
(4) Irregular work
(5) Unnatural work posture
(6) Excessive drinking and smoking
(7) Change in religious activities

One remarkable fact is the difference between the death of a spouse and the death of a parent is not significant. This cultural connection to one's parents as the central axis of life is being diminished rapidly these days in the young generation. Also, the culture of honoring parents which acted as the hub of all life is going to disappear rapidly.

The analysis of the difference in intensity of the sample response rate for stress factors between these items was not significant.

$H_o : P_i = P_j \ H_a : P_i \neq P_j$ (Where, i, j=1,2,...., 33)

Table 3. The normal testing result by significance level (α =0.05)

Item of sample rate	Test statistic	Rejection region
$P_1 = P_3$	4.71	$Z \geq \lvert \pm 1.96 \rvert$
$P_3 = P_4$	5.43	$Z \geq \lvert \pm 1.96 \rvert$
$P_4 = P_{14}$	2.07	$Z \geq \lvert \pm 1.96 \rvert$
$P_{14} = P_{22}$	2.08	$Z \geq \lvert \pm 1.96 \rvert$
$P_{22} = P_{27}$	2.14	$Z \geq \lvert \pm 1.96 \rvert$
$P_{27} = P_{33}$	3.33	$Z \geq \lvert \pm 1.96 \rvert$

Where P_1 = Death of spouse

P_3 = Death of close family members

P_4 = Changes in family member's health

P_{14} = Irregular work

P_{22} = Unnatural work posture

P_{27} = Excessive drinking and smoking

P_{33} = Change in religious activities.

4   DISCUSSION AND CONCLUSIONS

The industrial accident in the knowledge-based society is induced by complex factors not just one cause factor. Accordingly, one of the ways to eliminate industrial accidents is to analyze and evaluate stress factors thoroughly. It can prevent traditional accidents and eliminate industrial accidents from the fundamental complex causes. From the results of this paper, the following conclusions can be drawn for the fundamental strategy of industrial accident prevention.

First, we develop a Korean stress model fitting the real state of married workers for refining the national strategy considering job, life, and workplace stress.

Second, it should propose concrete and systematic safety management methods along the Eastern ideology for prevention of industrial accidents because Eastern culture and ideas in Korea are based on the principal axis of all life.

Third, it should systematically improve the organization of the company and the working environment in order to prevent industrial accidents in heavy industries resulting from troubles with the boss, and unpleasant working environment.

Fourth, it should address items that consider the mental and physical health of workers to prevent occupational accidents as well as occupational diseases.

Finally, if the development of a KSM is applied in education on safety management in the field, it provides decisive information to dramatically minimize fatal accidents as well as industrial accidents.

In the future, it is required to develop Korean stress models based on married workers and unmarried workers with a correlation analysis between stress factors and disabilities.

REFERENCES

Kang, Y.S., Yang, S. H., and Kim, T. G., (2011), "Development of the Korean Stress Model for a National Strategy –Focused on Unmarried Workers-", *Journal of the Korean Institute of Plant Engineering*, 16(4), 75-80.

Kang, Y.S., et al . (2008), *Modern Statistic*, Donghwa Technology Press.

Kim, D. S., et al., (2005), "Evaluation Management of Job Stress", *OSHRI*, 1-17, 2005.

KOSHA (2008-2011), *Statistics of the Industrial Accident (per years)*.

Yang, S. H. and Kang, Y. S., et al., (2007), *Work Analysis and Management*, Medical Korea.

Yang, S. H. and Kang, Y. S., et al., (2006), *Ergonomics*, Shin Gwang, 118-130.

Yang, S. H. and Kang, Y. S., et al., (2011), *Safety Management System*, Hyunmoon Press.

Jung, K. T., (2000), "A Method Considering Performances Shaping Factors in Quantitative Human Error Analysis", *Journal of the Korean Institute for Industrial Safety*, 12, 113-121.

Cho, J. J., et al., (2005), "A Study on Reliability Assessment and Accuracy of Measurement Instrument for Korean Job Stress", *OSHRI*.

ACHI Masatomo, *et al.*, (2001), "National Occupational Health Research Strategies", *International Journal of Industrial Health*, 39, 287-307.

Cooper, C. L., and Cartwright, (1997), "An Intervention Strategy for Workplace Stress", *Journal of Psychosomatic Research*, 43(1), 7-16.

Kang, Y.S., Hahm, H. J. (2008), Yang, S. H., and Kim, T. G., "Application of the Life Change Unit Model for the Prevention of Accident Proneness among Small to Medium Sized Industries in Korea", *Industrial Health*, 46(5), 470-476.

Kim, T. G., Kang, Y. S., and Lee H. W., (2011), "A Study on Industrial Accident Rate Forecasting and Program Development of Estimated Zero Accident Time in Korea", *Industrial Health*, 49(1), 56-62.

Larson, K. M., (1973), "Leadership Style, Stress, and Behavior in Task Performance", *Organization Behavior and Human Performance*, 9, 407-420.

T. Saaty, (1983), "Priority Setting in Complex Problem", *IEEE Transactions on Engineering Management*, 30(3), 140-155.

CHAPTER 35

Study the Variability of Anthropometric Measurements of Women's Feet Laser

Lima, Edmilson Gabriel de
Olle, Francisco A. da Luz
Okimoto, Maria Lucia L.R.

Federal University of Parana
Curitiba, PR, Brazil
edmilsonlimas@gmail.com
olledaluz@yahoo.com.br
lucia.demec @ufpr.br

ABSTRACT

This article aims to present the study of the variability of anthropometric measurements of women's feet laser. 30 samples were selected for convenience of a population of 60 university students and Region of Curitiba. The selection criterion was used in the body mass index. We selected 10 samples with a BMI less than 18,5 and samples with a BMI between 18,5 and 24,9 and 10 samples with BMI 25 to 29,9. Anthropometric variables were measured six feet by right and left calipers and four dimensional analog checker heights called direct method. For the method used is called an indirect three dimensional laser scanner manual. In total 288 were observed variables. Was applied in the analysis of variability between methods Linear Pearson Correlation obtaining the value near 0, 99 for the group with a BMI less than 18, 5 confirming that the methods correlate. For the group with BMI between the mean of 0,849 mm confirming that the data collected is accurate and will have similar quality, both in the question of exactly how much accuracy. Finally we applied ANOVA with a criterion to evaluate the differences between the mean values for samples with a "BMI between 25 and 29, 9", yielding a population ratio of 0, 05. Proving that there was no significant variation between the direct and indirect method.

Keywords: Anthropometry. Foot. Sampling. Laser Scanning. Three-dimensional

1 Anthropometric Measurement Methods

The traditional anthropometric measurement methods for physical contact with the object causing pressure on the epidermal tissue, hindering the orientation with respect to management of measurement may have parallax error; lack of calibration of the instrument can cause precision errors.

Regarding the method of indirect measurement without physical contact with the object through the technique of laser scanning three dimensional we can find many advantages such as: reliability of measures generated by the new technology, capturing an entire object in one continuous scan without physical contact, analysis of data high quality, obtaining pieces of formal geometry of high complexity; positioning itself, there is no need for bulky mechanical arms fixed position tripods or external positioning devices. The National Institute of Technology (INT) highlights the following advantages to this technology: a survey of thousands of points on the body surface in a few seconds, archiving of the scanned image of the body in 3D allows us to raise a large amount of anthropometric variables, independent of physical presence of the individual; the data obtained can be used to calculate surface areas, forms of body segments, body contours and other measures that cannot be obtained through traditional methods in anthropometry. Na figure 1 Shows the scanner 700Cx manufacturer Z Corporation used in this work and table 1 technical information:

Figure 1 – ZScanner 700 CX (ZCORP, 2011)

Table 1 - Technical Specifications

Description	Specifications
Applications:	Reverse Engineering, Medical Orthosis, Simulations.
Texture Resolution:	50 a 250 DPI
Sampling of Rate:	18.000 measurements per seconds
Number of Cameras:	3
XY accuracy up to 50 microns:	(up to 0.002 inches)
Z-resolution :	mm (0.004 inches Z)
Weight:	1,3 kg (2.85 lbs)

The feet of Brazilians are relatively shorter and more "fat in relation to European feet, which are thinner and longer. Like many for shoe manufacturing are based on European forms that explains the cases of tightness in the feet (Lacerda, 1984).

The footwear industry has skyrocketed in the last decade. Brazil is an exporter of shoes for many countries, according to the Brazilian Association of Footwear Industry Brazil is the third largest producer of shoes, with 800 million pairs per year, surpassed only by India with 900 million, and China, with 9 billion. The increase in exports was 15, 2% in the second half of 2010 (GLOBE 21, 2011).

1.2   Methods

We interviewed and obtained the variables weight and height of 60 females for convenience in   order to   reduce collection   time,   costs   and logistics. After reading the agree with the Terms of Consent (IC), were collected identification data such as age, height, weight, etc. After this was done analysis and calculation of BMI   and were   then   selected 30 individuals with BMI   below 18.5,   BMI between 18, 6 to   24, 9 and 25 to29, 9. Called G1, G2, G3. The following is a flowchart summarizing the steps proposed in this study applied:

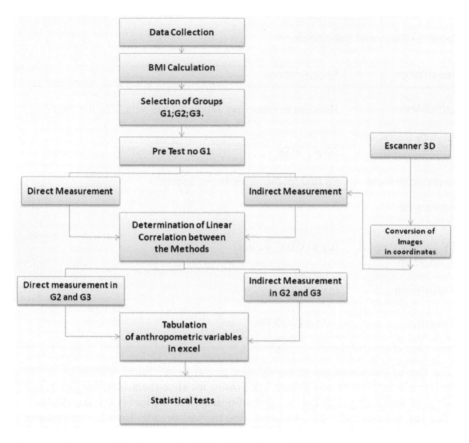

Figure 2 – Flowchart of Implementation Methodology (Author, 2011)

2 CONCLUSIONS

Analyzing the results of this study, it becomes possible to verify the relevance of the tests: Linear correlation Pearson, ANOVA test. Through the results in Table 2 can be seen that both the direct and indirect methods of collecting anthropometric data are accurate, since the group thinness, the linear correlation coefficient (r) approaches 1 (TRIOLA, 2008). For the healthy group the value of the t test of the population proportion is 0, 01 indicating a significant level of confidence as shown in table 2. For the group overweight ANOVA value indicates a population ratio of 0, 05 which is a significant level of confidence. Followed the same protocols for obtaining measurements presented in this study. After analysis of the three statistical tests described in Table 2, we observe that for each group in question, taking into account the six anthropometric variables the difference between the means of direct and indirect methods are: Group 1: 0,0001 mm . In group 2, 0,849 mm. And for group 3: 0,016 mm.

The absence of direct data collection favor the use of tables approach is not always reliable, Pheasant (1995) highlights that even a direct collection of anthropometric is virtually impossible to achieve accuracy less than 5 mm, and this error is negligible compared to the results found in tables and data approximation. He also reports that the daily practice of ergonomics anthropometric specifications require an accuracy of 25 mm, so below these errors do not compromise the results or its applicability. Therefore the work meets these values and international standards which is 5%. Regarding the direct and indirect methods, we note that the first is simpler, as a function of the difficulty of putting the meter on the same point, pressure on the epidermal tissue, parallax error, noting the dimension table, etc.., and the error probability of reading is greater. The second is more complex and precise, since it requires training, knowledge for working with images and extraction of anthropometric dimensions. It follows that the probability of error is smaller.

Noteworthy is the indirect method as the main contribution in this work since the manual scanner is versatile, lightweight, fast, accurate to 0,05 mm did not bring any harm to the object with total scan of the object can be imaged in the image geometry and item filed.

In relation to the proposed objectives verified after the analysis that the differences between the methods and variables to the same individual are not significant, as shown in table 2:

Table 2 - Result comparison between the methods and reliability (Author, 2011)

Objective	n	Group	Difference between the Means	Test	Results	Standard
Comparison between Methods and Reliability	10	G1	0, 0001	Pearson	0,99	1
Comparison between Methods and Reliability	10	G2	0, 849	t	p = 0,01	0,01 a 0,05
Comparison between Methods and Reliability	10	G3	0, 0016	ANOVA	p = 0,05	0,01 a 0, 05

As a contribution this study can add the ergonomic design of footwear targeted to a specific population with specific characteristics. We envision the following recommendations that may guide further studies:

• Expand and improve the method of scanning in the plantar region of the toes;
• study of the characteristics of the generated mesh in the 3DStudio MAX mesh geometry. (Volume, area, circumference, etc.)
• Applicable for research and product development for people with deformities of the feet as Hallux Valgus, Obese, Epis.
• Applicable to research and development of Prosthetics and Orthotics.

ACKNOWLEDGMENTS

To all friends and colleagues who somehow collaborated in data collection, and only then was it possible to perform this work. To all the volunteers, without them this research would not become reality.

REFERENCES

GLOBE 21. Leather and footwear. Available at.: http://www.global21.com.br/informessetoriais/setor.asp?cod=3. Accessed on 03 June In 2011.

Lacerda, D. F. Anthropometric Measurement of Foot. Dissertation. Rio de Janeiro, COPPE / UFRJ, 1984.

NATIONAL INSTITUTE OF TECHNOLOGY. Three-Dimensional Anthropometric Survey of the Brazilian Population. Ministry of Science and Technology, November 2005. http://wear.io.tudelft.nl/files/brasil05/PATPB.pdfavailable.

Pheasant, S. Body space - Anthropometry, Ergonomics and Design. 2. Ed London, 1996.

Triola, M. F. Introduction to Statistics. Rio de Janeiro: LTC, 2008.

ZCORP.COM. http://www.zcorp.com/en/Products/3D-Scanners/ZScanner-700-CX/spage.aspx

CHAPTER 36

Work Posture Assessment of Computer Users using CARULA

(Computer Aided Rapid Upper Limb Assessment)

Lakhwinder Pal Singh
Assistant Professor, Department of Industrial & Production Engineering,
Dr B R Ambedkar National Institutes of Technology (NIT), Jalandhar
Punjab, India
lakhi_16@yahoo.com

Amanpreet Singh
St. Soldier Institute of Engineering and Technology, Jalandhar
Punjab, India
goldeneye4u@gmail.com

Supreet Kaur
Lovely Professional University (LPU), Phagwara
Punjab, India
soni.supreet@gmail.com

ABSTRACT

It's the computer era and pcs (personal computers) are daily used by thousands of people. When people at work they used pcs (personal computers) minimum for four hours per day which cause a lots of pain in a various parts of the body .Computer Vision Syndrome (CVS), low back pain, tightness headaches and psychosocial pressure are some common problem faced by computer users. Rapid Upper Limb Assessment (RULA) is a method concerned with work related to upper part of the body which includes wrist position, upper arm position, lower arm position, neck position and trunk position. RULA done manual calculations by entering values into the work sheet equivalent to the position which is observed, if there are bulky numbers of working employees it means manual calculations are more, it can become monotonous if many activities and corresponding postures need to be analyzed and there are more chances of subjective biasness. In this paper, Computer Aided Rapid Upper Limb Assessment (CARULA) will help to decide the correct work posture analysis for industry. The present work focused on the calculation of RULA score with the help of introducing new algorithm CARULA which is quite efficient and precise than the manual calculations. CARULA has been computerized to calculations in a user friendly environment with pictorial representation using MATLAB. We have tested this CARULA by analyzing on 10 different work posture; CARULA is relatively much faster and significantly reduced the risk of subjective biasness as compared to manual RULA.

Keywords: Work Posture Analysis using CARULA, Reduced Subjective Biasness, CARULA.

Introduction and Literature Survey:

Ergonomically considered products have seemed in a variety of industries in reaction to the increasingly observable problems addressed by ergonomics. Ergonomically designed office furniture, baggage, lawn tools and computer products are growing in popularity. Ergonomics has many advantageous tools for examining risk in work posture. Rapid Upper Limb Assessment (RULA) is a method concerned with work related to upper part of the body which includes wrist position, upper arm position, lower arm position, neck position and trunk position (Microsoft Corp, August 2005).

RULA done manual calculations by entering values into the work sheet corresponding to the working posture which is observed, if there are huge numbers of working employees it means manual calculations are more, it can become monotonous if many activities and corresponding postures need to be analyzed and there are more chances of subjective biasness. Therefore, CARULA is developed as a computerized process as it would result effective and accurate assessment of postural analysis of various activities in industry. The inspiration behind CARULA is to computerize the manual work sheet of RULA so that it is more useful when carrying out postural analysis in the field. Some snapshots can also be taken of the activity and uploaded to help in making decision about the immobile posture in the activity. The user can choose the neck, wrist, shoulder and trunk position from the various options available. The posture score and final score are updated automatically as user chooses from the various positions. The Health and Safety Executive report that one-third of the injuries reported each year to the HSE and to local authorities is affected by manual handling. Heavy manual Labour, uncooperative postures, manual materials handling, and previous or existing injury are all risk factors implicated in the development of Musculoskeletal Disorder (MSD). The tool for assessment can be divided by broadly into 3 categories, checklists, graphical assessment of posture and film and video based. Several tools have been proposed for postural analysis (Martha J. Sanders). REBA is usually done by hand calculations by entering values into the work sheet corresponding to the posture activity observed. It can become tiresome if many activities and corresponding postures need to be analyzed. From now CAREBA is developed as a computerized procedure as it would result effective assessment of postural analysis of various activities in industry. The enthusiasm behind CAREBA is to computerize the manual work sheet in REBA so that it is more useful when carrying out postural analysis in the field. CAREBA is developed as standalone software in java and is equivalent to using a calculator with mechanism to calculate REBA score. CAREBA is divided in to 3 sections the score A, score B and the final REBA score. A photograph can also be taken of the activity and uploaded to help in making decision about the static posture in the activity. The user has to just choose the posture from a set of pictures but the main limitation with CAREBA is that once the user input the basic details, then user can proceed to calculate the score A. The user can choose the neck, legs and trunk position from the various options available. The posture score and final score A are updated automatically as user chooses the various positions. Rapid Entire Body Assessment (REBA) is a tool used in occupational ergonomics to access the risk factor associated with carrying out a particular task. The risk factor is determined from analyzing the various demands on the musculoskeletal system while performing the task (Preeti Sheba Hepsiba D and Darius Gnanaraj S, 2009).

"A comparison of posture and muscle activity means and variation amongst young children, older children and young adults whilst working with computers". Teenagers and young adults are now the most frequent users of computers; however computer-related posture and muscle activity differences between children and adults have not been previously examined. This study found children tended to use more spinal flexion and young children had greater spinal asymmetry when interacting with computers. The potential for these posture differences to result in greater risk of musculoskeletal discomfort and disorder may be offset by the tendency for children to have greater variation of posture and muscle activity. Despite these differences, similar prevention messages may be appropriate for children and adults including the encouragement of appropriate postures and appropriate task variation (Barbara Maslen and Leon Straker, 2008).

RULA:

The RULA method was first developed by Drs. McAtamney and Corlett of the University of Nottingham's Institute of Occupational Ergonomics in 1993. RULA is a screening tool that assesses biomechanical and postural loading on the whole body with particular attention to the neck, trunk and upper limbs. Reliability studies have been conducted using RULA on groups of VDU users and sewing machine operators. A RULA assessment requires little time to complete and the scoring generates an action list which indicated the level of intervention required to reduce the risks of injury due to physical loading on the operator. RULA is intended to be used as part of a broader ergonomic study. RULA (rapid upper limb assessment) is a survey method developed for use in ergonomics investigations of workplaces where work-related upper limb disorders are reported. This tool requires no special equipment in providing a quick assessment of the postures of the neck, trunk and upper limbs along with muscle function and the external loads experienced by the body. A coding system is used to create an action list which indicates the level of intervention required to reduce the risks of injury due to physical loading on the operator [L McAtamney, E Nigel Corlett,1993).

RULA is divided into four exposure classes: negligible, low, medium and high. Medium and high risk actions should be urgently addressed to reduce the level of exposure of risk factors (L P Singh, 2010).

CARULA and its Methodology:

CARULA is developed as a computerized process as it would result effective and accurate assessment of postural analysis of various activities in industry. The inspiration behind CARULA is to computerize the manual work sheet of RULA so that it is more useful when carrying out postural analysis in the field. Implementation of CARULA is done with the help of **MATLAB**. It can be used to do very simple as well as very sophisticated tasks. The reason for selecting MATLAB is an easy software package to use even without much knowledge (E. Pachepsky).

Environment of CARULA

CARULA is divided in to 3 sections the calculation of score A, score B and the final RULA score. CARULA calculate the angel and scores of upper limb which include neck, trunk, wrist, Arm positions lower arm and Upper Arm. The user can choose any one position from available options, by clicking on that position a radio button will be activated. User select an angel from the static image the angel will be calculated automatically and then score of that angel would also display on computer screen.

Advantages of CARULA

- Reduced Subjective Biasness- User chooses an angel from the static image the angel will be calculated automatically.
- Easy to understand and produce accurate results –because of computerized work it is easy to understand and reduce the chances of inaccuracy.
- Portable- it can run on Laptop, PDA and Tablet.
- User friendly interface – it is also used by those users who are not computer literate.
- Speedy results- there are no manual calculations.
- Used graphical user interface, enable different postures and provides complete report of the resulting scores.

Figure.1 CARULA output showing different calculations.

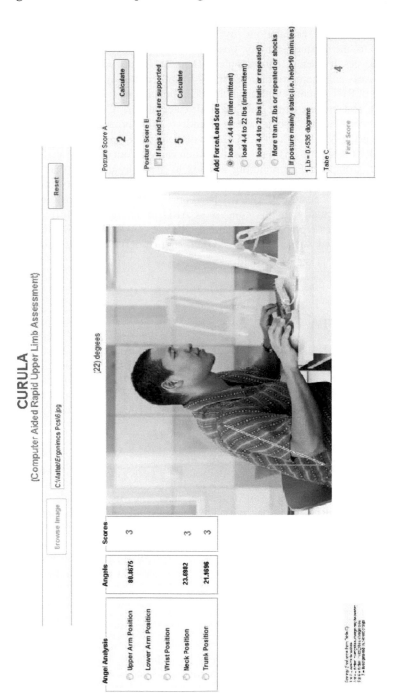

354

Result and Discussion:

RULA done manual calculations by entering values into the work sheet corresponding to the posture activity observed, if there are bulky numbers of working employees it means manual calculations are more which is very time consuming process and there are more chances of inaccurate results. CARULA do the calculation of RULA scores with the help of introducing new algorithm which is quite efficient and precise than the manual calculations. CARULA provides computerized calculations in a user friendly environment with pictorial representation using MATLAB. CARULA will help to computerize the manual work sheet of RULA so that it is more useful when carrying out postural analysis in the field. Here, Computer Programming is used to generate an action list which indicates the level of intervention required to reduce the risks of injury in upper limb due to wrong work posture. CARULA tried on 10 different work posture it has been approved it's geniality in this concern. CARULA is significantly reducing Subjective Biasness as compare to manual RULA. CARULA will help to computerize the manual work sheet of RULA so that it is more useful when carrying out postural analysis in the field.

Analysis of CARULA

Testing of CARULA is done with three different ways. One posture is performed on two different users. Calculations of manual RULA performed by a professional, a student and on CARULA. Resulted scores of them are not matched. CARULA generate the accurate results, in less time as compared to manual RULA and reduced the subjective biasness. Screen shot of CARULA is shown in Figure.1.

CONCLUSIONS

CARULA is quite efficient and precise than the manual calculations. Implementation of CARULA is done with the help of Matlab. CARULA produce accurate results as it was verified by manual RULA. Computer Programming is used to generate an action list which indicates the level of intervention required to reduce the risks of injury in upper limb due to wrong work posture. User select an angel from the static image the angel will be calculated automatically and then score of that angel would also display on computer screen.

REFERENCES

Microsoft Corp August 2005. The Importance of Ergonomic Input Devices in the Workplace, The Scope of Computer-Related Repetitive Strain Injuries and Methods for Their Prevention.

Martha J. Sanders .,"Ergonomics and Management of Musculoskeletal disorders",pg.214-218.

Preethi sheba Hepsiba D and Darius Gnanaraj 2009. COMPUTER AIDED RAPID ENTIRE BODY ASSESMENT (CAREBA) Department of computer science, sri rama Krishna institute of technology, pachapalayam, Coimbatore 641010 school of mechanical sciences, Karunya University, Coimbatore 641114, Tamil Nadu ,India. paper presented at International Ergonomics Conference Humanizing Work and Work Environment on December 17-19, 2009.

L McAtamney, E Nigel Corlett 1993. RULA: a survey method for the investigation of work-related upper limb disorders, Institute for Occupational Ergonomics, University of Nottingham, University Park, Nottingham NG7 2RD, UK.

Barbara Maslen and Leon Straker 2008. A comparison of posture and muscle activity means and variation amongst young children, older children and young adults whilst working with computers. School of Physiotherapy, Curtin University of Technology, Perth, WA, Australia.

L P Singh 2010. WORK POSTURE ASSESSMENT IN FORGING INDUSTRY: AN EXPLORATORY STUDY IN INDIA Department of Industrial & Production Engineering, DR B R Ambedkar National Institute of Technology Jalandhar (Punjab), 144011.

E. Pachepsky. Basic Matlab Tutorial www.pachepsky.com/MatlabTutorial.pdf

CHAPTER 37

Factors That Lead to an Early Decline of Road Traffic Fatalities in China

De-Yu WANG, Wei ZHANG

Department of Industrial Engineering, Tsinghua University, Beijing, China, 100084
wangdy11@mails.tsinghua.edu.cn

ABSTRACT

Road traffic accident has become a leading cause of death for human. Traffic fatality risk is closely related to the number of vehicles, traffic participants, traffic conditions, etc. Worldwide studies show that per capita income is a good indicator for road fatality risk. Kopits (2005) suggest that fatalities per population in a country tend to follow a Kuznets's Curve over per capita income. The peak of the curve is found to be at around $8,600. However, in comparison with other major economies in the world, China has an earlier peak of traffic fatality risk at $1,490.

This article investigates into societal, economical, individual factors to explain the early decline in China's traffic fatality risk. It is of great significance to other developing countries to find out what factors contribute to the decline of traffic fatality risk at an earlier stage of economic development.

Keywords: Kuznets's Curve, traffic fatality, economic development

1   INTRODUCTION

As a country develops, economic growth usually brings about improvement in various social aspects. In developed countries, motorization rates are usually high. Larger number of vehicles on road increases the chance of a traffic accident. However higher national income level also means more abundant funding for road infrastructures, law enforcement, emergency medical service, etc. In general, the traffic conditions in more developed countries are much safer than less developed countries. Consequently, the relationship between traffic fatalities and the economic development of a country is more complex than a monotonic function. Smeed (1949) discovered that the number of deaths in a country, the number of registered vehicles, and the population is correlated with each other. He proposed a formula to forecast the number of deaths as the product of vehicle quantity and population. However, statistics in more recent years show that the number of road fatalities declines when a society enters a certain stage of development. Kopits and Cropper (2005) discovered that the relationship between the number of fatalities and per capita income of a country follows a Kuznets's Curve. According to her analysis, the traffic fatality risks (fatalities/population) in major economies peak at around $8600 of per capita income, in 1985 international dollars.

The economic development patterns in different countries or regions are usually of various types. Furthermore, many other factors would affect the road traffic fatality risk. Statistical figures in recent years in China show that traffic fatalities per population peaks at 8.79 in 2002 and began to decline in 2004 (see Figure 1). In 2010, the fatalities per 100,000 persons dropped 44.37% to 4.89. In terms of domestic income, the per capita GDP (measured in US dollars by official exchange rates of current years) in 2002, 2003, and 2004 are $1135, $1274, $1490, respectively[1]. These figures are far less than the $8600 level given by Kopits.

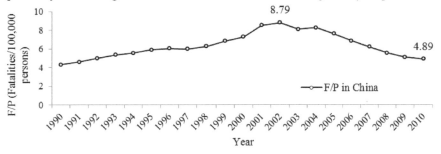

Figure 1 Traffic fatality risks in China from 1990 to 2010. Source: Ministry of Public Security of China, Traffic Management Bureau, Ministry of Public Security, *Statistical Yearbook of Road Traffic Accidents, 2010.*

It is of great importance that the turning point of traffic fatalities in China deviates from the statistical figures worldwide. China is currently the most populous country in the world, with 1.37 billion citizens[2], accounting for 19.32% of world population[3]. In terms of automobile industry, China has become the largest auto market since 2009, when 13.79 million vehicles were produced throughout the year[4].

---

[1] National Bureau of Statistics of China: 'China Statistical Yearbook 2010'.

[2] National Bureau of Statistics of China: 'Official Report of Major Statistics from the Sixth National Census, 2010'.

[3] United Nations: 'World Population Prospects: The 2010 Revision'.

[4] China Association of Automobile Manufacturers.

The production figure mounted to 18 million in 2010, giving China the world's second-largest vehicle population with 207 million units in operation[5]. Given its large amount of traffic participants and vehicles, China is an important country worthy of deeper investigation in the trend of traffic fatalities. Several factors lead to China's earlier decline of road fatalities risks. This paper examines some direct and vital reasons and tries to explain this phenomenon from various aspects including societal, economical and individual factors.

2   METHODS

Traffic fatality risk is closely related to the population of vehicles on road, driver's safety awareness, road conditions, vehicle types and safety standards, the level of law enforcement, the access to emergency medical service, etc. All these factors are promoted by the rise of income level of citizens. However, these factors have positive of negative impact on road traffic fatality risk. Comprehensively, as income level rises, the fatality risk tends to follow an inverted U-shaped Kuznets's Curve.

2.1. The Kuznets's Curve for Traffic Fatality Risk

Statistical figures in China show that traffic fatality risk reached its peak and began to drop when per capita income was only around $1400 by official exchange rate. Calculated by purchasing power, the per capita GDP (2005 PPP dollars) in 2004 is 3706[6]. Although the per capita income may vary according to the evaluation approach it follows, the figure is at least 30% less than the $8600 line given by Kopits. Different calculation approaches in evaluating a nation's income will be discussed in section 4.

In year 2002 and 2003, vehicle production in China increased for 39% and 37%. From then on, the annual production figure kept growing at a rate of around 20%. This implies that the downturn of fatality risk is not triggered by the decrease of traffic volume. In actuality, in the first decade of the new century, underdeveloped regions in China have been shrinking. In these areas, traffic participants are less protected than those in more developed regions, such as cities and towns. Besides, underdeveloped regions account for a high proportion in China. Consequently, once the traffic accidents and the corresponding fatalities in underdeveloped regions are well controlled, the total fatality figures would decrease. In the following section, we would find evidence and try to explain why China's traffic fatality risk begin to decline before the country reaches a high level of per capita income.

---

[5] Traffic Management Bureau, 'Statistical Yearbook of Road Traffic Accidents, 2010'
[6] According to United Nations Economic and Social Commission for Asia and the Pacific,
http://www.unescap.org/stat/data/swweb_syb2011/DataExplorer.aspx, retrieved December, 2011.

3 ANALYSIS

3.1. Human

The principal component of a transportation system is the people using it. In general, higher income level means higher human development level. Drivers, cyclers, pedestrians, and other traffic participants would be more aware of the safety of their own and others. Correspondingly, the fatality risk would be low.

3.1.1. Life Expectancy

Life expectancy reflects human development level in a country. Besides, as the total resources are limited, the government prioritizes its investment in public health in countries with short life expectancy (David Bishai, 2006). On the contrary, in other countries with longer life expectancy, more investment is available in traffic safety and accident rescue, etc. In the past decade, life expectancy in China has increased, sparing more money for advanced medical system.

Besides, among countries with similar life expectancy, China has a relatively low level of income (Table 1). Despite the economic condition, citizens in China have proper access to medical services. Health care development in China is in lead of the income level.

Table 1 Counties with similar life expectancy with China. a: 2004 estimated; b: Population is estimated from per capita GDP and total GDP; c: subtotal excludes China; d: Average per capita GDP is estimated from subtotals of total GDP and population; e: 2001 data; f: 2003 data; g: 2002 data. Average Per Capita GDP=Subtotal GDP/SubTotal Population = 915.27*1000/117.54=7787 USD. Source: Central Intelligence Agency, *The World Factbook*.

Rank	Country	Life expectancy[a] (year)	Per capita GDP ($)	Total GDP (billion $)	Population[b] (million)
100	Solomon Islands	72.66	1700	0.80[e]	0.47
101	Lebanon	72.63	4800	17.82[f]	3.71
102	Hungary	72.4	13900	139.80[f]	10.06
103	Mauritius	72.38	11400	13.85[f]	1.21
104	Turkey	72.36	6700	458.20[f]	68.39
105	**China**	**72.27**	**4700**	**5,989[g]**	**1274.26**
106	Malaysia	72.24	9000	207.80[f]	23.09
107	Saint Kitts and Nevis	72.15	8800	0.339[g]	0.04
108	Bulgaria	72.03	7600	57.13[f]	7.52
109	Panama	71.94	6300	18.78[f]	2.98
110	Antigua and Barbuda	71.90	11000	0.75[g]	0.07
	Subtotal[c]		7787[d]	915.27	117.54

3.2. Vehicle

Traffic participants utilize motor vehicles to commute from one place to another. In case of an accident, vehicle will protect drivers and passengers. In general, pedestrians are least protected, exposed directly to contact with objects. Cyclers have helmet, if they obey the regulations and laws, to provide adequate protection for their head in a collision. However, the capability of such equipment to lower the fatality risk is limited. Severe accident such as high speed crash is still fatal for bike riders. Passenger cars, including hatchbacks, sedans, vans, and buses, give the drivers and passengers the best protection in an accident.

3.2.1. Vehicle Category

Vehicles of different classes vary in their protection for the passengers. In general, motorized vehicles without proper bodywork are at the highest risk of fatality in accidents. These vehicles usually have less than four wheels and are excluded from passenger car category (category M)[7]. In China, according to government regulations, compact cars, sedans, SUVs, MPVs, vans, various types of buses are classified as passenger cars. Other civil vehicle such as motor cycles and scooters with two or three wheels, farm vehicles, etc. are excluded, and categorized as 'other vehicles'.

Since 2000, the increase in the volume of automobile market is mostly contributed by the growth of passenger car market (see Figure 2). As a result, the proportion of passenger cars on road kept growing in the last decade (see Figure 3).

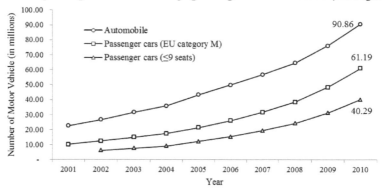

Figure 2 Vehicle population in China. Categorization was modified in 2004. Figures before 2004 are adapted from corresponding categories in previous regulations. Source: Traffic Management Bureau, Ministry of Public Security, *Statistical Yearbook of Road Traffic Accidents*, 2006-2010, *A Compilation of Road Traffic Accidents Statistics*, 2001-2005.

---

[7] European Automobile Manufacturers' Association, *Definition of vehicle categories*, http://www.acea.be/collection/regulation_and_standards_background/, retrieved December, 2011.

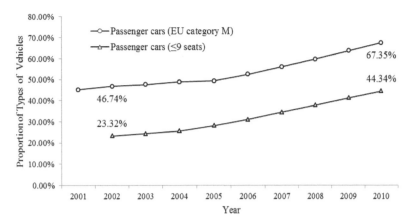

Figure 3 Proportion of vehicles in different categories. Passenger cars with no more than 9 seats account for nearly half of all vehicles in 2010. Source: Traffic Management Bureau, Ministry of Public Security, *Statistical Yearbook of Road Traffic Accidents*, 2006-2010, *A Compilation of Road Traffic Accidents Statistics*, 2001-2005.

Ceteris paribus, traffic participants exposed less proportion of time to lower class of vehicle would have lower fatality risk in accidents. Moreover, an increase of people's income level would lead consumers to upgrade their daily commuting tools. This decreases the chance of crashes between asymmetrically sized vehicles, where the fatality risk is highest (Tay, 2003).

3.3. Road

The road infrastructure is important in providing a safe environment for vehicles. High standard road is not only comfortable for commuters, but also safe for drivers. According to modern standards, it is required that a road is adequately paved and the surface provides enough grips under all weather. Lightings make sure that drivers have keen vision after nightfall. Such features prevent fatal accidents from happening and reduce the risk in travelling.

3.3.1. Leading Expressway Standard

The expressway system in China is one of the most advanced road infrastructures in the world. The system is built and maintained according to the national standard (Ministry of Transport, 2004). Defined by the standard, expressways are limit-access sections of road for automobile to travel in a smooth manner in multiple lanes. The traffic flow is physically separated with central reservation into two directions. Besides, the alignments are specially designed to provide adequate view for drivers. By 2011, nearly all sections of the expressway are toll ways.

3.3.2.   Proper Infrastructure ahead of Economic Development

Commuters travelling on low-standard road would suffer greater fatality risk. Renovate road infrastructure and construct fast and safe highways for inter-city travel are key to reduce the fatality risk. The first section of expressway in mainland China was completed in 1988. Afterward, the expressway network gradually came into being. In 1999, there were 11,605 km of international standard expressways in mainland China, who surpassed Canada to rank the second in terms of expressway mileage. By 2011, the expressway mileage in mainland China has reached 85,000 km (53,000 miles)[8], second to the United States with 95,000 km (59,000 miles) of expressways by 2009[9].

China National Highways are also high standard passage for long distance travel. In the last decade, most areas in the mainland have been covered with National Highways, or provincial ways. In general, roads above Class III in China are able to provide adequate road surface quality, lightings, alignment designs, and signs. Statistics showed that from 2000, the proportion of fatalities on roads of Class III or below has decreased from over 40% to 33%. Fatality risk in accident under better road conditions is more easily to control. On the contrary, fatalities under poor road conditions contribute more to the overall fatality risk. As the proportion of fatalities on roads of lower class decreased, the overall fatality risk is reduced.

3.4. Society

3.4.1. Urbanization

In the last decade, China has made great progress in urbanization. The urban population exceeds the rural population in 2011 (see Figure 4). This means that more people are living in the cities, where commuting environment is safer than in rural areas. In urban areas, road infrastructures are in better conditions, most intersections are controlled with traffic signals, pedestrians have sidewalks, crosswalks and overpasses, and commuters have public transit as a safer means to travel. Consequently, as more population shifts from underdeveloped rural to more developed urban area, the overall traffic fatality risk would decrease.

In terms of road infrastructure, to cope with the high traffic volume, intersections in city area are usually controlled with traffic signals. As the traffic flow is regulated to follow certain rules, the risk of traffic fatality is relatively low.

Adequate facilities for pedestrians also help in creating a safer traffic environment. In Beijing, there are 420 pedestrian overpasses and 200 underpasses[10].

---

[8] Ministry of Transport of the P.R.C., http://www.chinanews.com/gn/2011/12-31/3573689.shtml, retrieved December, 2011.
[9] U.S. Department of Transportation, Federal Highway Administration,
http://www.fhwa.dot.gov/policyinformation/statistics/2009/hm220.cfm, retrieved December, 2011.
[10] Beijing Bureau of Statistics, *Beijing Statistical Yearbook 2011*.

The separation of different means of transportation would dramatically reduce the chance of fatal accident as crashes involved with asymmetrical subjects are the most dangerous.

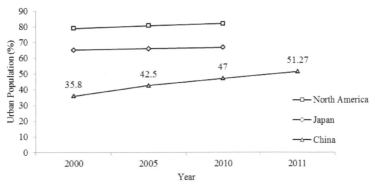

Figure 4 The proportion of urban population in major economies. Source: United Nations Economic and Social Commission for Asia and the Pacific

Public transport is another advantage for city dwellers. Buses, subways, trains run by the government or private entities provide much safer travel solutions. In cities like Beijing, over 40% of commuters choose public transit[11]. During the process of urbanization, more population would use public transit, helping reduce the fatality risk in road transportation.

3.4.2. Government and social awareness

Before year 2004, the traffic accident fatality risk kept rising and in 2003, annual fatality numbers exceeded 100,000 for the third year. From then on, the administration paid much more attention to the issue of road safety. Besides, the society began to realize the seriousness of this problem.

Road Traffic Safety Law is enacted in 2004. Several government agencies including Ministry of Health, Public Security, Transport, Education, and State Administration of Work Safety jointly issued a notice about road safety improvement. Local governments were required to publicize the newly-enacted law and to strengthen education about danger on road as well as knowledge of first aid. The fatality reduction plan required that fatality number should be lowered in each year thereafter. Local governments were to be checked at the end of each year for their effort in reducing the fatality risk. As a result, local administrations raised investment on road renovation and maintenance, on traffic management facilities, on public educations, and on law enforcement, etc. These approaches achieved significant results, effectively improved the road traffic conditions, and ultimately

---

[11] Beijing Municipal Commission of Transport,
http://www.cnr.cn/newscenter/gnxw/201108/t20110812_508360910.shtml, retrieved December, 2011.

push the fatality figure through the peak towards an early decline. They last till today and contribute to the steady decline of fatality risk.

4   DISCUSSION

4.1. Economy

There are several distinctive approaches in calculating national income level. In this, study, statistics of China come from the yearbook published by National Bureau of Statistics. In the yearbook, the GDP is calculated by production approach. Deeper investigation revealed that the income figures vary a lot when different calculation approaches are adopted.

The peak point given by Kopits when fatality risk begins to decline is 8,600, PPP 1985 international dollar. This figure is generated from data in Penn World Table (PWT) 5.6[12]. However, PWT 5.6 does not include China. The income data for China in Penn World Table 7.0 is 4,082, PPP, 2005 international dollar. Other institutes also gave various values of per capita GDP of China in 2004. Nonetheless, the income level of China in 2004 is doubtlessly lower than the average level when road traffic fatality risks begin to decline.

Table 2 Per capita income evaluated by purchasing power parity given by different institutes or programs.

Institute/Program	PPP	Year
United Nations Development Programme (UN)	4,115	2005
Economic and Social Commission for Asia and the Pacific (UN)	3,706	2004
World Economic Outlook Database (IMF)	5,299	2004
World Development Indicators (WDI), 2005 (World Bank)	6,760	2005
International Comparison Program (ICP) (World Bank)	4,091	2005
World Development Indicators & Global Development Finance (WB)	3,719	2004
World Factbook (CIA)	5,600	2004 est.

4.2. Underreport

The statistics in China showed an obvious decline in fatality risk after year 2004. Yet, it may have been pointed out that the figures do not necessarily reflect the reality. Data collected from local statistics agencies may have already been distorted. However, the possibility that such cases may exist does not impair the overall credibility of this study, given the following reasons.

First, the problem of underreport may not be as serious as many people think. Specifically, in major cities in China, statistics of road traffic fatalities are usually

---

[12] http://pwt.econ.upenn.edu/php_site/pwt_index.php, retrieved December, 2011.

more accurate and credible, as the more thorough reporting mechanisms are adopted. Statistics of 36 major cities, including Beijing, Shanghai, Tianjin, Guangzhou, etc., presented an obvious decline of fatality risk after 2004. Although this may not be a direct evidence for the accuracy of national figures, it still indicates that the descending trend of fatality risk in China after 2004 is believable.

Second, even if this underreport problem existed, the extent to which it would affect the reliability of the conclusion is limited. As the phenomenon of underreport widely exists in developing countries, the statistics are still effective when we base our conclusion on cross-country comparison. Nevertheless, the aim of this study is to provide possible solutions for developing countries to effectively control the fatality risk downward earlier.

5 CONCLUSIONS

As we investigated many social aspects in China, we draw the conclusion that although per capita income of China is relatively low, it made great achievements in education, vehicle manufacturing, road construction, legislation, public transit, mobile communication, etc. These factors contribute to the early decline of traffic fatality risk.

Similar to China, other BRICS countries, Brazil, Russia, India, and South Africa, are experiencing fast economic growth. The process of motorization has begun. In terms of population, BRICS countries have significant effect on Kuznets Curve of fatality risk versus per capita income. Consequently, the average level of income when the fatality risk begins to decline should be adjusted, given the cases in developing countries like BRICS, especially among them, China.

6 ACKNOWLEDGEMENT

REFERENCE

David Bishai, A. Q. P. J. a. A. G., 2006. National road casualties and economic development. Health Economics, Volume 15, pp. 65-81.

Kopits, E. & Cropper, M., 2005. Traffic fatalities and economic growth. Accident Analysis and Prevention, Volume 37, pp. 169-178.

Smeed, R. J., 1949. Some Statistical Aspects of Road Safety Research. Journal of the Royal Statistical Society. Series A (General), 112(1), pp. 1-34.

Tay, R., 2003. Marginal Effects of Changing the Vehicle Mix on Fatal Crashes. Journal of Transport Economics and Policy, 37(3), pp. 439-450.

United Nations Development Programme, 1990. Human Development Report 1990. 1st ed. New York: Oxford University Press.

United Nations Development Programme, 2010. Human Development Report 2010 20th Anniversary Edition The Real Wealth of Nations: Pathways to Human Development. 1st ed. New York: Palgrave Macmillan.

CHAPTER 38

The Effect on Physiology from Different LED Lighting

Chih-Ling Huang[1], Chih-Wei Lu[2]*, Chun-Yuh Yang[1], Hsiao-Wen Tu[3], Chun-Hsing Lee[3], Ya-Hui Chiang[3], Hung-Lieh Hu[3]

Department of Public Health,
College of Health Sciences,
Kaohsiung Medical University, 807 Taiwan

ABSTRACT

Because of the global energy tension, LED technology plays an important role in the recent years. According to the past studies, it could save huge energy if replace LED with traditional illumination, therefore the research and developments about LED products are very important. The effect on physiology or physiology from LED lighting is an important factor to product devise. There were 10 male subjects participated in this study. They read for 4 to 5 hours at 4 kinds of LED lighting environments in random order: 4500K and the illumination<30 lux, 2700K, 4500K, and 6500K. The Farnswrth Munsell 100 test and vision test were being used before and after the subjects' reading to evaluate their color discrimination and vision. Furthermore, body temperature was being measured during their reading at

[1] The Department of Public Health, College of Health Sciences, Kaohusing Medical University, R.O.C
[2] Department of Industrial and Systems Engineering, Chung Yuan Christian University, R.O.C
[3] Industrial Technology Research Institute, R.O.C

8 times. The result shows that the LED lighting at 2700K had the greatest effect on the subjects' vision. The body temperature was getting lower during reading at all 4 kinds environments, especially the third time (P=0.043) and sixth time (P=0.013) at 6500K lighting, and the last time at 2700K (P=0.036) and 6500K (P=0.002) lightings. These values have significant difference to the one was measured before reading. These findings could be some references to product development.

Keywords: Exemplary chapter, Human Systems Integration, Systems Engineering, Systems Modeling Language

INTRODUCTION

Human Factors Because of the global energy tension, renewable energy developing becomes an important part in technology research and development in the future. Solar energy technology is the most widely application topic in this field. By this situation, LED development and application plays an important role in the recent years.

As a consequence, good lighting has a positive influence on health, well-being, alertness, and even on sleep quality [1]. It also could save huge energy if replace LED with traditional illumination, therefore the research and developments about LED products are very important.

Many studies indicated that human physiology could be affected by lighting. Human performance on color discrimination in visual display terminals may be affected by illuminant colors [2]. Wenting Cheng find out the effects of light-emitting diode (LED) lighting on elderly people's color discrimination and preference in 2011 study, elderly people perform better in color discrimination with higher color-correlated temperature of LED light sources and it also increases with higher illuminance (30lx-1000lx) of LED lighting [3]. In Boyce's study, performance significantly increased with increasing illluminance under the 2700K and 2500K lamps [4].

Measure body temperature is one of the popular ways to evaluate sleepiness. Changes in the distal-toproximal skin temperature gradient (DPG) have been shown to correlate with sleepiness immediately after waking up [5]. By Maan Van De Werken's study, sleepiness was positively related to momentary distal skin temperature [6]. According to these studies, the effect on physiology or physiology from LED lighting is an important factor to product devise.

The aim of this study is to evaluate the the effect on Physiology from Different LED Lighting. There were 10 male subjects participated in this study. They read for 4 to 5 hours at 4 kinds of LED lighting environments in random order: 4500K and the illumination<30 lux, 2700K, 4500K, and 6500K. The Farnswrth Munsell 100 test and vision test were being used before and after the subjects' reading to evaluate their color discrimination and vision. Furthermore, body temperature was being measured during their reading at 8 times.

METHODS

Subjects

Subjects were recruited by advertisements at internet. 10 male subjects (mean age 26.4 years (between 19 to 33 years), mean height 172.5cm, mean weight 69 kg, mean BMI 23.12) were selected base on the following criteria: subjects did not have sleep disorders as assessed by questionnaires, and had to keep a sleep–wake schedule during the 7 days at home prior to participation.

This study was conducted in accordance with the principles outlined in the Declaration of Helsinki and was reviewed and approved by the Human Research Committee of the Industrial Technology Research Institute. Each subject provided written informed consent prior to starting the study.

Farnsworth Munsell 100

For visual performance measurements, Farnsworth Munsell 100 was used to study the hue discrimination of light sources and found out that lamp type (spectrum) and illuminance for more than 40 years [7]. It consists of 85 color discs arranged in four series. The color discs are randomly arranged beforehand. The subjects must put all the discs in proper sequence. An error score is introduced to evaluate how the visual task is accomplished [8]. In this study, each subject tested 2 times (before and after reading) to evaluate color discrimination error score.

Experimental design

This experiment consists 4 nights. Subjects stayed in individual bedrooms with no information about time of day. The room was lit by ceiling lighting resulting in an intensity of experimental lighting level (4500K and illumination<30 lux, 2700K, 4500K, and 6500K) measured at eye level. The condition arrangement was random expect the first day was set 4500K and illumination as baseline level.

RESULTS AND DISCUSSION

Color discrimination
Figure 1 is the subjects' mean color discrimination error score which tested after their reading by LED lighting types. Compare to the base line level (tested before their reading at the first day). The highest mean score is 4500K (both 4500K, and 4500K and illumination <30lux) and the positive effect happens at 6500K. Figure 2 is the subjects' mean color discrimination error score which tested after their reading by experimental days. The value is decreasing after the second day.

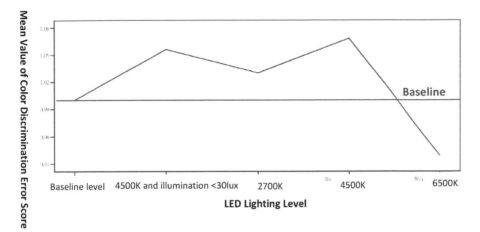

Figure 1. The subjects' mean color discrimination error score which tested after their reading by LED lighting types

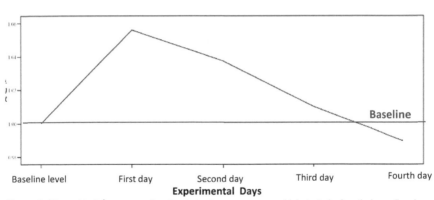

Figure 2. The subjects' mean color discrimination error score which tested after their reading by experimental days

Optometric assessment

Figure 3 to 5 are subjects' mean optometric assessment results (both eyes, right eye, and left eye). According to these figures, the worst value all at 2700K.

370

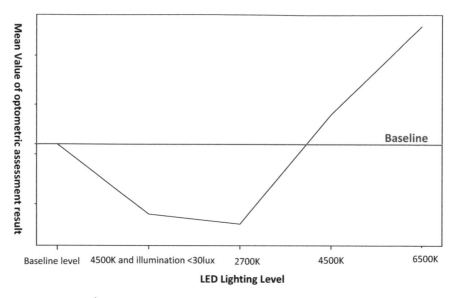

Figure 3. subjects' optometric assessment results (both eyes)

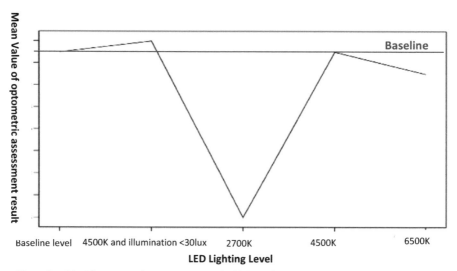

Figure 4. subjects' optometric assessment results (right eye)

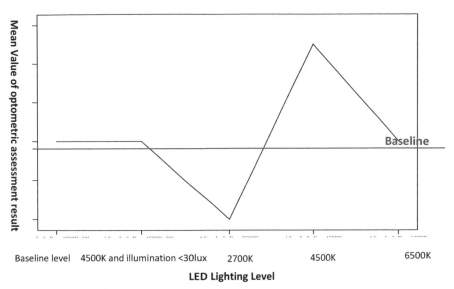

Baseline level   4500K and illumination <30lux   2700K   4500K   6500K

LED Lighting Level

Figure 5. subjects' optometric assessment results (left eye)

Body temperature

Figure 6 to 9 are the subjects' mean body temperature value by time at19:50, 20:20, 21:20, 21:50, 22:20, 22:50, 23:20 in 4 LED lighting condition. The decreasing trend could be found at all 4 situation s. In 6500K situation, the subjects' body temperature mean values which measured at 20:50, 22:20, and 23:20 have significant difference between the value which measured at 19:50 (P=0.043, P=0.013, P=0.002). The subjects' body temperature mean value at 23:20 has significant difference between 19:50 in 2700K situation (P=0.036). According to these results, these two LED lightings might related to subjects' sleepiness.

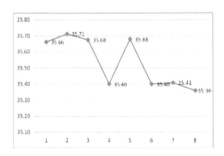

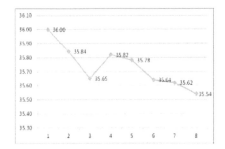

Figure6. the subjects' mean body temperature value by time (4500K,<30lux)

Figure7. the subjects' mean body temperature value by time (6500K)

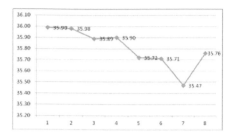
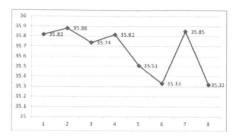

Figure8. the subjects' mean body temperature value by time (4500K)

Figure9. the subjects' mean body temperature value by time (2700K)

There are some limitation factors in this study. First, the decreasing trend could be found at the subjects' mean color discrimination error score which tested after their reading by experimental days. The learning-effect might happen in this study. Second, the subjects' were asked to take a rest every 30 minutes for 10 minutes to avoid visual fatigue. This could influence the color discrimination results. Third, the experiment was carry out during work days, so the subjects' work type might be a confounding factor in this study.

These findings could evaluate the employees' physiology effects in LED companies. Furthermore, these could be some references to product development.

ACKNOWLEDGEMENT

This study has been supported by national Science Council (NSC 99-2221-E-033-070-MY3). We thank the subject volunteers for their participation, Wen-Guei Lin (Tainan Hospital, Department of Health, Executive Yuan) for her help and advice. Apparatus support was obtained from College of Electrical Engineering and Computer Science, Chung Yuan Christian University (CYCU-EECS-9901).

REFERENCES

1. *Wout J.M. van Bommel*, Non-visual biological effect of lighting and the practical meaning for lighting for work, Applied Ergonomics 37 (2006) 461–466
2. Feng-Yi TSENG, Chin-Jung CHAO, Wen-Yang FENG, Sheue-Ling HWANG, Assessment of Human Color Discrimination Based on Illuminant Color, Ambient Illumination and Screen Background Color for Visual Display Terminal Workers, Industrial Health 48 (2010) 438–446

3. Wenting Cheng, Jiaqi Ju, Yaojie Sun, and Yandan Lin, The Effect of LED Lighting on Color Discrimination and Preference of Elderly People, Human Factors and Ergonomics in Manufacturing & Service Industries **00** (0) (2011)1–8

4. Boyce, P. R., & Cuttle, C., Effect of correlated_colour temperature on the perception of interiors and_colour discrimination performance. Lighting Research and Technology, 22(1) (1990) 19–36.

5. Krauchi, K., Cajochen, C. and Wirz-Justice, A. Waking up properly: is there a role of thermoregulation in sleep inertia? J. Sleep Res., 13(2004)121–127.

6. Maan Van De Werken , Marnic . Gimenez, Bonnie De Vries ,_Domien G . M. Beersma, Eus J . W. Van Someren, Marijke C. M. Gordigun, Effects of artificial dawn on sleep inertia, skin temperature, and the awakening cortisol response, European Sleep Research Society, J. Sleep Res., 19 (2010) 425–435

7. Boyce, P. R., & Simons, R. H., Hue discrimination_and light sources. Lighting Research and Technology, 9(3) (1997) 125–140.

8. Wenting Cheng, Jiaqi Ju, Yaojie Sun, Yandan Lin, The Effect of LED Lighting on Color Discrimination and Preference of Elderly People, Human Factors and Ergonomics in Manufacturing & Service Industries 00 (0) (2011) 1–8

CHAPTER 39

Effects of Evening LED Lighting of Different CS/P on Human Subjective and Objective Perceptions

Hsiao-Wen Tu[1]; Chun-Hsing Lee[1]; Hsiang-Chi Chung[1]; Ya-Hui Chiang[1]; Hung-Lieh Hu[1];Mu-Tao Chu[1]; Chih-Ling Huang[2]; Chih-Wei Lu[3]

[1] Industrial Technology Research Institute, R.O.C
[2] Graduate Institute of Environmental and Occupational Safety and Health, The Department of Public Health, Kaohusing Medical University, R.O.C
[3]Department of Industrial and Systems Engineering, Chung Yuan Christian University, R.O.C
HsiaoWenTu@itri.org.tw

ABSTRACT

Many literatures and the international lighting companies suggest that light can affect the secretion of cortisol and melatonin via visual and non-visual pathway. Moreover, it also can alter human circadian rhythms. Thanks for the easily-controlled light parameters (e.g., spectrum, correlated color temperature (CCT), color rendering index (CRI), circadian stimulus (CS), etc) of Light-emitting diode (LED) and its energy-saving property, it can provide the proper spectrum to fit the environments and seasons. This study attempted to determine the suitable illumination for evening activities by physiological and psychological assessments. The experimental light source is a RGBW-LED lamp with adjustable light parameters. The experimental light source was the RGBW LED lamp of adjustable lighting parameters, which can produce an LED frequency spectra of three different CS/P values at luminance of 500-lux. In addition, to provide conditions of almost no lighting interference for control, the experiment included a dim light environment (<30-lux). The objective biology data (e.g., heart rate (HR), galvanic skin response (GSR)) and the subjective questionnaire (e.g., heaviness, lightness, etc) from ten participants exposed in different light spectra at environment temperature $25^\circ C$

and the humidity 50% are analyzed. Our results show that CS/P ： 0.412 (2700K) light can make human comfortable, increase the time of concentration, and reduce the fidgety feeling and the heavy loading.

Keywords: LED, Spectrum, Galvanic Skin Response (GSR), Heart rate (HR), Human factors in lighting

1   INTRODUCTION

In recent years, energy shortages, abnormal climates, and other factors have resulted in increasingly higher public awareness of environmental protection. Carbon reduction and energy saving have become important issues at present. LED is highly profiled for its energy saving advantage. In the past, lighting equipment manufacturers focused on improvements of technological aspects, such as illumination performance. However, with rising awareness of Human factors in lighting, LED manufacturers have shifted focus from technological and performance indicators to quality illumination in order to satisfy consumers' human physiological and psychological needs. For example, in 2006, Philips launched dynamic lighting equipment, which can change in response to different situations of lighting (e.g., office lighting, industrial lighting, classroom lighting, etc) [1]. Moreover, lighting equipments that can change luminance along with time, season, and climate have been developed to create environments with a sense of delightfulness and comfort. In 2008, Osram developed lighting that can change CCT according to different physiological periods. The product concept is that white light with more components of blue light is required during the day, while white light with more red light ratio is more suitable for evening. The human perception of spatial luminance should simultaneously consider the perceptions of the luminance of the floor, ceiling, and walls, rather than using the luminance of light source to represent visual perception of lamination. Panasonic proposed the concept of Eco & Feu Lighting, which is applied to practical office lighting. Compared with traditional lighting environments, space modified by Feu assessment can better suit human perceptions to improve the sense of comfort [2].

The early studies on Human factors in lighting are limited in color lighting control technology, and can only apply a few changes in spectrum and CCT, in addition to different degrees of luminance, to observe the effects of light source on human physiological and psychological perceptions. Kruithof (1973) discussed the effects of light source CCT and luminance of traditional light sources on human physiological and psychological perceptions. The results showed that the different combinations of light source color temperature and luminance would result in different public evaluations. People prefer the low color temperature in a low luminance environment [3]. On the contrary, they prefer high luminance in high luminance environments. At present, LED has become the global replacement of traditional light sources for purposes of energy savings. The LED advantage is its ability to adjust lighting parameters (e.g., spectrum and CCT), which makes it possible to use LED for different lighting parameters for human factor lighting experiments. Thus, studies on ergonomic lighting can be conducted on different lighting parameters.

Scientists found that light affects human physiological periods [4-10] and the maximum wavelength affecting human physiological response is 464nm [11-15]. This circadian stimulus of the spectrum has been defined as a non-visual perception, and the physiological period response spectrum has been defined as CS. Non-visual perceptions can interfere with brain messages, which further inhibits the secretion of melatonin to affect the response of physiological periods and ultimately change the circadian rhythm [16-22]. According to previous studies, different lighting parameters, such as the CS spectrum and CCT, would affect human physiological and psychological perceptions. Studies on the effects of LED on humans have recently been conducted, but without definite results. Hence, it has become a trend to explore the effects of the lighting parameters of LED light sources on human physiological periods and psychological perceptions.

2 RESEARCH PROCEDURES AND ANALYSIS

This experiment selected four types of lighting parameters by focusing on changing the CS frequency spectrum proportions [23] (Table 1) in given luminance (lumen). The luminance measurement was at an approximate distance of 160 cm from the lamp. With the exception of luminance of 30- Lux in a dark environment, luminance was 500- Lux in other cases. Figure 1 shows the frequency spectra of the four types of light sources.

Table 1 Experimental settings of lighting parameters

CCT	2700K	4500K	6500K
X coordinate	0.44	0.36	0.31
Y coordinate	0.39	0.37	0.33
CS (lm)	236	383	527
P (lm)	575	637	672
CS/P	0.412	0.601	0.784

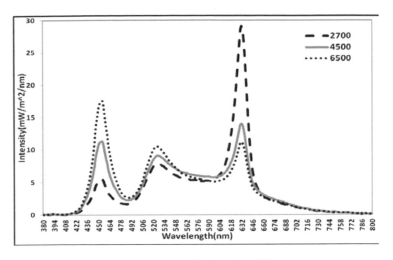

Figure 1　Spectrum distribution of the light source

The common lamp module available on the market was used in experiments. It was composed of a reflector, a fixed metal plate, a fixed metal frame, front cover, and heat sink (internal diameter is 200mm，outer diameter is 220mm) (Fig.2). A single lamp used 6 LED lights, typed Cree MC-E RGBW, and each had an LED light source of RGBW, which could be individually driven to generate different driving current proportions to produce different combinations of CCT, ranging from CS/P: 0.412 (2700K) ~ CS/P: 0.4120.784 (6500K).

Figure 2　LED lamp structure

Regarding environmental design, this study simulated a bedroom space of 270cm×300cm×240cm, where the height of lamp was 235cm from the ground. With a table, a chair, and a bed in the space, this experiment installed 9 lamps evenly throughout the experimental space. The room temperature was set at 25°C and relative humidity was 50%. The human physiological and psychological responses were measured in an environment of the above specifications.

378

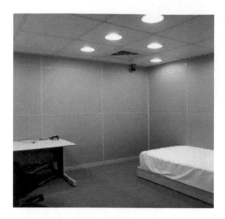

Figure 3 Experimental space

The subjects were 8 men, aged 18-32, with good physiological and psychological health and normal vision. The subjects were asked to wear an active watch to measure physical activity levels during the week prior to the experiment, in order to monitor their routines. On the day of the experiment, subjects were inhibited from drinking alcohol, tea, or coffee. The experimental process design is as shown in Figure 4. The experiment, including preparation to the completion of the questionnaire, required 6 hours. Subjects were asked to simulate daily life in the experimental space by reading books or resting. Sleep or use of display equipment was not allowed during the experiment. During the experimental process, physiological measurement instruments were used to record GSR and HR, while the subjective questionnaire was used to assess psychological perceptions, such as dizziness, headache, and mental conditions of the subjects.

18:00	18:10	18:30	19:00	19:50 - 23:30
Arrival Measurement weight, vision and Munsell	Eating	Bathe	Setup instrument	Recording biological data (HR、GSR)、Questionnaire (Don't sleep) 20min Read and 10min Rest (One cycle)
			Illuminating LED light, Different parameter of light (CS、CCT、CRI), process is about 5 hours 30 mins. (500 lx)	

Figure 4 Experimental process design

The subjective evaluation indicators of this experiment included five items of negative perceptions, dizziness, headache, irritability, heaviness, and nervousness, which were measured on a 5-point scale, ranging from 1 (strongly agree) to 5 (strongly disagree).

3   EXAMPLE VALIDATION AND DISCUSSION

Analysis of variance was conducted on the experimental data. The results suggested that, four types of lighting parameters have significant variance in the effects on the subject's objective physiological data GSR and HR ($p<0.05$). In addition, most items concerning the subjective perception showed significant differences.

GSR can reflect emotional fluctuations. A high GSR value indicates an excited state, while a low GSR value indicates a calm state. Figure 5 shows the averaged data of GSR, determined by dividing the sum of the original GSR data by 8. The data suggested that the subjects have mostly stable emotions in a dark environment, which is significantly different from the emotional state in the cases of three different settings. In the cases of three settings of lighting parameters, GSR is relatively low when the subjects are in lighting conditions of CS/P: 0.412 (2700K), and significant differences can be found from the other two types of lighting parameters.

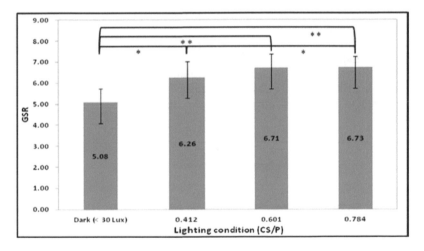

Figure 5 Average GSR data of 8 subjects (* : P<0.05, ** : P<0.01)

HR can reflect a state of nervousness or relaxation. In general, a high HR value represents an increasing sense of nervousness. Figure 6 shows the average data of the eight subjects, determined by dividing the sum of the original data by 8. The data suggested that HR is most stable when the subjects are in a dark environment, which is significantly different from that of the other three settings of lighting parameters. The subject's HR is relatively stable when they are in lighting conditions of CS/P: 0.412 (2700K), which is significantly different from the case of lighting conditions at CS/P: 0.601 (4500K).

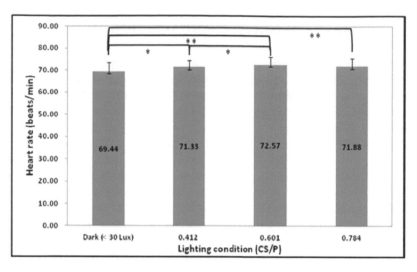

Figure 6 Average GHR data of 8 subjects (* ∶ P<0.05, ** ∶ P<0.01)

The average score was determined by adding the measurement data of the eight subjects, and dividing the sum by 8. Figure 7 shows the negative comments. A higher value indicates a higher level of disgreement of the subjects. As seen, it is relatively more unlikely to produce psychological perceptions of dizziness, headache, or irritability in the case of lighting parameters of CS/P: 0.412 (2700K); as compared with the dark environment and the other two settings of lighting parameters.

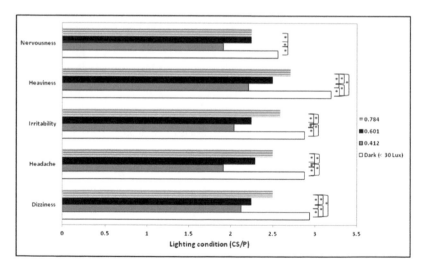

Figure 7 Average evaluation of psychological perceptions of the 8 subjects (* ∶ P<0.05)

4   CONCLUSIONS

Based on the results, this paper proposes the following conclusions. First, when in a dark environment, human physiological responses are relatively stable and peaceful, indicating emotional stability and comfortableness. Second, in the case of CS/P: 0.412 (2700K), GSR and HR are relatively more stable, with significant differences as compared with the other two settings of lighting parameters. The questionnaire survey found that, in the case of lighting conditions of CS/P: 0.412 (2700K), the subjects did not feel dizzy, headache, or irritable.

As humans cannot live and work in dark environments, and require lighting for evening activities, this paper proposes the most suitable lighting parameters of CS/P: 0.412 (2700K), which creates perceptions of environmental comfortableness, improved concentration, and general quality, while reducing irritability and feelings of heaviness. The experimental variables are lighting parameters. The statistical results can be used to propose changes to illumination parameters for the benefit of human perceptions, and can serve as references for the development of product applications in illumination.

REFERENCES

"Philips Dynamic Lighting",
    http://ambientenvironments.wordpress.com/2010/09/20/philips-dynamic-lighting/
"Panasonic co&Feu Lighting", http://denko.panasonic.biz/Ebox/eco-feu/feu.html
Λ.Λ.Kurithof (1941). Tubular luminescence lamps for general illumination. *Philips Tech. Rev. 6(3).*
John E. Pauly *et al* (1967). Circadian rhythms in blood glucose and the effect of different lighting schedules, hypophysectomy, adrenal medullectomy and starvation. *American Journal of Anatomy, Vol. 120, 627-636*
Richard H. Swade (1969). Circadian rhythms in fluctuating light cycles: Toward a new model of entrainment. *Journal of Theoretical Biology, 227-239.*
M Zatz and MJ Brownstein (1979). Inteaventricular carbachol mimics the effects of light on the circadian rhythm in the rat pineal gland. *Science, Vol. 203, 358-361.*
CA Czeisler *et al* (1986). Bright light resets the human circadian pacemaker independent of the timing of the sleep-wake cycle. *Science, Vol. 233, 667-671.*
AJ Lewy *et al* (1987). Antidepressant and circadian phase-shifting effects of light. *Science, Vol. 253, 352-354.*
Diane B. Boivin *et al* (1996). Dose-response relationships for resetting of human circadian clock by light. *Nature 379, 540-542.*
CA Czeisler *et al* (1996). Phase-shifting human circadian rhythms: influence of sleep timing, social contact and light exposure. *J Physiol Vol.495, 289-297.*
David S. Minors *et al* (1991). A human phase-response curve to light. *Neuroscience Letters Vol.133, 36-40.*
George C. Brainard *et al* (2001). Action Spectrum for Melatonin Regulation in Humans: Evidence for a Novel Circadian Photoreceptor. *The Journal of Neuroscience, 6405–6412.*
Victoria L. Warman *et al* (2003). Phase advancing human circadian rhythms with short wavelength light. *Neuroscience Letters Vol.342, 37-40.*

M.S. Rea *et al* (2005). A model of phototransduction by the human circadian system. *Brain Research Reviews 50 213–228.*

M.S. Rea *et al (2010).* Circadian light. *Journal of Circadian Rhythms.*

AJ Lewy *et al* (1980). Light suppresses melatonin secretion in humans. *Science, Vol. 210, 1267-1269.*

Armstrong SM *et al* (1986). Synchronization of mammalian circadian rhythms by melatonin. *J Neural Transm Suppl.*

Iain M. McIntyre *et al* (1989). Human Melatonin Suppression by Light is Intensity Dependent. *Journal of Pineal Research, Vol. 6, 149-156.*

Maija-Liisa Laakso *et al* (1993). One-hour exposure to moderate illuminance (500 lux) shifts the human melatonin rhythm. *Journal of Pineal Research, Vol. 15, 21-26.*

George C. Brainard et al (2001). Human melatonin regulation is not mediated by the three cone photopic visual system. *The Journal of clinical Endocrinology & Metabolism, Vol.86.*

WJM van Bommel Professor *et al* (2004). Lighting for work: a review of visual and biological effects. *Lighting Res. Technol. 36,4 (2004) pp. 255–269.*

Seithikurippu R. Pandi-Perumal *et al* (2007). Dim light melatonin onset (DLMO): A tool for the analysis of circadian phase in human sleep and chronobiological disorders. *Progress in Neuro-Psychopharmacology & Biological Psychiatry, 1 – 11.*

(2008). A new retinal photoreceptor should affect lighting practice. *Lighting Res. Technol, 373–376.*

Section III

Product Design and Evaluation

An Analysis Framework of User Oriented Systems for Call Center Services in Telecommunication

An-Che Chen, Hau-Wei Huang

Mingchi University of Technology
New Taipei City, TAIWAN
anche@mail.mcut.edu.tw

ABSTRACT

The growing global market of telecommunication services continuously promotes growing customer needs and customer cares. Despite the efforts to facilitate Web-based services for customer contacts, Call Center Services (CC) with Interactive Voice Response System (IVR) currently remain the primary channel for customer services in telecommunication industries. In reality, the development of the system structure for IVR and the subsequent Call Center Information System are mainly based on the perspectives from the internal function processes on the provider side, i.e., a rather technical- or business- function oriented approach. For most of the customer users who do not familiar with the technical or internal processes in telecommunication services, however, may have greater chances of wasting times or making errors while interacting with these service systems which are not designed from their perspectives. Therefore, a user oriented system is needed for better user experiences for such customer services.

This study aims to establish a template analysis framework for the system improvement towards user-oriented customer services, through conducting an empirical study in a major telecom company in Taiwan. Sampled system records in IVR logs are extracted and further linked with the corresponding transaction records which are routinely reported by customer service representatives for call handlings. In addition, individual interviews with customer service agents are also the other primary part of the system analysis. The interview results show that the problematic repair service dispatch policy and the personnel proficiency in business inquiries are critical to the quality of customer services. By further cross-referencing the results of system analysis and agent interviews, practical suggestions for system improvement towards a user-oriented customer service system as well as the implications for theoretical research are in further discussion.

Keywords: user oriented, customer service, telecommunication

CHAPTER 41

Investigating the Interface Usability of Bluetooth Earphone Controls

An-Che Chen, Yi-Hsiang Su, Chen-Chih Chen

Mingchi University of Technology
New Taipei City, TAIWAN
anche@mail.mcut.edu.tw

ABSTRACT

One of the recent trends of the design of electronic gadgets is consolidating multiple functions in one small device. This idea leads to adding more controls than ever to even smaller devices. Therefore such interface design is also becoming more complicated. Because of the growing popularity in using cell phones for music listening, the wireless Bluetooth earphones are also demanded to include the music control functions. The challenges of integrating more controls to the designers also bring even greater difficulties in using such devices for the end users. In fact, because of the front-worn design of such devices, most of the interface controlling is not visually guided. This study therefore seeks to investigate the controlling behaviors of such Bluetooth earphone devices, mainly through conducting stereotype survey and observing hand postures under different types of control configuration and device sizes in lab experiments as well.

Data analysis reveals that the gaps between user stereotypes and the real control layouts of the Bluetooth earphones on the market are not significant. Personal preference and device experience do not influence the stereotype of Bluetooth earphone controls. From the results of lab experiments, it finds that, unlike cross-button layout and single-button-joystick layout, significant differences in hand postures were observed with the separated double-button layout for different device sizes and different device orientations (front-hung vs. free-holding style). For free-holding style, hand postures were mostly observed with palm-inward and the device was put in the center of palm. Subjects tended to use thumb, index-finger and middle-finger for such control manipulation. For front-hung style, hand postures with palm-downward and using both thumb and index finger are the majority.

Regardless device orientations or sizes, on the other hand, most subject participants orient the control device on their fingers with their palms outwards and use their thumbs for controlling both cross-button layout and single-button-joystick layout of Bluetooth earphone sets.

Keywords: interface usability, compact controls

CHAPTER 42

Evaluation of Grip Strength in Children: the Ergonomic Design Used in the Development of Secure Lids for Packaging

Luis Carlos Paschoarelli, Laura Schaer Dahrouj

UNESP – Univ Estadual Paulista
Bauru, Brazil
paschoarelli@faac.unesp.br

ABSTRACT

Packaging design should provide safety, comfort, efficiency and aesthetic satisfaction, as well as consider user capacity, especially usability and accessibility when opening. Product access should be easy for users with limited capabilities (i.e., the elderly) and difficult for those restricted from using it (i.e., children). The purpose of this study was to verify manual grip (torque) strength while opening a container of bleach. A total of 102 subjects (51 females) between 2 to 5 years old participated in the study. A static torque screwdriver was placed inside three mock-up bleach containers. In all mock-ups, there was a significant increase ($p \leq 0.05$) in grip strength ability according to age. The three analyzed lids were of different dimensions and the results indicated that this variable had a strong influence on the application of torque. The recommended dimensions for the lids of products with toxic potential must be near those of the lid with the smallest diameter, whose design reduces children's capacity to apply torque. These results may contribute to better usability and security for packaging design.

Keywords: packaging design, lids, grip strength, ergonomic design

1   INTRODUCTION

Technological developments have improved the design of packages with respect to commercialization, preservation and the integrity of consumer products. Aspects such as efficiency, comfort and safety are equally important, but depend on understanding the user-package interface. When this interface is disregarded, usability is compromised and severe accidents can happen. One example that stands out is access to products with toxic potential by users who were not considered in the design of their packages. While colors and labels are designed to attract consumers, the sealing systems for household cleaning products do not prevent possible intoxication accidents, particularly those involving children.

Knowledge of the grip strength necessary for removing the lids of household cleaning products may result in reliable and safe systems that provide easy access for users with limited capacities (e.g. elderly individuals), and impede access for those who should not handle the product (e.g. children). The lack of appropriate and reliable biomechanical parameters for populations of all ages, including children, seems to be one reason that still prevents designers and the packaging industry from offering products with truly safe lids.

2   REVIEW

Everyday products should have packaging with high usability levels that guarantee comfort and safety. The Swedish Packaging Research Institute recognizes the following main usability problems in packages: the use of inappropriate information and handling difficulties (Berns, 1981). The latter may be associated with biomechanical limits related to the strength and ability of users.

In Canada, it is estimated that 67,000 people seek medical attention every year due to accidents involving package handling. Plastic containers are involved in 14% of these accidents, mainly the poisonings of children who ingest all or part of their toxic contents (Winder, 2002). Every year the American Poison Control Center registers more than 1.1 million cases of accidental poisonings involving children under five years old (Beirens, Beeck, and Dekker, et al. 2006).

Brazilian data from the SINITOX (Sistema Nacional de Informações Tóxico-Farmacológicas) indicated that in 2009 children in this age range represented 20.41% of the total number of poisoning cases, and the ingestion of household cleaning products was responsible for 4,817 cases (Sinitox, 2009).

Ozanne-Smith (2001) affirms that, after falls, poisoning is the main cause of accidents for children 4 years old and under, and the most vulnerable moment for this type of accident is during cleaning when the toxic product can be easily accessed by children, which demonstrates the great danger of unsafe packages for household cleaning products. According to CEATOX/SP (Centro de Atendimento Toxicológico de São Paulo), between August 2003 and March 2004, there were 483 cases of bleach poisoning, one of the main household cleaning products sold in Brazil. The problem is aggravated by the sale of artificially colored bleach in recycled containers (PET bottles), which make it more attractive not only to customers, but to children, who confuse it with soft drinks, which leads to increased occurrences of poisoning (Inmetro, 2007).

Child resistant packaging (CRP) involves lids whose handling mechanisms are above children's capacity and are obligatory for some products in certain countries. In the USA, CRP was introduced in 1970, and since that time it has contributed reductions in the number of child poisoning cases. Specific methods are used to evaluate the effectiveness of the packaging and its use by children, adults and elderly individuals (Pattin, 2003).

Thien and Rogmans (1984) report that in many countries industry and government are still reluctant to adopt CRP regulatory measures and that when this type of packaging is used the elderly face handling and access difficulties. Berns (1981) reports that child-resistant lids with a "press and turn" system are considered the most difficult to remove. This author also points out that users want CRP to present difficulties for children, not adults.

It should also be considered in the design of these containers that the less complicated the opening system of the packaging is, the higher the degree of acceptance by adults and the elderly. Accordingly, there is a demand to know the real hand grip strength capacity of children for opening packages.

In general, studies about the hand grip strength of children are developed with clinical objectives, since their results can describe the development of coordination, be useful for diagnosing neurological disorders and/or to identifying upper limb pathologies (Esteves, Reis and Caldeira, et al. 2005). The main variables analyzed are palm grip strength (PGS) and finger grip strength (FGS) can be influenced by different methods and/or particular characteristics of the samples.

There are still few studies about hand grip strength among extreme age groups, including children. Moreover, since they perform a number of activities of daily living, it is necessary to better understand their biomechanical capacities, including product handling.

The objective of the present study was to evaluate the hand grip strength torque (maximum voluntary isometric contraction) exerted by children between 2 and 5 years old during the simulated opening of a bleach container (household cleaning product), demonstrating parameters for the ergonomic design of safe lids.

3   MATERIALS AND METHODS

This was a cross-sectional study, and the procedures met the guidelines of Certified Ergonomist Deontology Code (Código de Deontologia do Ergonomista Certificado) - Norm ERG BR 1002 (Abergo, 2003). It was approved by the Research Ethics Committee of the involved institution (Prot. 032/07) and a free and informed consent form was signed by the legal guardians of all 102 subjects. The sample included 51 males and 51 females of the following ages: 2 (n=11), 3 (n=25); 4 (n=33) and 5 years old (n=33).

An advanced force gaugue (AFG-500, Mecmesin Ltd., UK) and a static torque screwdriver (ST10, Mecmesin Ltd., UK) were used to record maximum voluntary isometric contractions. The latter was placed inside real bleach bottles with the sensor fixed to the different lids in order to more closely reality.

Three different bleach containers, among the eight different brands sold in Brazil, were selected. The selection criteria were containers with screw-on lids with different morphological/dimensional characteristics. The lid of container "A" was 35 mm in diameter and 25 mm high, the highest among the brands analyzed; the lid of container "B" had a lower diameter (29 mm) and height (15 mm); the lid of container "C" was 19 mm high and 36 mm in diameter, the largest diameter of the analyzed brands.

The procedures were previously tested and the activity was carried out with the subject standing erect and free to handle the container as desired, including alternating the hands to hold the container and/or twist the lid (Figure 1). An arrow on the upper face of the lid indicated the direction (counterclockwise) it opened. Verbal encouragement was given during the task to obtain the best performance. The sequence of the containers was randomized, and a 60 s interval was set between the tasks.

Figure 1 Procedures adopted in data collection.

Two data collections were carried out and statistically significant differences were found ($P \leq 0.05$) in two ("A" and "B") of the three containers. Therefore, we opted to use the greatest grip strength for all containers in the data analysis.

The results were described statistically, and Student's t-test was applied to identify statistically significant differences ($P \leq 0.05$) in grip strength between hands (right and left), genders (male and female), age ranges (2, 3, 4 and 5 years old) and the three analyzed containers.

4   RESULTS

The mean values (N.m) and respective standard deviations of grip strength obtained with the three containers for each age, gender and hand are presented in Table 1.

Table 1. Means (N.m) and standard deviation for the different genders, hands and ages.

Age	G	N	Hand	A MEAN	s.d.	B MEAN	s.d.	C MEAN	s.d.
2	F	06	Right	0.543	0.047	0.294	0.092	0.425	0.116
			Left	0.494	0.176	0.356	0.109	0.464	0.114
	M	05	Right	0.720	0.373	0.470	0.146	0.636	0.255
			Left	0.732	0.259	0.367	0.181	0.527	0.235
	Total	11	Right	0.624	0.256	0.374	0.146	0.521	0.212
			Left	0.602	0.240	0.362	0.138	0.493	0.172
3	F	12	Right	0.766	0.155	0.519	0.127	0.669	0.155
			Left	0.775	0.193	0.435	0.114	0.621	0.213
	M	13	Right	0.839	0.211	0.542	0.072	0.726	0.234
			Left	0.805	0.174	0.505	0.148	0.705	0.221
	Total	25	Right	0.804	0.187	0.532	0.101	0.699	0.198
			Left	0.791	0.180	0.472	0.135	0.666	0.217
4	F	15	Right	1.026	0.296	0.651	0.222	0.767	0.206
			Left	1.111	0.346	0.668	0.206	0.780	0.318
	M	18	Right	1.239	0.310	0.810	0.245	0.989	0.311
			Left	1.339	0.379	0.843	0.198	1.078	0.351
	Total	33	Right	1.142	0.318	0.738	0.245	0.888	0.287
			Left	1.236	0.377	0.764	0.218	0.943	0.364
5	F	18	Right	1.333	0.327	0.756	0.244	1.097	0.272
			Left	1.327	0.368	0.767	0.260	1.097	0.291
	M	15	Right	1.400	0.370	0.905	0.265	1.154	0.322
			Left	1.626	0.461	1.001	0.306	1.184	0.446
	Total	33	Right	1.364	0.315	0.824	0.262	1.123	0.292
			Left	1.463	0.434	0.874	0.302	1.137	0.366
2-5	F	51	Right	1.017	0.380	0.616	0.247	0.820	0.311
			Left	1.036	0.418	0.612	0.253	0.818	0.348
	M	51	Right	1.134	0.382	0.737	0.266	0.936	0.341
			Left	1.228	0.494	0.757	0.316	0.961	0.412
	Total	102	Right	1.075	0.384	0.676	0.262	0.878	0.329
			Left	1.132	0.465	0.684	0.293	0.889	0.386

The general mean values (N.m) and respective standard deviations of the grip strength obtained with the three containers, for each hand, gender and age are presented in Table 2. This table also presents the significance values ("P") of the comparisons between the means.

No significant differences were observed between the maximum torque exerted by the right or left hands of the subjects for any of the containers. There were significant strength differences in favor of the males for all containers ($P \leq 0.05$). The means significantly increased ($P \leq 0.05$) with age.

Table 2. Statistical analysis between dependent and independent variables (Student's *t*-test).

		A MEAN (s.d.)	"p"	B MEAN (s.d.)	"p"	C MEAN (s.d.)	"p"
HAND	RIGHT	1.075 (0.384)	0.07	0.676 (0.262)	0.71	0.878 (0.330)	0.70
	LEFT	1.132 (0.466)		0.684 (0.294)		0.889 (0.387)	
GENDER	FEMALE	1.026 (0.398)		0.614 (0.249)		0.819 (0.328)	
	MALE	1.181 (0.442)		0.747 (0.291)		0.948 (0.376)	
AGE	2	0.613 (0.242)		0.368 (0.139)		0.507 (0.189)	
	3	0.798 (0.182)		0.502 (0.122)		0.682 (0.207)	
	4	1.189 (0.350)		0.751 (0.230)	0.03	0.916 (0.327)	
	5	1.413 (0.380)		0.849 (0.281)		1.130 (0.329	

In all other interactions: $p \leq 0.01$

The maximum grip strength values obtained for each container according to gender, hand and age range are presented in Table 3. The highest reference value was obtained by a five-year-old male subject with the left hand. Another analysis demonstrated that container "B" allowed less grip strength transference.

Table 3. Absolute maximum values (N.m) for different hands, genders and age ranges.

		A	B	C
HAND	RIGHT	2.126	1.522	1.672
	LEFT	2.412	1.732	2.072
GENDER	FEMALE	1.816	1.208	1.620
	MALE	2.412	1.732	2.072
AGE	2	1.234	0.640	0.906
	3	1.190	0.824	1.150
	4	2.104	1.366	1.838
	5	2.412	1.732	2.072
MAX. VALUE		2.412	1.732	2.072

5   DISCUSSIONS

Container lids have different functions (sealing, dosing, preservation, aesthetic value) including safety and are always based on usability principles.

High national and international rates of accidents and poisonings involving children underscore the need for further developments in lid design for products with toxic potential, particularly bleach.

The non-existence of child grip biomechanical standards for product design restricts the industrial sector regarding the development of effectively safe containers and lids and necessitates understanding of grip capacity in this population.

Studies evaluating the grip strength (PGS and FGS) exerted by children (Ager, Olivett, and Johnson, 1984; Fullwood, 1986; Peeble and Norris, 2000; Bear-Lehman, Kafko, and Mah, et al. 2002; Esteves, Reis and Caldeira, et al. 2005; and Moura, 2008) have presented specific clinical objectives and methodologies, which impedes applicability in product design.

The methods in the above-mentioned studies are not standardized, which complicates comparison of the results (Mathiowetz, 1986). In the present study, characterized by simulated daily activities, freedom of movement and posture naturally becomes necessary, as long as a standard is kept between individuals. The use of verbal stimulation can also affect performance (Mathiowetz, 1986); however, it is a fundamental procedure when dealing with children. Another aspect to be considered is the difference between populations, which was observed by Fullwood (1986) when comparing samples (7 and 8 years old) of American and Australian individuals.

Even though it was not possible to compare the results presented here with previous studies that have evaluated PGS and FGS, some tendencies could be observed, particularly regarding gender and age range.

Nevertheless, data collection was repeated in the present study and significant differences ($P \leq 0.05$) were found, thus the highest value was adopted. Voorbij and Steenbekkers (2002) also repeated data collection with adults, but found no differences ($P > 0.05$). This may be explained by the diversity of the age ranges, considering that the young age of the subjects would seem to require at least one repetition in evaluations of this nature. Furthermore, the differences occurred with containers whose lids had extreme dimensions, i.e., the greatest height ("A") and smallest diameter ("B"), which involve the highest and lowest torque values, respectively. Accordingly, the dimensions of the lids seem to have determined the ability to apply strength between the two collections, possibly due to the difference in surface contact with the subjects' hands.

No significant strength differences ($P > 0.05$) were found between left and right hands, even though the left hand was slightly stronger. Bear-Lehman, Kafko, and Mah, et al. (2002) report that laterality has not yet been established in children under 5 years old and that the ability to perform certain movements, such as turning a lid in a counterclockwise direction, seem not to be associated with either hand.

With respect to gender, the boys were significantly stronger ($P \leq 0.05$) than the girls. Studies with adults have also found the same result (Mital, 1986; Mital and Sanghavi, 1986; Shih and Wang, 1997; Imrhan and Jenkins, 1999; Kim and Kim 2000; and Paschoarelli and Santos, 2011). However, Peebles and Norris (2000) found no gender differences below 60 years old. Again, population differences (Fullwood, 1986) seem to be the most plausible explanation for this condition.

Significant differences were also found regarding age differences ($P \leq 0.05$), which were associated with an increase in capacity. This was also previously observed by Peebles and Norris (2000), who reported a gradual increase from 2 to 15 years old.

Morphologically speaking, the present study evaluated containers with lids of different dimensions. The highest mean values were reached with the higher diameter lids ("A" and "C"), a result that corroborates Peebles and Norris (2000), who found an increase in grip strength capacity proportional to the increase in lid diameter. On the other hand, the lid height probably determined the statistically significant differences ($P \leq 0.05$) between the three containers analyzed, particularly in the case of less strength transmission in "A" than "C". The higher lid of container "C" increased the contact surface with the subjects' hands, which probably determined the strength increase.

Even though lower strength transference values were observed for container "B", demonstrating its superior morphological composition (design and dimension) to the other analyzed CRP lids, other studies contradict this hypothesis.

Rohles, Moldrup, and Laviana (1983) recommend the use of lids with at least a 70 mm diameter for CRP, even though Peebles and Norris (2000) report that children continue to apply greater strength to this type than to those with smaller diameters. Ivergard (1979) also point out that although the elderly transfer more strength to lids with larger diameters, children apply less torque to them.

Another aspect to consider is the effect of the lid's surface finish (smooth or grooved) on the application of torque. The present study did not consider this variable, since only currently-available lids were used, all of which were grooved. Nevertheless, Peebles and Norris (2000) point out that the presence of this kind of finish affects torque, improving the capacity to apply strength due to friction.

The formal characteristics (design) of the packaging, including diameter, height, and grooves are another influence on torque, as well as any elements that interface with one hand while positioning the other on the lid. However, the differences in the morphology of the three containers do not seem to have interfered in the results, since the hands of the subjects were very small compared to the volume of the containers, which thus focused attention on variations in lid form.

6   CONCLUSIONS

Accidents involving the poisoning of children by household cleaning products seem to be related to usability problems found in the lids of the containers of these products. The present study proposed to analyze the hand grip strength of children from 2 to 5 years old and present biomechanical parameters for the ergonomic design of safe lids.

This study concluded that variables such as gender, age and product morphology influence the applied torque and should be considered in the development of CRPs. Even though the results indicate that dimensional variations in lid surface, optimized by increased diameter and height, interfere significantly in strength transmission, the best combination of characteristics for CRP is still not completely clear.

Therefore, the dimensions recommended for CRP lids should be close to those of container "B", which allowed the least torque to be applied by the children. In a real situation, this would complicate their access to the toxic product. The surface of these lids should also be smooth, contributing to less transmitted torque.

Finally, it is important to point out that the procedures adopted in the present study were towards a specific purpose, i.e., the torque applied by the children when opening the container. This biomechanical movement is also applied in other activities of daily living, and considering that this population interfaces with several products, the grip strength standards established could contribute to better usability and safety for other objects.

ACKNOWLEDGMENTS

This study was supported by FAPESP (Procs. 05/59941-2 and 07/53668-8) and CNPq (Proc. 303138/2010-6).

REFERENCES

Abergo. 2003. *Norm ERG BR 1002 – Code of Deontology of Certified Ergonomists* Accessed February 3, 2007. http://www.abergo.org.br/arquivos/norma_ergbr_1002_deontologia/pdfdeontologia.pdf. [in portuguese]

Ager, C. L., B. L. Olivett, and C. L. Johnson. 1984. Grasp and pinch strength in children 5 to 12 years old. *The American Journal of Occupational Therapy* 38 (2): 107-113.

Bear-Lehman, J., M. Kafko, L. Mah, L. Mosquera, and B. Reilly. 2002. An Exploratory Look at Hand Strength and Hand Size among Preschoolers. *Journal of Hand Therapy* 15(04): 340-346.

Beirens, T. M. J., E. F. Beeck, R. Dekker, J. Brug, and H. Raat. 2006. Unsafe storage of poisons in homes with toddlers. *Accidents Analysis and Preventions* 38: 772-776.

Berns, T. 1981. The Handling of Consumer Packaging. *Applied Ergonomics* 12(3), 153-161.

Esteves, A. C., D. C. Reis, R. M. Caldeira, R. M. Leite, A. R. P. Moro, and N. G. Borges Junior. 2005. Grip force, laterality, sex and hand anthropometry of children in scholar age. *Brazilian Journal of Kinanthropometry and Human Performance* 7(2): 69-75. [in portuguese]

Fullwood, D. 1986. Australian Norms for Hand and Finger Strength of Boys and Girls Aged 5-12 years. *Australian Occupational Therapy Journal* 33(1): 26-36.

Imrhan, S. N. 1994. Muscular strength in the elderly – Implications for ergonomic design. *International Journal of Industrial Ergonomics* 13: 125-138.

Imrhan, S. N. and G. D. Jenkins. 1999. Flexion-extension hand torque strengths: applications in maintenance tasks. *International Journal of Industrial Ergonomics* 23: 359-371.

Inmetro. Bleach – Product and Safety of the Package. Accessed February 4, 2007, http://www.inmetro.gov.br/consumidor/produtos/agua_sanitaria2.asp?iacao. [in portuguese]

Ivergard, G. 1979. Handleability of consumer packaging - Observation technique and measurement of forces: Appendices Report No. 14 16 46. Stockholm: Swedish Packaging Institute.

Kim, C. and T. Kim. 2000. Maximum torque exertion capabilities of Korean at varying body postures with common hand tools. *Proceedings of the International Ergonomics Association*. San Diego: IEA.

Mathiowetz, V., D. M. Wiemer, and S. M. Federman. 1986. Grip and pinch strength: norms for 6 to 19-year-olds. *The American Journal of Occupational Therapy* 40 (10): 705-711.

Mital, A. 1986. Effect of body posture and common hand tools on peak torque exertion capabilities. *Applied Ergonomics* 17 (2): 87-96.

Mital, A. and N. Sanghavi. 1986. Comparison of maximum volitional torque exertion capabilities of males and females using common hand tools. *Human Factors* 28 (3): 283-294.

Moura, P. M. L. S. 2008. *Study of the hand grip strength in diferente age groups of human development* [Master Dissertation]. Brasilia: UnB. [in portuguese]

Ozanne-Smith, J. 2001. Childhood poisoning: access and prevention. *Journal Pediatric Child Health* 37: 262-265.

Paschoarelli, L. C. and R. Santos. 2011. Usability Evaluation of Different Door Handles. In: *Advances in Cognitive Ergonomics* eds. D. Kaber and G. Boy, Miami: CRC Press, p. 291-299.

Pattin, C. A. 2003. Child Resistant Packaging – Regulations and Effectiveness. *Safety Brief* 23(03): 1980-2002.

Peebles, L. and B. Norris. 2000. *Strength data for design safety – phase I*. Nottingham: Institute of Occupational Ergonomics.

Rohles. F. H., K.L. Moldrup, and J.E. Laviana. 1983. *Opening jars: An anthropometric study of the wrist-twisting strength of children and the elderly*. Report no. 83-03. Kansas: Institute for Environmental Research.

Shih, Y. C. and M. J. J. 1997. Evaluating the effects of interface factors n the torques exertion capabilities of operating handwheels. *Applied Ergonomics* 28 (5): 375-382.

Sinitox. 2009. Cases of poisoning for bleach by federated unit and age group, registered in 2009. Accessed January 5, 2012, http://www.fiocruz.br/sinitox_novo/media/tab02_domissanitarios_2009.pdf [in portuguese]

Thien, W. M. A. and W. H. J. Rogmans. 1984. Testing child Resistant Packaging for Access by Infants and the Elderly. *Accident Analyses & Prevention* 16 (3): 185–190.

Voorbij, A. I. M. and L. P. A. Steenbekkers. 2002. The twisting force of aged consumers when openinig a jar. *Applied Ergonomics* 33:105-109.

Winder, B., K. Ridgway, A. Nelson, and J. Baldwin. 2002. Food and drink packaging: who is complaining and who should be complaining. *Applied Ergonomics* 33: 433-438.

CHAPTER 43

Perception of Products by Different Levels of Integration: A Study with Pruning Shears

Lívia Flávia de Albuquerque Campos, Jamille Noretza de Lima Lanutti, Liara Mucio de Mattos, Douglas Daniel Pereira, Elen Sayuri Inokuti, Luis Carlos Paschoarelli

São Paulo State University
Bauru, Brazil
liviaflavia@gmail.com

ABSTRACT

Six different models of pruning shears with eight subjects were assessed using semantic differential (SD) applied at four levels of integration: see the product image, see the real product, touch the product, and finally, use the product. The results showed awareness of the variables changed at different levels of interaction. The perception of the subjects changed after using the product.

Keywords: Ergonomic Design, Pruning shears, Semantic Differential, Levels of integration.

1 INTRODUCTION

The current methods of design have considered the inclusion of typical needs of a conscious user of the cognitive and emotional aspects involved during use, such as the meanings assigned to products, the pleasure they can give or the emotions and feelings that they evoke (Vergara et al., 2011). However, some of the studies that have sought to evaluate these variables, only the visual intervention has been used and, sequentially, subjective opinions are achieved by the semantic differential test

(SD) (Hsu et al, 2000; Chuang and Ma, 2001; Yannou and Petiot, 2004; Mondragón et al, 2005; Cheng and Chang, 2009). Given this observation, Vergara et al. (2011) developed a study aiming to understand the influence of the level of interaction (see pictures, see the real product, touch and use it) about the perception of products with SD tests. The results showed that the level of integration (LI) modifies the perception of the factors that are linked to physical interaction, for the case of products (hammers) evaluated, which represents important information for the study of methods of ergonomic evaluation. The aim of this study was to determine whether in pruning shears, LI is also able to affect significantly their perception of SD tests, in order to contribute to the study of products evaluation methodology, considering the different sensory interactions.

2 MATERIALS AND METHODS

2.1   Ethical questions

The procedures were based on ethical recommendations (Cns, 1996) and the ethics parameters were based in "Norm ERG BR 1002 – Code of Deontology of Certified Ergonomists" (Abergo, 2003). All subjects signed an informed consent statement previously approved by the local ethics committee (240/2010, Feb, 24, 2011-CEP-USC).

2.2   Subjects

Eight subjects participated, 2 (two) females and 6 (six) male. The mean age was 20.63 years old (s.d. 1.19 years).

2.3   Materials and equipments

We used six different models of pruning shears (Figure 01A). For the execution of activity bamboo stakes (40 cm in length and diameter≈ 1.5 cm)   were used, arbitrarily marked for cutting (Figure 01B).

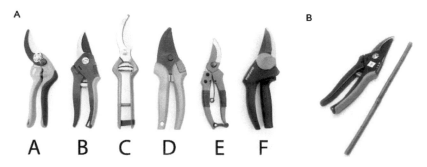

Figure 01A Pruning shears used in the study. Figure 01B Bamboo used in the fourth phase.

We also applied: Statement of Free and Informed Consent (SFIC) (CNS, 1996); recruitment and identification protocol; assessment protocols for scissors (SD).

2.4   Definition of the semantic space

The semantic space, to prepare the SD protocol consisted of 24 pairs of bipolar adjectives, general and specific (according to the cable or on the cut) (Table 1). Those adjectives were selected from the collection of words used in advertising of pruning shears found on the websites of manufacturers, distributors and shops. The order of presentation of terms was randomized so adjectives with positive or negative characteristics were not in only one side of the scale.

Table 1 Words used in the SD scales.

Translated words	Original words used (Portuguese)
Confortable / Unconfortable	Confortável / Desconfortável
Heavy / Light	Pesado / Leve
Multipurpose / Limited	Polivalente / Limitada
Inaccurate / Precise	Imprecisa / Precisa
Unrestricted / Restricted	Irrestrito / Restrito
Effective / Ineffective	Eficaz / Ineficaz
Simple / Sophisticated	Simples / Sofisticada
Safe / Dangerous	Segura / Perigosa
Fragile / Strong	Frágil / Resistente
Durable / Provisional	Durável / Provisória
Hard drive / easy drive	Acionamento difícil / Acionamento fácil
Ergonomic / No ergonomic	Ergonômico / Não ergonômico
Large / Small	Grande / Pequena
Weak / Powerful	Fraca / Potente
Modern / Classical	Moderna / Clássica
Unpleasant / Pleasant	Desagradável / Agradável
Beautiful / Ugly	Bonita / Feia
Amateur / Professional	Amadora / Profissional
Adherent Cable / Sliding Cable	Cabo Aderente / Cabo Deslizante
Hard Cable / Soft Cable	Duro / Macio
Easy Cut / Difficult Cut	Corte Fácil / Corte Difícil
Unstable Cut / Firm Cut	Corte Instável / Corte Firme
Precise Cut / Inaccurate Cut	Corte Preciso / Corte Impreciso
Hard Cut / Soft Cut	Corte Duro / Corte Macio

2.4   Procedures

The test was conducted in four phases with different LI. In Phase 1, subjects viewed the image of scissors in a multimedia projection. Still with the visualization of the projection being shown, the subject answered questions on tests of SD. In Phase 2, the scissors were presented and subjects were not allowed to touch them, while answering SD tests for this phase. In the third phase the subjects were able to touch the scissors, and answer the SD test. In Phase 4, the activity was performed and each subject used the scissors individually to cut the bamboo stakes. It was allowed to cut more pieces if they wanted to be sure of the answer. After finishing the task, the subject answered the SD test questions.

This activity was repeated for each of the six scissors. The order of evaluation of the scissors was randomized in each phase. The SD scale had seven points so the closer it was from the adjectives, the greater is the agreement about the meaning of the terms. Each of the phases took about 20 minutes, except the fourth stage, which took about 40 min. It was applied a break of at least one week between the phases. All the phases were performed in group, however, the subjects were asked not to exchange opinions among each other not to show any reaction of acceptance or rejection of the product.

2.5   Statistical analysis

Statistical analysis was performed in STATISTICA ® software. For each point in the range, it was assigned a value (from 1 to 7) and the levels of the averages and standard deviations were obtained. We applied the Wilcoxon test ($p \leq 0.05$) to check significant differences between phases for each pair of adjectives, for each object.

The pairs of adjectives were divided into groups according to the characteristic of the analyzed variable. We identified three groups: Style Variables (Simple / Sophisticated; Modern / Classic; Beautiful / Ugly); quality / robustness Variables (Heavy / Light; Weak / Powerful; Fragile / Strong; Durable / Provisional; Large / Small); Usage Variables (Effective / Ineffective; Hard Drive / Easy Drive; Easy Cut / Difficult Cut; Unstable Cut / Firm Cut; Comfortable / Uncomfortable; Inaccurate / Precise; Amateur / Professional, and Adherent Cable / Sliding Cable; Ergonomic / No ergonomic; Unrestricted / Restricted; Multipurpose / Limited; Unpleasant / Pleasant; Safe / Dangerous; Inaccurate Cut / Precise Cut; Hard cut / Soft cut; Hard Cable / Soft Cable).

3 RESULTS AND DISCUSSIONS

Variables of style Simple / Sophisticated and Modern / Classic showed no significant difference when comparing the results of the levels of integration, only one of these three style variables (Beautiful / Ugly) presented significantly different ($p \leq 0.05$) between phases of two scissors (Figure 02). Both scissors (C and F) were evaluated as more beautiful after use.

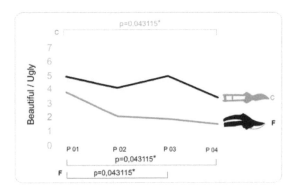

Figure 02 Results of the Wilcoxon test (p ≤ 0.05) for variable style (Ugly / Beautiful).

Among the variables of quality and robustness of the object, two of them (Heavy / Light; Weak / Powerful), showed a significant difference among the results of phases of three shears evaluated (Figure 03). Shears B and E were evaluated in the initial phases as heavier than in the phases in which it was possible to touch and use the products. The C shear was evaluated as lighter at the beginning and after use, as heavier. The subjects evaluated the shears A and E as the most powerful in the early stages and more fragile in the last stages. The opposite was observed for the shear B.

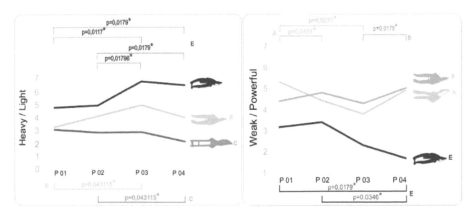

Figure 03 Results of the Wilcoxon test (p ≤ 0.05) for variables Robustness / Quality.

For two variables (Weak / Strong; Durable / Provisional) results showed significant differences among phases for two shears, and the variable (Large / Small) did not show significantly different results among phases for any scissors (Figure 04). Shears A and E were considered more durable in the more superficial levels of interaction. In the same way they were considered more fragile in the levels of greater tactile interaction with the product.

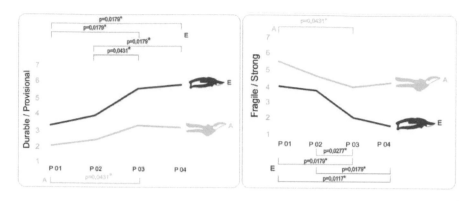

Figure 04 Results of the Wilcoxon test (p ≤ 0.05) Robustness / Quality variables

The variables of use (Effective / Ineffective; Difficult Drive / Easy Drive; Easy Cut / Difficult Cut, Unstable Cut / Firm cut) showed significantly differences (p ≤ 0.05) among phases for most shears (Figure 05).

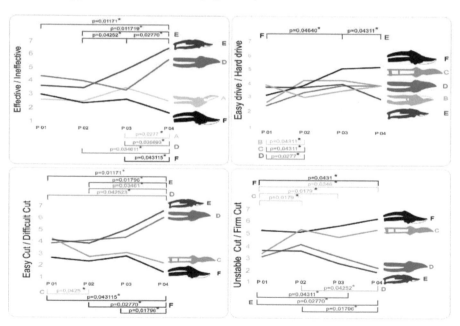

Figure 05 Results of the Wilcoxon test (p ≤ 0.05) variables of Use.

The E and D shears were evaluated in the initial phases as more effective than after use. The A and F shears were considered more effective at this stage. Talking about the ease of driving, there are significant differences between phases P1 and P2 for three shears. C and D were best assessed at the initial phase (P1) compared to the second phase (P2). The opposite occurred for the B shear. The F shear also received better evaluation at the initial phase (P1) compared to phase P3, in the same phase the E shear received the worst evaluation, compared to the final stage (P4). The cut of the F shear was considered the easiest after use, and C comparing P1 and P2. The cut of D and E shears were considered hard to cut while LI increased. Those shears were also analyzed as the most unstable cut when the level of integration increased. F and C shears were considered the most firm cut with the increase.

For the other four variables of use (Comfortable / Uncomfortable; Inaccurate / Precise; Amateur / Professional; Adherent Cable / Sliding Cable) they showed significantly differences ($p \leq 0.05$) among phases for half of the shears (Figure 06).

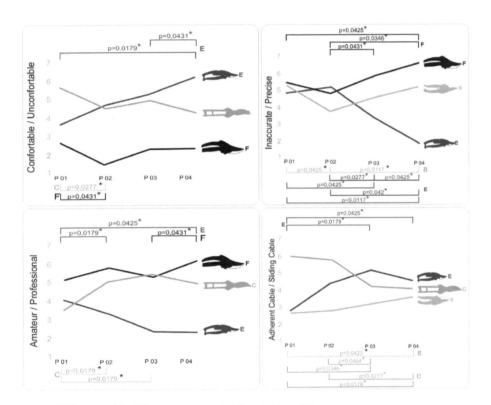

Figure 06 Results of the Wilcoxon test ($p \leq 0.05$) variables of Use.

C and F are considered the most uncomfortable when visualized in phase P2, compared to the initial phase (P1). The opposite occurred with E, which was rated as the most comfortable in the initial phases. B and F were rated as the most accurate with the increase of LI and the opposite occurred to E. C was considered the most professional when the LI increased, F and, particularly after being used (P4). The opposite was observed for E. The cable C was considered the most sliding

406

in initial stages after use. We found the opposite for E and B.

For the other seven variables (Ergonomics / No Ergonomic; Unrestricted / Restricted; Multipurpose / Limited; Unpleasant / Pleasant; Safe / Dangerous; Inaccurate Cut / Precise Cut; Hard Cut / Soft Cut) significant differences (p≤ 0.05) between phases of two shears were observed (Figure 07).

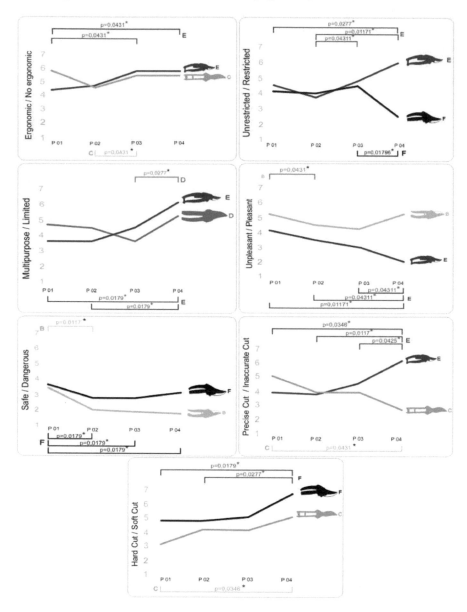

Figure 07 Results of the Wilcoxon test (p ≤ 0.05) variables of Use.

The C and E shears were evaluated as least ergonomic when the level of interaction increased. The E shear was considered the most restricted, the most limited, the least pleasant, and the most inaccurate cut after being used. The F shear, the most unrestrained, the safest and softest cut, according to the interaction increasing. The B shear was considered more unpleasant and more secure when comparing P1 to P2. The C shear was evaluated as the most precise and soft cutting, when the level of integration increased. Finally, we found significant differences between phases of only one scissors for the variable Hard Cable / Soft Cable (Figure 08). The subjects assessed the B cable as the hardest after using it.

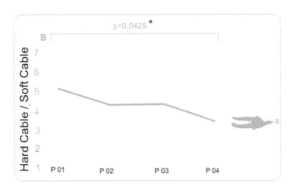

Figure 08 Results of the Wilcoxon test (p ≤ 0.05) variables of Use.

CONCLUSION

The analysis showed that the perception variable changed in different LI. We found that the usage variables are the most affected by LI, either by the number of shears where these differences were observed or by the number of affected variables. Vergara et al. (2011) also found that the "physical variables" were the most affected. This study showed that in pruning shears, as in the case of hammers (Vergara et al., 2011) LI also influences the perception of users. This information is important in the specification of the methodology of analysis of products with the application of SD tests.

ACKNOWLEDGMENTS

This study was supported by FAPESP (FAPESP 2010/21439-9).

REFERENCES

ABERGO. (2003). *Code of Deontology of Certified Ergonomists.* Norm ERG BR 1002. Accessed December 20, 2009, available at: http://www.abergo.org.br/arquivos/norma _ergbr_1002_deontologia/pdfdeontologia.pdf [In portuguese].

CNS. National Health Council. (1996). *Resolution No. 196,* October 10, 1996. Available at: <http://conselho.saude.gov.br/docs/Resolucoes/Reso196.doc>. Accessed: April, 8, 2010. [In Portuguese].

Cheng, H.-Y., Chang, Y.-M., 2009. Extraction of product form features critical to determining consumers' perceptions of product image using a numerical definition-based systematic approach. *Int. J. Ind. Ergon.* 39 (1), 133 e 145.

Chuang, M.C., Ma, Y.-C., 2001. Expressing the expected product images in product design of micro-electronic products. *Int. J. Ind. Ergon.* 27 (4), 233 e 245.

Hsu, S.H., Chuang, M.C., Chang, C.C., 2000. A semantic differential study of designers' and users' product form perception. Int. J. Ind. Ergon. 25 (4), 375e391.

Mondragón, S., Company, P., Vergara, M., 2005. Semantic differential applied to user-centred machine tool design. Int. J. Ind. Ergon. 35 (11), 1021e1029.

Petiot, J.F., Yannou, B., 2004. Measuring consumer perceptions for a better comprehension, specification and assessment of products semantics. *Int. J. Ind. Ergon.* 33 (6), 507 e 525.

Vergara, M.; Mondragón, J. L. S.; Company, P.; Agost, M. 2011. Perception of products by progressive multisensory integration. A study on hammers. *Applied Ergonomics,* (42) 652 – 664.

CHAPTER 44

Evaluation of Manual Torque Forces in Simulated Activities with Brazilian Adults of Different Genders and Age Groups: Ergonomic Design of Faucets

Lívia Flávia de Albuquerque Campos, Luis Carlos Paschoarelli

São Paulo State University
Bauru, Brazil
liviaflavia@gmail.com

ABSTRACT

This paper presents an evaluation of torque strength and on the perceptual aspects of the use of faucets. We evaluated 180 subjects (18–29 years old = 30M and 30F; 30–55 years old = 30M and 30F; >55 years old = 30M and 30F). A digital dynamometer, a static torque transducer, five different models of handles (faucets) and other complementary tools were used for data collection. The procedures were standardized, based on ethical recommendations and biomedical criteria. The data analysis was based on descriptive statistics and included the application of tests for verification of significant differences. The results show the male performed torque significantly higher ($p \leq 0.05$) than the female subjects and the forces applied by the age group above 55 years old were the lowest ($p \leq 0.05$). With the lever-type handle was possible to achieve the highest magnitude of torque to all groups of subjects and these differences were significant ($p \leq 0.05$) in all comparisons. The lever-type handle was considered the easiest to use and the spherical handle was the hardest.

Keywords: Ergonomic Design, Hand Tools, Faucets.

1   INTRODUCTION

The faucets handles marketed and used in Brazil already have a high level of technology and design. However, problems of usability, performance and security are still observed, especially when different individuals in different age groups and genders (male and female) use this product. Hence the importance of ergonomic knowledge related to the capacity of the musculoskeletal system in generating forces for manual manipulation of these objects.

Muscle strength is defined by Chaffin et al. (2001) as "[...] maximum force that a muscle group can develop on prescribed conditions" (p. 101). His knowledge was treated by Mital and Kumar (1998, p. 102) as extremely important, because by that "[...] it is possible to design devices which will conform to the physical capabilities of humans, and prevent musculoskeletal injuries".

The amplitude of force varies substantially in an adult (normal and healthy) population, (Chaffin et al., 2001) and as influence we have: gender and age. Iida (2005, p.99) explains that "[...] women have a muscle capacity of approximately two thirds of men." For the hand grip, for example, there are reports that women perform strength between 50% and 74% of the strength of men (Crosby et al., 1994; Mamansari; Salokhe, 1996; Caporrino et al., 1998; Imrhan, 2003 ; Edgren et al., 2004). For the manual torque, studies indicate that the female performs around 49.12% to 66% of the strength of males (Imrhan; Jenkins, 1999; Kim, Kim, 2000; Shih, Wang, 1996, 1997). To Morse et al. (2006) the mean manual torque were also higher for males than for females and Matsuoka et al. (2006) found that men generated torque twice higher than women. Paschoarelli (2009) observed that male subjects showed significantly greater torque than the female subjects, ranging from 43% to 77%.

In addition to gender, age is another variable of influence, because over the years, there is "[...] a gradual loss of strength and mobility, making muscle movement weaker, slower and lower in amplitude" (Iida , 2005, p.100). According to some studies, the time that you get the maximum force is in the range of 25-29 years old (Montoye, Lamphiyer, 1977; Voorbij; Steenbekkers, 2001) and the mark to the beginning of loss of muscle strength is related to age located between 50 and 55 years old (Montoye, Lamphiyer, 1977; Mathiowetz et al., 1986; Hanten et al., 1999; Voorbij; Steenbekkers, 2001). According to other studies of strength in different age, muscle strength seems to be higher at around 30 years old, with a decline after this period (Asmussen; Heeboll-Nielson, 1962; Chaffin, Herrin, et al., 1977). Paschoarelli (2009) found that individuals of 30-55 years old had the highest mean torque.

Imrhan and Jenkins (1999) specified that, among other things, the characteristics of the object such as dimensions, materials, or design, are also variables of influence on management activity which involves the manual torque. The object should be designed to maximize comfort, performance of tasks and the contact area between the palm and the handle, in order to provide better distribution of power and reduce pressure on the hand. The proper design of management is considered a major influence on "User x Product" system performance (Napier, 1985; Iida, 2005, Kong et al., 2007) and on the investigation of this variable, there are several studies on the relationship of the handling drawing and the capacity of the human musculoskeletal system (Deinavayagam; Nagashima; Konz, 1986, Review, 1988, Adams, Peterson, 1988; Imrhan; Loo, 1989; Imrhan; Jenkins, 1990 and 1999; Imrhan et al., 1992; Peebles, Norris, 2000, 2003, Kong et al., 2007), but few results are applicable in the design of products, either because of research situations, which do not always match the reality of product use, or because data do not correspond to different age groups (Peebles, Norris, 2000, 2003).

Reviews and physical analyzes of activity manuals are still scarce in Brazil, and despite some occasional advances, information about the Brazilian population is still incomplete, which requires further studies; mainly because it is impractical to apply data from studies conducted outside Brazil in the design of local produce. Thus, it is necessary to know the stress levels in the management of these products to both genders and different age groups, and the kind of design that facilitates/hinders the manipulation.

2 MATERIALS AND METHODS

2.1 Subjects

The sampling was based on the theory of statistical inference and the principle of independence of the sample about the population (Triola, 1999, p. 149). We defined 30 subjects per group: 18 to 29 years old (30M and 30F), 30 to 55 (30M and 30F)> 55 (30M and 30F). The general characteristics are shown in Table 1.

Table 1 General characteristics of the sample

Variables			Age (years)	Weight (kg)	Height (m)
Male	18-29	$\bar{x}$	22,9	73,3	1,77
		d.p.	3,45	13,5	0,07
	30-55	$\bar{x}$	42,8	82,7	1,69
		d.p.	5,86	14,1	0,15
	>55	$\bar{x}$	63,4	77,9	1,63
		d.p.	6,99	12,2	0,15
Female	18-29	$\bar{x}$	22,0	59,8	1,61
		d.p.	2,53	14,5	0,07
	30-55	$\bar{x}$	44,6	67,9	1,57
		d.p.	7,73	17,1	0,28
	>55	$\bar{x}$	67,4	71,5	1,53
		d.p.	12,1	20,2	0,23

2.2 Ethical questions

Due to the fact of involving experimental procedures on humans, it was submitted and approved by the Ethics Committee in Research of USC / Bauru - SP (Protocol 005/09 of 20.02.2009) and meets the resolution and 196/96-CNS-MS "Standard ERG BR 1002," "Code of Ethics of the Certified Ergonomist" (Abergo, 2003).

2.3 Materials and equipments

To evaluate the torque, we used an AFG - Advanced Force Gauge (digital dynamometer) AFG 500 (Mecmesin Ltd., UK) and Static torque trasducer - STT (Mecmesin Ltd., UK), Model ST 100-872-003. The STT was set at a base to actuate with standardized height of 1.00 m from the floor (simulating the opening of a shower). We used a graphic scale to collect perceptual data, with five spaces (from easiest to hardest) to organize the faucets handles after the simulation of use. Five faucets handles commercially available in Brazil were also used (Figure 01). To obtain weight (kg) and height (m), we used a scale with an anthropometer. Protocols were also used: Term of Free and Informed Consent (TFIC) (CNS, 1996), protocol of recruitment and identification, and registration protocol of strength and perception.

H01   H02   H03   H04   H05

Figure 01 faucets handles group selected for analysis.

2.4 Procedures

The subjects, after the objectives of the study being explained, they read, filled out and signed the TFIC and then filled out the identification. The collection of weight (kg) and height (m) was performed. The use simulations occurred with five models of faucets handles, for shower command. The subjects were instructed to perform maximal voluntary strength (Voorbij; Steenbekkers, 2001; Chaffin et al., 2001, Edgren et al., 2004, Dias et al., 2010), in the upright position (standing). The handles were driven in a clockwise direction with the right hand, regardless of the dominant hand, based on Paschoarelli (2009). A random sequence was determined for each subject. Each model was driven, respecting an interval of 60 seconds. The procedures were standardized and based on biomedical criteria and recommendations (Caldwell et al., 1974; Chaffin, 1975, and Kroemer et al., 1994). The procedures did not result in pain or damage to the organism of subjects.

2.5 Statistical analysis

In the gender influence analysis, the Student t test (p≤ 0.05) was applied to the cases of normal variances, otherwise, the Mann-Whitney test (p ≤ 0.05) was applied. For the different age groups we used the Kruskal-Wallis test (p ≤ 0.05) and the *pos hoc* test of nonparametric multiple comparison (Zar, 1999, cited Paschoarelli, 2009). To identify significant differences (SD) in the comparison of five faucets handles we used the analysis of variance (ANOVA - Multivariate tests, p≤ 0.05) for repeated measures, for cases of normality performance and *post hoc* test of *Bonferroni* to verify groups in which occur the SD, otherwise, the Friedman test (p ≤ 0.05) and *post hoc* of *Dunn*. The perception data values refer to score assigned to each faucet handle based on its position (1st position = 1, 2 position = 2 ...). The graphic scale ranged from 1 (easiest) to 5 (hardest). For the analysis of SD we used the *Wilcoxon* test (p ≤ 0.05).

3 RESULTS AND DISCUSSIONS

3.1 The influence of gender in the manual torque

The results (Figure 02) show that the male subjects performed a significantly higher torque (p ≤ 0.05) than females.

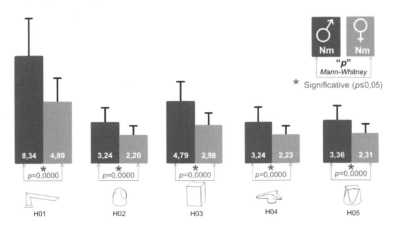

Figure 02 Comparison between genders (means, standard deviations and "p" value).

The male subjects had greater magnitudes of torque with H01, with which they surpassed the average torque of females 42.44%. In general, the torque reached by females ranged from 57.56% to 68.83% of the torque achieved by the male subjects

414

corroborating with other studies that talk about gender differences in this type of analysis (Shih, Wang, 1996, 1997; Imrhan; Jenkins, 1999; Kim, Kim, 2000; Paschoarelli, 2009).

3.1 The influence of age on manual torque in sanitary metal handles

The results (Figure 03) showed that the strength registered to H02, by the group of subjects between 18-29 years old was significantly higher than the strength achieved by the subjects aged over 55 years old.

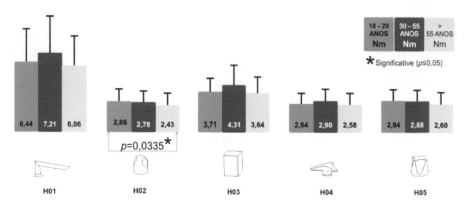

Figure 03 Comparison between age groups (means, standard deviations and "p" value).

The forces applied by the age group over 55 years old were the lowest for all handles, and the subjects aged from 30 to 55 years old accounted for the largest force magnitudes for most of the handles, although most of these differences were not statistically significant (p ≤ 0.05).

The percentage of torque for the age group from 30 to 55 years old was higher than about 9.32% to 15.95% of the strength of subjects aged over 55 years old. These results are in agreement with other studies (Asmussen; Heeboll-Nielson, 1962; Chaffin, Herrin, et al., 1977).

3.3 The influence of the design of faucet handles on manual torque

The results (Figure 5) showed significant differences (p ≤ 0.05) in all age groups and genders, for all comparisons to H01. H03 showed significant differences in most comparisons, except for H02, H03 and H04 for the female subjects and those aged from 18 to 29 years.

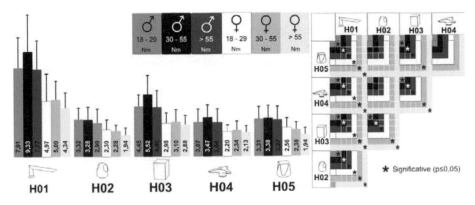

Figure 05 Comparison of gender and age groups (means, standard deviations and "p" value)

H03 presents geometric form and according to Iida (2005) this type of management has little contact surface, by distancing itself from the human anatomy focusing points of tension in the hands and transmits less force. Nevertheless, had the second highest torque for all groups of subjects. This is probably due to the fact that their size allows greater contact area, although the geometric shape.

Additionally, their faces acted as a lever. Already H02, H04 and H05 showed similar performance, we have not verified SD for the comparisons. These interfaces have smaller forms and therefore with little contact area, affecting the distribution of forces (Napier, 1985; Iida, 2005, Kong et al., 2007) and consequently the magnitude of force generated in the activation of these interfaces (Adams; Peterson, 1988).

3.4   Perceptual aspects

Statistical analysis showed significant differences in all comparisons (*Wilcoxon* - $p \leq 0.05$), as shown in Table 2.

Table 2 Values of "p" for the Wilcoxon test ($p \leq 0.05$)

	H05	H04	H03	H02
H01	p=0.0000*	p=0.0000*	p=0.0000*	p=0.0000*
H02	p=0.0000*	p=0.0003*	p=0.0000*	
H03	p=0.0004*	p=0.0361*		
H04	p=0.0000*			

The results, in which all comparisons presented DS, suggest that H01 was considered easier to use and the handle H02 was the hardest to use (Figure 06). The handles H03 and H04 have morphological structure that, in theory, could also provide further facilitated prehensions, especially due to the design with different support points. However, the presence of edges and corners, may have contributed to significant pressure on the palm side of the hand, making the user experience more uncomfortable. The handle V02 provided a hard perform, probably due to its spherical shape, which, despite reducing the pressure points with the hand, it does not allow a management that facilitates the turn, because it does not have support points. These results are similar to those reported by Paschoarelli (2009).

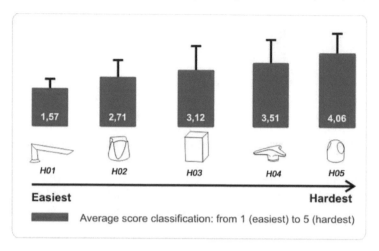

Figure 06 Results (mean and standard deviation) of the perceived ease / hard of use.

4. CONCLUSION

We observe the influence of gender in driving faucets handles. The female subjects performed a significantly lower torque than the male. We have found that gender differences have changed according to the design of the interface. The highest torques and the largest differences were seen in the drawing with a lever, and the lowest torques and the smallest differences among genders were found in the interfaces with lower contact surface. Regarding differences among age groups, the elderly subjects performed the lowest torque in most cases, especially when compared with subjects aged from 30 to 55 years old, although the majority was not statistically significant, this significance was observed in spherical handles.

We note the importance of considering the characteristics of individual users. The interfaces design must allow an appropriate prehension to the involved operation and require muscle actions, which fit the limits and capabilities of different groups of subjects. We can say that the lever-type handles can be considered those with the best performance. It is recommended to use this type of handle configuration, however the design should prioritize anthropomorphic forms, compared to geometric designs. The handles should prioritize the increasing of contact area, because larger dimensions allow better performance. Finally, the spherical designs should be avoided and, if not essential, they must have a larger size, more robust forms and finishing with textures to prevent the hands slippage.

ACKNOWLEDGMENTS

This study was supported by FAPESP (FAPESP 2009/02125-0, 2010/21439-9).

REFERENCES

Abergo - *Code of Deontology of Certified Ergonomists.* Norm ERG BR 1002. Accessed December 20, 2009, http://www.abergo.org.br/arquivos/norma_ergbr_1002_deontologia/pdfdeontologia.pdf [In portuguese].

Adams, S. K.; Peterson, P. J. (1988). Maximal voluntary handgrip torque for circular electrical connectors. *Human Factors,* 30 (06), 733-745.

Asmussen, E., Heeboll-Nielson, K. (1962). Isometric Muscle strength in relation to age in men and women. *Ergonomics,* 5 (1), 167 - 169.

Caldwell, L. S., et al. (1974). A proposed standard procedure for static muscle strength testing. *American Industrial Hygiene Association Journal,* 35, 201-206.

Caporrino, F. A. et al. (1998). Estudo populacional da força de preensão palmar com dinamômetro Jamar®. *Revista Brasileira de Ortopedia,* 2 (33), 150-154.

Chaffin D. B. (1975). Ergonomics guide for the assentment of human strength. *Amer. Ind. Hyg. J,* 36, 505 – 510.

Chaffin, D.B.; Anderson, G.B.J.; Martin, B.J. (2001). *Biomecânica Ocupacional.* Belo Horizonte: Ergo.

Chaffin, D.B.; Herrin, G.D.; Keyserling, W.M.; Foukle, J.A. (1977). *Preemployment strenght testing.* Cincinnati: Health, Techinical Report No. 77 - 163 of National Institute for Occupational Safety.

CNS. National Health Council (1996), *Resolution No. 196,* October 10. Available at: <http://conselho.saude.gov.br/docs/Resolucoes/Reso196.doc>. Accessed: 8, Apr, 2010. [In Portuguese].

Crosby, C. A.; Wehbé, M. A.; Mawr, B. (1994) Hand strength: normative values. *The Journal of Hand Surgery,* 4 (19A), 665-670.

Deinavayagam, S.; Weaver, T. (1988). Effects of handle length and bolt orientation on torque strength applied during simulated maintenance tasks. In: Aghazadeh, F. (Ed.) *Trends in Ergonomics / Human Factors.* Amsterdam: Elsevier, 827-833.

Dias, J. A. et al. (2010). Força de preensão palmar: métodos de avaliação e fatores que influenciam a medida. *Revista Brasileira de Cineantropometria & Desempenho Humano,* 3 (12), 209-216.

Edgren, C. S.; Radwin, R. G.; Irwin, C. B. (2004). Grip force vectors for varying handle diameters and hand sizes. *Human Factors,* 2 (46), 244-251.

Hanten, W. P. et al. (1999). Maximum grip strength in normal subjects from 20 to 64 years of age. *Journal of Hand Therapy,* 12, 193-200.

Iida, I. (2005). *Ergonomia:* projeto e produção. 2ª. ed. Rio de Janeiro: Edgard Blucher.

Imrhan, S. N. (2003). Two-handed static grip strengths in males: the influence of grip width. *International Journal of Industrial Ergonomics,* (31), 303-311.

Imrhan, S. N.; Jenkins, G. K. (1999). Flexion-extension hand torque strengths: applications in maintenance tasks. *International Journal of Industrial Ergonomics,* 23 (04), 359-371.

Imrhan, S.N.; Jenkins, G.K. (1990). Hand turning torques in a simulated maintenance task. In: Das, B. (1990). *Advances in Industrial Ergonomics and Safety II.* Londres: Taylor e Francis, 437-444.

418

Imrhan, S. N.; Jenkins, G. K.; Townes, M. (1992). The effect of forearm orientation on wrist-turning strength. In: Kumar, S. (1992). *Advances in Industrial Ergonomics and Safety IV*. Londres: Taylor & Francis, 687-691.

Imrhan, S. N.; Loo, C. H. (1989). Modelling wrist-twisting strength of the elderly. Ergonomics, 31 (12), 1807–1819.

Kim, C.; Kim, T. (2000). Maximum torque exertion capabilities of Korean at varying body postures with common hand tools. In: International Ergonomics Association, 14, *Proceedings of the International Ergonomics Association*, San Diego: IEA.

Kong, Y.K.; Lowe, B.D.; Lee, S.J.; Krieg, E.F.(2007).Evaluation of handle design characteristics in a maximum screwdriving torque task.*Ergonomics*, 50 (9), 1404–1418.

Kroemer K. H., Kroemer H. and Kroemer-Elbert K. (1994). Ergonomics: how to design for exertions. *Human Factors*, 12 (3), 297 – 313.

Mamansari, D.U.;Salokhe, V.M.(1996).Static strength and physical work capacity of agricultural labourers in the central plain of Thailand. *Applied Ergonomics*,(271),53-60.

Mathiowetz V., Kashman N, Volland G., Weber K., Dowe M. and Rogers S. (1985). Grip and pinch strength: normative data for adults. *Archives of Physical Medicine and Rehabilitation*, (66), 69-74.

Matsuoka, J; Berger, R.A.; Berglund, L.J.; An, K.N. (2006). An analysis of symmetry of torque strength of the forearm under resisted forearm rotation in normal subjects. *The Journal of Hand Surgery*, 31A (5).

Mital,A.; Kumar, S. (1998). Human muscle strength definitions, measurement, and usage: Part I - Guidelines for the practitioner. *International Journal of Industrial Ergonomics*, 22, 101-121.

Montoye, H. J.; Lamphiyer, D.E. (1977). Grip and arm strength in males and females, age 10 to 69. *The Research Quarterly*, 1(48), 107-120.

Morse, J.L; Jung, M.C.; Bashford, G.R.; Hallbeck, M.S. (2006). Maximal dynamic grip force and wrist torque: the effects of gender, exertion direction, angular velocity, end wrist angle. *Applied Ergonomics*, (37), 737–742.

Nagashima, K.; Konz, S. (1986). Jar lids: effect of diameter, gripping material and knurling. *Human Factors Society - 30th Annual Meeting*, 30, pp. 672-674.

Napier, J. (1985). *A mão do homem: anatomia, função e evolução*. Rio de Janeiro: Universidade de Brasília.

Paschoarelli, L. C.(2009). *Design Ergonômico: Avaliação e Análise de Instrumentos Manuais na Interface Usuário X Tecnologia [Tese de Livre Docência]*. Bauru: UNESP.

Peebles, L.; Norris, B. (2003). Filling 'gaps' in strength data for design. *Applied Ergonomics*, 34, 73 - 88.

Peebles, L.; Norris, B. J. (2000). *Strength Data for Design Safety – Phase I*. Londres: Department of Trade and Industry.

Shih, Y.C.; Wang, M.J.J. (1997). Evaluating the effects of interface factors in the torque exertion capabilities of operating handwheels. *Applied Ergonomics*, 5(28), 375-382.

Shih, Y. C.; Wang, M. J. J. (1996). Hand/tool interface effects on human torque capacity. *International Journal of Industrial Ergonomics*, (18), 205-213.

Triola, M.F. (1999). *Introdução à Estatística*. Rio de Janeiro: Editora LTC.

Voorbij, A.I.M.; Steenbekkers, L.P.A. (2001). The composition of a graph on the decline of total strength with age based on pushing, pulling, twisting and gripping force. *Applied Ergonomics*,(32), 287-292.

CHAPTER 45

An Ergonomic Look at the Open Plan Office

Ana Paula Lima Costa; Vilma Villarouco

Universidade Federal de Pernambuco, BRASIL

aplimacosta@gmail.com; villarouco@hotmail.com

ABSTRACT

The open-plan layout is a form of physical occupation for an office used in Brazilian public sector agencies to maximize the occupation of physical space. However, the work environment should not only be adapted to the structural and cultural requirements of the organization, but also to employees' needs. It should be understood that, when designing work environments, complementary approaches should be used to meet physical and behavioral needs. Ergonomic assessments were made in the offices of public sector companies in accordance with the systemic approach of the Ergonomic Methodology for Evaluating Built Space – MEAC (in Portuguese) (Villarouco, 2009), which analyses the physical space by linking physical-spatial evaluations closely to tools for identifying how the environment is perceived. The results showed that using open plan offices without observing these premises leads to users suffering various kinds of discomfort.

Keywords: organizational culture, built environment, architecture

1   INTRODUCTION

In recent years, the adoption of open-plan offices has been questioned in view of the high rate of complaints from users of such space. This configuration was proposed so that communication flows and administrative rationality might hold sway over the status of its members. However, open-plan offices are commonly employed to reduce set-up costs, since, by integrating a suite of rooms, the effective cost per square meter and infrastructure services are reduced, which makes the cost per square meter of a company with open-plan areas much smaller than that of a company with enclosed spaces. However, in an open-plan office, many people from different administrative areas work together in the same space and need to create rules for such shared use so as to administer personal differences that arise from it.

Bearing in mind that, in order to propose a physical occupation model for offices what must be identified are the behavioral attitudes, the work and the tasks and their influences within the organization, with regard to this physical occupation, it is important to take into account the needs of those who will use the environment. Thus, ergonomics, a scientific discipline that focuses on man in situations of real work, is seen to be an important tool to aid understanding of the built environment and its influence on the user. With this intention, ergonomic assessments of the built environment were conducted in the offices of two public sector companies in Brazil which had made their spaces for staff uniform by using the open-plan layout, but prior to this they had not undertaken studies on the activities to be performed in such environments in order to verify the suitability of these spaces for their users.

2   CULTURAL ORGANIZATIONAL CULTURE AND THE WORK ENVIRONMENT

Organizations are structured based on patterns of authority, the division of labor, control methods, forms of internal communication, etc. The first theories about office occupancy arose from the need to control employees' activities, and thus adopted a rigid fixing of the environments by positioning jobs as in a factory assembly line. Open-plan offices began to be set up from 1950 with the emergence of theories of human relations and the need for users to inter-relate with each other, which led the lay-out to following the geometry of the flows, thus gaining organic contours. Formal space started to be dissolved in order to show a "harmony of interests" within the company, and a hierarchical stratification began to be expressed by using partition panels and elements that set the territoriality of each user. The use of workstations was consolidated over the following years due to the economy of space occupied by stations grouped in relation to isolated desks, which reduces the space required and the cost of furnishing when setting up offices. (Andrade, 2007)

The work environment should not only be adapted to the structural and cultural requirements of the organization, but also to employees' needs (Asselbergs, Schreibers and Voordt, 2008). A pleasant atmosphere, with fixtures and fittings that satisfy its users has been proved to be important for work, because, according to Gifford (2005), satisfying the users of the built-space is important since its occupants spend significant parts of their lives there. Therefore, the environment must fit the needs and activities of its occupants by ensuring it is inhabitable. Conforming with such notions may, according to this author, increase an office worker's productivity and improve the social ties between people. According to Bins Ely (2003), in a work environment, besides the organizational aspects, the environmental aspects, the spatial conception of the environments, the layout and environmental comfort are also important tools to improve the conditions under which work is performed.

Physical comfort is the necessary condition if the built-environment is to be for living in, and the physical environment itself can affect the performance of those occupying it (Wilson, 1990). Despite the diversity of the physical elements of the environment such as temperature, lighting and sound, people work in the "total" environment, which makes the wholeness of the fit between all its elements important (Parsons, 1990). Gots (1998) states that, although users of a building are aware that the environment can affect their health, which leads them to associating the use of the environment with catching diseases, such perceptions have different degrees of precision. According to Gifford (et al, 2002), individuals disagree about the value of a building because they weight the physical characteristics of the environment differently, the result of the inference of different cognitive properties.

Gots (1998) states that some symptoms that workers may associate with the environment are very often of several natures, it being necessary to determine one diagnosis for the employee and another one for the building, and then integrating the two. This involves multidisciplinary considerations. Villarouco (2008) proposes an ergonomic assessment of the built environment based on a systemic approach, covering variables of the areas involved in the built space and in which the prime element is the user of this space and his/her environmental perceptions, because it is the user who absorbs the impacts that the environment transmits.

3   CASE STUDY

To check the suitability of using the open-plan layout in companies that have used it only with the aim of integrating its employees without checking its appropriateness for the activities they perform, ergonomic assessments were conducted in two workspaces of public sector enterprises by means of the Ergonomic Methodology for Assessing the Built Environment – MEAC (in Portuguese), proposed by Villarouco (2009). Based on the need to link methodologies for physical-spatial assessment to tools that identify environmental perception when applying ways to analyse built environments ergonomically, MEAC aims to set bases for systematizing the analysis of space, which rest on non-negotiable elements of the ergonomic approach, such as the focus is on the user, the systemic approach and usability. MEAC comprises four analytical steps: 1) Global Analysis of the Environment, 2) Identifying the Environmental Configuration, 3) Assessing the Environment in use in the Performance of the Activities and 4) Environmental Perception. MEAC rounds off the analysis of the environment with the Ergonomic Diagnosis of the Environment and Ergonomic Propositions.

3.1.   Global Analysis of the Environment

The first step of MEAC consists of conducting a comprehensive analysis of the spatial configuration, characterized by identifying if there are problems.

Packed into the work-room of office A (Figure 1) are 57 workers who undertake accounting auditing services in public sector companies. In the work-room of office B (Figure 2), services for preparing payroll are performed and there are 45 people on its staff list. These activities require to be done in isolation because outside distractions affect the concentration of staff and hence how services are performed. It is noted that the lack of a place to store work objects does not create a good impression of the environment, as can be seen in Figures 1 and 2. In these environments, people move around and talk all the time.

Figure 1- General view of office A

Figure 2 – General view of office B

3.2.   Identification of the Environmental Configuration

In this stage, the physical and environmental conditioning factors are identified by collecting data collected from the environment.

3.2.1. Layout

In this stage, the distribution of the work-room environments is checked. How the open-plan layout is used is identified in the floor-plans of Figure 3. There are no distinctions between the various work teams, only of the management area. The workstations are grouped into islands.

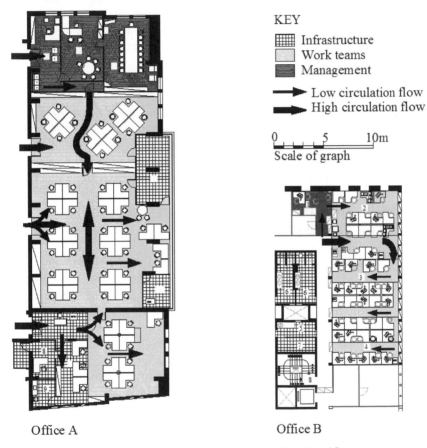

KEY

▦ Infrastructure
☐ Work teams
▩ Management

→ Low circulation flow
➡ High circulation flow

0   5   10m
Scale of graph

Office A   Office B

Figure 3 – Floor plan of the layout and circulation flows of Offices A and B

3.2.2. Circulation Flows

In this stage, the circulation flow of the users and the rational sequence for carrying out the activities are checked. The lack of zoning the activities in the office sees to it that the flow of people circulating through the rooms is close to staff's desks, which has an adverse effect on the conduct of services because the attention of staff is distracted (Figure 3). The lack of controlling access to the room makes it easy for outsiders to have entry to the services area.

3.2.3. Measurements

Working and environmental conditions should be adapted according to the workers' psychophysiological characteristics by providing comfort and safety. The assessments of environmental comfort parameters were conducted in accordance with Brazilian parameters (Brasil, 1978).

The measurements indicated the illumination levels of the ambiances were below the minimum levels laid down in regulatory standards; the values measured of temperature were above the maximum permitted limits and the values measured of noise in the ambient were above the maximum ones deemed as comfortable.

3.2.4. Work posts

The furniture in the offices was not sized according to the activities that users perform. The lack of partition panels between workstations makes it possible for work material to be mixed among the desks, and gives evidence of the lack of privacy. There is no place to store work and personal material. The team leaders' desks are not differentiated nor is there an area where their team can be attended to.

3.3. Evaluation of the Environment in Use in the Performance of the Activities

The step of assessing the environment in use when activities are being carried out aims to identify to what extent the environment facilitates or inhibits how activities are undertaken in areas set aside for them. In these offices in which the activities require to be done in isolation, it is observed that the open-plan environment integrates staff which affects their concentration on the tasks to be conducted, because, regardless of what the users wish, there are always people circulating through the space and there is constant talk.

The lack of defining work teams creates conflict in the management of tasks because each team works at a different pace. The lack of space to talk with interlocutors sees to it that conversations close to work desks adversely affect the concentration of those who are not taking part in the conversation.

3.4. Analysis of the User's Perception

Having completed the first block of assessments, the fourth step of the methodology begins. This consists of identifying variables of a more cognitive character by applying and analysing the Constellation of Attributes (Schmidt, 1974). Observations are made on the symbolic images and perceptions that staff have of the working environment they use. In the stage of assessing the individual's symbolic image vis-à-vis the environment, the user is asked to list the attributes that they long for in the working environment. In the step of assessing individual's perception vis-à-vis the environment, the user is asked to describe the environment he/she uses. In order to distinguish what is objective from what is subjective in the user's perception, the user's subjective impressions and objective impressions are compared and contrasted.

Users' examples of an idealized environment referred to one that is pleasant, comfortable and spacious. When they commented on the actual environment that they used, users cited the negative aspects, such as precarious fittings in their building, a noisy and cluttered environment. This reflected the discomfort they felt in the workplace. When comparing the users' subjective impressions with their direct impressions of the environment, it is seen that in addition to their longing for a place with good facilities, the users yearn for management solutions, such as cleanliness in their environment, dividing it into sectors as per the work teams, and a good environment for collegial fellowship.

3.5.   Ergonomic Diagnosis and Propositions

After conducting the analyses, an ergonomic diagnosis is constructed, which presents the information needed for an overall understanding of the situation so as to generate interventions on issues that interfere negatively in the performance of the system.

Table 1- Table of ergonomic diagnostics and propositions

Attribute	Ergonomic Diagnosis	Propositions
Layout	The lack of zoning and the small dimensions of the environments adversely affect and are not compatible with the use of the open-plan layout.	Redimensioning and zoning environment so as to have enough space to accommodate the activities.
Circulation Flows	The lack of zoning of activities creates a circulation flow close to work stations.	Zoning the activities so as to avoid a heavy flow of people close to the work stations.
Work stations	The dimensions of the work stations are not suitable for doing the activity. There is a lack of cabinets in which to place work material. Managers' work stations are not suitable for attending to the teams.	Sizing the posts based on the ergonomic analysis of the activity. Providing stations with vertical partitions which individualise the post. Providing lockers to store work objects. Differentiating managers' work areas.

Measurements	The luminosity, noise and temperature levels are not in accordance with the recommended indices.	Carrying out acoustic improvements to eliminate excessive noise, changes in the AC system to make the climate more comfortable and making adjustments to the lighting system to provide adequate illumination.
Environment in use	The environment with no partitions between departments causes distractions that adversely users' concentration.	Dividing the environment between the work teams, isolating them from each other. Signaling managers' work spaces.

The physical space was shown to be unsatisfactory for accommodating the users such that they can have comfort, security and privacy when performing their tasks. Confirming the data analyzed, the use of the open-plan lay-out was shown to be ergonomically unsuited to being used in locations that require concentration and isolation so as to perform tasks, and this is aggravated because the size of the environment is incompatible with the number of staff using the physical space.

4   RECOMMENDATIONS FOR IMPLEMENTING OPEN-PLAN OFFICES

The open-plan office was designed to encourage inter-relationships within the company; so this setting should be used only when this integration is in accordance with the organizational culture of the company.

The adoption of the layout should be preceded by an analysis of the architectonic space, to check if space is of sufficient size and has the spatial conditions to adopt this configuration. The environment should be zoned in such a way that it avoids an undesirable flow close to the work stations.

The adoption of an open-plan layout should be preceded by an ergonomic analysis of the tasks of the users of space in order to assess the suitability of this configuration, so that the work activity space may be sufficient to undertake the activity without restrictions (Boueri, 2008).

It should be noted if appropriate furniture is used, with panels and elements that set the limits of the territoriality, a behavior, which, according to Hall (1990), is part of human nature.

5   CONCLUSION

The open plan office was created based on administrative theories of human relations, in which giving value to teamwork turned the workplace itself into a means of achieving productivity, but the open plan system is being used in public sector offices because it economizes on the area occupied in relation to that occupied by individual rooms.

The ergonomic analyzes conducted demonstrated that using this type of occupation as there had been no prior observation of labor relations and the conditions of the physical environment needed to accommodate a large number of people, an environment had been generated that had an adverse effect on how activities were carried out.

This makes the decision to open or enclose the environment depend on the nature of the jobs and the need for there to be a flow of activities that necessarily need to be on display to each other. Any shortfall in observing this factor sees to it that the spatial configuration of open plan offices is not in accordance with the characteristics of the work, thus making them areas with no specificity and unfit for any type of productive activity.

REFERENCES

ANDRADE, Cláudia Miranda Araújo de. A história do Ambiente de Trabalho em Edifícios de Escritórios: Um Século de Transformações. São Paulo, C4, 2007

ASSELBERGS, F.; SCHREIBERS, K.; VAN DER VOORDT, J. M. Theo. Kantoorgebouwen. In.: Arbo Jaarboek 2008, Alphen a/d Rijn: Kluwer, p. 333-346. 333-346. 2008.

BINS ELY, Ergonomia + Arquitetura: buscando um melhor desempenho do ambiente físico. *In:* Anais do 3º ERGODESIGN Congresso Internacional de Ergonomia e Usabilidade de Interfaces Humano-Tecnologia: Produtos, Programas, Informação, Ambiente Construído, 2003

BOUERI FILHO, José Jorge. Projeto e dimensionamento dos espaços da habitação - Espaços de atividades. E-book- Livro II, Estação das Letras e Cores, São Paulo, 2008

BRASIL. Ministério do Trabalho e Emprego - NR -17 – Ergonomia (117.000-7), 1978

GIFFORD, Robert O papel da psicologia ambiental na formação da política ambiental e na construção do futuro. Psicologia USP, 16(1/2), 237-247. São Paulo, 2005.

_____; HINE, Donald W.; MULLER-CLEMM, Werner; SHAW, Kelly T. Why architects and laypersons judge buildings differently: cognitive properties and physical bases. Journal of Architectural and Planning Research, Chicago, 2002

GOTS, Ronald E. Investigating health complains. *In:* O´Reilley, J. T.; Hagam, P.; Gots, R.; Healge, A. Keeping buildings healthy: how to monitor and prevent indoor environmental problems. New York, 1998

HALL, Edward T. The Hidden Dimension – Anchor Books Editions, New York, 1990

PARSONS, Kenneth C. Ergonomics assessment of thermal environments. In: Evaluation of Human Work. A Practical Ergonomics Methodology. Second edition. Edited by John R Wilson and E Nigel Corlett. University of Nottingham. 1990

SCHMIDT, J. L. La percepción del habitat. Barcelona: Ed. Gustavo Gili, 1974

VILLAROUCO, Vilma. An ergonomic look at the work environment. *In:* Anais do 17th World Congress on Ergonomics, Beijing, China, 2009

_____. Construindo uma metodologia de avaliação ergonômica do ambiente - AVEA. *In:* Anais do XV Congresso Brasileiro de Ergonomia – ABERGO- Porto Seguro, 2008

WILSON, John R. A framework and a context for ergonomics methodology. *In:* Evaluation Of Human Work. A Practical Ergonomics Methodology. Second edition. Edited by John R Wilson and E Nigel Corlett. University of Nottingham. 1990

CHAPTER 46

Pulling Strength with Pinch Grips: A Variable for Product Design

Bruno Montanari Razza, Luis Carlos Paschoarelli, Danilo Corrêa Silva,
José Alfredo Covolan Ulson, Cristina do Carmo Lucio

Laboratory of Ergonomics and Interfaces, UNESP – Univ. Estadual Paulista
Bauru, Brasil
bmrazza@gmail.com

ABSTRACT

Pulling with pinch grips is an action frequently used in either occupational or daily living activities, especially in situations where the object is too small, the access to the object is restricted, the use of tools is prevented and situations such as pulling strips out of long-life packaging, removing seals from flask lids, tearing a plastic bag, etc. Excessive demands of strength in those actions may limit the user's access to certain activities and even lead to injury. The objective of this study was to collect valid data of pulling strength with pinch grips to be applied to the design of safer and more comfortable products and tasks. Three handles of different heights (1 mm, 20mm and 40mm) were used in the study and three types of pinch grips were measured with both hands: pinch-2, chuck pinch and lateral pinch. The study included 30 men and 30 women, all right-handed and healthy, and the maximum voluntary isometric contraction was measured in the standing posture. The results showed that the type of pinch grip used influenced more greatly the pulling strength than the size of the handles. This study provides biomechanical data of an action often performed in many daily living activities and also in several occupational tasks, but has been still little explored.

Keywords: Pinch Grip, Isometric Strength, Pulling Strength, Biomechanics, Ergonomic Design

1. INTRODUCTION

Despite the increasing automation in the industrial environment, many tasks still have great demand of manual efforts, such as: maintenance tasks, manual material handling, patients transportation in hospitals, among others (Imrhan, 199; Kim and Kim, 2000). Tasks and products that require inappropriate application of manual strength are considered risk factors for the development of occupational diseases (Kattel et al., 1996), and are responsible for much of the total injuries in industry (Aghazadeh and Mital, 1987).

Pinch grips, particularly, have been associated with high rates of occupational diseases (Armstrong and Chaffin, 1979; Eksioglu et al., 1996), being considered a risk factor in ergonomic evaluation (Keyserling et al., 1993). In addition to accidents and occupational diseases, problems in products of daily consumption related to the application of manual strength are also reported (Crawford et al., 2002; Voorbij and Steenbekkers, 2002). These records could be even greater, but many individuals avoid purchasing certain products because they already know the difficulty of use (Imrhan, 1994). Although there are products to help people with reduced muscle capacity, these devices are considered to marginalize people and to be little practical and difficult to utilize, besides being only short-term solutions (Ivergard et al., 1978 apud Imrhan, 1994; Imrhan, 1994).

For these reasons, several studies have investigated pinch grips strength, having already been established that it is significantly influenced by:

- Gender (Swanson et al., 1970; Mathiowetz et al., 1985; Imrhan and Loo, 1989; Hallbeck and McMullin, 1993; Crosby et al., 1994; Dempsey and Ayoub, 1996; Hefferman and Freivalds, 2000; Shih and Ou, 2005);
- Age (Imrhan, 1994; Imrhan and Loo, 1989; Peebles and Norris, 2003; Mathiowetz et al., 1986);
- Anthropometry (Imrhan and Sundararajan, 1992; Crawford et al., 2002; Imrhan and Loo, 1989);
- Laterality (Mathiowetz et al. 1985; Imrhan and Loo, 1989; Crosby et al., 1994);
- Grip type employed (Dempsey and Ayoub, 1996; Imrhan, 1991; Ager et al., 1984, Imrhan and Rahman, 1995; Swanson et al., 1970; Kraft and Detels, 1972; Mathioweitz et al., 1985; Imrhan and Loo, 1989; Imrhan, 1991; Fernandez et al., 1991);
- Body Posture (Catovic et al., 1991; Swanson et al., 1970);
- Wrist Deviation (Kraft and Detels, 1972; Imrhan, 1991; Fernandez et al., 1991; Hallbeck and McMullin, 1993; Dempsey and Ayoub, 1996; Shih e Ou; 2005; Lamoreaux and Hoffer, 1995);
- Grip spam (distance between the fingers that make up the grip) (Imrhan and Rahman, 1995; Dempsey and Ayoub, 1996; Shih and Ou, 2005; Heffernan and Freivalds, 2000); among others.

Although several studies have been conducted on pinch grips, few studies have investigated them associated with the pulling strength (Fothergill et al., 1992; Imrhan and Sundararajan, 1992; Peebles and Norris, 2003). This action is often used in occupational activities, especially when the object is too small, the access to the object is restricted or when the use of tools is prevented. In the activities of daily life, pulling with a pinch grip is also widely used in situations such as pulling plastic or paper strips from long-life packs, removing seals from flask lids, ripping plastic bags open, tearing vacuum sealed packages , opening drawers, etc...

The aim of this study is to collect biomechanical data from Brazilian healthy adults for the design of products in which pulling strength with pinch grips is used.

2. METHODS

The procedures for this study were approved by the Committee of Ethics in Research (CEP-FMB-UNESP n. 373/2005) and the recommendations of the National Health Council (Resolution 196-1996) and the Brazilian Association of Ergonomics (1002 ERG BR) for research involving humans were met.

2.1 Subjects

Sixty right-handed unpaid volunteers participated in the experiment, being 30 women in the mean age of 21.60 years (SD 3.05), ranging from 18 to 30 years, and 30 men, in the mean age of 21.83 years (SD 2.46), ranging from 18 to 28 years. None of the subjects showed any history of musculoskeletal disease in the upper limbs in the last year. The subjects' written consent was obtained and all procedures were widely explained to them. The Edinburgh Inventory (Oldfield, 1971) was applied to certify that the whole sample was right-handed.

2.3 Apparatus

The record of strength was performed by a digital dynamometer AFG500 (Mecmesin Ltd., England), with maximum capacity of 500N, accuracy of 0.1% (full scale), analogical communication interface +4 ... 0 ... -4V full scale, digital communication interface RS-232 and maximum sample rate of 5000 Hz. The data were obtained by a personal computer with a Windows XP operating system (Microsoft®, version 2002), and a specific software (SADBIO - System of Acquisition of biomechanical data, Labview 7.0, National Instruments®, England) was developed for this study.

The subjects took the strength measurements of three different handles, representing three objects with different heights, one being 40 mm high (40.0 x 40.0 x 40.0 mm), another one 20 mm high (20.0 x 40.0 x 40.0 mm) and the last one having an extension in cloth 1 mm thick (40.0 x 40.0 x 1.0 mm). In the 20 mm and 40 mm handles, fabric was applied all over the surface in contact with the hands, for a standardization of texture in the hand-object interface; besides, corners were made round to avoid pressure concentration on the hands of the subjects (Figure 1).

2.4 Procedure

The procedures of this study followed recommendation of related literature (Daams, 1993; Hook and Stanley, 1986; Mathiowetz et al., 1984; Caldwell et al., 1974; Chaffin and Andersson, 1990; Mital and Kumar, 1998). Subjects in this experiment were asked to exert their maximum pulling strength (maximum isometric voluntary contraction) with each hand, with three types of pinch grip. The grips used were: pinch-2 (thumb opposed to index finger), three jaw chuck pinch (index and middle finger opposed to thumb) and lateral pinch (thumb opposed to the lateral side of index finger; Also called key-pinch). Altogether, the study consisted of 18 variables (3x3x2), made of 3 handles, 3 pinch grips and 2 hands.

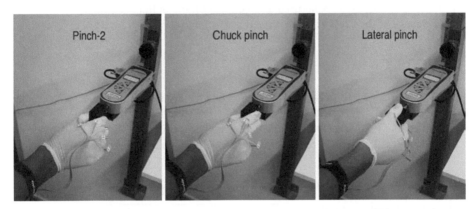

Figure 1: Evaluation equipment and the types of pinch grip evaluated.

For the evaluation of maximum voluntary isometric contraction in pulling, each subject was asked to remain in the standing posture facing the equipment, the elbow of the upper limb flexed 90°, the forearm in neutral position – horizontally aligned – and the wrist positioned freely, according to the preference of the subject. The equipment was positioned at the height of the subject's elbow. The subjects were asked to keep the fingers which were not active in the grip flexed to the palm of the hand. In the pulling strength measurement for pinch-2 and chuck pinch, the wrist remained in extension and slight ulnar deviation. In all measurements, the subjects used rubber gloves on the hands.

The force was measured in an interval of 5 seconds, the first and the last seconds being discarded; the maximum strength obtained in the remaining interval of 3 seconds was taken as the result. This type of measurement was previously employed in other approaches and was shown to be effective in ensuring more homogeneous results. In order to prove the data collection validity, a second measurement was taken and it was observed whether the difference between the values obtained in the two measurements did not vary more than 10%. When such variation was less than 10%, the highest value was considered as a result, whereas if the variation was more than 10%, a third measurement was performed; in this case, the greatest value was used as a result.

The subjects were instructed to exert their utmost strength as soon as they heard the beep, with no sudden movements, in an interval of approximately 1 second, and hold that contraction until they heard the beep again (for a 5-second measurement). A positive and general feedback was offered, but no information on the performance of the subjects was provided, and there was not any presence of spectators in the study environment.

The order of measurement of the variables was random, in order to avoid influences of unknown external variables in the study. It was offered an interval of 30s to 1min between the measurements and longer intervals were performed when signs of fatigue were recognized.

In all the research results, descriptive statistics analysis was utilized. Student's t-test was employed to determine significant influences of test variables (type of pinch grip, handle size, hand used and gender) on the pulling strength. The level of significance of the t-test was determined at 5% ($\alpha \leq 0.05$).

3. RESULTS

The results of pulling strength with pinch grips are shown in Table 1. The male gender showed significantly greater strength than the female gender in almost all variables analyzed, except for right-hand pinch-2 performed with the 40 mm and 1mm handles. The female gender reached 76.96% of male strength, on average.

Table 1. Pulling strength.

Handle	Hand	Grips	Total (kgf)		Female (kgf)		Male (kgf)	
			Max	S.D.	Max	S.D.	Max	S.D.
40 mm	Right	Pinch-2	4,45	1,50	4,01	1,11	4,89	1,70
		Lateral pinch	6,17	2,17	5,03	1,51	7,30	2,15
		Chuck pinch	5,56	1,64	4,90	1,47	6,22	1,56
	Left	Pinch-2	4,06	1,28	3,66	1,04	4,45	1,40
		Lateral pinch	5,89	2,14	4,84	1,58	6,94	2,12
		Chuck pinch	5,20	1,57	4,63	1,46	5,76	1,48
20 mm	Right	Pinch-2	4,32	1,12	3,94	0,96	4,70	1,15
		Lateral pinch	6,38	1,93	5,30	1,40	7,45	1,79
		Chuck pinch	5,27	1,43	4,65	1,23	5,88	1,37
	Left	Pinch-2	4,01	1,13	3,69	0,92	4,33	1,24
		Lateral pinch	6,06	1,85	4,95	1,42	7,17	1,56
		Chuck pinch	4,82	1,28	4,41	1,24	5,23	1,21
1 mm	Right	Pinch-2	3,95	1,28	3,57	1,07	4,32	1,38
		Lateral pinch	6,04	1,86	5,23	1,53	6,85	1,84
		Chuck pinch	4,64	1,28	4,18	1,18	5,10	1,23
	Left	Pinch-2	3,56	1,33	3,09	0,92	4,03	1,52
		Lateral pinch	5,61	1,75	4,80	1,42	6,41	1,70
		Chuck pinch	4,27	1,29	3,75	1,10	4,79	1,27

Considering grip span, the 40 mm handle was the one which generated the highest strengths, followed by the 20 mm and the 1 mm one, except for lateral pinch, in which the 20 mm handle generated the highest measures of strength. However, this difference was only significant for some grips, as it can be seen in Table 2. Figure 2 shows the variation of strength magnitude for male and female gender, respectively. For pinch-2 and chuck grips, there is a tendency of increase in strength when there is an increase in grip span. For lateral pinch, however, the behavior was of increase and decrease.

Table 2. T-test results (Student) for the handle size.

Gender	Hand	Grips	Handle size		
			40mm x 20mm	40mm x 1mm	20mm x 1mm
Female	Left	Pinch-2	0,870811*	0,105641	0,171121
	Left	Lateral pinch	0,655282	0,638469	0,853058
	Left	Chuck pinch	0,348642	0,068174	0,133525
	Right	Pinch-2	0,796370	0,010304	0,013600
	Right	Lateral pinch	0,715405	0,754158	0,688832
	Right	Chuck pinch	0,429908	0,012524	0,032686
Male	Left	Pinch-2	0,554120	0,236908*	0,247092
	Left	Lateral pinch	0,779257	0,369341	0,203428
	Left	Chuck pinch	0,282181	0,005982	0,023770
	Right	Pinch-2	0,801555*	0,169147*	0,183322*
	Right	Lateral pinch	0,433290*	0,468798*	0,066764*
	Right	Chuck pinch	0,170941	0,014770	0,173900

* Data analyzed via the Mann-Whitney test

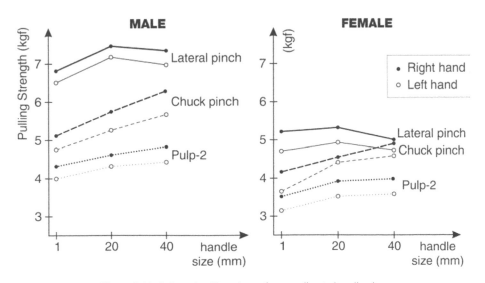

Figure 2. Variation of pulling strength according to handle size.

Table 3 presents the t-test results (Student) for the pinch grips. Statistical analysis showed that lateral pinch is significantly stronger than pinch-2 in all situations tested. Chuck pinch was also significantly stronger than pinch-2. Lateral pinch was significantly stronger than Chuck pinch, but not for all variables. Pinch-2 and chuck pinch represent 67.4% and 82.4% of lateral grip strength on average, respectively. A greater strength magnitude of the right hand over the left hand was obtained for both genders. This behavior had already been expected because the sample was right-handed. However, these differences were not significant.

Table 3. T-test results (Student) for the types of pinch grip.

Gender	Handle size	Hand	Grips		
			Lateral pinch X Chuck pinch	Lateral pinch X Pinch-2	Chuck pinch X Pinch-2
Female	40 mm	Right	0,622174	0,002569	0,010253
		Left	0,615954	0,000945	0,006684
	20 mm	Right	0,047829	0,000093	0,040515
		Left	0,076836	0,000083	0,012543
	1 mm	Right	0,008875*	0,000007*	0,018930
		Left	0,000672	0,000000	0,007571
Male	40 mm	Right	0,073629*	0,000030*	0,004069*
		Left	0,020007	0,000003	0,001366
	20 mm	Right	0,000673*	0,000000*	0,001259
		Left	0,000001	0,000000	0,012687
	1 mm	Right	0,000247	0,000001*	0,020950*
		Left	0,000083	0,000000*	0,013655*

* Data analyzed via the Mann-Whitney test

4. DISCUSSION

In manual activities evaluation, the size of the handled object has been extensively studied, and in many studies it has been shown as a determinant of task execution ease. In our study, an influence of the height of the grip on manual strength was observed, but this variable was not significant, in general. It can be supposed that the size of the handle has an influence on the joint structure of the hand executing the grip, but the strength employed by the arm and shoulders to execute the traction must be less influenced by the size of the handle.

It becomes apparent, however, a tendency of strength increase with the increase of the handle size for pinch-2 and chuck pinch. Other authors have also observed this behavior for these grips (Imrhan and Rahman, 1995; Dempsey and Ayoub, 1996; Shih and Ou, 2005; Peebles and Norris, 2003). As for lateral pinch, the highest strength values were obtained with the 20 mm grip span, showing a behavior of increase and decrease of strength as the grip span increases. Imrhan and Rahman (1995) and Dempsey and Ayoub (1996) also observed this same behavior for lateral pinch. According to Imrhan and Rahman (1995), the reason for this lateral pinch behavior can be explained by possible damages imposed by the grip span on the length-tension relationship of muscles that control the thumb and also a possible loss of lever in what concerns these joints.

The type of pinch grip employed was identified as one of the most influential factors in manual strength. In our study, it was observed that the strongest grip is the lateral one, followed by chuck pinch and then by pinch-2. The same results were also found in other studies (Dempsey and Ayoub, 1996; Imrhan, 1991; Peebles and Norris, 2003).

As for the proportion of strength magnitude, the grips also varied. In the study of Imrhan and Sundararajan (1992), chuck pinch and pinch-2, respectively, showed 60.3% and 39.6% of the pulling strength in lateral pinch, on average. In our study, considering only the handle of cloth (similar to the one from previous study), chuck pinch and pinch-2, respectively, made 77.7% and 59.3% of the lateral pinch pulling strength, on average. Imrhan and Sundararajan (1992) state that the shoulder muscles may have contributed with 45-56% of the strength applied to the manual traction associated with pinch grips, especially for lateral pinch, which may have contributed to the results of major magnitude for this grip.

In this study, it was observed that the left hand performs 92.5% of the strength of the right hand, on average. These results are consistent with several studies that investigated the influence of dominance in manual strength (Imrhan and Loo, 1989; Crosby et al., 1994; Imrhan and Sundararajan, 1992).

5. CONCLUSION

From all the factors that may affect the pulling strength with a pinch grip, the type of pinch grip is certainly one of the most influential. The reduction imposed on the strength when pinch-2 is utilized, for example, shows a greater magnitude than the task variables (size of the handled object), or individual characteristics (dominance). The use of pinch-2 can result in a reduction of 33.5% in the strength, on average, compared to the best condition (lateral pinch). Gender is another factor that has great influence on manual strength, having to be considered within the design of products and tasks where there is a participation of the female gender.

One recommendation for implementing the results of this study is that the use of lateral pinch in the design of products and tasks must be prioritized, rather than the use of pinch-2 and chuck pinch. This will make the activity more comfortable and easier for the user, since this type of grip allows the application of greater strength. For this, the area available for the fingers must be slightly larger, because this grip requires more space to be performed and the thickness of the object should be approximately 20 mm, a condition in which the strength generated was greater.

Therefore, the main relevance of this research is the generation of manual strength parameters that may help, as explained, for many areas of scientific and technological knowledge, always seeking to enhance comfort, usability and safety in widest range of human activities. Some examples would be: jobs and tasks in industry; the design of products for everyday use (packs, manual tools, etc.); helping you compose a database of normal values for more accurate diagnosis of musculoskeletal disease, among others.

436

ACKNOWLEDGEMENTS

This study was supported by FAPESP (Proc. FAPESP 05/58600-7).

REFERENCES

Ager, C. L. and Olivett, B. L.; Johnson, C. L. (1984) Grasp and pinch strength in children 5 to 12 years old. The American Journal of Occupational Therapy, 38 (2): 107-113.

Aghazadeh, F. and Mital, A. (1987) Injuries due to hand tools: results of a questionnaire. Applied Ergonomics, 18: 273-278.

Armstrong, C. A.; Chaffin, D. B. (1979) Carpal tunnel syndrome and selected personal attributes. Journal of Occupational Therapy, 21 (7): 481-486.

Caldwell, L. S.; Chaffin, D. B.; Dukes-Bobos, F. N.; Kroemer, K. H. E.; Laubach, L. L.; Snook, S. H.; Wasserman, D. E. (1974) A proposed standard procedure for static muscle strength testing. American Industrial Hygiene Association Journal, 35: 201-206.

Catovic, E.; Catovic, A.; Kraljevic, K.; Muftic, O. (1991) The influence of arm position on the pinch grip strength of female dentists in standing and sitting positions. Applied Ergonomics, 22 (3): 163-166.

Chaffin, D. B. and Andersson, G. B. J. (1990) Occupational Biomechanics, 2nd Ed., New York: John Wiley and Sons.

Crawford, J. O.; Wanibe, E. and Laxman, N. (2002) The interaction between lid diameter, height and shape on wrist torque exertion in younger and older adults. Ergonomics, 45(13): 922-923.

Crosby, C. A.; Wehbé, M. A.; Mawr, B. (1994) Hand strength: normative values. The Journal of Hand Surgery, 19A (4): 665-670.

Daams, B. J. (1993) Static force exertion in postures with different degrees of freedom. Ergonomics, 36 (4): 397-406.

Dempsey, P. G. and Ayoub, M. M. (1996) The influence of gender, grasp type, pinch width and wrist position on sustained pinch strength. International Journal of Industrial Ergonomics, 17: 259-273.

Eksioglu, M.; Fernandez, J. E. and Twomey, J. M. (1996) Predicting peak pinch strength: Artificial neural network vs. regression. International Journal of Industrial Ergonomics, 18: 431-441.

Fernandez, J.E., Dahalan, J.B., Halpern, C.A. and Viswanath, V. (1991). The effect of wrist posture on pinch strength. In: Proceedings of the Human Factors Society and 35th Annual Meeting, Human Factors Society, Santa Monica, CA, pp. 748-752.

Fothergill, D. M.; Grieve, D. W.; Pheasant, S. T. (1992) The influence of some handle designs and handle height on the strength of the horizontal pulling action. Ergonomics, 35(2): 203-212.

Hallbeck, M.S. and McMullin, D.L., 1993. Maximal power grasp and three-jaw chuck pinch force as a function of wrist position, age, and glove type. International Journal of Industrial Ergonomics, 11: 195-206.

Heffernan, C. and Freivalds, A. (2000) Optimum pinch grips in the handling of dies. Applied Ergonomics, 31: 409-414.

Hook, W. E.; Stanley, J. K. (1986) Assessment of thumb to index pulp to pulp pinch grip strengths. Journal of Hand Surgery (Br), 11 (1): 91-92.

Imrhan, S. N. and Loo, C. H. (1989) Trends in finger pinch strength in children, adults, and the elderly. Human Factors, 31 (6): 689-701.

Imrhan, S. N. and Rahman, R. (1995) The effect of pinch width on pinch strengths of adult males using realistic pinch-handle coupling. International Journal of Industrial Ergonomics, 16: 123-134.

Imrhan, S. N. and Sundararajan, K. (1992) An investigation of finger pull strengths. Ergonomics, 35 (3): 289-299.

Imrhan, S. N. (1994) Muscular strength in the elderly – Implications for ergonomic design. International Journal of Industrial Ergonomics, 13: 125-138.

Imrhan, S. N. (1991) The influence of wrist position on different types of pinch strength. Applied Ergonomics, 22 (6): 379-384.

Ivergard, G.; Hallert, I; Mills, R. (1978) Handleability of consumer packaging – Observation technique and measurement of forces. Report no. 40-14-16-46, Swedish Packaging Institute, Stockholm, Sweden.

Kattel, B. P.; Fredericks, T. K.; Fernandez, J. E.; Lee, D. C. (1996) The effect of upper-extremity posture on maximum grip strength. International Journal of Industrial Ergonomics, 18: 423-429.

Keyserling, W. M.; Stetson, D. S.; Silverstein, B. A.; Brouwer, M. L. (1993) A checklist for evaluating ergonomic risk factors associated with upper extremity cumulative trauma disorders. Ergonomics, 36 (7): 807-831.

Kim, C.; Kim, T. (2000) Maximum torque exertion capabilities of Korean at varying body postures with common hand tools. In: International Ergonomics Association, 14., San Diego. Proceedings of the International Ergonomics Association. San Diego: IEA, 2000, 4p. 1 CD-ROM.

Kraft, G. H. and Detels, P. E. (1972) Position of function of the wrist. Archives of Physical Medicine and Rehabilitation, 52: 272-275.

Lamoreaux, L. and Hoffer, M. M. (1995) The effect of wrist deviation on grip and pinch strength. Clinical Orthopaedics and Related Research, 314, 152-155.

Mathiowetz, V.; Kashman, N.; Volland, G.; Weber, K. and Dowe, M. (1985) Grip and pinch strength: normative data for adults. Archives of Physical Medicine and Rehabilitation, 66: 69-74.

Mathiowetz, V.; Weber, K.; Volland, G. and Kashman, N. (1984) Reliability and validity of grip and pinch strength evaluations. The Journal of Hand Surgery, 9A (2): 222-226.

Mathiowetz, V.; Wiemer, D. M. and Federman, S. M. (1986) Grip and pinch strength: norms for 6 to 19-year-olds. The American Journal of Occupational Therapy, 40 (10): 705-711.

Mital, A.; Kumar, S. (1998) Human muscle strength definitions, measurement, and usage: Part I – Guidelines for the practitioner. International Journal of Industrial Ergonomics, 22: 101-121.

Oldfield, R. C. (1971) The assessment and analysis of handedness: The Edinburgh inventory. Neuropsychologia, 9: 97-113.

Peebles, L. and Norris, B. (2003) Filling 'gaps' in strength data for design. Applied Ergonomics, 34: 73-88.

Shih, Y. C. and Ou, Y. C. (2005) Influences of span and wrist posture on peak chuck pinch strength and time needed to reach peak strength. International Journal of Industrial Ergonomics, 35: 527-536.

Swanson, A.B., Matev, I.B. and de Groot, G. (1970). The strength of the hand. Bulletin of Prosthetics Research, 9: 387-396.

CHAPTER 47

Evaluation of an Interactive Graphic Interface for Mobile Financial Services

Ya-Li Lin, Huei-Ting Yang

Tunghai University
Taichung, Taiwan
yllin@thu.edu.tw

ABSTRACT

The usability of interactive graphic interface for mobile financial service is evaluated using an orthogonal array experiment. The effects of design factors including frame size (Large--468×326 pixels, Small--231×331 pixels), interactivity (Yes/No), data series (multiple, single), grid (Yes/No), and luminance contrast (0, 0.33, 0.67, 0.99) on the visual performance and visual fatigue are examined. Response variables including response time, critical flicker fusion (CFF), user interface satisfaction rating, and overall workload would be collected. Five types of financial tasks would be assigned to each participant including exact value, maximum, correlation, comparison, and trend tasks. Users' experience will help us to evaluate service efficiency and user interface satisfaction with mobile financial services. The results indicate two-factor interactions of interactivity × grid and interactivity × luminance contrast would significantly affect total response time. In addition, the design with interactivity has higher satisfaction than ones without interactivity.

Keywords: mobile financial service, usability evaluation, visual performance, critical flicker fusion (CFF)

1 INTRODUCTION

Mobile phones have provided an unprecedented opportunity for financial development and access, and are set to become a common tool for conducting financial transactions in the near future. Apple's initial success with the iPhone and the rapid growth of mobile phones based on Google's Android operating system has led to increasing use of applications downloaded to the mobile device. The growth of these services depends on the presence of a viable business model, customer demand, and an enabling business environment. Personal financial apps help you monitor your bank accounts, credit cards, household bills, and tracking the foreign exchange rate and online stock trading for global stock market. The best part of going mobile is that you don't have to be at a computer to do all this (Mint.com, 2012).

Studies in graphical perception, both theoretical and experimental, provide a scientific foundation for constructing statistical graphics (Cleveland and McGill, 1984). After that, they continued to propose a general discussion of graphical perception and dynamic graphical methods (Cleveland and McGill, 1987, 1988). Statistical graphs and alphanumerical displays of quantitative data are a major component in software applications, e.g., statistical packages, spreadsheets, and decision support systems (Meyer, Shinar, and Leiser, 1997). Line graphs are used if readers need to determine the rate of increase in the means of the dependent (criterion) variable as a function of changes in the independent (predictor) variable (Gillan, Wickens, Hollands, and Carswell, 1998). In addition, for identifying the trend, the line-type is significantly better than others (Meyer, Shinar, and Leiser, 1997). They claimed the trends could easily be read from line graphs because the slope of a line equals its trend and can be directly perceived. The traditional aim of graphics was to show as much as possible in one display but not display it all at once. Information is hidden to make the plots easily readable and understandable. Only information fundamentally necessary for interpreting the data view is included by default. Any additional information required by the user can be directly interrogated from the graphics. Interactive queries triggered at graphical elements give information on the data. Which information do we expect to gain by querying? The general strategy for a graphical element is to show the most important information connected to it (Wilhelm, 2003).

This study would like to examine and provide an optimal user-centered statistical graphic interface for mobile financial services (MFS). Some principles of user interaction in statistical graphics have been extracted from the study of Wilhelm (2003) as follows. First, detailed information can be made available on demand by responding to interactive user queries. Second, resizing the frame can be implemented in such a way that whenever the size of the frame is altered, the plot size is changed as well. Third, there is no need to immediately show the particular values of plot axes, since our interest lies more in the overall pattern. Fourth, the background color of the frame does not bear any information but is usually chosen to let the data points stand out. Different background colors might be more appropriate to make the patterns more visible (Wilhelm, 2003). This study differs from previous studies in that it concentrates on the effects of interactivity, frame size, grid, data series, and luminance contrast that affect user's performance and visual fatigue. In addition, understanding the needs of statistical graphs for mobile financial services (MFS) will help improve user experience and increase the service's usability.

The details of the construction of a graph would determine what visual process we must employ to decode the information. The construction is successful only if our visual systems perform this graphical perception accurately and efficiently. In this paper, we discuss interactive statistical graphs interface that arise in using mobile financial services. We first review Wilhelm's interactive statistical graph principles, then present our experimental design based on a simulated interface including a description of the interfaces evaluated. It helps Apps designers provide an optimal user-centered interface for MFS. We follow with a description of our research methodology, and then present the results. The paper concludes with a discussion of the design implications followed by future work.

2 RESEARCH METHODOLOGY

In this study, having decided to present the data in a graph, we followed the guidelines for presenting quantitative data proposed by Gillan, Wickens, Hollands, and Carswell (1998). For example, use a line graph to represent a continuous independent variable and use a scatter plot if readers need to determine the degree of correlation between two variables.

2.1 Participants

Sixteen undergraduate and graduate students from Tunghai University voluntarily participated in the experiment. The participant pool consisted of eight females and eight males (mean age of 24 years old and standard deviation of 1.36 years). They all had experience with Statistics and had normal vision or corrected vision above 0.8.

2.2 Experimental Design

The usability of the interactive graphic interface was evaluated using an orthogonal array experiment. The effects of design factors including frame size (S), Large (468×326 pixels) and Small (231×331 pixels), interactivity (I), Yes (value-label added) and No, data series (D), Multiple and Single series per page, grid (G), Yes (grid added) and No, and luminance contrast (LC) with 0, 0.33, 0.67, and 0.99 on the visual performance and visual fatigue of interactive graphs would be examined. A simulated interactive statistical graphic interface based on the combinations of design factors is developed in advance to illustrate the mobile financial system. Figures 1-3 illustrate a part of the simulated interface applied in the experiment. Five financial tasks will be assigned to each of the sixteen participants based on the definitions of Meyer, Shinar, and Leiser (1997), including: (1) reading the exact value of a single point (Exact), (2) identifying the maximum value from a specific data series (Maximum), (3) identifying the strength of linear correlation for two variables (Correlation), (4) comparing two points that belong to the same data series but have different values on the x axis (X-comparison) , and (5) identifying the trend of a data series (Trend).

Response variables including response time and NASA-TLX task load for each task, critical flicker fusion (CFF) before- and after-experiment (CFFB and CFFA), and user satisfaction rating would be collected. Response time is defined as the time to complete the assigned task consisting of correctly perceiving the target and correctly making a decision. The researcher's hypotheses proposed below will be discussed. Does the two-factor interaction influence visual performance, visual fatigue, user satisfaction, and overall workload? The objective is to propose the design guidelines of the optimal interactive graphic interface for stock marketing data of the mobile financial system.

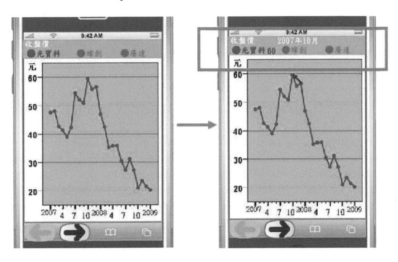

Figure 1 Illustration of interactive design for statistical graph using a line graph.

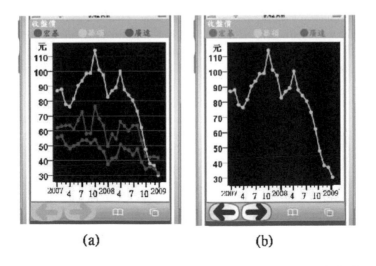

(a)   (b)

Figure 2 illustration of the types of data display: (a) multiple, (b) single display.

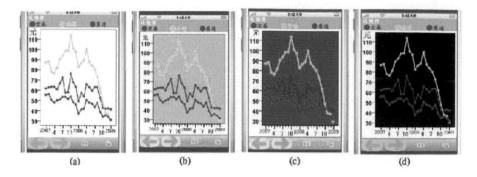

Figure 3 Illustration of luminance contrasts (LC): (a) LC=0, (b) LC=0.33, (c) LC=0.67 and (d) LC=0.99.

2.3   User Interface Satisfication and NASA-TLX Rating

This study uses a questionnaire for user interface satisfaction (QUIS) as a structured subjective assessment of usability (Chin, Diehl, and Norman, 1988). International standard ISO 9241 part 11 defines usability as 'The extent to which a product can be used by specified users to achieve specified goals with effectiveness, efficiency and satisfaction in a specified context of use (Benyon, Turner, and Turner, 2005).' An after-experiment questionnaire was filled out by the participants when they had completed the experiment. In addition, NASA-TLX (National Aeronautics and Space Administration Task Load Index) was used to evaluate task load including mental demand, physical demand, temporal demand, performance, effort, and frustration level for each task (Hart and Staveland, 1988). The overall workload (OW) is computed using the mean weighted score for the mental demand, physical demand, temporal demand, performance, effort, and frustration level for each task.

2.4   *Experimental Procedure*

Participants took part in the experiment individually. First, operation instructions were presented on the computer screen explaining the experimental tasks and the corresponding stock marketing data for the mobile financial service. The sitting posture was freely adjusted by each participant. Statistical graphic interfaces were randomly assigned to each participant. He or she entered a question bank in the first step. This presentation was self-paced. Each participant had to randomly perform three trials for each of five tasks. Performance data was collected in terms of computing response time defined as the time to complete the task. The summation of response time for all five tasks is termed the total response time in the following analysis. In addition, QUIS and NASA-TLX task load ratings will be implemented following the experiment.

3   RESULTS

The response time of five tasks for each participant will be simultaneously recorded with the aid of Stream Author 4.0. Based on the records, the response time will be computed after the experiments.

3.1   Descriptive Statistics

The descriptive statistics of response time for the five tasks are summarized in Table 1. For reading an exact task, it indicates the design with interactivity is better for improving the mean response time (88.0 seconds). However, for identifying a trend task, the design without interactivity is better for improving the mean response time (62.63 seconds). For the exact value, correlation, and maximum value tasks, the lowest luminance contrast has the fastest mean response time, but for trend and comparison tasks, the highest luminance contrast has the fastest mean response time. In addition, for comparing the mean response time of five tasks, Table 1 showed the exact value task has the longest mean response time (120.2 seconds) but the maximum task has the fastest mean response time (53.9 seconds).

For meeting the applications of statistical graphs in practice, total response time is used for the following analyses and is defined as the summation of response times over five tasks. The descriptive statistics of total response time are summarized as listed in the table, but are omitted here. It shows mean total response time varied among different interface sizes, interactivity, grid, data display, and luminance contrasts for overall tasks. Statistical graphs with interactivity have faster total response time on average (374 seconds) than without interactivity (411 seconds). Statistical graphs with a large frame size have faster response time on average (381 seconds) than smaller ones (404 seconds). Statistical graphs without a grid design have a faster response time on average (380 seconds) than with grid ones (405 seconds). Statistical graphs with single data series per page have a faster response time on average (362 seconds) than with multiple data series (423 seconds). Statistical graphs with the lowest luminance contrast have a faster response time on average (343 seconds) than the other higher luminance contrast ones.

3.2   Analysis of Visual Performance using total response time

It is important to screen the possible interactive effects from the total of ten combinations. After a series of model selections, the final solution identifying two-factor interactions I×G and I×LC are included in the model and shown in Table 2. It indicates both the two-factor interaction effects between interactivity and grid (I×G) and interactivity and luminance contrast (I×LC) reach statistical significance (F=9.46 with p-value=0.037 and F=6.80 with p-value=0.048, respectively). The model fitting is good enough because the coefficient of determination is over 90% (R^2 =91.74%) and the coefficient of adjusted determination is about 70%

(R_{Adj}^2 =69.03%). The model adequacy checking based on the residuals indicates the residuals are satisfactory with normal distribution for the basic assumption of the ANOVA model (Kolmogorov-Smirnov normality test=0.081, p-value>0.150). For the design guidelines, the combination of interactivity and no grid is recommended since it has the fastest total response time on average (316.3 seconds), on the other hand, the combination of interactivity and the lowest luminance contrast 0 is also recommended since it has the fastest total response time on average (241.5 seconds). Due to the limitation of the experimental sample sizes, we could not provide sufficient evidence to support the inference of three-factor interaction among interactivity, grid, and luminance contrast. This is worth investigating in further studies.

Table 1 Summary statistics of response time for five different tasks (unit: second)

Factor	Task Level	Exact Mean	STD	Maximum Mean	STD	Correlation Mean	STD	comparison Mean	STD	Trend Mean	STD
I	Yes	88.0	27.2	54.5	27.8	60.3	26.9	76.1	25.3	94.9	55.5
	No	152.4	39.7	53.3	10.8	64.8	19.3	78.3	28.2	62.6	19.1
S	Large	119.8	55.5	54.1	15.1	63.5	15.9	65.4	15.8	78.5	33.6
	Small	120.6	40.0	53.6	25.8	61.5	29.3	89.0	29.5	79.0	54.0
G	Yes	106.4	23.4	59.6	24.8	.	.	81.1	24.6	84.4	56.6
	No	134.0	60.9	48.1	14.2	.	.	73.3	28.2	73.1	27.8
D	Multiple	112.1	46.4	51.3	13.7	61.4	21.8	65.9	12.1	71.3	28.0
	Single	128.3	48.8	56.5	26.2	63.6	25.2	88.5	31.6	86.2	56.0
LC	0.00	105.0	57.1	45.5	14.2	59.5	21.6	71.5	34.9	61.8	9.4
	0.33	120.0	47.8	61.0	35.8	66.8	39.9	80.0	25.5	112.3	74.1
	0.67	131.5	31.3	50.8	11.0	61.3	17.5	88.5	23.2	88.0	29.8
	0.99	124.3	62.5	58.3	16.6	62.5	13.9	68.8	25.0	53.0	9.9
Task		120.2	46.8	53.9	20.4	62.5	22.8	77.2	25.9	78.7	43.4

"." Denotes no such design is provide.

3.3   Analysis of Visual Fatigue using DCFF

Visual fatigue was measured by the difference in critical fusion frequency (DCFF). It is defined as the difference between CFF before-experiment (CFFB) and CFF after-experiment (CFFA) and expressed as DCFF=CFFA-CFFB. After a series of model selections, the final solution identifying two-factor interactions I×G and I×LC are not significant but still included in the model and shown in Table 2. This indicates the main effects of interface size (S), data series (D) and luminance indicates the main effects of interface size (S), data series (D) and luminance contrast reach statistical significance (F=27.76 with p-value=0.006, F=21.68 with p-value=0.010, and F=29.86 with p-value=0.003, respectively). The model fitting is good enough because the coefficient of determination is over 97% (R^2=97.64%) and the coefficient of adjusted determination is over 90% (R^2_{Adj} = 91.16%). Large interface, multiple data series, and the lowest luminance contrast (LC=0) are highly recommended, however, small interface size, single data series, and the highest luminance contrast (LC=0.99) should be avoided.

3.4   Analysis of NASA-TLX and UIS

The rankings of overall workload from high to low for five tasks are in the order Exact, Comparison, Trend, Correlation, and Maximum. In contrast to visual performance, on average, participants took the longest response time to complete the task of reading an exact value; meanwhile, they overtook the most serious overall workload. In addition, on average, participants took the shortest response time to complete the task of finding a maximum; meanwhile, they overtook the least serious overall workload. The result of ANOVA indicates the effect of I×D is statistically significant on overall workload (F=6.40 with p-value=0.039). Both the designs of multiple data series without interactivity and single data series with interactivity are highly recommended, however, the design of multiple data series with interactivity should be avoided. In addition, the effect of interactivity (I) is statistically significant for the UIS score (F=7.10 with p-value=0.029). The design of interactivity tends toward a higher UIS scoring on average.

Table 2 Analysis of variance for total response time and DCFF

Response variable	Total response time			DCFF	
Source of variation	DF	F	P-value	F	P-value
Interactivity (I)	1	1.65	0.268	2.15	0.216
Frame Size (S)	1	0.59	0.484	27.76	0.006**
Grid (G)	1	0.75	0.436	0.06	0.821
Data Series (D)	1	4.41	0.104	21.68	0.010*
Luminance Contrast (LC)	3	2.39	0.210	29.86	0.003**
I×G	1	9.46	0.037*	5.45	0.080
I×LC	3	6.80	0.048*	6.34	0.053
Error	4				
Corrected Total	15				

*p<0.05; **p<0.01

4   DISCUSSION

Which objects of a data display can receive user-interaction requests? In principle, all objects could be interactively accessible but not all objects are directly visible. How should the user directly manipulate objects that cannot be seen? It seems reasonable to restrict direct manipulation to objects that have a graphical representation. Therefore, Wilhelm proposed his first postulation, "Axiom1: Only objects that are graphically represented can be directly manipulated." Depending on the level of a data display that is concerned with user interaction, our study indicates two-factor interactions between I×G and I×LC are statistically significant on total response time. This partly agrees with Wilhelm's postulations that for a single frame changing the entire frame and changing the size and color of the frame are the only user interactions that are useful. The MFS interface with the design of interactivity, no grid, and the lowest luminance contrast 0 is highly recommended because it has the fastest response time on average. However, from the point of view of visual fatigue, our study indicates the effects of interface size (S), data series (D) and luminance contrast (LC) all reach statistical significance. It is fully consistent with Wilhelm's postulations of Axiom 1 (Wilhelm, 2003). On the other hand, many of the user interactions that aim at the model or sample population level force the type and the frame level to update correspondingly. When we change a model, we like to see the display react immediately. In summary, an internal linking structure is needed to control the transmission of the interaction message between the graphic levels within one display. Therefore, Wilhelm proposed his second postulation, "Axiom2: Interactive user request can be passed to objects without graphical representation using the internal linking structure." The design of interactivity and data series in our study makes as much information available as possible but does not display it all at once. It is also consistent with Wilhelm's postulations of Axiom 2. Except the choice of graph type, two-factor interaction between I×D is shown to be an important factor affecting the overall workload. Multiple data series with interactivity would result in a higher workload over the five tasks. This is similar to the study of Meyer, Shinar, and Leiser (1997) that the complexity of the data would affect the tasks difference, as prior familiarity with the display did. For the tasks of Comparison and Trend, the highest luminance contrast (LC=0.99) achieves the best visual performance. It is consistent with the study of Snyder (1988) that luminance contrast is better to set at a level higher than o.67.

5   CONCLUSIONS

User interactions are an essential element of mobile financial services. The results of this study have implications for interactive statistical graphs of mobile financial services in general. Most MFS users prefer using interactivity of value-label to no value-label. For NASA-TLX, the interaction of multiple series and interactivity of value-label would result in the highest workload over the five tasks, but the interaction of single series and interactivity of value-label would result in a lower workload. For reading the exact value of a single point, the results show both the mean response time (120.2 seconds) and overall workload (5.25) are higher than other tasks; however, in the design of interactivity of the value-label, mean response time was reduced to 88 seconds. Depending on the task differences, it is better to use the design of interactivity with no value-label for the trend-identifying task (mean=62.63 seconds). The interactions of I×G and I×LC are statistically significant in terms of total response time but only frame size, data series, and Luminance contrast would affect DCFF. It should not be recommended for small size, single series, and highest LC (0.99) for serious visual fatigue.

ACKNOWLEDGMENTS

This work was supported by the National Science Council, Taiwan under Grant NSC 100-2221-E-029-021.

REFERENCES

Benyon, D., P. Turner, and S. Turner. 2005. *Designing Interactive Systems: People, Activities, Contexts, Technologies*, England: Pearson Education Limited.

Cleveland, W. S., McGill, R., 1984, Graphical perception: theory, experimentation, and application to the development of graphical methods, *Journal of the American Statistical Association* 79: 531-554.

Cleveland, W. S., R. McGill. 1987, Graphical perception: the visual decoding of quantitative information on graphical displays of data. *Journal of the Royal Statistical Society, A (General)* 150: 192-229.

Chin, J. P., V. A. Diehl, and K. L. Norman. 1988. Development of an instrument measuring user satisfaction of the human-computer interface. *Proceedings of SIGCHI '88*, 213-218, New York: ACM/SIGCHI.

Gillan, D. J., C. D. Wickens, and J. G. Hollands, et al. 1998. Guidelines for presenting quantitative data in HFES publications. *Human Factors* 40: 28-41.

Hart, S. G. and L. E. Staveland. 1988. Development of NASA-TLX (Task Load Index): Results of Empirical and Theoretical Research. In. *Human Mental Workload*, eds. P. A. Hancock and N. Meshkati. 239-250.

Meyer, J., D. Shinar, and D., Leiser. 1997. Multiple factors that determine performance with tables and graphs. *Human Factors* 39: 268-286.

Wollman, D. "Manage Your Money Anywhere." Accessed February 29, 2012, http://www.mint.com/.

Snyder, H. L. 1988. Image quality. In. *Handbook of Human-Computer interaction, eds.* M. Helander. Amsterdam: Elsevier.

Wilhelm, A. 2003. User interaction at various levels of data displays, *Computational Statistics and Data Analysis* 43: 471-494.

CHAPTER 48

The Design and Evaluation of a Radio Frequency Identification (RFID) Enabled Inpatient Safety Management System

Ta-Wei Chu [1], Chuan-Jun Su [2]

Department of Industrial Engineering & Management
Yuan Ze University
135, Yuandong Rd., Zhongli City, Taoyuan County 32003, Taiwan (R.O.C)
taweichu@gmail.com [1]
iecjsu@saturn.yzu.edu.tw [2]

ABSTRACT

Clinical work in modern hospitals is characterized by a high degree of mobility, frequent interruptions, and much ad hoc collaboration between colleagues with different areas if expertise. Hospitals are faced with higher patient loads and smaller budgets, necessitating greater efficiency. Greater productivity could be partly achieved through improvements in hospital information systems. However the medical care for patients differs from traditional workplaces. The primary concern of doctors and nurses is the patient, not the information system, and thus they put a strong premium on usability. In this paper we describe the design of a Radio Frequency Identification (RFID)-enabled inpatient safety management system (RISMS) and how it can offer a viable solution for the point of care, achieving a high degree of usability through a user-centered design process. The performance evaluations as well as the expected benefits of the system are also discussed.

Keywords: patient safety, Radio Frequency Identification (RFID)

1 INTRODUCTION

Recent medical reports show that, on an average day shift, hospital nurses in Taiwan attend to an average of seven to thirteen patients per shift, while the average for night shift nurses is twelve to thirty patients per shift – an average patient load several times higher than that of hospital nurses in the United States. Not only are nurses in Taiwan responsible for providing regular care to patients, but they also have to support surgical procedures, and occasionally contribute to emergency room care. Moreover, during rounds, nurses are required to take detailed hand-written records of patients' vital signs. This work is regularly interrupted by emergency calls or other distractions, and the nurses then need to be able to pick up where they left off perhaps hours later, often resulting in incomplete records and inappropriate care. Exhausted, distracted nurses are much more likely to make mistakes, including providing the wrong medication or inappropriate dosages.

Missteps in simple hospital management procedures are often the cause of unnecessary medical crises. Automating certain aspects of patient care and administration work can help avoid such situations and can directly contribute to improved patient safety. The research presented in this paper aims to reduce the workload of nurses and doctors through simplifying access to patient documentation and automating the measurement and recording of patients' vitals. Reducing the workload on medical staff would help promote patient safety and reduce the incidence of medical problems in hospitals.

RFID (Radio Frequency Identification) is seen by many as a critical technological development with potentially unlimited applications. RFID allows for large-scale automated identification, tracking and positioning over the wireless spectrum. When combined with an electronic product code network, and regulated by EPC Global standards, RFID data can be turned into practical knowledge useful for decision making. RFID has already been widely used in veterinary medicine, largely for identifying animals. Other research has shown that RFID can be used in the identification and tracking of medical materials and case histories, and also for identification and tracking in remote care (Accenture, 2004).

Most hospital information management systems are broken down by department, but this structure often creates barriers to the exchange of information. Without taking a holistic view of information management, miscommunication and delays of information transfer result in poorer-quality patient care and increased costs. Therefore, when adopting a new technology such as RFID, care must be taken not to repeat the mistakes of the past. A few medical institutions in Taiwan are applying RFID systems in trials, but they seldom integrate procedures and structures, thus severely limiting the benefits the technology can bring to the quality of patient care.

2.   RFID APPLICATIONS AND HEALTH RISKS IN MEDICAL INDUSTRY

2.1. RFID Applications

This paper considers three potential applications for RFID in the medical industry. First, the technology can be used to track material and supplies including equipment, reusable materials, and raw materials, thus improving production efficiency and reducing human resource requirements. Second, it can be used to improve production processes through continuous tracking of asset locations and updating status data, thus allowing for the automated management of resources and error detection, contributing to higher-quality end products and services. Third, it can contribute directly to patient safety through monitoring and verifying drug use by patients.

2.2. Health Risks of RFID

Today there are hundreds of thousands of RFID systems in use, all using EM fields to detect and scan tags. A report by the International Commission on Non-Ionizing Radiation Protection (ICNIRP) examined the effects of EM radiation on humans (ICNRP, 2004) and described the mechanisms of thermal and non-thermal interaction between EM fields and biological systems. Thermal interaction is the heating of tissue which can cause damage. The most notable non-thermal interaction is membrane stimulation, in which membrane potentials may be altered at a cellular level and might have effects on the nervous system. The report states that the high frequencies of Electronic Article Surveillance (EAS)/RFID systems produce no heating or thermoregulatory stress (ICNRP, 2004). However, EAS and RFID devices may interact with and disrupt medical devices such as pacemakers. The report recommends further study and that device manufacturers should provide information needed for health risk assessments. There is also a need to continue to collect exposure data, especially for occupational groups, differentiated, if possible by the frequency of exposure.

MRI incompatibility is perhaps the most serious issue. An MRI machine uses powerful magnetic fields coupled with pulsed radio frequency (RF) fields. According to the FDA's Primer on Medical Device Interactions with Magnetic Resonance Imaging Systems, electrical currents may be induced in conductive metal implants that can cause "potentially severe patient burns".

3.   RFID-ENABLED INPATIENT SAFETY MANAGEMENT SYSTEM (RISMS) DESIGN

Mindful of the importance of patient safety, many countries have established professional institutions to conduct patient-safety related research and policy consulting. However, technological developments (such as RFID and barcode applications, etc.) should also be considered a critical aspect of improving patient safety. Many hospitals (i.e. Massachusetts General Hospital, Germany Saarbrucken Hospital, etc.) have successfully used RFID technology to reduce the medical errors and improve patient safety. Our ultimate objective is to build an RFID-enabled inpatient safety management system (RSIMS) to reduce the occurrence of medical errors and improve patient safety.

3.1. Case Study and System Modeling for RISMS

The sheer volume of case studies on a given subject doesn't necessarily indicate the quality of the underlying research. While individual research projects focus on a specific case, multiple case reviews search for generalities across cases. The choice of appropriate research methods and design is determined by the particular topic, whether it focuses on individuals, departments, information units or an entire organization.

The hospital used for this case study began opened in 1996 with 800 beds as the regional teaching hospital in northern Taiwan. In 1997, the hospital received regional accreditation as a Class I teaching hospital. The hospital comprised 31 specialized medical departments including internal medicine, surgical, gynecology, etc. The emergency care operation ran year-round, and offered holiday and evening clinical services. At the time of writing, 603 beds are currently occupied. The hospital is located in a mixed residential, industrial area which has a higher than average proportion of elderly residents. Development plans focus on expanding regional medical services and establishing the hospital as a regional medical center offering high quality and efficient medical services to the people of the region.

The hospital was selected according to the following criteria:

1. As a regional teaching hospital, patient safety is central to the hospital's mission, as is constant improvement to the quality of nursing
2. Patient selection was facilitated by the relative simplicity of patient conditions in the OB/GYN department. The majority of patients were expectant mothers or experiencing gynecological problems, with one patient per room
3. Patient data followed a comprehensive fundamental structure, and the hospital has excellent wireless coverage.

In order to improve the nursing quality and patient safety, it is important to first understand the various nursing tasks, grouped by shift:

A: Morning shift 1 (08:00-12:00, 13:00-17:00)
B: Morning shift 2 (08:00----16:00)
C: Night shift (16:00--24:00)
D: Evening shift (16:00--21:00)
E: Late shift (24:00----08:00)

During ward rounds, the doctor traditionally is engaged in carrying, searching, and browsing medical histories, recording patient conditions, and giving or changing medical instructions. Nurses' ward rounds traditionally include updating medical histories and recording vital signs including temperature, blood pressure, pulse, etc. These activities require extensive paperwork, with the most important records listed below:

I.   Pharmaceutical Treatment Record:.
II.   Parental Treatment Record.
III.   Vital Signs Record.
IV.   Body Temperature Chart.
V.   Progress Note.

3.4   System Architecture

At present, nurses rely on visual cues and personal memory to identify a patient on the ward, which frequently results in patient misidentification, which could lead to incorrect medication prescription, especially when the patient is out of her bed or room. RFID tags on the patient's wrist, combined with an RFID reader, would enable the nurse to accurately identify each patient, displaying the patient's profile on a flat panel. Such a system would reduce human error, especially those made by medical trainees and new nurses. We designed the architecture of RISMS composed of two sub-systems: Ward Round and Nursing Station.

● **Ward Round Sub-system**

The Ward Round subsystem performs two important functions for inpatient safety management:

1.   Identity verification: with a portable RFID reader, nurses can automatically verify a patient's identity from the patient's wrist tag, immediately calling the patient's profile (including doctor's instructions) to a portable display. By eliminating the possibility of mistaken identity, the system ensures that each patient gets the correctly-assigned medication and treatment.

2.   Ward round automation: when the system retrieves a patient's information, the nurses can start their ward rounds, during which they are primarily concerned with administer medication, recording vital signs (blood pressure, respiration, pulse, etc.) and making progress notes. In administer medication, the list of required medicines and dosages would appear on the ward round system via wireless internet, allowing nurses to verify the medication and dosages, and to timestamp the patient's receiving the medication. The system will also display the patient's vital signs for the past 24 hours, allowing for at-a-glance comparison and freeing nurses from having to carry paper records on rounds. Data entry is performed through touchpad/tablet, allowing for handwritten recordkeeping and easier data management. The handwriting recognition system would standardize handwritten notes into legible text. Nurses will not be able to overwrite records written by others, and all input/modification activity would be logged in a central database for review purposes.

● **Nursing Station Sub-system**

The main functions of the nursing station are:

1.   Admission/discharge administration: The system is used to verify patient identity and to create a wrist tag upon admission. The nurse orally confirms patient identification with the patient or admitting party. The identity of transfer patients is

double-checked upon receipt from the Hospital Information System (HIS) and the transfer patient would also receive a wrist tag. For example, a patient with ID number A123456789 would receive a wrist tag ID 5678 which would remain constant throughout the patient's entire hospital stay. The patient's photography will appear on the UI for verification. When the patient is discharged, the tag will be recycled by removing the association of the patient's ID card and the tag ID. The tag's data would be wiped clean, ready to be written again for another patient.

2.   Search, modification and maintenance of medical records: The accuracy, accessibility and timeliness of medical records requires that records be centrally maintained at the nurses' station and modified based on the ward rounds system, which allows nurses at the nursing station to retrieve and modify information from ward rounds. Privacy is ensured by placing restrictions on various levels of authorization access.

3.   Reports generation: At the nursing station, nurses can print out Progress Notes, physiology information and medical histories directly onto pre-formatted hospital charts and tables, obviating the need for hand-drawn tables.

3.5   RISMS Implementation

Based on the abovementioned architecture, we developed a prototype using the following hardware and software components: 1) Nurse station workstation: Intel Pentium 4 CPU with 1GHz RAM personal computer with Microsoft Windows XP Professional Version; 2) Portable Devise Cart: the cart consisted of a tablet PC, (Acer N200, with Intel Pentium 4M CPU and 512MB RAM), an RFID reader (with built-in antenna) and APC Uninterruptible Power Supply (UPS). The passive-type 13.56 MHz RFID reader was attached to the back of the tablet PC via CF card. The tablet PC and RFID reader were powered by the UPS battery; 3) RFID Wristband: The wristbands used for this test were based on the Tag-it passive tag chip with 13.56 MHz frequency band. The wristbands used were pre-programmed with fictitious patient information; 4) Wireless Network: An 802.11b Cisco access point (AP) with a 50mA signal provided wireless connectivity. The access point was also configured to act as a DHCP server to allow the association of wireless clients; 5) Server with HIS Interface: we used an HP dl-385 as the server, with a software interface, developed by The Taiwan Electronic Data Processing Co. (TEDPC), to the hospital's Patient Information System (PAS), The HIS software interface was setup using a Microsoft Windows Server 2003 Standard Edition OS and connected directly to the hospital's network through an Ethernet network switch.

To increase ease of use, the prototype's integrated development environment (IDE) was built on Microsoft Visual Studio.NET 2003, which features a GUI that can be easily customized by the user. To encourage quick uptake by medical staff most UIs are modeled on conventional forms. For the server, we used Oracle 9i to simulate a hospital database. Moreover, the Oracle Client must be installed on each workstation and tablet PC for assessing and storing data for the server. In this study, we used an object linking and embedding database (OLEDB) to connect the proposed application with the database. In addition, we also used Crystal Reports to access the requisite records and design the forms for reports. All reports are used as substitutes for hospital paperwork involved in the ward routine.

4   SYSTEM PERFORMANCE EVALUATIONS

Two types of performance evaluation were conducted for RISMS: 1) clinical evaluation and 2) RISMS system evaluation.

4.1   Clinical Evaluation

We established a standard format for a patient's nursing record via applying RISMS, including patient physiological, psychological, medication and health education through real data collection and references research. The records were collected in the obstetrics and gynecology ward after a process of test, redesign, education and training. The results are illustrated below:

4.1.1   Evaluation Sample Basic Attributes

The basic attributes of samples used in the clinical evaluation is shown below. Twenty staff members were asked to complete our questionnaire with a 100% return rate. The average age was 26.1 years old, with 55% of the respondents between 20-25 years old. Most of the nursing staff (75%) had a college degree Forty percent had worked three years or less, 35% had worked 3-5 years. Thirty-five percent held an N2 nursing staff rank, while 25% held N1. All had experience using desktop computers, but only 10% had worked with tablet PCs.

In the first part of the study, we are interested in finding out whether the system improved the nursing work. We therefore focused on simple delivery and DRG (Diagnosis Related Group) cases. The nurses included in the study all participated in the system design and were familiar with the system operation.

4.1.2   Record Integrity

During ward rounds, the nurses used the RISMS for obstetric patients and created new nursing records including general patient data, discharge notes, records for vital signs and a drugs check list. Four nurses, including the head nurse, created 36 nursing records. No duplicate or missing records were noted. Having participated in the system's design, the nurses had no difficulty learning to operate the system. Compared to traditional hand write chart, Record integrity was 100%, and system use increased the integrity of nursing records by 55.4%.

4.1.3   Shortening Completion Time of Medical Records

In using the RISMS, the average time to complete a medical record decreased from 190 seconds to 60 seconds. Time required for record creation was reduced by

125 seconds per record for normal deliveries, and 150 seconds per record for cesarean deliveries. Generating patient progress notes was sped up by 54 seconds per record.

- Average writing time: The nurses participated in the design of the RISMS and were well trained, resulting in a 110 second time savings in record generation per record.
- Normal spontaneous delivery records: 55 seconds per record, a savings of 125 seconds per record.
- Cesarean records: 80 seconds per record, a savings of 150 seconds per record.
- Daily nursing progress note: 47 seconds per record, a savings of 54 seconds per record.

4.1.4 Staff Satisfaction towards Applying RISMS on Nursing Operations

Of the 20 staff members surveyed 88.6% stated that the system had a positive effect on the creation of nursing records, while 11.4% felt the system offered no particular benefit and preferred the previous paper-based system. Those who felt the system offered significant benefits particularly pointed to the system's ability to improve the ease and integrity of transferring records from one shift to the next.

4.2 RISMS System Evaluation

The system evaluation adopted the Likert 5-point system: 5 = strongly agree, 4 = agree, 3 = no comment, 2 = disagree, and 1 = strongly disagree.

4.2.1 Performance Evaluation of RISMS in Nursing Operations

The performance evaluation of the system among nurses was 3.10 + 0.34, with each topic scoring between 2.65 and 3.90. The strongest incentive among nurses to use the system was to improve "service quality" (3.21), and the weakest was improving "operating procedures" (2.96). Following up on this, we found that nurses found the system to be favorable in terms of "reducing errors caused by misidentification of patients" (3.60), "increasing interaction with patients and understanding patients" (3.50), and "reducing costs due to medical error" (3.55). Low degrees of acceptance were found for "reducing nursing workload" (2.65) due to paper-based and RFID-based systems used in parallel, "improving nursing shifts" (2.65) and "reducing human resource redundancies" (2.70).

Simple t-test and one-way ANOVA tests were used to analyze the performance of RISMS in nursing operations. If a significant difference was found, the Scheffe test was applied for further analysis. Nurses over the age of 31 had a higher acceptance rate than those under 25 (3.88±0.56 vs. 3.00±0.73; p<0.05), and acceptance increased with the nurse's rank (3.90±0.55 vs. 2.86±0.59; p<0.05) and with their length of service (3.88±0.56 vs. 2.78±0.62; p<0.05). However, education level seemed to be insignificant in acceptance, as was experience in using desktop and tablet PCs.

4.2.2 The Satisfaction Survey of Using the RISMS in Nursing Operations

The average satisfaction score among the nursing staff applying the RISMS was 3.20+0.78, with topic scores running from 2.75 to 3.70. Satisfaction with "facility"-related factors averaged 3.32, while "problem solving" areas only averaged 2.78. Upon further investigation, we found that nurses were most satisfied with "the ease of retrieving RFID tags" (3.70), "the stability of tablet PC" (3.55) and "the ease of interface use" (3.40). Disappointment was evident; however, in "the difficulty of problems solved" (2.75), "the speed of problem resolution" (2.80) and "reliability of roaming" (2.95).

4.2.3 Correlation of Nursing Staff Acceptance of and Satisfaction with the RISMS

We analyzed the correlation with the Pearson correlation test to examine acceptance of and degree of satisfaction of using RISMS. The results revealed no correlation between the acceptance of the system and degree of satisfaction (r=0.25). This indicates that, although the nursing staff had a positive attitude towards accepting the system, the degree of satisfaction is still low. At the same time, we found that there is positive correlation among satisfaction with "operating procedure", "service quality" and "medical costs" (r=0.610, p<0.05; r=0.627, p<0.05; r=0.601, p<0.05), indicating that the higher a nurse's expectation of the system's effect on nursing quality, the higher the degree of acceptance. Overall, satisfaction with the system was high (3.20+0.78), but we could only find a positive correlation between "concept design" and "facility"(r=0.467, p<0.05), but with "problem solving". If modifications to the system can improve "problem solving", we expect it will further raise overall satisfaction.

5   CONCLUDING REMARKS

In recent years, hospitals in Taiwan have attempted to improve the quality of medical treatment and patient safety through using RFID technology to accomplish specific tasks, including newborn care, detection of SARS patients and emergency room treatment. In most the technology was used to verify patient identification, to manage patient care, or to ensure proper medication. However, the technology has only very rarely been applied to the routine ward processes. Our on-site survey found a high incidence of medical errors resulting from faulty ward routines, and this study aims to apply RFID technology into ward routines to improve inpatient safety. The proposed RISMS improves the quality of inpatient care and reduces the workload of nursing staff. It not only reduces the incidence of medical errors, thus improving improve patient safety, but also promotes the development of EMR.

Via a pilot trial of our RISMS, we found a significant reduction in patient misidentification, leading to more reliable and accurate access to patient information, along with a reduction of human errors and administrative workload.

REFERENCE

Brock D.L. 2002. *Smart Medicine the Application of Auto-ID Technology to Healthcare.* Auto-ID Center.

Accenture. 2004. "Accenture Helps Form RFID Industry Group to Evaluate Technology's Value in Pharmaceutical Industry." Accessed February 28, 2012, http://newsroom.accenture.com/article_display.cfm?article_id=4081.

Brewin, B. 2004. "RFID gets FDA push." Accessed February 28, 2012, http://fcw.com/articles/2004/11/14/rfid-gets-fda-push.aspx.

Thompson, C. A. 2004. Radio frequency tags for identifying legitimate drug products discussed by tech industry. *American Journal of Health-System Pharmacy* 16(14):1430-1431.

EPCglobal Web site. Accessed February 28, 2012, http://www.epcglobalinc.org

ICNRP. 2004. ICNIRP Statement Related to the Use of Security and Similar Devices Utilizing Electromagnetic Fields. *Health Physics* 87(2):187-196.

Wallin, M., T. Marve, and P. Hakansson. 2005. Modern wireless telecommunication technologies and their electromagnetic compatibility with life-supporting equipment. *Anest Analg.* 101(5):1394-400.

Young, J. 2005. "RFID applications in the healthcare and pharmaceutical industries." Accessed February 28, 2012, http://www.radiantwave.com/whitepapers/healthWP.doc.

Chen, H. P. 2007. An RFID- enabled Inpatient Safety Management System [Master Thesis], Yuan Ze University.

CHAPTER 49

The First Time We Never Forget: Product Evaluation in the Context of First Usage

Caio Márcio Almeida e Silva¹, Maria Lúcia Okimoto²,

¹Universidade Federal do Paraná | Lactec
Curitiba, Brasil
caiomarcio1001@yahoo.com.br

² Universidade Federal do Paraná
Curitiba, Brasil
lucia.demec@ufpr.br

ABSTRACT

This paper presents a research that approached the intuitive use starting from the first use of a product. Being like this, the objective was of evaluating the intuitive use starting from the first use, as well as in a new use context. They announced eight voluntary subjects. With the results, it was ended that the fact of putting the participants in a context of different use influenced in the execution of the tasks, as well as in the perception of the product.

Keywords: intuitive use, first usage, lift platform.

1   INTRODUCTION

This paper reports a study with emphasis on intuitive use from the experience with the product from the first use. Intuitive use, in turn, comes if showing as an important area to be considered in product design. However, it is still little what if writes about the association of intuition in the relationship between people and products. As an example, we have the following rearches: Rutter, Becka, Jenkins (1997); Frank e Cushcieri (1997); Thomas e Van-Leeuwen (1998); Okoye (1999) e Blackler e Popovic (2003). And this sphere of "understand intuitively" is therefore the starting point for the research in question. In this way, the understanding and use intuitive get along when the observer/user comes into contact with a new product to your repertoire and it can to perform tasks of intuitive way, without the use of manuals or help from others. For both, we assume that the intuition helps in the process of interaction with a product.

In Silva (2012) it presented some approaches for the intuitive use. Some identified ones in other authors, and another developed experimentally. One of them, is the intuitive use with a new product in an use atmosphere. The author suggests that the fact of using the product for the first time facilitated the novelty The other unpublished aspect was the subject's contextualização in a new scenery.

2   METHODOLOGY

Starting from the foundations, it was developed a protocol that orientated the experiment. They participated in that experiment 8 people. They didn't present physical deficiency. For the accomplishment of the same, the participants had to fill out a term of free and illustrious consent. Soon afterwards, the participants had to accomplish the following task: seating in a wheel chair (simulating wheelchair) to go until the lift platform of the station-tube of Curitiba-PR, to identify which the appropriate commands to elevate the platform and then, to lower her/it at the level of the sidewalk. Finally, the participants had to leave the platform using the wheel chair.

The participants were people that had never used a wheel chair nor the lift platform. Like this, they could evaluate starting from the experience of the first use, without addictions nor use habits. At the end, the participants had to fill out a questionnaire. The questionnaire treated of themes, as: technological familiarity, usability, use of controls and influence of the sensorial incentives in the activity. The experiments lasted, on average, 30 minutes. The technique of used research was the interview. As metric, the solemnity-report "evaluation of specific attributes" was used. The experiment was driven outdoors, in the station-tube located in the Federal University of Paraná, Brazil.

The questionnaire counted with the following subjects:

1. Is the experience of using the lift platform similar with some other experience lived by you (or done live in the daily)?
    ( ) Yes   ( ) No

    If yes, what: _____

2. How was it made the choice do(s) command(s) it goes use the platform?
    ( ) For the disposition of the controls in the panel
    ( ) For the format of each one of the controls
    ( ) For the symbolism of the color
    ( ) For the written indication of each button
    ( ) Other:

3. In your opinion, which is the degree of it influences of the format of the controls for the success of the use?
    ( ) It influences strongly
    ( ) It influences
    ( ) It influences, nor it doesn't influence
    ( ) It doesn't influence
    ( ) It doesn't influence strongly

4. In your opinion, which is the degree of it influences in the way of working the controls for the success of the use?
    ( ) It influences strongly
    ( ) It influences
    ( ) It influences, nor it doesn't influence
    ( ) It doesn't influence
    ( ) It doesn't influence strongly

5. In your opinion, which is the degree of it influences of the color of the controls for the success of the use?
    ( ) It influences strongly
    ( ) It influences
    ( ) It influences, nor it doesn't influence
    ( ) It doesn't influence
    ( ) It doesn't influence strongly

6. Do mark with a "x" in which of those controls you would find more suitable for the use in the panel of the platform?
    ( ) button "turns on-turns off"
    ( ) Interrupting
    ( ) Keyboard

( ) Rotative button
( ) Discreet button
( ) Lever
( ) Crank
( ) Voter
( ) Pedal "turns on-turns off"
( ) Simple pedal

7. For you, which sensory input predominant in the use of the platform? (How many alternatives mark are necessary)

( ) Vision
( ) Audition
( ) Touch
( ) Sense of smell
( ) Palate

8. For you, which sensory could be more explored in the use?

( ) Vision
( ) Audition
( ) Touch
( ) Sense of smell
( ) Palate

3 RESULTS

The results were organized and verified the percentile of the answers. Soon afterwards, it was applied the proportion test that indicated that there was not influences of the acquired repertoire, starting from other products, in the use of the platform. Another identified aspect was the lack of influence of the design of the product in its usability.

Is the experience of using the lift platform similar with some other experience lived by you (or done live in the daily)?

(0) Yes   (7) **No**

If yes, what: _____

How was it made the choice do(s) command(s) it goes use the platform?

(1) For the disposition of the controls in the panel
(0) For the format of each one of the controls
(4) For the symbolism of the color

(4) For the written indication of each button
(0) Other:

In your opinion, which is the degree of it influences of the format of the controls for the success of the use?
(3) It influences strongly
(3) It influences
(2) It influences, nor it doesn't influence
(0) It doesn't influence
(0) It doesn't influence strongly

In your opinion, which is the degree of it influences in the way of working the controls for the success of the use?
(5) It influences strongly
(3) It influences
(0) It influences, nor it doesn't influence
(0) It doesn't influence
(0) It doesn't influence strongly

In your opinion, which is the degree of it influences of the color of the controls for the success of the use?
(6) It influences strongly
(2) It influences
(0) It influences, nor it doesn't influence
(0) It doesn't influence
(0) It doesn't influence strongly

Do mark with a "x" in which of those controls you would find more suitable for the use in the panel of the platform?
(3) button "turns on-turns off"
(3) Interrupting
(0) Keyboard
(0) Rotative button
(0) Discreet button
(2) Lever
(0) Crank
(0) Voter
(0) Pedal "turns on-turns off"
(0) Simple pedal

For you, which sensory input predominant in the use of the platform? (How many alternatives mark are necessary)
(7) Vision
(4) Audition

(4) Touch
(0) Sense of smell
(0) Palate

For you, which sensory could be more explored in the use?
(3) Vision
(3) Audition
(0) Touch
(0) Sense of smell
(0) Palate

For the participants' great majority (seven), the experience of using the platform was not shown similar the any experience before lived by them. When such participants went to choose the command to work the platform, four of them orientated its choice for the color, and four for the written indications.

When we ask on which was the influence of the format of the controls for the success of the task, we had as answer that six participants thought it influenced somehow. You gave six, three thought it influenced strongly, and three that influenced. For that questioning, six participants pointed that the format of the command doesn't influence.

Was asked which was the influence in the way of actuation of the controls for the success of the task, we had as answer that all the participants thought it influenced somehow. You gave, five pointed that it influences a lot and three that influences.

When we questioned the participants concerning the degree of influence of the color of the control for the success of the use, we had that seven participants pointed that the color influences in some way. You gave seven, five analyzed that it influences strongly, and two that simply influence. In that item, a person didn't answer.

It was still questioned, with the participants, which the most appropriate controls for the actuation of the platform. In that item, three people indicated that the most appropriate control was the button league-turns off. Other three pointed that the most suitable control was the switch. And, finally, two participants pointed that the most suitable control is the lever.

In that experiment, it was also verified the influence of the sensorial incentives that they were involved in the moment of the use. The vision was the visual incentive of larger frequency, with seven indications. The audition and the touch were also shown influential for the participants. Both had four indications. The sense of smell and the palate were not suitable. Already when it was questioned which could be explored better in the use of the incentives, three people answered the vision and other three indicated the audition.

4 FINAL CONSIDERATIONS

This paper presented a research that approached the intuitive use starting from the first use of a product. Being like this, the objective was of evaluating the intuitive use starting from the first use, as well as in a new use context.

The experience of accomplishing a test with people that never used a product, as we were working in some of the other experiments, it points an approach possibility for the intuitive use. In that case, besides being the first use, it is still the first time in that the user meets inserted in a scenery while a wheel chair user. That influenced in the attention, in the compromising with the experiment, and in the construction of a real situation, on the part of the participants. Another important aspect of being considered is that that experiment was accomplished outdoors, giving a more real character to the experiment.

For future works, we suggested that the research is redone with a larger number of people properly selected. Another given suggestion, it is that that approach is made for products of daily use, participants of m new use context.

ACKNOWLEDGMENTS

To CAPES and Fundação Araucária for the financial support to the research.

REFERENCES

A. V. Cardello and P. M. Wise. Taste, smell and chemesthesis in product experience, In: Product Experience, Oxford: Elsevier (2008).

B. G. Rutter, A. M. Becka and D. A. Jenkins. 'User-centered approach to ergonomic seating: a case study' *Design Management Journal* Vol Spring (1997) 27–33.

B. Thomas and M. Van-Leeuwen. 'The user interface design of the fizz and spark GSM telephones human factors in product design' in W S Green and P W Jordan (eds) *Current Practice and Future Trends*, Taylor & Francis, London (1999) pp 103–112.

C. M. A. e Silva and M. L. Okimoto. Considerando a intuição no uso de produtos. In: Anais do 11º Congresso Internacional de Ergonomia e Usabilidade de interfaces humano-tecnologia: produtos, informações, ambiente construído e transporte. Manaus, 2011.

C. M. A. Silva, M. R. L. Okimoto. Experiência com o produto a partir do uso intuitivo. Dissertação de mestrado em design), Progama de pós-Graduação em Design, Universidade Federal do Paraná, Curitiba, 2012.

D. A. Norman. O Design do dia-a-dia. Rio de janeiro, Rocco, 2006.

D. A. Dondis. Sintaxe da linguagem visual. 2ª edição. São Paulo, Martins Fontes, 1997.

H. C. Okoye. Metaphor mental model approach to intuitive graphical user interface design, College of Business Administration thesis, Cleveland State University, Cleveland (1998)

H. N J. Schipperstein and P. Hekkert. *Product Experience*. Elsevier, 2008.

H. N J. Schipperstein; M. P. H. D. Cleiren. *Capturing product experiences: a split-modality approach*. In: Acta Psychologica 118. Elsevier, p. 293–318, 2005.

H. T. Neefs. On the visual appearance of objects. In: *Product Experience.* Oxford, Elsevier, 2008.

J. Nielsen. Usability engineering. Boston, Academic Press, 1993.

K. Krippendorff. The semantic turn. Boca Raton, Taylor & Francis Group, 2006.

L. Lidwell, K. Holden. B. Jill. Princípios Universais do Design. Porto Alegre, Bookman, 2010.

M. H. Sonneveld and H. N. J. e SCHIFFERSTEIN, H. N. J. The tactual experience of objects. In: Product Experience. Oxford: Elsevier, 2008.

M. Van Hout. (2004). Interactive Products and User Emotions. Dissertação de mestrado. Twente.

M. Van Hout. Interactive Products and User Emotions. Dissertação de mestrado. Twente.

M. Wong. Princípios de Forma e Desenho. São Paulo: Editora Martins Fontes.

R. Van EgmondVAN. The experience of product sounds. In: Product Experience, Oxford: Elsevier, (2008).

R. Arnheim. Arte e percepção visual, Nova versão, São Paulo, Pioneira, 2005.

T. Frank and A. Cushcieri. 'Prehensile atraumatic grasper with intuitive ergonomics' *Surgical Endoscopy* Vol 11 (1997) 1036–1039.

W. Cybis and A. Bertiol. Ergonomia e usabilidade: conhecimentos, métodos e aplicações. São Paulo, Novatec Editora, 2007.

CHAPTER 50

Parameters for Evaluating the Usability of the Platform Lift

Cristiana Miranda, Maria Lúcia Ribeiro Leite Okimoto, Caio Marcio Almeida e Silva

Universidade Federal do Paraná
Curitiba, Brasil
cmiranda3001@gmail.com

ABSTRACT

The purpose of this article is to present the process of defining parameters for the assessment of usability of the bus station platform lift interface, in Curitiba City, South of Brazil. The platform lift is a device that allows access for people with mobility restrictions on public transportation in Curitiba. This product features Facility of Operation as its main attribute of usability (Ease of Operation). The good performance of operating effectiveness (effectiveness of operation) enables the use of such equipment in public transport systems in other cities.

Keywords: method, usability, platform lift

1   INTRODUCTION

City planning, including urban transportation system, stands out in Curitiba, and it is considered a development model for other cities throughout Brazil and worldwide (WHO, 2011). This process started in the 1970s and it is on continuous improvement. In 1991, a new concept for urban public transportation system was implemented in Curitiba, the "Hotline", which amongst other innovations developed a boarding and disembarking station, the "Tube Station", featuring raised flooring, with the same height as the bus floor, just like a train station. To ensure access for people with mobility restrictions to the tube station, it was introduced a system named platform lift (Urbs, 2011).

Figure 1 - Sequence of photos simulating the movement of the tilting platform lift (Miranda, C. archive).

The equipment has had three versions; the latest is the tilting platform lift, which has been in use since 2008. Nevertheless there are no tests or usability parameters defined for this model. Therefore, the objective of this article is to present the process of defining the parameters of usability of the platform lift. We begin this article with some thoughts on usability and Ease of Operation. Afterwards, we introduce the development stages of the process, which followed the guidelines of ISO 20282-2: 2006 – Ease of operation of everyday products - Part 2: Test method for walk-and-use products. Finally, we present the results of a pilot test, involving two subjects, and final thoughts on the developed process.

2   USABILITY OF THE WALK-UP-AND-USE PRODUCT

Usability has many definitions (Santos, 2008; Tullis and Albert, 2008). According to the International Standards Organization (ISO 9241-11, item 3.1), usability is the "extent to which a product can be used by specified users to achieve specified goals with effectiveness, efficiency and satisfaction in a specified context of use". Other attributes in addition to effectiveness (accuracy and completeness with which users achieve specified goals), efficiency (resources expended in relation to the accuracy with which users achieve specified goals) and satisfaction (freedom from discomfort) may be added to usability, like learnability, legibility and ease to operation (Santos, 2008). Usability of a product is a variable of context of use, i.e., metrics such as task success, task time, and mistakes may have different values according to, for example, the user experience (beginner, with training or expert). This possibility is crucial to the assessment system used in professional activities. In the case of products of daily life such as walk-up-and-use products, which provide a service to the general public, usability has as its main attribute the ease of operation in the user interface. (ISO9241-11:2007).

Ease of operation means "usability of the user interface of an everyday product when used by the intended users to achieve the highest goal(s) supported by the PRODUCT. (ISO 20282-1:2006, item 3.4) high success rate (effectiveness of operation), acceptable task times (efficiency of operation) and acceptable level of satisfaction with operation. The condition of first use is based on a person who is not familiar with the product and has to use it right away without prior training or help from others. Therefore, the primary metrics of usability of a walk-up-and-use product is the effectiveness of operation (ISO 20282-1:2006).

The usability evaluation should follow a development method. Initially, one should understand the objects of study and the user goals, and then select the right metrics in order to obtain a feedback as reliable as possible (Tullis &Albert, 2008). Thomas and Bevan (1996) proposed the following procedure for the elaboration of a usability test: product description, context of use, critical usability factors, context of evaluation, evaluation plan and usability measurements. So it has been agreed that the best procedure to set parameters of usability of the platform lift would be to develop a complete method of usability testing of walk-up-and-use product.

3   TEST METHOD FOR WALK-UP-AND-USE PRODUCT

The procedure for definition of interface usability parameters of the platform lift has been developed from the guidelines of ISO 20282-2: 2006 – Ease of operation of everyday products – Part 2: Test method for walk-and-use products together with the contributions of some other authors. The steps are: 1) identify product; 2) identify context of use; 3) verify that the product is suitable to user characteristics; 4) define participants and scenarios; 5) define parameters of usability; 6) develop the test protocol; and 7) pilot test.

3.1 Identifying the product

The research methods and data gathering used in this step have been: document-based and rapid ethnography (Bevan, 2009, Usability Methods, 2011).

Table 1 – Product Identification and description

Item	Description
Name	Tube Station Platform Lift
Model	Tilting Model.
Goal	Transporting people with limited mobility from the sidewalk to the tube station elevated floor and vice versa, allowing access to the public transportation system.
Technical Specifications	Equipment built in tubular structure and steel plate, incorporating electro-mechanical trigger device.
Dimensions	

A= 4'- 7"
 (1400mm)
B= 3'- 1"
(940mm)
C= 3' – 3"
 (1000mm)
D= 2' – 4"
 (720mm)
E= 2' – 10"
 (860mm)

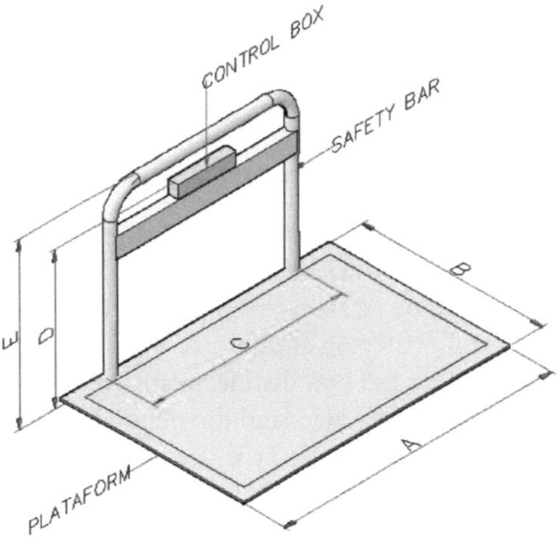

Figure 2: Illustration platform-lift (C. Miranda personal archive)

Interface	The control trigger is positioned in the intermediate sidebar. The lid must remain raised by the user while the power button is pressed. The red button moves the platform lift upwards, the green button moves it downwards and a larger button, also red, activates the emergency shutdown, as Figure 3

Figure 3: Detail of the control box - (C. Miranda personal archive).

3.2 Identifying context of use

To identify the context of use the following methods have been applied: rapid ethnography, observation of users and theoretical surveys (Bevan, 2009; Usability Methods, 2011). The subtopics of context of use are: physical environment, main goal, user types, user characteristics and task.

Physical environment: The platform lift is installed on the sidewalk next to the tube station at ambient temperature and at streetlight. The environmental noises are the transiting cars and people, when the platform is in use, a siren is triggered. There are no usage time restraints, and when the equipment is triggered, it has a nominal speed of 0.15 m / s. Users might get stressed due to bad weather (rain, wind, sun) and technical problems with the equipment.

Main goal: Entering and exiting the tube station, using the platform-lift, with autonomy.

User types: The platform lift has been designed to be used by people with mobility restrictions, such as: in wheelchairs, with difficulty in climbing / walking down stairs (knees or hips issues, respiratory insufficiency and / or heart failure, obesity, etc.); and carrying stroller or bags in general. In the case of users in wheelchairs, there are two distinct realities: the active wheelchair users, who use the platform lift on their own, and the dependent wheelchair users, which are aided by assistants who perform the task. Regarding all users, the condition of mobility restrictions may be temporary or permanent.
For the process of usability parameters definition two user groups have been determined: 1) wheelchair user: comprising people who autonomously use wheelchairs; 2) user on foot: consisting of people who use the platform-lift in the following situations: pushing a wheelchair or stroller or carrying various bags; with difficulty in climbing /walking down stairs.

User characteristics: Users older than 12 years of age are able to operate the platform lift autonomously. There are no restrictions regarding the dimensions of the user's body, and the equipment supports a nominal weight of 551.16 lb (250 kg). Consider a user without experience in similar equipment and without prior instruction to operation. Consider the possibility of user who does not read in Portuguese (tourist) or illiterate.
Disabilities: the listed impairment conditions may be temporary or permanent, in process of deterioration or reestablishment of the physical functions. Consider users with mild cognitive disabilities and mild to moderate visual abilities. As to auditory abilities, consider all possibilities of disability. For the description of Biomechanical abilities, it is necessary to divide the disabilities into two groups: wheelchair user and standing up user.

Table 2 – Biomechanical abilities of users group

Wheelchair User	Standing Up User
1- lower limbs and / or hips immobilization; 2- paraplegic; 3- tretaplegic, column levels C5, C6, C7 e C8 (McKinley, 2011); 4- Amputations: lower and / or superior limbs, hands and fingers, total or partial.	1- no physical restrictions and having at least one of the hands employed holding something; 2- with balance restrictions; 3- with mobility restrictions; 4- Amputations: lower and / or superior limbs, hands and fingers, total or partial. 5- 2/3/4 cases with the use of crutches, canes or prosthetics.

Task: There are two separate but similar tasks: the first one is to enter the tube station; the second one is to exit the tube station. In both tasks the lift platform is used autonomously. The activities to operate the equipment are simple, but there are two actions that are performed by the tube station collector, as outlined in Table 3. The user activities are flagged with * and the collector ones with**.

Table 3 – Task description

Task	Actions
Enter	1*;move from the sidewalk to the platform lift base; 2* lift the lid of the control box, keeping it this way; 3* press the green button (the platform goes up), keeping it pressed until the equipment reaches the same level as the tube station; 4* remove the hand from the control box; 5* move from the base of the platform lift to the floor of the tube station; **the tube station collector should trigger the command alongside the ratchet so that the platform lifts down to the sidewalk level
Exit	** the tube station collector should trigger the command alongside the ratchet so that the platform lifts up to the tube station level; 1* move from the tube station to the base of the platform lift; 2* lift the lid of the control box; 3* press the red button (the platform goes down), keeping it pressed until the equipment reaches the same level as the sidewalk; 4* remove the hand from the control box; 5* move from the base of the platform lift to the sidewalk

3.3 Verifying that the product is suitable to user characteristics

By analyzing the functional and physical characteristics of the platform lift interface relative to context of use, we believe the equipment complies with certain user characteristics, such as, the control box is positioned in the recommended height by the NBR 9050: 2004 (ABNT, 2004) for wheelchair users. Specific issues of ease of operation should be evaluated in usability tests.

3.4 Defining participants and scenarios

The usability test context of evaluation will be made of real components: users, tasks, environment organization, technical environment, physical environment (Thomas and Bevan, 1996).

Users: the evaluation will be based on a selection (n=20) of the general public. Participants in the usability tests should sign the Term of Free and Informed Consent, in compliance with Resolution 196/96-CMS-MS and Standard ERG BR 1002 - ABERGO. They may be male or female, over 18 years old, with no experience using the equipment, either as user or as observer. This selection is justified by two reasons: 1) any person can push a wheelchair due to illness or accident, temporarily or permanently, 2) providing context for first use, which is not possible if participants are equipment users.

Task: Participants should enter the tube station, using the platform lift, pushing an empty wheelchair. The task script is the following: 1) the participant remains behind the wheelchair, at a predetermined site and receives instructions; 2) the beginning of the task is authorized; 3) the participant finishes the task, when the wheelchair is completely outside the platform lift.

Environment Organisation: Three moderators will be required for the usability test. Moderator 1: operates the camera and authorizes the beginning of the task. Moderator 2: makes notes in a specific form and guides the participant should he be unable to operate the equipment. Moderator 3: welcomes the participant, explains the purpose of the test and orients the task, distributes post-task survey. Participants will be instructed to operate the platform lift without seeking help, but if necessary they can question moderator 2. The tests will be videotaped for subsequent analysis and data comparison.

Physical environment: Wheelchair, digital camcorder, tripod, electric power cable, chronometer, forms to be filled out by the moderator and the participant, pens. The starting location of the task will be marked on the sidewalk surface (where the wheelchair will stay) as well as the camera positioning, focusing on the platform lift control box.

3.5 Defining parameters of usability

Because the platform lift is a walk-up-and-use product, the main attribute of usability is the effectiveness of operation, with the following measurements: 95% success rate on first use, without help from others. For this usability test, the success of the task will be decomposed into a binary system: 1) successfully completed the task without help from others; 2) task completed successfully, but with help from others.

In order to complete the information on ease of operation of the equipment, the task times and the operation mistakes will also be recorded, and classified into types: mistake due to misinterpretation of use information; mistake due to misinterpretation of the interface, and mistake due to distraction, which will be recorded by frequency. The task times will be broken down into two moments, in case the participant needs help to complete the task: Phase 1: from the beginning of the task until the requesting assistance moment; Phase 2: total time. Data on task times and mistakes might contribute to the platform lift interface analysis, in case the successful task results are found to be outside specifications.

3.6 Developing test protocol

The form for Moderator 2 will be prepared so it can be filled out quickly. Below you find a model.

Name:		Age:	
Date:		Time:	
( ) Task success without help		( ) Task success with help	
Time spent:		Phase 1:	Total:
Interpretation Mistakes		Mistakes due to distraction	
Information	Interface		

Figure 4: model form for Moderator 2.

3.7 Pilot test

Two pilot tests have been carried out: the first test has been conducted with a volunteer, who had used the platform lift as assistant of a wheelchair user and the second one with a voluntary participant, who had never used the equipment. The experienced user data will be used as base of comparison in the effective test. The pilot test results are shown in Table 4 bellow:

Table 4 – Pilot test results

Mensuration	Experienced Participant	Inexperienced Participant
Task success without help	X	
Task success with help		X
Time spent on Phase 1		39 seg.
Total time spent	40 seg.	85 seg.
Mistakes		
Information Interpretation		2
Interface Interpretation		2
Distraction		

4   FINAL CONSIDERATIONS

The purpose of this article has been to present the process of defining parameters for the assessment of usability of the platform lift interface, a walk-up-and-use product. This procedure has been based on the guidelines of ISO 20282-2:2006 – Ease of operation of everyday products – Part 2: Test method for walk-and-use products. Stages involved have been: 1) identify product; 2) identify context of use; 3) verify that the product is suitable for user characteristics; 4) define participants and scenarios; 5) define parameters of usability; 6) develop the test protocol; and 7) pilot test.

The usability test method as proposed by the ISO 20282-2:2006 has been efficient and effective for the formulation of usability parameters for the platform lift. The main attribute of usability of the platform lift is the ease of operation, taking under consideration as relevance measurement the effectiveness of operation, based on the success rate of first use.

The implementation of the pilot test has allowed some operational details to be corrected, such as camcorder positioning choice and camera zoom. The results of a test with a participant without experience are not the basis for a discussion on the usability of the equipment, but in this particular case, it may be assumed that the equipment has readability issues.

ACKNOWLEDGMENTS

The authors would like to acknowledge the CAPES – Coordenação de Aperfeiçoamento de Pessoal de Nível Superior (Coordination for the Improvement of University Graduates).

REFERENCES

ASSOCIAÇAO BRASILEIRA DE NORMAS TÉCNICAS (BRAZILIAN ASSOCI-ATION OF TECHNICAL STANDARDS)- ABNT. NBR 9050:2004 – *Acessibilidade a Edificações, Mobiliário, Espaços e Equipamentos Urbanos. (Accessibility Guidelines to Edifications, Furniture, Urban Spaces and Equipment.)*Rio de Janeiro, 2004

BEVAN, N. 2009. *Criteria for selecting methods in user-centred design.* Acessed January, 15, 2012, http://www.nigelbevan.com/papers/Criteria_for_selecting_methods_in_user_centred_design .pdf

INTERNATIONAL STANDARDS ORGANIZATION – *ISO 20282-1:2006(E) Ease of operation of everday products – Part 1: Design requirements for context of use and user □characteristics.* Switzerland, 2006.

_____. *ISO 20282-2:2006(E) Ease of operation of everday products – Part 2: Test method for walk-up-and-use products.* Switzerland, 2006.

_____. *ISO 9241-11: 2007. Ergonomic requirements for office work with visual display terminals (VDTs). Part 11 — Guidelines for specifying and measuring usability.* Switzerland, 2006.

JORDAN, Patrick.1998. W. *An Introdution to Usability.* London: Taylor & Francis.

McKINLEY, W. "Funcional Outcomes per level of spinal cord injury". Accessed November, 23, 2001, http://emedicine.medscape.com/article/322604-overview

NIELSEN, J. 1993. *Usability Engineering.* San Francisco: Morgan Kaufmann.

PETRIE, H. and N. BEVAN. 2009. *The evaluation of accessibility, usability and user experience.* The Universal Access Handbook, ed C. Stepanidis.

SANTOS, R.C. "Systems Usability Evaluation Metrics Review." Paper presented at GBATA Global Business And technology Association Conference, Madri, 2008.

THOMAS, c. and N. BEVAN. 1996. *Usability Context Analysis: A Pratical Guide.* Version 4.04. Teddington, Middlesex, UK: National Physical laboratory.

TULLIS, T; ALBERT, W. 2008. *Measuring the user experience: collecting, analyzing and presenting usability metrics.* Burlington: Morgan Kaufmann.

URBANIZAÇÃO CURITIBA S.A. URBS. "História do transporte coletivo de Curitiba." Accessed June, 05, 2011, ("The History of Public Transportation in Curitiba) http://www.urbs.curitiba.pr.gov.br/PORTAL/historiadotransportecoletivo.php

("Usability Methods", accessed December, 28, 2011, http://www.usabilitybok.org/methods

WORL HEALTH ORGANIZATION – WHO. 2011. *World Recort on Disability.* Accessed January, 15, 2012, http://whqlibdoc.who.int/publications/2011/9789240685215_eng.pdf

CHAPTER 51

Usability Testing as a Systematic Tool for Product Performance Valuation During Design Process of an Ischial Support

Luz Mercedes Sáenz, Andrés Valencia

Industrial Design Faculty, Pontificia Bolivariana University, Medellín-Colombia
luzmercedes.saenz@upb.edu.co, andres.valencia@upb.edu.co

ABSTRACT

This study aims to show the particular features of the usability evaluation process, with an emphasis on the analysis models and the way that tests were conducted using protocols and instruments for recording information. The paper also aims to show how the participation of users can be regarded as a substantial tool, not only in achieving a more accurate final result, but also the manner in which the functional analysis process is facilitated. The completion of this project has shown that a detailed analysis of usability at different stages of the design process is a fundamental tool in determining which objects in a public space adapt adequately to their environment and to their potential users.

Keywords: usability testing, design process, ischial support.

1   INTRODUCTION

Since several years ago the Medellín Metro has adopted a strategy of improving the conditions of well-being for users of the Metro system. This strategy has focused on the resting times in the waiting areas of the stations of those users who experience reduced mobility due to permanent or temporary disability. It was acknowledged that the system had not provided an adequate solution for those users without causing unnecessary physical discomfort and for that reason it was decided to explore the theme of physical support for the body.

This strategy was developed with the support of the Faculty of Industrial Design at the Pontificia Bolivariana University. The University carried out a research and development project which culminated in a design for an ischial support that was adaptable to the needs that were identified for each group of users. Within these groups, the most important were the elderly, people of short stature, expectant mothers and those that required technical aids – including orthotics and prosthetics – to support mobility.

A methodological process of 'User Centered Design' was utilized during the project development. A set of technical tests were carried out which assessed the structural safety of the product through analysis of stresses and strains in critical points. In addition, usability tests with both low and high fidelity prototypes were carried out with actual users. This helped to define the final formal, material and functional attributes of the support as well as other conditions – including anchoring systems and installation and maintenance procedures – in order that the support should function correctly in the Metro stations. These tests were essential in obtaining information for the design of the ischial support and highlighted the importance of the conception and application of these tests when Ergonomics and Design were used in product configuration.

2   FRAMEWORK

Metro de Medellín and the design team considered the following concepts as important: ischial supports, reduced mobility and temporary disability situations. Besides of them, the next aspects were taken into account for the project execution:

2.1 User-Centered Design

User-Centered Design (UCD) is a methodological tool that aids the design process. Following on from (Norman, 1988), a design process supported by UCD should:

- Make it easy to determine what actions are possible at any moment (make use of constraints).
- Make things visible, including the conceptual model of the system, the alternative actions, and the results of actions.
- Make it easy to evaluate the current state of the system.
- Follow natural mappings between intentions and the required actions; between actions and the resulting effect; and between the information that is visible and the interpretation of the system state.

In other words, make sure that the user can figure out what to do, and the user can tell what is going on.

Also, it was considered the Nigel Bevan approach (Bevan, 2009), which states that the context of UCD, typical user experience concerns include understanding and designing the user's experience with a product: the way in which people interact with a product over time: What they do and why. Thus, the assessment of design proposals by a group of users is considered user-experience.

A user-centered object should maximize usability and accessibility in proportion to the user experience is facilitated to the greatest extent possible, in light of some established safety constraints. To achieve this, it is necessary not only to know in depth the relationship between user, product and context, but also interact directly and participation (participatory design) with all the stakeholders who will be involved both in the design process and in the different situations of use that will happen with the object. (Sáenz & Valencia, 2012)

2.2 Ergonomics

"Ergonomics (or human factors) is the scientific discipline concerned with the understanding of the interactions among humans and other elements of a system, and the profession that applies theoretical principles, data and methods to design in order to optimize human wellbeing and overall system performance.

Practitioners of ergonomics, ergonomists, contribute to the planning, design and evaluation of tasks, jobs, products, organizations, environments and systems in order to make them compatible with the needs, abilities and limitations of people" (IEA Council 2000, s.f.)

2.3 Usability

'Usability' pertains to how easy a product can be used. The ISO (International Standards Organization) defines it as "...the effectiveness, efficiency and satisfaction with which specified users can achieve specified goals in particular environments (ISO DIS 9241-11) (Jordan, 1998).

'Usability' and 'ergonomics' are compatible because they share a common goal – the generation of optimum conditions for users.

It does not necessarily follow that a product that is usable for one person will be usable for another. However, objects do possess a number of characteristics that can indicate usability (Jordan, 1998).

This project, considered the conception about Designing for Usability by taking a "user centered" approach throughout the design process (Jordan, 1998). Specifying User-Product-Context characteristics and considering methods for Usability evaluation.

2.4 Ergonomics and design integration

Considering that the project involved a design process and as a fundamental requirement of the Company was to include a special condition of a user group:

Reduce Mobility Persons (RMP), the team considered important to articulate the configuring process of ischial support and Ergonomics from the following approaches:

"Ergonomics and design can be regarded as intervention/application disciplines; both use systematic procedures that include observation, analysis, diagnosis, and presentation of proposals that materialize into products, procedures, and environments. They also establish a methodological relationship, complementing each other through their common interests, objectives, and procedures. Both disciplines provide elements that are required for the understanding and application of criteria that support the user-product-context relationship" (Sáenz, 2011).

"When the elements of ergonomics and design are integrated, it is also considered important to include a detailed account of the characteristics, capacities, and limitations of the user, the product's requirements, and the conditions—both environmental and social—that relate to the context in which it is being used, and may influence the use and acceptation of the designed product" (Sáenz, 2011).

3. METHODOLOGY

Development of the project began with a methodological proposal by the Ergonomics Research Division of the Design Studies Research Group (GED) at Pontificia Bolivariana University (Sáenz, 2006). Their vision of ergonomics is anthropocentric (the starting point is the individual and subsequent conditions of well-being, health and security), systemic (they observe, analyze and draw conclusions about the conditions that optimize the User-Product-Context nexus) and interdisciplinary (the support of other disciplines leads to a more integrated vision of the situation under analysis).

From a starting point of the User-Product-Context nexus, the proposal includes thematic units, themes and priority components for analysis and application in the design process. In addition, the proposal generates certain activities and 'moments' that future designers can develop in a parallel manner, coherent with the vision of ergonomics.

The methodology focused on the requirements of the users and took into account the existing space (the Metro System), thus establishing relationships of use that were coherent with the principles of ergonomics and User-Centered Design (Sáenz and Valencia, 2012).

The specific variables considered for the information achievement process, besides, the indicators for the user's, product and context observation during the valuation testing were:

User: Reduced Mobility Person, Platform habits, Measurements relationships

Product: Material and form criteria

Context: San Antonio station, Accesses, Circulation-Movements

Also were taken into account: precedents, legislations and regulations, framework, analysis and diagnosis about actual situation, opportunities and requirements of design.

A systematic approach for the project development was proposed in the form of a clear ordered and methodical set of phases with linked activities which allow integrate several disciplines and external nonprofessional actors to the process (users), besides of articulate all the best evaluated valuation and analytical procedures to the object design. The Project was developed in five phases. Each one has linked a set of activities necessary for its implementation. Figure 1 shows schematically how this phases and its respective activities were proposed and executed.

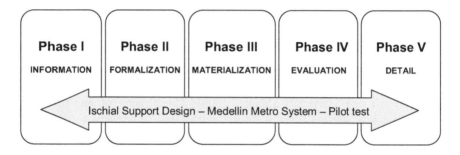

Figure 1 Phases in the project

During the execution of each phase it was noted that the phases I and V needed a lot of user contact with the design team. The users were invited to participate in two ways: participatory and non-participatory with the aim of get relevant information for:

- To identify the main users features and requirements during the waiting time in stations.
-To characterize the use context.
-To configure the object in the different design proposals
-The develop of functional tests
-To obtain and validate the technical data for the structural testing

The phases II, III and IV were of specialized technical character. They required the support of technical staff to concept and fabricate all the proposals.

In table 2 to 6 are presented all the activities associated to each development phase emphasizing in moments were the actions focused in the user characterization, use valuation and the design requirements definition were developed.

3.1 PHASE I - INFORMATION

Activity, methods and results

- Documentation: State of the art about ischial supports, Documental review with a basic functional analysis from images taken from internet.
- User lecture – Inquiry Method: Observation and register of the user typologies separated by ages, gender and mobility possibilities.
- Context lecture - Inquiry Method: Photographic record and non-participatory observation of the main architectural features of the stations, especially in platforms, accesses, circulation and permanence patterns, and ways of use of the existing furniture.
-Diagnosis and analysis - Inquiry Method: Documental search, photographic record and non-participatory observation, about disability situation in the Medellín region, the existing furniture elements in Metro system, the permanence dynamics and mobility patterns (flows) within the stations and the existing accessibility conditions.

Tools and Results from this phase were document information and graphic recording defining the habits of people when they are in platform and the opportunities that they represent for the design process. Some of these content and images were included in the first presentation showed to Planning Department of Metro de Medellín.

3.2 PHASE II – FORMALIZATION

Activity, methods and results

-Design proposals: All proposals were developed from the information got in the phase I, including the ergonomics, technical and aesthetics requirements.

-The proposals selection process was do it in two moments. The first was carried out by the design team based on the information of the phase I. The second was do it by the Metro de Medellín personnel who analyzed all the institutional aspects of each design like the anchorage possibilities, the maintenance requirements and coherence with the institutional language.

Tools and results in this phase were: sketching, formal and technical approximation including renders was presented to Planning Department of Metro de Medellín.

3.3 PHASE III - MATERIALIZATION

Activity, methods and results

-Low fidelity prototypes: Several low fidelity prototypes in non real materials were developed to decide which solution to each aspect was the best.

-Digital CAD modeling: Using CAD software several models were developed specially to confirm that the structural requirements were fulfilled. To do that a Finite Elements Analysis (FEA) was used.

-High fidelity prototype: One test prototype with all the features designed was developed with the help of the manufacturer. This prototype was installed in a station for its valuation.

3.4 PHASE IV - EVALUATION - FUNCTIONAL AND OPERATIVE TESTING

Activity, methods and results

-Low fidelity prototype testing – User Expression Protocol: Participatory Design: The real users were invited to evaluate all the features of the design using it for the time they define convenient. The evaluation was made it based on the user's commentaries and the analysis of the photos. One special work in this part of the project was developed with the low stature users.

Non structured survey and a photographic record were used. From the information got it decisions about anthropometric support height dimensions were taken.

-With the information taken from the initial user's test, the almost definitive form and shape of the ischial support were defined. Two testing high fidelity prototypes will be constructed with this information and placed in the most used Metro station.

-Structural Testing: Before the fabrication of the high fidelity prototypes, a Finite Elements Analysis was developed over a digital model using Siemens Solid Edge software. This analysis guaranteed that the ischial support comply with the three structural design criteria: strength, stiffness and stability (Sáenz & Valencia, 2012), and with that optimize the mechanical behavior of the system.

-Besides the structural criteria, the technical analysis includes the verification of the static safety factor for the system. To do that both, the stresses and the strains, were evaluated in all the critical points over the entire object and the Von Misses criteria was applied (Valencia, 2007).

When the security of the system was guaranteed, optimization techniques were applied to reduce the weight of the support. Latter, the definitive mechanical plans were obtained and sent to the manufacturing people for the construction of high fidelity prototypes.
-Second group of tests with users: One month after the high fidelity prototypes were installed in the station, a second group of tests with users was developed.

Non-participatory and participatory observations with photographic record of all the use situations.

Surveys to measure the user's satisfaction with emphasis in comfort perception, aesthetics and functionality of the support.

The results of this work were traduced in formal changes over the prototype, which allow enhance the usability of the ischial support.

3.5 PHASE V - DETAIL DESIGN AND TECHNICAL INFORM

Activity, methods and results

-Second set of functional test: From the second set of functional tests, the design team made a functional behavior evaluation inform, which included conclusions related with the final form and shape of the support, the users perception about the aesthetic, utility and comfort and all the technical details needed to the support manufacturing.

A technical inform was presented to Planning Department of Metro de Medellín.

-Mass production: From the last months of the 2010, the Metro de Medellín has installed more than 30 ischial supports in several of its stations.

4. RESULTS/CONCLUSIONS

The main result of the project was the ischial support itself, but all the usability testing protocol appear as a very interesting result since allows validate the design methodology used by the design team. The Project finalized when the definitive plans were given to the Metro de Medellín for the manufacturing of the ischial support for all Metro stations. From the date when the first high fidelity prototype was installed, there not are received any requirements from the metro de Medellín about the performance of the product.

However there are several situations that have been important to analyze the way in that a project like this is transferred to a city community. Some of them depend of the Metro de Medellín and its policies, and others depend on the Medellín citizens' culture. The main of this situations area presented in the following:

-      In some stations a Handicapped logo was located below the ischial supports. This respond to an internal policy of the Metro de Medellín, but have generated some confusion between the users since there are a high respect intentions for this kind of objects within the Metro culture and could be a transitory disability person who need to use the system but don't do it because the logo.

-      Some people, especially adults with no one disability situation, adopt non adequate postures when use the ischial support. A sitting posture is adopted by people who don't know or don't understand that this type of furniture must be used in a semi/sitting posture.

This information expresses the opportunity to continue the project with an accompaniment to the Metro de Medellín in a direct work with user in a way that allows users stay being part of a process that doesn't end with the installation of the ischial support. The project gain a future work line because the diversity of the users and the different users perceptions about a furniture element in mass transportation system.

The usability tests not only generate information useful for the design process, but feedback other stages of the use process of everyday life objects, and can orient to improvements for the performance of the system. Likewise, usability valuation can contribute with the culture creation with respect to the use and understanding of urban furniture proposals.

ACKNOWLEDGEMENTS

The authors wish to thank all the members of the work team for their assistance in the development of project activities: Camilo Andrés Páramo V., Industrial Designer and teacher in Industrial Design Faculty(UPB), Juan Esteban Vélez V., Mechanical Engineer (UPB), Juan Sebastián García C., student in Industrial Design Faculty(UPB), Liliana Sanín S. coordinator in University-Industry Projects Office in Industrial Design Faculty(UPB) and Adriana Arcila and partners in Metro de Medellín Planning Office.

REFERENCES

Bevan, N., 2009. What is the difference between the purpose of usability and user experience evaluation methods ?. Children, pp. 2005-2008.

IEA Council 2000, 2000. International Ergonomics Association. [En línea] Available at: http://www.iea.cc/01_what/What%20is%20Ergonomics.html [Último acceso: 19 August 2011].

Jordan, P., 1998. An Introduction to Usability. United Kindomg: Taylor & Francis.

Norman, D., 1988. The design of everyday things.. New York: Basic Books.

Sáenz, L. M., 2006. Methodological proposal for learning, research and application in ergonomics and products design. Maastricht, Netherlands., Elsevier Ltd..

Sáenz, L. M., 2011. Integration of Ergonomics in the Design Process: Conceptual, Methodological, and Practical Foundations. En: Human Factors and Ergonomics in Consumer Product Design, Methods and Techniques. Section1: Methods for Consumer Products Design 1. Boca Ratón: CRC Press Taylor & Francis Group, pp. 155-175.

Sáenz, L. M. & Valencia, A., 2012. Ergonomics and design in a ischial support proposal for the Medellin metro, Colombia.. Work: A Journal of Prevention, Assessment and Rehabilitation, Volumen 41, Suplement 1, pp. 1323-1329.

Valencia, A., 2007. La estructura: un elemento técnico para el diseño. Medellín: Universidad Pontificia Bolivariana.

CHAPTER 52

The Development of a Device Selection Model for Wireless Computing Devices in High Consequence Emergency Management

Pamela Bush[1], Susan Gaines[1], Arturo Watlington[1], Mohammed Jeelani[1], Lewellyn Curling[2], and Phillip Armbrister[2]

[1]University of Central Florida; Orlando, Florida
[2]College of The Bahamas; Nassau, Bahamas
Pamela.McCauleyBush@ucf.edu

ABSTRACT

The use of wireless technology is growing globally at an exponential rate. In regard to emergency management, wireless technology provides many advantages such as portability and resilience, which can play a pivotal role in improving information exchange. The official implementation and optimized use of wireless technology during emergencies is crucial in emergency management planning for the United States, as well as internationally. Developing island and Caribbean nations, such as The Bahamas, which face unique geographical, infrastructural, political, and cultural hurdles, should be especially focused on the communication needs during high consequence emergencies. This research focused on developing a human-centered methodology for the assessment of hand-held communication devices for use in emergency management. Interviews with Subject Matter Experts and surveys of both Emergency Management Officials and civilians were conducted. This model considered the usability factors associated with hand-held communication devices, and considered the weighted priority of each selection factor based on input provided by a team of subject matter experts through the use of AHP analysis. The most commonly used hand-held communication devices in The Bahamas were identified by the Bahamian National Emergency Management Agency (NEMA) and then tested against the model to prove its effectiveness. In addition, data was collected at the 2011 Florida Governors Hurricane Conference and a comparable model was developed. The most commonly used hand-held communication devices identified by the emergency management officials at the Governors' Conference were then tested against the model to prove its effectiveness.

Keywords: Usability, emergency management, mobile communication

1 INTRODUCTION

The use of wireless technology is growing at an exponential rate and has revolutionized communication in the modern day. According to the International Telecommunication Union (2009), 67% of the world's population, or over 4 billion people, are cellular phone users. Wireless technology is proving to be the most resilient forms of communication during emergency situations (U.S. Department of Justice, 2002). Unlike other communication devices, hand-held communication devices operate using multiple communication methods, utilizing both voice and data networks. Wireless communication can be employed to locate individuals who are in need of aid as well as to provide civilians with critical information during the emergency, such as where they need to go to receive aid, how to treat themselves for injuries, and what precautions to take until help arrives. Wireless technology can also be used to help civilians affected by emergency situations reunite with family members who are also in the affected areas, as well as communicate with family members abroad. With the use of wireless technology increasing in developing societies, hand-held communication devices are in a position to play a pivotal role in emergency management. Due to the lack of official implementation of these devices and the lack of the establishment of standard guidelines for device selection, the use of hand-held communication devices in emergency management is yet to be optimized. Island nations such as The Bahamas, which face unique challenges in regard to emergency management due to geographical, infrastructural, political, and cultural hurdles, can especially benefit from the optimized implementation of hand-held communication devices in emergency management. Since humans are the end users of these hand-held communication devices and considering that "the psychological, physiological and cognitive states of individuals are increasingly stressed, leading to the introduction of new, unfamiliar and possibly unidentified human factors related stressors" (McCauley Bell, et al. 2008) during high consequence emergencies, it is critical that the human factors issues associated with the use of these devices in such conditions are considered during device selection.

2   LITERATURE REVIEW

Zingale, Ahlstrom, and Kudrik (Zingale et al, 2005) prepared a technical report which provides human factors guidance for the use of handheld, portable, and wearable computing devices. According to this study, understanding the needs and goals of the user is critical when optimizing the selection and use of equipment for a specific job function. For a device to be used with minimal training, major features and functions must be easily accessible and visible. Devices that have good legibility and color contrast are generally easier to learn, more effective, and more readily acceptable to users. A set of criteria must be established in order to determine whether or not a device will be adequate for the user's task performance expectations. Other factors to be considered should include portability, appropriate human-computer interfaces, accommodation to environmental conditions, and durability.

The Department of Defense (1995) released a guide for human engineering design considerations and included guidance for handheld test equipment. According to the Department of Defense, handheld equipment should allow the user to attach the device to his or her clothing without interfering with its use or task performance. Handheld equipment should have a non-slip surface and should be shaped so that it does not slip out of the user's hand. Handheld equipment should also be small and lightweight. In addition, portable equipment should feature rounded corners and edges.

In the U.S. commercial market, less expensive cell phones are typically larger, heavier, and have fewer features than high cost cell phones, and cost under $100 (Windle, 2002). Moderately priced cell phones are smaller and lighter than low cost phones, feature extended-life batteries, and range in cost from $100-$300. High-end cell phones offer the latest features, the smallest designs, and cost over $300. There are currently three standard mobile phone batteries: nickel cadmium (NiCad), nickel metal hydride (NiMH), and lithium ion (Li-ion). NiCad is an older technology and has known problems such as being subject to memory effects, or damage due to charging repeatedly without being fully discharged. NiMH is a newer technology which does not suffer from memory effects like NiCad batteries and holds its charge longer. Li-ion is a long lasting and light battery type which does not suffer from memory effects and is the most expensive of the three standard battery types. Talk time and standby time should also be considered when selecting a mobile device. Talk time is the amount of time a battery can power a phone when it is being used to make or receive calls. Standby time is the amount of time a battery can power a phone when it is on but not being used.

According to Arif and Wolfgang (2009), the standard QWERTY keyboard is the fastest of all text entry layouts. The multi-tap phone keypad is considered the slowest text entry method. Amongst QWERTY type keyboards, the mini-QWERTY text entry method is the second fastest alternative. The size of keyboard layout does not have a noticeable impact on performance. Soft text entry is faster than text entry using a multi-tap phone keypad, but not as fast as text entry using QWERTY and mini-QWERTY keyboards.

In Nielson's model of usability, usability is a component of usefulness (Leventhal, 2008). If a system is not useful, then the usability of the system will not matter. Factors aside from usability, such as reliability, can impact whether or not a system is considered useful. Nielson's model specifies that five dimensions are important to usability: whether the device is easy to learn, efficient to use, easy to remember, induces few errors, and is subjectively pleasing. Nielson does not weight the dimensions in this model since the relative importance of each dimension is dependent on the project.

3 METHODOLOGY

3.1   Summary of Methodology

The objective of the methodology was to identify and quantify device selection factors which address the equipment needs that were determined to be the most relevant during emergency management situations. The device selection factors were chosen with consideration only to features which were publically available on end user hand-held communication devices at the time this study was conducted. Upon identification of the subject population, the methodology consisted of the following activities:

1. Development of survey instrument
2. Data collection
    a. Survey of commercial and scientific literature
    b. Bahamas – National Emergency Management Agency (NEMA)
    c. U.S. – Florida Governors Hurricane Conference
3. Data Analysis
    a. Summary statistics
    b. Determination of devices utilized
    c. Identification of device selection factors
4. Determination of weighted priorities of selection factors
5. Development of device selection model
6. Model evaluation

3.1.1   Development of survey instrument

The paper-based survey was developed to collect data in order to establish a baseline for current wireless communication use by Bahamian emergency management officials as well as to identify equipment deficiencies experienced by participants. The survey consisted of 28 multiple choice questions as well as 5 free response questions. The questions verified information about the emergency management officials and their personal communication devices.

On the survey, questions were grouped under the associated topics of:

background, handheld communication experience, device performance, usability, other communication devices, and suggestions. In the background section, personal questions were asked to acquire information regarding demographics, emergency management position, types of tasks performed by emergency management officials, years of employment, and formal training. Other questions were used to determine the type and service capabilities of the specified handheld communication device; the skill level of the user, frequency of use, and related tasks were also determined in this section. Questions geared toward device speed, reliability, battery life, and durability were used in the device performance section of the survey. In the usability section, civilians rated their devices on their ease of use, size, weight, and accuracy of text entry. Officials were also asked if they had problems with their devices slipping out of their hands.

The five usability factors that were included in the model for the Bahamian study were taken from Nielsen's model of usability. The usability section was comprised of questions about the visual clarity, audio clarity, lighting, and interference with device caused by outside factors. In the section of other communication devices, officials were asked to specify any other communication devices to which they have regular access; such devices included satellite phone, landline phone, radio, and personal computers. A series of phase analysis questions was also included in the survey which asked about the types of communication devices utilized by the emergency management officials during the various phases of emergency management; however, that data was not used in this analysis and will be used in a future study. The final section on the survey called for suggestions regarding design improvements for handheld devices and ways the devices can be used to improve their effectiveness during emergency management.

Initially, the plan was to use the same survey for the U.S. SME's. However, after the data collection with the Bahamian Emergency management officials, it was determined that adaptations needed to be made to the survey. Specifically, the SME's stated that there issue of portability was essentially universal for all of the mobile devices being evaluated. In other words, when it pertains to these devices since they are all extremely portable and the differences are only relative, thus there is no need to collect this information. Due to the importance expressed in the area of safety and utility in the disaster management environment, it was determined that these factors should be include in the usability analysis. Thus the survey was adapted to evaluate these factors from a usability perspective.

3.1.2 Data Collection

The first phase of this project utilized knowledge acquisition and data collection techniques in order to determine the communication needs in The Bahamas as related to emergency management, thus Bahamian Emergency Management Professionals with the National Emergency Management Agency (NEMA) were the subject matter experts. The knowledge acquisition tools which were used included a literature review, an interview analysis of Bahamian emergency management officials acting as subject matter experts, and a survey on Bahamian officials.

The participating 31 Bahamian emergency management officials answered demographic, skill level, and functionality questions related to the use of hand-held communication devices to support emergency related activities including those directed toward preparation, mitigation, and response. These emergency management officials provided invaluable input based upon their practical experience in high consequence emergency situations. A comprehensive list of communication needs was identified from these sources and each need was appropriately classified as infrastructural, organizational, or equipment needs.

Once initial equipment needs were identified, this phase of the project focused on developing a human-centered methodology for the assessment of hand-held communication devices for use in emergency management. This model considers both the usability factors associated with hand-held communication devices, and the weighted priority of each selection factor based on input provided by the subject matter experts. The proposed human-centered methodology for assessing hand-held communication devices for use in emergency management can be used by emergency management officials when assessing potential devices to be purchased for use in emergency management.

The next phase of the study involved an evaluation of mobile device perspectives of U.S. Emergency Management Professionals. In order to obtain a broad cross section of EM professionals, the research team attended the Florida Governor's Hurricane conference. A data collection environments was provided that allowed the team to collect survey data and explain the research to the Subject Matter Experts. This aspect of the study included 44 participants. After subject matter expert consultation, the survey used in the Bahamian data collection activity was slightly adapted.

The same methodology was utilized in each phase to analyze the data, including prioritizing device selection factors. The Analytical Hierarchy Process (AHP) was use to obtain relative weights. The results of the surveys from U.S. Subject Matter Experts were compared with those from the Bahamian experts. This comparison produced modest differences in features considered most critical by SMEs, most often accounted for by strategic differences in infrastructure and geographical issues associated with Island Nations as compared to the U.S. Mainland.

3.1.3 Data Analysis

The surveys were analyzed using statistical methods of categorical data analysis and correlations were identified. Several communication needs which were categorized as infrastructure, organizational, and equipment needs as well as a hierarchy of device selection factors in regard to the use of hand-held communication devices during emergency management situations were identified.

Device selection factors were selected which addressed the equipment needs that were determined to be the most relevant during emergency management situations according to the subject matter experts. The device selection factors were chosen with consideration only to features which were publically available on end-user hand-held communication devices at the time this study was conducted. As mobile technology advances and as more innovative features are incorporated in the designs of future devices, this list of device selection factors is expected to be altered in order to better reflect the latest technological advancements.

3.1.4 Device Selection Factors

The following 10 device selection factors, characterized as first level factors, were identified: durability, battery life and type, accommodation to environmental lighting, text entry method, grip, screen size, portability, audio clarity, usability, and unit cost. In addition to these, the following second level factors coming under the more complex of the first level factors identified: battery life and type was identified as being a product of battery type, standby time, and talk time; portability was considered to be dependent on device weight and volume; and finally, five secondary components of usability were identified based on Nielson's model of usability: ease of learning, efficiency of use, ease of remembering, frequency of errors, and subjective pleasure. Figure 1 depicts the hierarchy of the identifieddevice selection factors.

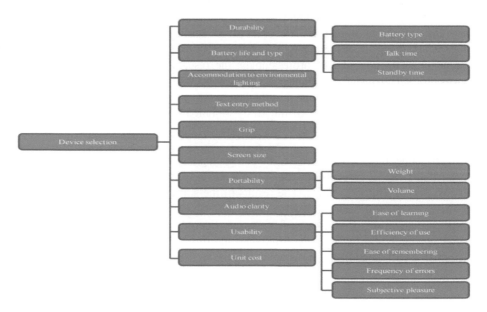

Figure 1: Hierarchy of device selection factors

3.1.5 Weighted Priorities of Device Selection Factors

The weighted priorities of the device selection factors for Bahamian and US emergency management profesionals which were calculated using AHP analysis. Bahamian and U.S. emergency management officials answered a series of pairwise comparison questions. Their answers were used to determine the relative, weighted priority of each factor with the use of the Expert Choice 11 software. Weighted priorities for second level selection factors including battery life and type considerations, portability factors, and usability considerations were also determined.

3.1.6 Rating Methodology for Device Selection Factors

The rating methodology in Table 6 is to be used in order to rate each of the device selection factors and second level factors on a scale of 1-3. A combination of physical measures (yellow/green), subjective opinions from potential users (blue), and operator-use measures (pink) are incorporated in this methodology.   A laboratory study was also conducted to further assess usability however the details of this study are beyond the scope of this article.

Table 1: Rating scales for selection factors

	3	2	1
Durability [a]	Designed for rugged use and is submersible in water	Designed for rugged use, but is not submersible in water	Designed for standard use only
Battery life and type			
Talk time [a]	Equal to or greater than 8h	Greater than 4h but less than 8h	Less than 4h
Standby time [a]	400h+	200h-400h	Less than 200h
Battery type [a]	Li-Ion	NiMH	NiCad
Accommodation to environmental lighting [b]	Extremely well	Well	Poorly

Text Entry [a]	Mini-Qwerty physical text entry with large buttons	Mini-Qwerty physical text entry with small buttons	Mini-Qwerty soft text entry or limited key physical text entry
Grip [b]	Highly Adequate	Adequate	Inadequate
Screen size [a]	320x480 or larger	176x220 to 320x480	176x220 or smaller
Portability			
Weight [a]	Under 3.0 oz	3.0-6.0 oz	Over 6.0 oz
Volume [a]	Under 5.0 in^3	5.0-6.0 in^3	Over 6.0 in^3
Audio clarity [b]	Extremely clear	Clear	Unclear
Usability			
Ease of learning [b]	Extremely easy	Easy	Difficult
Efficiency of use [c]	Under $(\bar{x}-.5s)$ task time	$(\bar{x}-.5s)$- $(\bar{x}+.5s)$ task time	Over $(\bar{x}+.5s)$ task time
Ease of remembering [b]	Extremely easy	Easy	Difficult
Frequency of errors [c]	Less than $(\bar{x}-.5s)$	$(\bar{x}-.5s)$- $(\bar{x}+.5s)$	More than $(\bar{x}+.5s)$
Subjective pleasure [b]	High	Medium	Low
Unit Cost [a]	Over $300	$100-$300	Under $100

LEGEND: Method used to obtain ratings for parameter
Derived from manufacturer's spec (a)
Subjective from Subject Matter Experts (b)
Derived from laboratory experiments (c)

The rating of the physical measures involves analyzing the manufacturer's specifications for each device. In order to rate the subjective measures (marked in blue on Table 6), a survey was employed which instructed potential operators to perform a series of tasks using each device, following which they were required to give their opinions in answers to a series of questions. In order to rate the operator-use measures (marked in pink on Table 6), measures in regard to efficiency of use and frequency of errors were taken on participants during the device testing session.

3.1.7 Development of the Device Selection Model

Rating scales, which included ratings ranging from 1-3, were developed for each device selection factor. It was determined that a combination of the physical analysis, subjective opinions from potential operators, and the operator-use measures was necessary in order to fully evaluate the appropriateness of hand-held communication devices for use during emergencies. A methodology was prescribed for determining a score for each factor, belonging to one of these three categories. The factor's score for a given device is an indication of how well a given device ranks in terms of that factor, was prescribed. A device selection model was then developed which consider the ratings and weighted priorities for all of the selection factors in order to provide an overall score for a given device, indicating its appropriateness during emergency situations.

A human-centered device selection model was proposed which allows for the calculation of a score for a given hand-held communication device which indicates how appropriate it is for use during emergency situations in The Bahamas. This model is truly human-centered, given that the identification of device selection factors evolved from user input, the model incorporates Nielson's model of usability, and subjective opinions and operator-use assures are integrated into the methodology. The parameters are show in Table 2 through 5 below.

Table 2: Overall Device Selection Model Factors

Overall Model Factors	Parameter	Relative significance parameter
Audio clarity	f1	a1
Usability	f2	a2
Portability	f3	a3
Accommodation to light	f4	a4
Battery life	f5	a5
Unit cost	f6	a6
Text entry	f7	a7
Grip	f8	a8
Screen size	f9	a9
Durability	f10	a10

Table 3: Usability Factors

Usability Sub-Model Factors		
Remembering	g1	b1
Learning	g2	b2
Ease of Use	g3	b3
Error Frequency	g4	b4
Pleasure	g5	b5

Table 4: Portability Factors

Portability Sub-Model Factors		
Weight	h1	c1
Volume	h2	c2

Table 5: Battery Factors

Battery Sub-Model Factors		
Talk time	i1	d1
Stand by	i2	d2
Battery type	i3	d3

A human-centered device selection model was developed which allows for the calculation of a score for a given hand-held communication device which indicates how appropriate it is for use during emergency situations in The Bahamas. The AHP results are being combined to create relative weights that represent the

Bahamian and US emergency management professionals. The model is human-centered in that the identification of device selection factors involved user input, the model incorporates Nielson's model of usability, and being that subjective opinions and operator-use measures are integrated into the methodology. A mathematical version of this model can be seen below:

$$Z = f1a1 + f2a2 + f3a3 + f4a4 + f5a5 + f6a6 + f7a7 + f8a8 + f9a9 + f10a10$$

The following equations can be used to determine the values of f1, f2, f7, and, f9, which are dependent on second level factors:

f1= g1b1+g2b2

f2= h1c1+h2c2+h3c3

f7= i1d1+i2d2

f9= j1e1+j2e2+j3e3+j4e4

where

1) Z = overall score
2) fi = rating for each device selection factor
3) a j = weighted priority for each factor
4) gi = rating for each durability factor
5) bk = weighted priority for battery life and type consideration
6) hi = rating for each durability factor
7) cl = weighted priority for battery life and type consideration
8) ii = rating for each durability factor
9) dm = weighted priority for battery life and type consideration
10) ji = rating for each durability factor
11) en = weighted priority for battery life and type consideration

The following equation is a comprehensive equation which integrates the five equations listed above:

$$Z = (g1b1+g2b2)a1 + (h1c1+h2c2+h3c3) a2 + f3a3 + f4a4 + f5a5 + f6a6 + (i1d1+i2d2)a7 + f8a8 + (j1e1+j2e2+j3e3+j4e4) a9 + f10a10$$

4 MODEL EVALUATION

In order to test the functionality of the proposed model, five of the most commonly used devices among Bahamian emergency management officials and civilians were tested against the model. The most commonly used devices were selected based on the survey results on Bahamian emergency management officials and civilians from the previous phase of this study. These devices included a BlackBerry Torch, a BlackBerry Curve, a Motorola Bravo, a Nokia 2330, and an iPhone 3GS. Five of the most commonly used devices, as reported by U.S. emergency management officials on the surveys were also selected and analyzed. These devices included iphone 3G and 4G, Blackberry Torch and Curve, and HTC EVO. Factor scores and overall scores were calculated for each of device in order to determine the appropriateness of each device in terms of use during emergency situations in for both The Bahamas and the U.S. The current model evaluation provides an overview of how the mathematical model is to be utilized and the ease of use. However, the complete analysis of the model will entail include ding a comparative assessment of the devices from subject matter experts. This aspect of the validation is ongoing and results will be reported in future studies.

5 CONCLUSION

A human-centered methodology for the assessment of hand-held communication devices for use in emergency management allows for the rating of hand-held communication devices in terms of appropriateness in emergency management situations, with the needs of users as a main priority. A hierarchy of device selection factors was developed based on results from the knowledge acquisition techniques.. AHP was used in order to determine the weighted priority of each device selection factor. This process incorporates physical analysis, subjective opinions from potential users, and operator-use measures in order to rate devices, ensuring the model is human-centered. By developing and testing the proposed methodology, it was proven that a mathematical model can be developed to holistically represent human factors' issues associated with the use of hand-held communication devices in emergency management was proven to be correct

6 FUTURE AREAS OF RESEARCH

The research methodology, specifically the AHP analysis can be performed by subject matter experts from other countries in order to develop tailored models applicable to specific geographical, infrastructure, and political considerations. In addition, the proposed methodology can be altered in order to be suitable in industries other than emergency management. Emerging device functionality such as video and photo capability will be incorporated into future the models.

ACKNOWLEDGEMENTS

The authors would like to acknowledge the National Emergency Management Agency (NEMA) of the Bahamas as well and the Florida Governor's Hurricane Conference 2011 for providing access to outstanding subject matter experts for this research. Appreciation is also extended to the National Science Foundation for funding this research project.

REFERENCES

Ahmad Arif, and W. Stuerzlinger, "Analysis of Text Entry Performance Metrics". (2009) *IEEE Toronto International Conference-Symposium on Biomedical Engineering.* pp 100-105

Department of Defense. (1995). *Human engineering design guidelines.* (MIL-HDBK 759C). Navy Publishing and Printing Office, Philadelphia, PA

International Telecommunication Union. (2009) *"The World in 2009: ITC Facts and Figures."* http://www.itu.int/ITU-D/ict/material/Telecom09_flyer.pdf.

Laura Leventhal, and J. Barnes (2008). *Usability Engineering. Upper Saddle River: Pearson*/Prentice Hall.

Pamela McCauley-Bell, S. Durrani, M. Jacobson, A. Hemphill, and S. Vaughn. (2008) Human factors and ergonomic issues in large scale disaster management" Proceedings of the Institute of Industrial Annual Conference and Industrial Engineering Research Conference, pp 1429-1432

Pamela McCauley-Bush, M. Jeelani, S. Gaines, L. Curling, P. Armbrister, A. Watlington, R. Major, L. Rolle, S. Cohen, "Assessment of communication needs for emergency management officials in high-consequence emergencies" Journal of Emergency Management, January/February 2012; pp 15-25

U.S. Department of Justice, "Guide for the Selection of Communication Equipment for First Responders" (NIJ Guide 104-00). *National Institute of Justice Publication*, Vol. I, February 2002.

Dale Windle (2002). *The Role of Wireless Technology in Disaster Recovery.* Issue brief. Binomial International. Retrieved from www.binomial.com/resources/**wireless**_in_dr.pdf.

Wireless Guide. (2007). *How To Choose Your Cell Phone. Web.* http://www.wirelessguide.org/phone/choosing-phones.htm

Carolina Zingale, V. Ahlstrom, and B. Kudrick, B. (2005). Human factors guidance for the use of handheld, portable, and wearable computing devices (DOT/FAA/CT-05/15). Atlantic City International Airport, NJ: Federal Aviation Administration William J. Hughes Technical Center.

CHAPTER 53

Assessment Tests for Mattresses and Pillows as Tools of the Ergonomic Conditions Diagnosis

Luz Mercedes Sáenz., Ana María Lotero,

Martha. Arias de L., Emilio Cadavid

Facultad de Diseño Industrial, Universidad Pontificia Bolivariana,
Medellín, Colombia
Empresas Públicas de Medellín.
Medellín, Colombia
luzmercedes.saenz@upb.edu.co , ana.lotero@upb.edu.co,
marta.arias@upb.edu.co, emilio.cadavid@epm.com.co

ABSTRACT

Comfort when sleeping is something that we all hope for and cherish; for that reason, comfort is valued by both users and manufacturers of mattresses and pillows. The aim of this paper is to present an account of a set of evaluation tests which sought to identify the ergonomic conditions of a mattress and pillow brand. The project was conducted by research professors from the Ergonomics Research Division – Design Studies Research Group (GED) at Universidad Pontificia Bolivariana (UPB) in Medellín.

The conceptual foundation of the project – in terms of Ergonomics and Design, is based on the systemic User – Product – Context relationship, which helps to identify the product's quality criteria. Thereafter it supports the formulation of the methodology and the design of evaluation tests that identify the ergonomic conditions of the product. As a result of the conceptual and methodological support – and taking into account the test findings – it was possible to identify mattress and pillow design requirements geared towards the optimization of ergonomic conditions.

Keywords: Ergonomics and Design, Evaluation Tests, Ergonomic Conditions, Mattresses and Pillows

1. CONCEPTUAL FRAMEWORK

The definition of ergonomic evaluation criteria was supported by the disciplines of Ergonomics and Design; these disciplines consider the characteristics, requirements and relationships which are established between the components that constitute the User – Product – Context system; and the conception and measurement of comfort on an objective level (quantitative measurement) and on a subjective level (a user's particular perceptions and tastes).

1.1. Ergonomics and Design

The ergonomics – design relationship helps to define the variables (characteristics and requirements) of each component of the ergonomic system (user- product - context). From each of these components, a series of indicators were established based on: physical and cognitive dimensions and definition of ergonomics per the International Ergonomics Association's(International Ergonomics Association, 2000); other criteria discernible in methods and techniques of design; a list of attributes, industrial products' values of use(Fornari, 1989); In addition, it helps define the basic features that determine a product's relationship of use and a user's perception of well-being and comfort. It helps explain how factors such as well-being and comfort interact with objective factors of ergonomics associated with product quality vis-à-vis usability, functionality, prevention and promotion of health and security. Each of these terms are briefly explained below:

Comfort: Vink, Overbeek & Desmet define comfort as an individual's personal experience with an environment or product, and influenced by multiple subjective (experiences and particular conditions), and objective (characteristics of the product or environment) factors. Comfort is evaluated by the user; therefore it is the user's perception that should be assessed, referring to amongst other things, sensory, thermal, acoustic, visual and physiological features. (Vink, 2001). It means that, as Crowely explains, comfort can vary from person to person thus making the identification of criteria for the standardization of consumer products problematic (Crowley, 2001). Nevertheless, it is possible to identify key factors that should be considered during the design process of products that generate comfort. These factors can include the relationship between ergonomics and comfort, and the relationship between functionality, usability and comfort.

Functionality, Usability: Denis Cohelo argues that for a product to generate comfort, it should provide functionality. This can be expressed, amongst other things, in the physical and mechanical characteristics that the product requires to fulfil its function. These characteristics can include size and ability to adapt to the posture and demands of the user. Cohelo also points to conditions of usability as determinants of comfort: one of the conditions that results in product comfort is ease of use and/or the product's ability to be used as and when it is required (Cohelo, 2009).

Functionality is supported by the recognition of formal requirements, the relationship that is established between the different parts to form a system and the adaptability of the psycho-physical features and requirements of the user. These factors determine the use of the product and conditions of utility which allow the product to be used in situations of well-being and efficiency in accordance with the user – product – context system. (Pontificia Bolivariana University, 2005)

1.2.   Conceptualization of Variables and Indicators

The features (variables and indicators) that were considered in the user evaluation tests, in terms of ergonomics and design, were determined by the conceptualization of the components in the User - Product - Context system. As a basis for the analysis of ergonomic conditions geared towards the design and the evaluation of the mattresses and pillows, the basic design dimensions of the products are used as a reference and supported in the disciplinary model of the Industrial Design Faculty at the UPB: functional, aesthetic-communicative and morph-productive. (Universidad Pontificia Bolivariana, 2005).

Characteristics and details related to each of the indicators were established, important features and opportunities were analyzed, and product design components were considered, based on the Disciplinary Model and the Educational Project – part of the Curriculum at the Faculty of Industrial Design at the UPB. These included:

"The functional-operative features i.e. technical features relating to shape and material, the product' utility (what it is used for) and the Human-Object relationship i.e. the criteria that facilitate the product's adaptability to the users, taking into account their characteristics, capacities and limitations.

The aesthetic-communicative features which include, among others, symbolic features: what the product expresses according to its appearance and aesthetics, and the signification it provokes in the different users.

Technical-productive features i.e. those related to material, technological and productive dimensions of the product, and the nature of the manufacturing process" (Saenz, et al., 2012).

This stage comprised the moment for evaluation, analysis and contact between human and object in the designated context, and to determine the way that the relationship of use is established.

2. CONTEXTUAL FRAMEWORK

An approach was made to the Company with three objectives: knowing the brand; identifying those product characteristics that could be ergonomically evaluated; assimilating information about the context in which design opportunities can be defined which utilize the Company's materials and processes.

The Company that requested participation in the project is a Colombian company which uses cutting edge German technology to manufacture flexible polyurethane foam and other products designed for rest and relaxation - including mattresses and pillows. During a visit to the factory installations, engineers and design and production staff were consulted about manufacturing processes and materials. It was then possible to identify the characteristics of three types of mattresses that would be evaluated in the study. (See Table 1)

Table 1.Characterization of Mattress Types

Mattress	Composition	d (v/cm3)	h (cm)	Covering Material	Firmness	Guarantee (years)
R1	Ultra flexible foam Sensaflex quality	34	18	American Jackard or cotton. Nylonsewing.	Semi Firm	7
R2	Penta soft extra flexible foam	26	30	Belguim Jackard. Nylon sewing	Firm	10
	Firm flexible foam block	40				
R3	Penta soft extra flexible foam	26	19	American Jackard or cotton. Nylon sewing	Extra-firm	10
	Extra flexible foam	26				
	Cassata	120				
	Extra espuma flexible	26				

The Company's website provided additional information about the Company and its products. Attributes that the Company emphasizes in advertising and promoting its mattresses and pillows were documented and the clarity of the information that the Company provides to its potential users – vis-à-vis the terms that have been referred to in this project – were analysed.

3. METHODOLOGY

It is important to describe the methodology used in the design of the ergonomic evaluation tests of the mattresses and pillows. It should be pointed out that the evaluation tests are understood as an integral concept that includes methodological design as well as the tools designed and utilized for the measurement. In order to achieve a precise evaluation, the measurement tools should be supported by the design of accurate experimental conditions, the formulation of hypotheses, the identification and operationalization of the variables and indicators, an accurate processing of information and finally the formulation of an analysis plan through the application of relevant statistical tests.

3.1. Methodological Design

Study Approach. Three approaches were applied in the evaluation of the mattresses and pillows: descriptive, comparative and relational.

Descriptive level, designed to obtain a subjective and objective evaluation of the products, as indicated below.

Comparative level, with the objective of making a comparison between the comfort levels obtained through the subjective evaluation based on the sensory perception of the subjects, and the levels of comfort obtained through the objective evaluation using a technological tool.

Relational level, designed to contrast the relationship hypothesis between the user, the product and the context.

Techniques and Tools for Collecting Information. The evaluation of the mattresses focussed on the measurement of comfort. Two types of tests were applied to each subject.

Subjective Test. A tool was designed and applied through a survey which gauged the user's subjective perceptions of comfort when in contact with the mattress. Evaluation related to the user's perception of comfort, and the firmness of the mattress and was analysed in accordance to each of the variables and indicators in Table 2.

Objective Test. An electronic rug of Canadian origin – known as the X-Sensor – was acquired by the Company and used as a tool for the objective evaluation.

Variables and Indicators- The variables and indicators used in the evaluation of the mattresses and pillows were identified through the conceptualization presented in the theoretical framework. As explained above, this led to the conclusion that the evaluation of the product should incorporate an evaluation of the user, product, context system. (See Table 2)

Scales of Subjective Comfort. The subjective perception of comfort was measured on a scale of quartiles. Using a straight line divided into quarters, each subject was asked to indicate their level of comfort on a scale of 0 to 10. This translated into a percentage scale of four levels of comfort: $0 - 25, 25 - 50, 50 - 75, 75 - 100$.

Scales of Objective Comfort. The Rug measures comfort levels of a mattress on a scale of 0 to 1. Measurement is assessed in terms of pressure points and support area when the user makes contact with the Rug. Visual information is also provided through means of a colour coded diagram that highlights different levels of pressure intensity that the user exerts on the surface of the mattress.

Table 2. Variables that were analysed with their criteria and indicators

Variable	Criterion or Dimension	Indicators
USER	Population Information	Gender, Age, Weight, Height, IMC, Social stratum, Educational Level, Profession, Extra-curricular activities
	Physical Features of the User	Sleep patterns, Sleeping position, Sleep disorders; Sleep interruptions – frequency/reasons; Sensations when waking up
MATTRESS PRODUCT	Mattress Uses and Preferences	Length of time mattress has been in use, Mattress size, Mattress material, Motives for renewing the mattress, Features that are deemed important when choosing a mattress (ideal mattress),
	The Brand	Brand awareness, Recognized brand products, Other recognized mattress brands
CONTEXT	Sleeping locations and customs	Sleep partners, Other activities conducted in bed, Garments worn, and objects used when sleeping, Siesta: frequency, place, Surface of mattress used, Mattress care: Mattress protectors, bed linen, frequency of change and turning over of mattress
PILLOW PRODUCT	Customs, uses, preferences	Length of time pillow has been in use, Pillow material, Number of pillows used when sleeping, Other uses of the pillow, Frequency of pillow-case change, Features that are deemed important when choosing a pillow – temperature, firmness, appearance, price, antiallergenic, Perception of pillow firmness and temperature, Ideal pillow features

Population and Sample Test. For the practical stage of the research, the User was defined as someone who buys and/or uses those Company products (mattresses and pillows) relevant to the Project. Potential clients and the Company's sales staff were also included. Evaluation tests were conducted using 149 users of different ages and social strata to identify their relationship with the formal and material qualities of the product.

Objectives and Hypothesis. The evaluation tests that examined the ergonomic characteristics of the mattresses began with the following hypothesis:

H: There are no significant differences between comfort levels obtained by the subjective evaluation that measured the sensory perception of the subjects, and comfort levels obtained by the objective evaluation that used the Rug.

This hypothesis arises from the conclusions drawn from the conceptual framework and taking into account the relationships in the user, product, context system. It is also important to understand that in addition to an individual's personal experience with an environment or product, subjective comfort is influenced by numerous subjective and objective factors.

Hypotheses concerning the relationships shown below in Table 3 were also contrasted.

Table 3. Complementary Hypotheses. User - Product - Context Relationships

Hypothesis	Relationships of Interest
Criteria for mattress use differs to the criteria for pillow use	Mattress time – Pillow time Mattress material – Pillow material
The relationship of mattress comfort and pillow comfort is related to material	Mattress material – feeling when getting up Pillow temperature – pillow material Sensation when getting up – mattress material
The relationship of the user with the mattress is related to the user's social stratum	Mattress time – social stratum
Habits associated with the place of purchase are related to social stratum	Place of purchase – social stratum
The relationship of the user with the brand is related to age, gender and social stratum	Brand awareness and its relationship with: age, gender and social stratum Knowledge of the brand's products and the relationship with age and social stratum
Usage habits and mattress care (of the bed) are related to age	Activities conducted in bed – age Use of a mattress protector – age Objects used when sleeping – age
Sleep comfort is related to cleanliness of bed clothing	Sleep disorders - frequency that bed clothing is changed
Sleep comfort is related to adopted position while sleeping	Comfort Evaluation - adopted position while sleeping

3.2. Procedure

Each subject carried out both subjective and objective tests using each of the three Company mattresses on a random basis. The tests were conducted under the following experimental conditions:

Condition 1: A mattress supplied by the Company. Three types were used: firm, semi-firm and extra-firm.

Control condition: A board that acted as a control surface (this had the appearance of a mattress, and was covered by the same material used to cover the other mattresses).

A selection of mattresses was used at random to guarantee the arbitrariness of the test. In other words, each subject had the same probability of using any one of the mattress types. In order to guarantee identical conditions of temperature, humidity and cleanliness for the subjective evaluation of the mattresses, a spacious and well-ventilated room was utilized in all experimental conditions. Each of the subjects was asked to lie for some minutes on each of the three surfaces. They were then asked to grade each surface using a scale of 1 to 10 according to comfort levels and the firmness of the mattresses; results were then classified in quartiles. In addition, each subject carried out an objective test using the Rug and Company mattress.

Processing of Information. After the information from the test and the subjective and objective evaluations had been collected, data was processed in accordance with the variables and indicators identified in Table 2:

First stage: information obtained from the subjects' responses to the survey and the results from the objective and subjective comfort evaluations were entered on to an Excel spreadsheet.

Second stage: additional Excel applications were used to produce frequency distribution tables that classified the 149 respondents in to categories according to their responses.

Third stage: double-entry frequency tables were produced, in line with the relationships of interest between variables, in order to establish the statistical validity of the hypotheses (see Table 3).

Fourth stage: comparative tables of comfort evaluation of the mattresses were produced, taking into consideration both the subjective evaluation in terms of perception and the objective evaluation with the Rug.

Analysis Plan. In order to analyse the information, descriptive, comparative and relational statistical tests were used that led to the following objectives being achieved: Characterization of the population; Development of a diagnosis that looked at the way in which the mattresses and pillows that were analysed fulfilled ergonomic conditions; Contrasting of the hypotheses; The definition of design requirements, understood as characteristics that the Company's products should possess, and taking into account their desire to include ergonomic conditions as a selling point that could strengthen their market position.

4. RESULTS. Statistical Analysis of the Information

Statistical analysis considers three types of analysis consistent with the processing stated above: descriptive, comparative and relational. The process and results are presented below.

4.1.   Descriptive Statistical Analysis

Excel applications were used to produce simple frequency distribution tables – relative and absolute. In other words, the number and percentage of the 149 respondents classified into categories according to each of the questions, and taking into account the relationship of the user with the mattresses and pillows.

The response trend for the qualitative variables were emphasized using the Mode as a central trend measurement, and the Range as a measurement of dispersion. The response trend for the quantitative variables were defined using the Average and Mode as central trend measurements, and the Range as measurement of dispersion.

A *confidence level of α= 0.10* was utilized for the estimation of population parameters, with a 90% certainty in the conclusions that are presented and a margin of error d = 0.63.

These results were classified: In accordance with the categories listed in Table 2: User, Product, Context, and with Comfort Evaluation (The average subjective and objective evaluation of comfort was established).

4.2.   Comparative Analysis

A comparative analysis between the subjective evaluation and the objective evaluation of the mattresses was conducted – using the *Student's t-test*, with a level of *α= 0.10* – on each of the Company's three mattress types. The difference between the objective evaluation and the subjective evaluation is significant, only for R1 mattress, with a certainty of 99%; the subjective evaluation is significantly higher.

4.3.   Statistical Relational Analysis

In order to contrast the relational or independence analysis with the variables, presented in Table 3, the *chi-squared test* was employed through the Epi-info application. The hypotheses is analysed with a *confidence level of α= 0.10*, with a 90% certainty in the conclusions. Results let conclude the relationships that were deemed significant with 90% certainty between *Comfort subjective evaluation* and the following variables: weight; sleep patterns; sleep disorders -wake up during their hours of sleep-; hours of sleep per day -It is more common to wake up feeling tired or unenergetic after a lot of sleep (9 hours or more) or very little sleep (4 hours), and wake up feeling rested and energetic after 7 to 8 hours of sleep adopted position while sleeping; the degree of firmness of the mattress and pillows which were deemed preferred.

5. CONCLUSIONS AND DISCUSSION OF THE RESULTS

With regards to the hypotheses in the study, only the R1 semi firm mattress showed significant differences between the subjective and objective evaluations of comfort. This relates to the results of the survey that asked respondents to point to those characteristics of a mattress that they deemed most important. A significant group stressed the level of firmness (56%). Of this group, the highest percentage (69%) prefers soft, some prefer firm (15%) and a small number (3%) prefer very firm.

Measurements of comfort of the board, as a control surface, were made with the Rug and subjective perception. In both cases, comfort levels of the board were significantly lower than those of the mattress types in the study.

Because it is a subjective concept, some authors attest that it is not possible to identify the exact causes of comfort and discomfort. This research has concluded that through ergonomics, it is possible to identify key factors that can be considered in the design of products that generate comfort.

The evaluation tests constituted a tool which was used to assess the way in which the Brand incorporates ergonomic conditions into its products i.e. the characteristics and requirements of the users and the way in which a product's features are determined. It includes criteria for the context of use for the mattresses and pillows, and seeks to optimize the User – Product – Context system for well-being, health and security.

The evaluation tests were designed in response to the specific requirements of the Company. However, considered in the broader sense of methodological design, these tests constitute a methodological support for the ergonomic evaluation of mattresses and pillows in general, and other design products in particular, providing that the necessary adaptations to variables and indicators that guide the design process are made (Table 2). This helps define the design of the tools, in accordance with the interests of each particular company in mind.

Based on the conceptual and methodological foundation of the Project, and supported by the concepts of ergonomics and design, indicators for the evaluation of a product's (in this case mattresses and pillows) use situation are derived.

REFERENCES

Cohelo, D., 2009. Confort and Pleasure. En: *Pleasure with Products : Beyond the Usability* . London: Taylor and Francis.

Crowley, J., 2001. The Invention of Comfort: Sensibilities and Design in the Early Britain and in the Early America. . En: Baltimore, London: The Johns Hopkins University Press, p. 361.

Fornari, T., 1989. Las Funciones de la Forma. En: T. Editores, ed. México: Universidad Autónoma Metropolitana Azcapotzalco, pp. 41-89.

International Ergonomics Association, 2000. Definición de Ergonomía. En: *Consejo Mundial de Ergonomía.* s.l.:s.n.

Sáenz, L. M., Lotero, A. M., Cadavid, E. & Arias, M., 2012. Analysis os ergonomics conditions of a brand of mattress and pillows. University-Industry project. En: M. M. S. a. K. Jacobs, ed. *IEA 2012: 18th World congress on Ergonomics - Designing a sustainable future.* Medellín: IOS Press, pp. 1281-1287.

Universidad Pontificia Bolivariana, 2005. Proyecto Educativo del Programa Diseño Industrial. En: Medellín: Universidad Pontificia Bolivariana.

Vink, P. e. a., 2001. Comfort Experience. En: The Invention of Comfort: Sensibilities and Design in the Early Britain and in the Early America. Baltimore, London: The Johns Hopkins University Press, p. 361.

CHAPTER 54

Usage Problems and Improvement Needs of the Clinical Upper Extremity Rehabilitation Devices in Taiwan: A Pilot Study

Lan-Ling, Huang [1]; Chang-Franw, Lee [2] and Mei-Hsiang, Chen [3]

[1,2] National Yunlin University of Science and Technology
[1,2] Yunlin, Taiwan
[1] g9630806@yuntech.edu.tw
[2] leecf@yuntech.edu.tw

[3] Chung Shan Medical University/Chung Shan Medical University Hospital
[3] Taichung, Taiwan
[3] cmh@csmu.edu.tw

ABSTRACT

There are many existing upper extremity rehabilitation devices (UERD) for various purposes, not all of them meet the practical needs of rehabilitation therapy. This study aims to survey the UERD used in Taiwan hospitals. The types, usage problems and improvement needs of the existing devices are categorized and thus possible design improvements proposed. Semi-structured interviews and field observations were conducted surveys to collect data. Field interviews with 10 senior professional occupational therapists in 10 medical institutions, and to observe the usage status of patients. The results can be summarized as follows: 1) 10 UERD are the most commonly used in hospitals in Taiwan. They are used for movements in the shoulder and elbow positions. 2) The main usage problems of the 10 existing UERD are the Base unstable, the operable components are easily damaged, and the components are not adjustable to fit user's operating height; the main improvement needs are all functions adjustable, and improved device durability. This study combines the clinical use of upper extremity rehabilitation device and professional status of occupational therapists with the views of patients, improve the design put forward recommendations to provide to the therapist, rehabilitation device designers, manufacturers and other experts during the rehabilitation development of device design or re-design reference.

Keywords: occupational therapy, upper extremity rehabilitation device, usage problem, design improvement needs, user satisfaction

1   INTRODUCTION

Stroke is one type of the cerebrovascular disease threatening people in modern societies. The most common and widely recognized impairment caused by stroke is motor impairment. Occupational therapy is one way to help stroke patients to restore their upper extremity motor function. Up to 85% of stroke patients experience hemiparesis immediately after stroke, and between 55% and 75% of survivors continue to experience motor deficits associated with diminished quality of life (Saposnik et al., 2011). Many daily living tasks are performed by the upper extremity; therefore, rehabilitation treatment of upper extremity is very important for stroke patients. Rehabilitation device is an essential tool in the process of rehabilitation therapy. With it, the therapist plans a series of activities for the treatment, and the patient performs the assigned activities accordingly. Advantages and disadvantages of its use may affect the work quality of the therapist and the effectiveness of treatment for the patient. Therefore, rehabilitation devices must be designed with users in mind. Such products can increase acceptance, and improve quality of life (Jacobs, 2008).

Many studies in the literature have focused on issues of the need and use of assitive devices for daily life at home after stroke (Moreland et al., 2009; Lui and Mackenzie, 1999; Sonn and Grimby, 1994). However, thus far, a review of the literature on assessing user requirements found that little published work exists on the medical device development (Martin et al., 2008). The following are two studies related to the ergonomics on assistive device. One study is a new design approach called user-centered design on a personal assistive bathing device for hemiplegia (Ma, et al., 2006). The researchers proposed three assistive devices for bathing. The results by user-based assessment showed the devices can help individuals clean the body parts, such as backs and armpits that they were previously unable to clean independently. The second study is a new user-centered design approach, proposed a hair washing assistive device design for users with shoulder mobility restriction (Wu et al., 2009).

In recent years, the fields of ergonomics, occupational therapy and physical therapy have grown increasingly intertwined as professional knowledge and skills are blended to advance applications that optimize human well-being and performance. Ergonomics is not solely confined to the workplace, products and environments should match the abilities, needs, and perceptions of the user (Jacobs, 2008). High quality, well-designed medical devices are necessary to provide safe and effective clinical care for patients as well as to ensure the health and safety of professional and lay device users. In order to design good medical equipment, capturing the requirements of users and incorporating these into design is an essential component of it (Martin et al., 2008).

Furthermore, in the process of rehabilitation device development, the product designers should fully aware of the user's needs, where users include not only patients and therapists, but also peripheral users such as caregivers, hospital stuff and technicians. In other words, the user's needs encompass more than just clinical effectiveness. The peripheral users, in this context, functions in the provision of clinical status and knowledge to the designer. Designer's responsibility, Based on considerations of clinical status and user needs, the designer's responsibility is then to design the rehabilitation equipment to meet the needs of all those concerned. Therefore, the usage assessment of the UERD by therapists is very important for the improvement and redesign of existing UERD for better function as well as performance.

The current clinical use of upper extremity rehabilitation devices (UERD) vary in types and use patterns, and several problems concerning the use of UERD and needs for improvements have also been identified (Huang et al, 2009). This points a need to further improve existing UERD to meet the practical needs of the therapist's in treatment and to enhance the effectiveness of rehabilitation patients as well. In order to effectively improve the existing device, we must be collected usability data from the users of existing products. Features and functions that have been well received may be retained while the redesign effort focuses on solutions to problems that users have identified as being the most serious (Cushman & Rosenberg, 1991). Therefore, a comprehensive understanding of the current situation of the devices in use is necessary. The objective of this study is to survey the types, usage problems and improvement needs of the existing UERD by occupational therapists at hospitals in Taiwan. The usage problems and needs of the existing UERD are categorized and thus possible design improvements proposed.

2   METHODS

Subject

Purposive sampling method was used to select 10 hospitals in Taiwan with rehabilitation department of facilities based on their location, size, and type. From each hospital, a therapist with practical experience of at least 5 years was contacted and briefed about the purpose of this research.

Measures

Semi-structured interviews and field observations were conducted surveys to collect data. In semi-structured expert interviews, therapists were asked about the clinical use of existing UERD: the types, frequency of use, usage problems encountered, and design improvement needs. In field observations, patients were observed while using these UERD without interference.

Expert interviews with occupational therapists include two main parts: 1) basic information, including the therapist's gender, age, the hospital, and years of work experience; and 2) the current use of UERD, including the type of UERD, the frequency of use, use pattern, problems encountered in use, and need of device improvement. Photographs of the devices currently used were shown in a list followed with a series of question as mentioned above. The use pattern here focuses on how the same device would be assigned to patients at different stage of recovery for different patterns of operation in order to fully take advantage of the design features of the device for different needs of the patients.

Procedure

A convenient time for interview was then arranged, and the researcher conducted the expert interview and field observation accordingly. During each expert interview, the therapist was asked to firstly see a list of photos of UERD, and check out those he/she had ever used in treatment, and then answer specific questions for each device. The therapist was asked to demonstrate the basic use pattern of each checked device, and how the patients in treatment would be asked to operate this device, and explain any problems encountered during use and improvement suggestions. If any device in used were not found in the gives list, the researcher would take photograph of it and ask the therapist about the same questions as those for the items in the list.

During field observation, the researchers were guided by the therapist to observe patients using the devices. In addition to questioning the therapist about how the devices in use are related to the symptoms of the patients, the researchers also observed the process of devices being operated, independently by the patients or with help of the therapist. The researchers spent 15-20 minutes to observe a patient using the devices. The researchers chose to stand either next to or behind the patient in order not to interfere or cause any adverse effect on his operation of the device. Each field trip of observation took about 90-120 minutes.

From the point of view of ergonomic design, the researcher would also take note of the usability of the devices, with special emphasis on the following points:
1) Is the UERD easy to operate?
2) Is it stable and firm during the operation?
3) Can it be operated independently by the patient?
4) Does the operation cause any unnatural movement or awkward posture of the body?
5) Does it require unduly exertion from the user?

The general features of the product design were also observed, such as:
1) Is it easy to break down? Is it stable and firm during the operation?
2) Are the sizes of the parts appropriate for the hand?
3) Any inconveniences due to manufacture inadequacy?
4) Any other design deficiencies.

Based on the data gathered from the field trips of observation, together with notes and photographs taken during the observation, the research findings were thus synthesized.

Data analysis

The collected most typical UERD were listed, without other products with similar types of operation but only different in shapes and sizes. To analyze the interview data, the recording was firstly transcribed to verbatim. Then, according the opinions of each device is thus summarized separately for further comparison. Similar opinions are combined and described in a more general way, so that each statement may contain several occurrence similar opinions and observed situations.

3   RESULTS AND DISCUSSION

Result of the expert interview and field observation can be summarized in 5 main points: 1) Data of surveyed subjects, 2) types of UERD in clinical use, 3) frequency of use, 4) use pattern, 5) usage problems and improvement needs, as described in the following:

1) Data of surveyed subjects

There were 10 therapists participated the in-depth interview, 5 male and 5 female, with average age 35.9 years (SD = 4.6 years), and average work experience of 12.9 years (SD = 4.8 years).

2) Types of UERD in clinical use

The most commonly used UERD in hospitals in Taiwan are shown in Table 1, where only the most typical products were listed, without other products with similar types of operation but only different in shapes and sizes. The clinical use of UERD in hospital, according to treatment objectives, can be divided into two major categories: proximal type and distal type. A proximal type UERD is defined to be one that is used for movements in the shoulder and elbow positions. The main function is to train a wide movement range for the upper extremity to restore its muscle strength and movement function. Most stroke patients in the beginning stage (with upper extremity movement recoverage in the Brunnstorm stages 2-4) need to use this type of device for rehabilitation. Examples of this type of devices are D1-D14 (Table 1).

On the other hand, a distal type UERD is one that is used for movements in the wrist and finger positions. Its main function is to train the dexterity of the hand and fingers or to restore its movement functions in daily life applications, such as turning a key to open the door. This type of devices is suitable for stroke patients in final stage (with upper extremity movement recoverage in the Brunnstorm stages 4-6). Examples of this type of devices are D15-D21 (Table 1).

Among the 21 rehabilitation devices listed in Table 1, there are 17 products found in all 10 surveyed hospitals; of which 10 (D1-D10) are for proximal rehabilitation and seven for distal rehabilitation (D15-D21). Each of the rest items (D11, D12, D13, D14) was only used in one hospital respectively.

3) Most frequently used UERD

Responses from the interviewed therapists concerning the frequently used UERD in their hospitals had a very consistent pattern. There were three devices (D1, D2, D3) indicated by all 10 therapists, and three other items (D6, D7, D9) indicated by 8 out of the 10 therapists. Interestingly, all these highly frequently used UERD belong to the proximal type, i.e., devices for rehabilitation of the movements in the shoulder or elbow positions. Furthermore, it is noteworthy that D9, a newly developed device in recent year, has such a high frequency of use and so well accepted by both the therapists and the patients. Based on the principle of recovery of upper limb function after stroke, patients tend to recover the shoulder and elbow motor functions first, so that the arm can have a greater range of motion and have enough power to assist the distal actions. The proximal type rehabilitation devices, therefore, are those that almost every stroke patient has to use. Hence, it is reasonable that these devices have a much higher frequency of use in hospitals.

The result shows that many types of UERD are used, of which the proximal devices are more preferable and more frequently used in the treatment process. Therefore, this study deals with the rehabilitation device for upper extremity, which include shoulders and elbows, the discussions of the use pattern, usage problem, and design improvement need and also the items for survey in the second phase would all be focused on devices belonging to the proximal type.

4) Use pattern of UERD

This survey item aims to see how therapists apply a variety of rehabilitation devices to meet the different needs of patients in treatment. The survey result of the usage problems and improvement needs indicates that the existing UERD facilities need to be redesigned, which is an important task for both industrial designers and ergonomic designers. The clinical use patterns of the UERD have to be considered in order to redesign the devices to better meet the need of their users. Discussions of the survey result of use pattern, usage problems, and improvement needs will be presented in the following.

Table 1 The most widely used UERD in clinics in Taiwan hospitals

Proximal type UERD

| D1. Exercise arm skate | D2. Exercise hand skate | D3. Vertical tower | D4. Horizontal tower |

| D5. Climbing board and bar | D6. Incline board | D7. Stacking cones | D8. Single curved shoulder |

| D9. Curamotion exerciser | D10. Upper bike | D11*. Oblique incline board | D12*. Shuttle mini press |

| D13*. Bilateral rotary wrist machine | D14*. Rotary wrist machine |

Distal type UERD

| D15. Graded pegboard | D16. Purdue pegboard test | D17. Finger extension remedial game | D18. Finger extension remedial game |

| D19. Clothes lacing activity | D20. Two-tiered horizontal bolt board | D21. Graded pinch exerciser |

*The devices marked were only used in one hospital respectively.

As to the use patterns of the devices, there are three main patterns: a) the patient use the affected upper extremity to operate the devices with assistance of the therapist, b) the patient uses the affected upper extremity to operate the device with help of the unaffected upper extremity, and c) the patient autonomously operates the device. Due to lack of adjustability of the existing devices to meet these use patterns, inevitably the therapist has to seek aid with additional attachments or auxiliary fixtures in order to make the device operatable by the patients. A few examples of these are:

a) Sandbags may be tied to the devices for increasing the weight and making the movement more stable. Some therapists suggested that although this can increase the stability of operation, the attached device may easily slip away or drop down during the operation.

b) Suspension frame is used to support the weight of the affected elbow, when its supporting force is insufficient.

c) Bandage is used to assist to grip the operation component when the finger grip strength is insufficient;

d) Splints are used to assist the affected side extension and control the movement of the activity range; and

e) C-type fixture is used to strengthen the base of the fixed device, making it more stable. Therapists also responded that excessive force from the C-type fixture often leads to damage to the table surface, and sometimes the fixture may even interfere with the patient's feet and cause uncomfortable feeling.

5) Usage problem, and improvement needs of UERD

As to the usage of the devices, the majority of therapists consider that the main problems with the existing UERD are: the lack of a firm standing base for each device, the operable components are easily damaged, and the components are not adjustable to fit user's operating height. Therapists have also suggested design improvement needs for existing UERD. In general, the proposed design improvements include: to make the device base more stable, all functions adjustable, and improved device durability, etc. Furthermore, the need of UERD for treatment need includes: a) to record the patient movement data in each treatment so as to provide reference for the next treatment for the physician, b) being able to change or adjust operable components and operation types (such as shape, size, resistance, etc) so as to meet different treatment needs.

When asked about their opinions of what is the patients' general psychological reaction to the use of these devices during treatment, many therapists responded that the patients seemed easily to get bored, feel lonely, lack of interest, and loss of motivation due to the monotonous repetition of movement without much interaction or feedback from the devices. This provides a piece of insightful information for improvement need, that is, a better device should provide some sort of interaction or feedback to its users, such as that much used in video gaming devices.

Based on the results obtained, it is clear that the existing UERD still have many problems concerning their structure, usage, function, and psychological aspects, which all need to be dealt with in their further improvement designs.

3   CONCLUSIONS

This study conclusions 5 design improvement features for existing UERD are following: 1) to make the device base more stable, 2) all functions adjustable, 3) improved device durability, 4) to record the patient movement data in each treatment so as to provide reference for the next treatment for the physician, 5) being able to change or adjust operable components and operation types (such as shape, size, resistance, etc). These improvement features improve the design put forward recommendations to provide to the therapist, rehabilitation device designers, manufacturers and other experts during the rehabilitation development of equipment design or re-design reference, make the device more in line with the rehabilitation goals and needs treatment, rehabilitation and better quality.

In addition, the results also offer some degree of demand related to treatment in patients with the psychological needs of the clinical effectiveness of rehabilitation therapy equipment, rehabilitation of the peripheral device design. Most therapists are still not based on clear clinical upper limb rehabilitation of existing equipment, the effectiveness of treatment, little research is also discussed, therefore, this study will focus on the future of clinical upper limb rehabilitation equipment to explore the efficacy of treatment. A rehabilitation device is used will affect the results after the rehabilitation treatment equipment design, and is also necessary to consider the design process elements.

ACKNOWLEDGMENTS

This study is supported by the National Science Council of the Republic of China with grant No: NSC99-2221-E-040-009.

REFERENCES

Cushman, W. H., and D. J. Rosenberg. 1991. *Human factors in product design*. New York.: Elsevier Science Publishers B. V.

Huang, L. L. and D. Cai. 2009. Rehabilitation apparatus design and development using ergonomic and physical principles. *Proceedings of Design Rigor & Relevance-IASDR2009*: 9 pp.

Jacobs, K. 2008. *Ergonomics for therapists (third edition)*. USA.: Mosby Elsevier.

Lui, M. H. L. and A. E. Mackenzie. 1999. Chinese elderly patients' perceptions of their rehabilitation needs following a stroke. *Journal of Advanced Nursing* 30 (2): 391-400.

Ma, M.Y., F.G. Wu. and R. H. Chang. 2006. A new design approach of user-centered design on a personal assistive bathing device for hemiplegia. *Disability and Rehabilitation* 29 (14): 534-541.

Martin, J. L., B. J. Norris, E. Murphy, and J. A. Crowe. 2008. Medical device development: the challenge for ergnomics. *Applied Ergonomics.* 39: 271-283.

Moreland, J.D., V. G. Dincent, A. L. Dehueck, S. A. Pagliuso, D. W. C. Yip, B. J. Pollock, and E. Wilkins. 2009. Need assessments of individuals with stroke after discharge from hospital stratified by acute function independent measure score. *Disability and Rehabilitation.* 31 (26): 2185-2195.

Sonn, U. and G. Grimby. 1994. Assistive devices in an elderly population studied at 70 and 76 years of age. *Disability and Rehabilitation.* 16 (2): 85-92.

Warlow, C., J. V. Gijn, M. Dennis, J. Wardlaw, J. Bamford, G. Hankey, P. Sandercock, G. Rinkel, P. Langhorne, C. Sudlow, and P. Rothwell, 2008, *Stroke: practical management (third edition).* USA. Blackwell Publishing.

Wu, F. G., M. Y. Ma, and R. H. Chang. 2009. A new user-centered design approach: A hair washing assistive device design for users with shoulder mobility restriction. *Applied Ergonomics.* 40: 878-886.

Saposnik, G., M. Mamdani, J. Hall, W. Mcllroy, D. Cheung, K. E. Thorpe, L. G. Cohen, and M. Bayley. 2010. Effectiveness of virtual reality using Wii gaming technology in stroke rehabilitation: a pilot randomized clinical trail and proof of principle. *Stroke.* 41: 1477-1484.

CHAPTER 55

Usability of Automotive Integrated Switch System using Face Direction

- Comparison of Usability between Integrated Switch System Using Face Directions and Traditional Touch-Panel Switch System -

*Atsuo Murata*1, Makoto Moriwaka*1, Takuya Endoh*1, Takehito Hayami*1, Shinsuke Ueda*2, and Akio Takahashi*2*

*1Graduate School of Natural Science and Technology, Okayama University
Okayama, Japan
murata@iims.sys.okayama-u.ac.jp
2 Department 2, Technology Research Division 8, Honda R&D Co., Ltd.,
Automobile R&D Center
Tochigi, Japan
Akio_Takahashi@n.t.rd.honda.co.jp

ABSTRACT

The development of switch system which leads to a faster response and less frequent visual off road would be useful for the purpose of enhancing safety and avoid inattentive driving. An attempt was made to evaluate the usability of newly developed integrated switch system using face direction. The usability was compared between the integrated switch system using face direction and the traditional touch-panel interface.

Keywords: usability, visual off road, integrated switch system, face direction

1   INTRODUCTION

In-vehicle equipments are becoming more and more complicated in order to make automotives more attractive (Murata et al., 2009a, Murata et al., 2009c).

While in-vehicle information systems such as IHCC system and ITS surely support driving activities, the operation of such systems induces visual and physical workload to drivers. It must be noted that in-vehicle information system has both positive and negative impacts on safety driving. With the development of by-wire technologies, automotive interfaces that control a display using computers have increased more and more. Due to the widespread of such systems, a variety of switch systems can be installed to in-vehicle equipments. The usability of switches are, in general, affected by many factors such as ease to operate, frequency of visual off road, and physical workload while operating, etc. (Murata et al., 2007, Murata et al., 2008, Murata et al., 2009b, Murata et al., 2011). The improvement of such factors leads to the enhanced usability of switches, and eventually contributes to the safety driving.

Therefore, the development of switch system which leads to a faster response and less frequent visual off road would be useful for the purpose of enhancing safety and avoid inattentive driving. An attempt was made to evaluate the usability of newly developed integrated switch system using face direction. The usability was compared between the integrated switch system using face direction and the traditional touch-panel interface.

2   METHOD

2.1   Participants

A total of 20 male participants licensed to drive took part in the experiment. Ten belonged to the group of young adults aged from 21 to 24 years. Ten were older adults aged from 65 to 72 years. All had a driver's license. They hand no orthopaedic or neurological diseases.

2.2   Apparatus

Two switches were used in the experiment. One was a traditional touch-panel switch, and the other was a developed integrated switch which can realize many functions by making use of facial directions. This system enabled the participant to select one of many functions by changing the facial direction and pressing a "Confirmation" key placed around a steering wheel. Using a three-dimensional magnetic-type location measurement system (POLHEMUS, 3-SPACE Fastrak), the switch system automatically recognized the facial direction (central, left, right, middle-left) and the combination of the recognition and the pressing of "Confirmation" key enabled the participant to carry out a pre-determined switch operation such as the open and close of left-side window. In the simulator, according to Japanese and British standard, the participant location was on the right (The driver's seat was located on the right). The experimental setting is shown in

Figure 1. EMG activities were acquired with measurement equipment. A/D converter (A/D instrument PowerLab 8/30) and a bio-amplifier (ML132) were used. Surface EMG was recorded using A/D instrument silver/silver chloride surface electrodes (MLAWBT9).

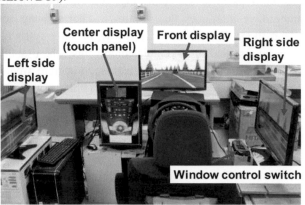

Figure 1 Outline of experimental setting.

2.3 Experimental task

Viewing the road in Figure 1, the participants were required to carry out the simulated driving task. The straight road was used in the simulation (The curved road was not used). The participants were also required to carry out the switch operation task using either the touch-panel interface (See Figure 1) or the integrated switch using facial directions. The switch operation included the following tasks: (1) open/close of a window on the driver's side, (2) open/close of a front passenger's seat, (3) operation of a door mirror on the driver side, (4) operation of a door mirror on the front passenger side, (5) on-off of a hazard ramp, (6)adjustment of a seat, (7) adjustment of an air conditioner, (8) operation of an audio system, (9) operation of a car navigation system, (10) on-off of a door mirror light on the driver side, (11) on-off of a door mirror light on the front passenger side, and (12) information presentation to HUD (Head Up Display). It must be noted that the operation (12) was carried out only by the integrated switch operation using facial directions.

The operation principle of the integrated switch using facial directions is briefly explained below. The three-dimensional magnetic-type location measurement system used for recognizing facial direction is shown in Figure 2(a). Five areas shown in Figure 2(b) were discriminated using the measurement system. The discrimination was carried out on the basis of the yaw angle output from the three-dimensional magnetic-type location measurement system (Fastrak). When the participant moved his head toward the left door mirror and the system recognized this, he could carry out the switch operations (2), (4), and (11). When the head movement to center panel was recognized, the participant could carry out the switch operations (6), (7), and (8). When the system recognized that the head was directed to the center of a seat, the participant could conduct the operations (5) and (12). The system recognition of the head movement to the right door mirror enabled the participant to carry out the switch operations (1), (3), and (10). After selecting the function, the participant can move his head freely and continue and complete the switch operation using a steering-wheel mounted switch. Thus, a lot of functions can be realized by making use of facial directions.

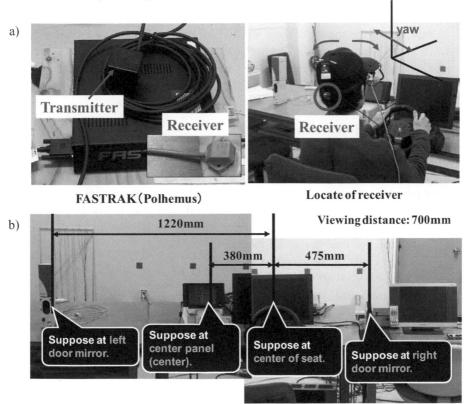

Figure 2 (a) Three-dimensional magnetic-type location measurement system used for recognizing facial direction and (b) areas that are discriminated using the measurement system.

The contents of switch operations (1)-(12) were orally presented to the participant. In case of the touch-panel interface, the participant carried out switch operation using the touch-panel shown in Figure 1.

2.4 Design and procedure

The switch type was a within-subject factor, and the age was a between-subject experimental factor. For each switch, four sessions of experiments were conducted.

One session consisted of 7-minute travel (drive) of a straight road. The twelve switch operations (1)-(12) randomly appeared once during the 7-min travel. The participants were required to carry out simultaneously the main tracking task and the secondary switch operation task such as the operation of CD and turning on of the hazard ramp. The duration of one experimental session was about 7 min. After the first and the fourth sessions for each switch was exhausted, the participant was required to evaluate workload using NASA-TLX. The following subjective ratings on usability were also carried out at the completion of each experimental session: general usability, usability of switch itself, number of functions, learnability, and usability for each functions (1)-(12) described above.

Separately from the experiment above, EMG was measured when using each switch in order to evaluate muscle workload. In EMG measurement, the electrodes were attached to the trapezius muscle. The EMG measured during the switch operation was converted into %MVC (Maximum Voluntary Contraction). In this case, the simulated driving task was not imposed on the participant.

2.5   Evaluation measure

Making use of the output data from the three-dimensional magnetic-type location measurement system (Fastrak), the frequency of visual off road and the visual time off road were obtained. When the head movement was out of the center area in Figure 2(a), this was regarded as visual off road. The following evaluation measures were used to compare the usability between two types of switches:
(1)Task completion time in switch operation
(2)Percentage correct in switch operation
(3)Frequency of visual off road
(4)Visual time off road
(5)Workload evaluation by NASA-TLX (WWL: Weighted Workload Score)
(6)%MVC when operating each switch

3   RESULTS

In Figure 3, the task completion time in the switch operation task is shown as a function of switch type and age. As for young adults, the developed integrated switching system using face directions led to faster response than the traditional touch-panel interface. On the other hand, this tendency was not observed for older adults. The task completion time of the developed system was longer than that of the traditional touch-panel interface. As a result of a two-way (age by switch type) ANOVA conducted on the task completion time, a significant main effect of switch type ($F(1,12)$=4.761, p<0.05) and a significant age by switch type interaction ($F(1,12)$=26.277, p<0.01) were detected. For each age group, a one-way (switch type) ANOVA was conducted. For both age groups, a significant main effect of switch type was detected (young adults: $F(1,9)=10,837$, $p<0.05$, older adults: $F(1,9)=9,837, p<0.05$).

In Figure 4, the percentage correct in the switch operation task is plotted as a function of switch type and age. As a result of a two-way (age by switch type) ANOVA conducted on the percentage correct, only a significant main effect of age ($F(1,12)=6.209, p<0.05$) was detected.

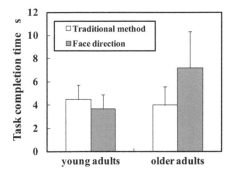

Figure 3 Task completion time in switch operation task as a function of age and switch type.

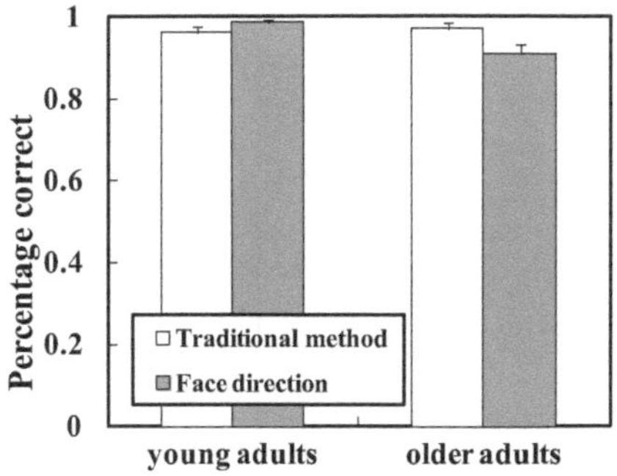

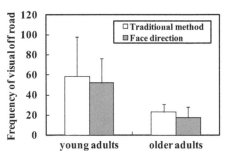

Figure 4 Percentage correct in switch operation as a function of age and switch type.

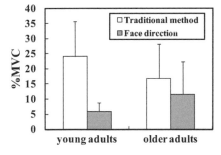

Figure 5 %MVC as a function of age and switch type.

Figure 6 Frequency of visual off road as a function of age and switch type.

Figure 5 compares the results of EMG measurements (%MVC) between the integrated switch system using face directions and the traditional touch-panel interface, and between young and older adults. A two-way (age by switch type) ANOVA carried out on the %MVC revealed a significant main effect of switch type ($F(1,13)=16.170$, $p<0.01$). A significant age by switch type interaction ($F(1,13)=4.890$, $p<0.05$) was also confirmed. A one-way (switch type) ANOVA was carried out on the %MVC for each age group. A significant main effect of switch type ($F(1,9)=28.875$, $p<0.01$) was confirmed for young adults. The result indicates that the physical workload is smaller when using the integrated switch than when using the touch-panel interface for both age groups.

In Figure 6, the frequency of visual off road is plotted as a function of switch type and age. A two-way (age by switch type) ANOVA conducted on the frequency of visual off road revealed only a main effect of age ($F(1,18)=11.591$, $p<0.01$). The developed switch system contributed to the reduction of frequency of visual off road for both age groups. As for older adults, the observation of the visual time off road revealed a similar result to this. On the other hand, concerning young adults, the visual time off road did not differ between the developed switch and the touch-panel interface.

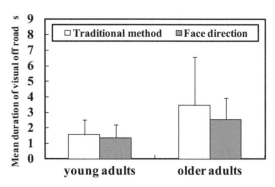

Figure 7 visual time off road as a function of age and switch type.

(a) young adults   (b) older adults

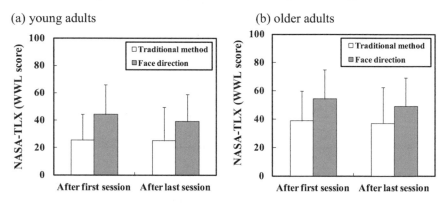

Figure 8 NASA-TLS (WWL) score as a function of experimental session and switch type.

In Figure 7, the visual time off road is plotted as a function of age and switch type. A two-way (age by switch type) ANOVA conducted on the visual time off road revealed only a main effect of age ($F(1,18)=5.536$, $p<0.05$). The visual time off road for the integrated switch using facial direction tended to be shorter.

In Figure 8, the NASA-TLS (WWL) score is plotted as a function of switch type and experimental session for both age groups. As a result of Mann-Whitney U test, significant main effects of age and switch type were confirmed. For both age groups, it tended that the workload was higher when using the integrated facial-direction switch than when using the touch-panel interface.

In Table 1, the results of subjective ratings on usability are summarized. The subjective rating on general usability improved with the elapse of experimental session. The usability of the touch-panel interface tended to be evaluated highly for the open/close of a window ((1) and (2)) and the operation of a door mirror ((3) and (4)), while the evaluation of usability of the integrated switch using facial directions tended to be high for (5) on-off of a hazard ramp and (9) operation of a car navigation system.

Table 1 Summary of subjective rating on usability.

	Type of interface	Young vs. older adults
(1) General usability	n.s.	n.s.
(2) Usability	n.s.	n.s.
(3) Number of functions	n.s.	n.s.
(4) Learnability	n.s.	$p<0.01$
(5)-1 Open and close of window	$p<0.01$	$p<0.05$
(5)-2 Open and close of door mirror	$p<0.01$	$p<0.01$
(5)-3 ON/OFF of Hazard lamp	$p<0.01$	n.s.
(5)-4 Seat adjustment	n.s.	$p<0.01$
(5)-5 Adjustment of air conditioner	n.s.	$p<0.05$
(5)-6 Operation of audio system	n.s.	$p<0.05$
(5)-7 Operation of car navigation system	$p<0.05$	$p<0.01$
(5)-8 ON/OFF of door mirror lamp		$p<0.01$
(5)-9 Display of HUD		n.s.
(6) Less frequent visual off road by face directional interface		n.s.

☐ Face direction < Traditional touch-panel
▨ Face direction > Traditional touch-panel

4   DISCUSSION

The task completion time of young adults was shorter when using the integrates switch using facial direction than when using the touch-panel interface (See Figure 3,) The %MVC measure was reduced to a larger extent when operating with an integrated switch using facial directions (See Figure 5). The frequency of visual off road for an integrated switch using facial direction decreased as in Figure 6. As far as young adults are concerned, the effectiveness of the developed switch system with facial direction was, as a whole, verified from the viewpoint of the task completion time, the number of visual off road, and the physical workload when operating the switch.

528

By contrast, the task completion time of older adults was mot improved by using the proposed integrated switch. The reason can be inferred using Figure 9. The cognitive information processing of both switches differs in processing denoted by the surrounded frame in Figure 9. Although it is sure that such processing leads to the reduction muscle workload and the decrease of visual time off road, it might

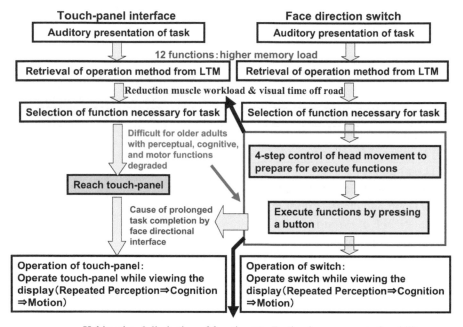

Habituation & limitation of function ⇒ Further improvement of usability

Figure 9 Explanation of cognitive information processing for a touch-panel interface or an integrated switch using face direction (face direction switch).

take longer for older adults with degraded perceptual, cognitive, and motor functions to execute the cognitive processing. The difficulty in the processing might cause longer task completion time. In this experiment, the participant was required to execute the switch operation using a total of 12 functions. This must impose more heavy memory load on older adults. It is possible that habituation to the operation and the limitation of function lead to further improvement of usability.

Comparison of data between young and older adults in Figures 6 and 7 shows that the strategy for visual information processing apparently differs between young and older adults. We generally execute visual information processing by controlling the number of fixations and the fixation duration (Murata, 2005). We interpret, from Figures 6 and 7, that young adults must adopt a strategy to minimize the visual time off road by increasing the frequency of visual off road. As for older adults, a strategy to minimize the frequency of visual off road and increase the visual time off road must be adopted, because they cannot adopt the same strategy with young adults due to the declined visual processing function. From the viewpoint of safety driving, the strategy to minimize the visual time off road would be desirable. Such difference should be taken into account to further enhance the usability of the proposed switch system.

Table 1 suggests that we should make proper use of both types of switches. Each switch has advantages and disadvantages. It is desirable that both switches are installed to the automotive cockpit. The touch-panel interface should be used for the open/close of a window ((1) and (2)) and the operation of a door mirror ((3) and (4)), whereas the integrated switch using facial directions is proper for (5) on-off of a hazard ramp and (9) operation of a car navigation system.

In future research, the usability of the proposed switch should be verified in real-world driving environment. Moreover, the learning characteristics of the proposed switch system should be investigated.

REFERENCES

Murata,A. 2005. Relationship between Display Features, Eye Movement Characteristics, and Reaction Time in Visual Search, *Human Factors* 46(3): 598-612.

Murata,A. and Moriwaka,M. 2007. Applicability of Location Compatibility to the Arrangement of Display and Control in Human-Vehicle Systems -Comparison between Young and Older Adults-, *Ergonomics* 50(1): 99-111.

Murata,A. and Moriwaka,M. 2008. Evaluation of Control-Display System by means of Mental Workload, *Proceedings of 4th International Workshop on Computational Intelligence & Applications*: 83-88.

Murata,A., Moriwaka,M. and Shugwang,W. 2009a. Development of Thumb-Operated Dial-Type Integrated Switch for Automobile and its Effectiveness, *Proceedings of 5th International Workshop on Computational Intelligence & Applications*: 330-335.

Murata,A., Tanaka,K. and Moriwaka,M. 2011. Basic study on effectiveness of tactile interface for warning presentation in driving environment, *International Journal of Knowledge Engineering and Software Data Paradigm* 3(1): 112-120, 2011.

Murata,A. Uchida,Y. and Moriwaka,M. 2009b. Fundamental Study for Constructing A System to Assist The Left Visual Field of Older Drivers -Effectiveness of The Alternative of The Left Front Side-view Mirror by The Central Visual Field-, *Proceedings of 5th International Workshop on Computational Intelligence & Applications*: 320-325.

Murata,A., Yamada,K. and Moriwaka,M. 2009c. Design Method of Cockpit Module in Consideration of Switch Type, Location of Switch and Display Information for Older Drivers, *Proceedings of 5th International Workshop on Computational Intelligence & Applications*: 258-263.

CHAPTER 56

Effects of Automotive Switch Type and Location on Usability and Visual Time off Road

Atsuo MURATA, Makoto Moriwaka and Takehito Hayami

Graduate School of Natural Science and Technology, Okayama University
Okayama, Japan
murata@iims.sys.okayama-u.ac.jp

ABSTRACT

The effects of switch type and its location on the in-vehicle task performance, the duration and the frequency of visual off road and the subjective evaluation on usability and workload were explored. The task completion time of the number selection-type switch mounted around the steering wheel was shorter than that of other conditions. The cursor control-type switch, irrespective of installation locations, also led to shorter switch operation time. The non-integrated switches were, as a whole, superior to the integrated switches from the viewpoints of task completion time and visual time off road.

Keywords: switch type; installation location; visual time off road; duration and frequency of visual off road

1 INTRODUCTION

The operation of in-vehicle equipments is becoming more and more complicated with the sophistication of in-vehicle information system. While in-vehicle information systems such as IHCC system (Intelligent Highway Cruise Control System), car navigation system, and ITS (Intelligent Transportation System) support driving activities, operating these systems induces visual and physical workload to drivers. Therefore, in-vehicle information system has both positive and negative impacts on safety driving.

Recently, a lot of studies have been conducted to investigate the effects of automotive switch functions and its installation location on the driving performance. Dukic et al. (2005) examined the effects of push switch location on driver's eye movement characteristics and driving performance such as steering wheel deviation, and safety perception. They showed that the visual time off road increased as the angle increased between the normal line of sight and push switch location (eccentricity) for the five switches placed on the central stack. In Dukic (2006, 2007), only a simple switch pressing was imposed on the participants. The participant did not conduct a real-world interactive task where a switch pressing like adjusting the temperature of air conditioner is carried out while interacting with information on the display system.

Murata and Moriwaka (2007) investigated the effect of age on the applicability of the location compatibility principle to the design of display and control system, and showed that the design of display and control taking into account the location compatibility principle enhanced the usability especially for older adults. Murata, Yamada, and Moriwaka (2009) explored the effects of switch type, location of switch, and display information on the primary driving task and the secondary switch operation task. Young adults were better than older adults at both abilities on processing displayed information and operating switches. The integrated switch placed around a steering was found to lead to high driving performance and shorter switch operation time.

However, these studies (Murata, Yamada, and Moriwaka, 2009) did not use eye movement characteristics for the evaluation of switches. Not only performance measures such as percentage correct and task completion time of a secondary task and primary tracking error, but also eye movements should be used to evaluate the usability of a switch system combined with a display system. Murata, Moriwaka, and Wang (2009) proposed a desirable conditions of torque and diameter for the design of a thumb-operated dial-type integrated switch on the basis of not a dual-task experiment in which both primary driving task and secondary switch pressing task were conducted as in a real-world driving environment, but a single-task experiment in which only a switch pressing was carried out. The desirable condition of torque and diameter should be explored using a dual-task experiment corresponding to a real-world driving environment.

The effectiveness of switch type and its installation location was verified on the basis of not only performance measures and psychological rating on workload and usability, but also drivers' eye movement characteristics (visual time off road, and frequency of visual off road) during a simulated driving task.

2 METHOD

2.1 Participants

A total of 20 participants took part in the experiment. All were male graduate and undergraduate students aged from 22 to 28 years, and licensed to drive from 1-5 years. The visual acuity of participants was matched and more than 20/20. They hand no orthopaedic or neurological diseases.

2.2 Apparatus

Experimental system consisted of (1) a tracking system equipped with a personal computer (mouse computer, 0602Lm-i211B), a projector (EPSON, EMP-S4), and a steering wheel (Logitech, MOMO), (2) a switch system equipped with a personal computer (HP, NX6120), a digital I/O card (Interface, PIO-24W (PM)), and a LD display, (3) an eye mark recorder (Nac Image Technology Inc.,EMR-9) connected with a personal computer (NEC, PC-VX1006F), and (4) a three-dimensional magnetic-type location measurement system (POLHEMUS, 3-SPACE Fastrak).

2.2.1 Switches used in this study

Two kinds of integrated switch and two kinds of non-integrated switch were used in the experiment (See Figure 1). Non-integrated switches included (a) number selection-type non-integrated switch, and (b) cursor control-type non-integrated switch. Integrated switches included (c) thumb-operated dial-type integrated switch, and (d) traditional dial-type integrated switch.

532

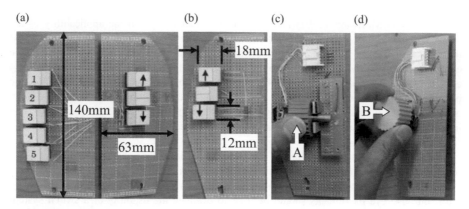

Figure 1   Switches used in the experiment. (a) Number selection-type non-integrated switch, (b) Cursor control-type non-integrated switch, (c) Thumb-operated dial-type integrated switch, (d) Traditional dial-type integrated switch.

As for non-integrated switch systems, each switch consisted of a hinge-type tactile switch (OMRON, B3J-1000). Pressing a 12mm X 12mm square surface to which an arrow or a number is attached, the cursor is moved or the function is selected or determined. The number selection-type non-integrated switch (See Figure 1(a)) consisted of eight keys (five number keys, upper and lower arrow keys, and an enter key). The cursor control-type non-integrated switch (See Figure 1(b)) consisted of upper and lower arrow keys and an enter key. According to Murata, Moriwaka, and Wang (2009), integrated switches were produced using encoders (Panasonic Electronic Device,11-type GS(EVER)). The rotation torque of encoder was fixed to 12.9mN · m, and a rotation dial made of PVC material was attached to the encoder. The diameter and the thickness of rotation dial were fixed to 30 mm and 17 mm, respectively, and thirty ditches with a width of 2.5 mm were set to the rotation dial with an equal interval between ditches. The integrated switch system, in general, needs both rotation and push operations. The thumb-operated dial-type integrated switch makes the switch rotate using only a thumb, and terminate a task by pressing A in Figure 1(c). When using the traditional dial-type integrated switch, the participant grabs the switch with a thumb and an index finger and rotates it in order to move the cursor. The switch operation is terminated by pressing B in Figure 1(d) along the rotation axis of the switch using a thumb.

2.2.2   Detection of eye-gaze location and method for judging the state of visual off road

According to Nonaka (2003), the eye-gaze was defined as a vector that connected an eye and an eye mark measured by an eye mark recorder. An eye mark recorder (Nac image technology, EMR-9) and a three-dimensional magnetic-type location measurement system (POLHEMUS, 3-SPACE Fastrak) were used to measure eye movements and head location, respectively. Using the data from these two measurement systems, the direction of an eye-gaze was calculated.

EMR-9 consists of a head unit and a controller. A head unit is equipped with a camera to film an eye and a camera to film objects within the visual field. Eye-gaze locations are detected using a relative distance between the location of a cornea reflective image (Purkinje image) which is obtained by near infrared LED radiation to an eye, and the center of a pupil. EMR-9 can output eye-gaze data (x and y coordinates in the 640×480 pixel coordination

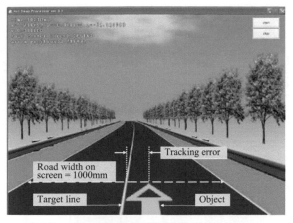

Figure 2 Display of tracking task.

system) with a maximum sampling frequency of 60 Hz via a serial port. The horizontal visual angle of a camera to film objects within the visual field corresponds to 44 degrees. Fastrak consists of a transmitter producing magnetic field and a receiver that detects the change of magnetic field. The relative location of a receiver to a transmitter (X, Y, Z) and the angle data (yaw, pitch, roll) can be obtained by Fastrak. Only angle information (yaw, pitch, roll) was used, because only a posture change of head is treated in this study. Only one receiver was used with a sampling frequency of 60 Hz. Serial data outputs from EMR-9 and Fastrak were synchronized and sent to a personal computer. According to the following procedure, the direction of eye-gaze was derived to judge the state of visual off road (whether the participant looked aside or not).

(1)Nine-point calibration

(2)Calculation of direction of eye-gaze

(3)Definition of visual off road

In this study, the state of visual off road (looking aside) was defined as follows. The state where the eye-gaze went out of the tracking screen was regarded as looking aside. The state where θ_{gx} and θ_{gy} satisfied Eq.(1) was judged to be the state of visual off road (looking aside).

$$\theta_{gx} > 12.1° \text{ or } \theta_{gx} < -12.1° , \theta_{gy} > 10° \text{ or } \theta_{gy} < -7.5° \text{ (1)}$$

The data for the left eye were used to judge the state of visual off road (whether the participant looked aside or not). The duration of one visual off road (looking aside) was defined as the duration of visual off road. The total time of visual off road per one trial was defined as visual time off road.

2.3   Task

The participants were required to carry out simultaneously a main tracking task and a secondary switch operation task.

2.3.1   Tracking task

A three-dimensional driving simulator was made using Hot Soup Processor 3.2. The simulation system was run with a sampling frequency of 50 Hz. The resolution of the simulator display was set to 1024×768 pixel. The refresh rate of the display was set to 60 Hz. The simulator display includes a one-lane road, a shoulder of road, a guardrail, and trees (See Figure2). The arrow and the center line of road corresponded to an operation object and a moving object, respectively.

534

a)
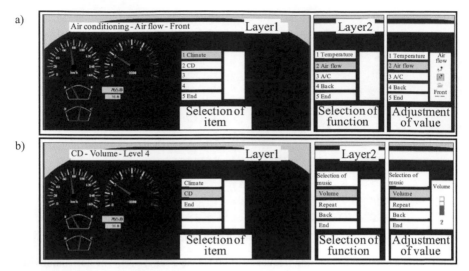

b)

Figure 3 Examples of display for switch operation task. (a). in case of using number selection-type nonintegrated switch, (b). in case of using cursor control-type non-integrated switch and both type of integrated switches

In the driving simulation program, the width of road was assumed to be 3.25 m so that the velocity of own virtual vehicle corresponded to about 35 km/h. The relation between the steering angle α and the yaw angle ω of the virtual vehicle in the simulation program is given by Eq.(2).

$$\omega = 0.078\alpha \text{ (deg/sec)} \quad (2)$$

The road in the driving simulation system was smoothly curved according to a sinusoidal function with a frequency of 0.05Hz. No other cars existed in the driving simulator display. The participant was required to operate a steering wheel so that the controlled arrow coincided with the changing center line as accurately as possible. The error between the controlled arrow and the center of a road was obtained every 0.1 s, and the mean value of this error was defined as a mean tracking error.

2.3.2   Switch operation task

The menu structure of the secondary switch operation task is demonstrated in Figure 3. The participant was required to carry out a switch operation task such as a control of an air conditioner (for example, set temperature to 24 deg.) and CD operation (for example, set repeat function on). Although (a) and (b) are very similar, these differ in that (a) is numbered on the surface and used only for the number selection-type non-integrated switch. Figure 3 (b) is for other types of switch. The display consisted of two layers. On the first layer, the participant was required to select either a control of an air conditioner or a CD operation. On the second layer, the participant was required to select temperature, air flow, or direction and adjust parameters of these items in case of a control of an air conditioner, and to select music, volume, or repeat function and adjust parameters of these items in case of a CD operation. When the participant selected "End" in Figure 3, one trial was completed. For one experimental condition (combination of switch type and installation location), the participant was required to conduct 10 trials. At the start and the end of a trial, beep sound was presented to the participant. In order to mask a sound produced by pressing a switch, a sound was output every time a switch was pressed. The minimum number of switch presses for one trial was the same for all of 10 trials. The "Back" function was used so that the participant could correct an error operation. The time from the presentation of a task content until the completion was recorded to a computer file.

2.4   Design and procedure

Switch type (four levels: number selection-type non-integrated switch, cursor control-type non-integrated switch, thumb-operated dial-type integrated switch, and traditional dial-type integrated switch) and switch installation location (three levels: steering, upper left side (front), and lower left side (lateral)) were within-subject experimental factors.

First, the participant was asked to adjust a seat so that the task could be comfortably performed and the switches could be pressed by reaching a hand naturally. Before the experimental task, the contents of primary and secondary tasks were explained to each participant. The driving history (years), age and the statue of each participant were checked. The participant was allowed to practice primary and secondary tasks before performing experimental tasks. When the experimenter judged that the participant clearly and fully understood how to perform primary and secondary tasks, the experiment was started. Before entering an experimental session, the eye mark recorder (EMR-9) was calibrated.

For each of 12 experimental conditions (switch type (four levels) by its installation location (three levels)), the experimental task above was conducted. Every time each experimental condition was exhausted, questionnaires on workload and usability were answered by the participant. The order of performance of the 12 experimental conditions was randomized across the participants. The participant was permitted to state their comments or opinions orally after the experiment was finished.

2.5   Evaluation measures

The following evaluation measures were used:
(A) Mean task completion time of a secondary switch operation (pressing) task
(B) Mean percentage correct of a secondary switch pressing task
(C) Mean tracking error
(D) Mean visual time off road
(E) Mean frequency of visual off road
(F) WWL (Weighted Workload) score of NASA-TLX
(G) 5-point subjective rating on usability: 1: very low usability, 5: very high usability

3   RESULTS

3.1   Tracking error

A two-way (switch type by installation location) ANOVA conducted on the tracking error revealed a significant main effect of switch type ($F(3,19)=6.58$, $p<0.01$). The tracking error of (b) cursor control-type non-integrated switch tended to be smaller than that of other switches. For each switch type, multiple comparisons (Fisher's PLSD) were conducted between arbitrary combinations of installation locations. For (d) traditional dial-type integrated switch, a significant difference ($p<0.01$) was detected between the steering and the upper left side (front) installation conditions.

3.2   Task completion time of secondary switch pressing task

Figure 4 shows the mean task completion time as a function of switch type and

536

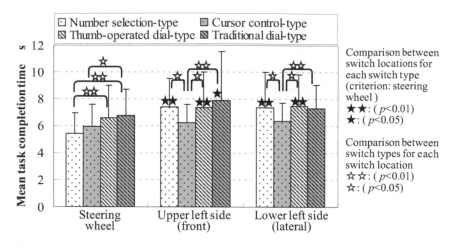

Figure 4 Mean task completion time as a function of switch type and switch location (error bar: standard deviation).

installation location. As a result of a two-way (switch type by installation location) ANOVA carried out on the task completion time, main effects of switch type ($F(3,57)$=4.606, p<0.01)and installation location ($F(2,38)$=19.537, p<0.01) and a switch type by installation location interaction ($F(6,114)$=3.019, p<0.01) were significant.

The task completion time installed around the steering tended to be shorter than that installed on locations other than the steering. In particular, the task completion time of (b) number selection-type non-integrated switch was the shortest. The cursor control-type non-integrated switch (c) installed on the locations other than the steering led to shorter task completion time than other switch types installed on the locations other than the steering. The task completion time of the cursor control-type non-integrated switch (c) installed on the locations other than the steering was nearly equal to that of the cursor control-type non-integrated switch (c) installed around the steering.

3.3 Percentage correct of secondary switch pressing task

A two-way (switch type by installation location) ANOVA carried out on the percentage correct revealed a main effect of switch type ($F(3,57)$=12.523, p<0.01). Multiple comparisons between arbitrary combinations of switch installation locations revealed no significant differences for each switch type.

The cursor control-type non-integrated switch installed around the steering switch and on the upper left side led to the percentage correct of 100%. The percentage correct of traditional dial-type integrated switch led to a lower percentage correct, irrespective of the installation location. The percentage correct of number selection-type integrated switch installed on the upper left side was low.

3.4 Visual time off road and frequency of visual off road

In Figure 5, the mean visual time off road is shown as a function of switch type and installation location. A two-way (switch type by installation location) ANOVA conducted on the visual time off road revealed significant main effects of switch type ($F(3,57)$=10.636, p<0.01) and installation location ($F(2,38)$ =6.767, p<0.01) and a significant switch type by

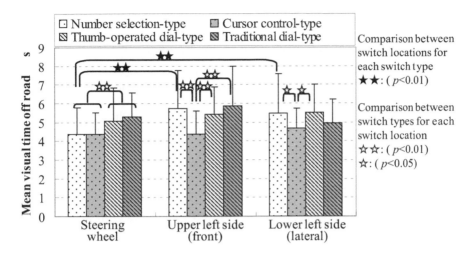

Figure 5 Mean visual time off road as a function of switch type and switch location (error bar: standard deviation).

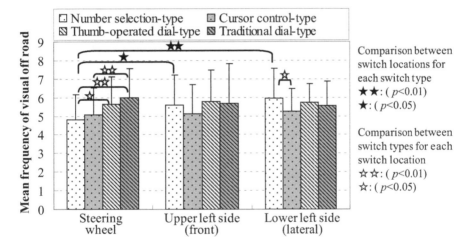

Figure 6 Mean frequency of visual off road as a function of switch type and switch location (error bar: standard deviation).

installation location interaction ($F(6,114)$ =3.047, p<0.01). In Figure 6, the mean frequency of visual off road is plotted as a function of switch type and installation location. As a result of a two-way (switch type by installation location) ANOVA conducted on the frequency of visual off road, a significant main effect of switch type ($F(3,57)$=3.966, p<0.05) and a significant switch type by installation location interaction ($F(6,114)$=2.196, p<0.05) were detected.

As for the visual time off road, the same tendency with the task completion time (see Figure 4) was observed. The visual time off road for (a) number selection-type non-integrated switch installed around the steering tended to be shorter. The visual time off road for (b) cursor control-type non-integrated switch tended to be shorter irrespective of installation locations.

3.5   WWL score and subjective rating on usability

WWL scores were less than 50 points for the four switch types installed around the steering, and for the cursor control-type non-integrated switch installed on the upper left (front) and the lower left (lateral) sides. Subjective rating scores on usability were more than 3 points for the four switch types installed around the steering, and for the cursor control-type non-integrated switch installed on the upper left (front) and the lower left (lateral) sides. In particular, the subjective rating scores on usability were more than 4 points when the cursor control-type switch was installed around the steering.

4   DISCUSSION

4.1   Effects of switch type and its location on secondary task performance

Operation errors occur frequently for the non-integrated switches, while the problem of over-shooting occurs for the integrated switches. From the viewpoint of the task completion time and the percentage correct, the cursor control-type non-integrated switch was, as a whole, superior to other switches. This must be caused due to the following reasons. (1) As the number of switches is few, the task completion time of the cursor control-type non-integrated switch is shorter than that of the number selection-type non-integrated switch according to Hick law. (2) The switches are gathered together, and no problem of over-shooting like integrated switches occurs.

As for the number selection-type non-integrated switch, the task completion time differed significantly between the steering and other installation conditions (Figure 5), Moreover, when installed around the steering, the task completion time of the number selection-type non-integrated switch was shorter than that of other switches. The reason why the task completion time and the percentage correct of the number selection-type non-integrated switch was higher around the steering than on other installation sites must be due to the fact that both hands are used only for this type of switch (See Figure 2). Operating switch with both hands according to the principles of motion economy must lead to higher efficiency.

4.2   Effects of switch type and its location on eye movement

As the switch operation in this study is complicated to some extent, it is impossible for the participant to complete the switch operation task within one visual off road (looking aside). Therefore, the participants had to allocate their cognitive resources properly between the primary tracking task and the secondary switch operation task. The mean duration of visual off road was nearly constant among switch types and among installation locations. Therefore, irrespective of switch type and its installation location, we infer that the participant adopted the following strategy. Once the participants direct their attention to the secondary switch operation task, and a constant period of time passes, they are sure to direct their attention again to the primary tracking task.

As for the cursor control-type non-integrated switch and the two types of integrated switch, the participants did not need to gaze at a switch once the switch operation task was undertaken. Therefore, the effect of installation location on the visual workload such as the visual time off road was not so remarkable as in Dukic et al.(2005) or the number selection-type non-integrated switch.

The reason why the two types of integrated switch led to the longer visual time off road must be due to the over-shooting when using the integrated switch and larger visual workload to confirm the change of operational state accompanied by the switch operation. As the non-integrated switches, on the other hand, have no problems like this, it can be inferred that the visual time off road was relatively shorter.

4.3   Effects of switch type and its location on workload and subjective rating on usability

The steering installation led to lower workload and higher usability as compare with the upper left and the lower left installation. This result reflects the tendencies that when the switch was placed around the steering, the task completion time was short (Figure 5), the visual time off road was short (Figure 6), and the frequency of visual off road was few (Figure 7). Concerning the psychological evaluation on workload and usability, the cursor control-type non-integrated switch was superior to other types of switch irrespective of installation locations. This result also reflects the tendency that the performance (task completion time and percentage correct of a secondary task) was overall higher when the cursor control-type switch was used.

Future work should overcome the problem of over-shooting when operating an integrated switch under more realistic driving environment, and the usability of integrated switch should be further improved.

REFERENCES

Dukic, T., Hanson, L., Holmqvist, K. and Waternberg, C., 2005, Effect of button location on driver's visual behavior and safety perception, *Ergonomics* 48: 399-410.

Dukic, T., Hanson, L. and Falkmer, T., 2006, Effect of drivers' age and push button locations on visual time off road, steering wheel deviation and safety perception, *Ergonomics* 49: 78-92.

Murata, A. and Moriwaka, M., 2007, Applicability of location compatibility to the arrangement of display and control in human-vehicle systems: Comparison between young and older adults, *Ergonomics* 50, 1-13.

Murata, A., Yamada,K., and Moriwaka, M., 2009, Design method of cockpit module in consideration of switch type, location of switch and display information for older drivers, *Proceedings of 5th International Workshop on Computational Intelligence & Applications*: 258-263.

Murata, A., Moriwaka, M., Wang, S., 2009, Development of Thumb-operated Dial-type integrated switch for automobile and its effectiveness, *Proceedings of 5th International Workshop on Computational Intelligence & Applications*: 330-335.

Nonaka, H., 2003, Communication interface with eye-gaze and head gesture using successive DP matching and fuzzy interface, *Journal of Intelligent Information System* 21: 105-112.

CHAPTER 57

Development of Similarity Measures by Extracting Design Features of a Shoe Last

Kazumasa Nozawa, Yoitsu Tahakashi, Yukio Fukui,

Jun Mitani and Yoshihiro Kanamori

University of Tsukuba
Tsukuba, JAPAN
nozawa@npal.cs.tsukuba.ac.jp

ABSTRACT

We propose a method of extracting the design features of a shoe last, and a quantitative estimation method of the difference between two lasts based on the design features. The design features are originally generated to manufacture a shoe last from a foot shape. They are defined as design information of a last, and are composed of shapes and lengths of contour lines of specific cross-sections of a last. Manufacturing close fitting shoes to specific person's feet requires lasts of specific shape. Making a newly designed last is expensive and time-consuming. Our idea is to select a most suitable last out of the stock instead of making one if and only if the feature discrepancies are acceptable level between the required and the measured features. Therefore extracting the design features and quantitative comparison methods of them are important to obtain close fitting shoes to specific person's feet. We studied the empirical design process of a last from a foot shape, where many steps of deciding the positions of cross-sections dimensions of the contour curves on cross-sections of a last. Our methods are developed from tracing the design process of a last and are implemented on a computer. Through experimental calculation we find they work well.

Keywords: design feature, shoe last, shape comparison

1   INTRODUCTION

As the quality of life has become more important in daily life, the desire of well fitting wearing goods are becoming more anticipated for the healthy lifestyle. Among wearing goods shoes are very important in the sense that they must be well fitting to the feet because they receive the whole body's weight, and shear force when the body moves. Wearing ill fitting shoes for a long time might cause deformation of foot bones called hallux valgus. However for the minority people who have nonstandard foot shapes it is very difficult to find well fitting shoes among the mass-produced ones designed for 90 or 95 percentile majority people. In order to obtain well fitting shoes the minority must purchase custom-made ones. It is expensive and takes time because it starts from making shoe last for the specific foot shape.

If the shoemaker could pick up a suitable last from their last stock instead of making one, the total cost and time would be drastically decreased. Then how can the shoemaker pick up the most fitting last if the design information of individual last is lost? We propose here a new method of extracting the last design features and a similarity measure between two lasts based on their design features. Though there are many design methods of shoe last around the world, we have referred the Kagami method (Kagami, F, 2000) because it has established numerical steps of design process of a last, so the methods are easily computerized.

2   RELATED WORK

J. Niu et al. (Niu and Salvendy, 2007) reviewed form comparison techniques into two categories depending on the usage of anatomical landmarks. These landmark-based methods are not applicable to shoe lasts which have no landmarks themselves, nor the statistical methods of landmark-free ones because no statistical methods are effective to individual person's shoe last which is different from person to person, and no common statistical form features exist. Other methods cannot handle with design features reflected onto last shapes. As for designing a last, Luximon et al. (Luximon and Luximon, 2009) developed a computer software which simulates the process of manufacturing a shoe last, and therefore user can design a last shape easily on a computer. However although it can design a last, it cannot compare the shape difference between lasts. On shape comparison Adamek et al. (Adamek and O'Connor, 2003) made a shape comparison method of 2D figures, focusing attention on their contour lines. Their method superposes a contour figure onto another figure roughly minimizing the crossover area, then calculates the average distance and its standard deviation between two contour curves. Using the average distance and its standard deviation the method calculates the Dissimilarity Measure. We developed a similarity measure partly referring to this Adamek's method. R Grimmer et al. (Grimmer, et al., 2011) compared and categorized human foot shape. Basically they define similarity measure as the volume difference between the volumes of two aligned point clouds. They propose another method to compare feet on basis of foot features. They extract foot dimensions such as length of the foot, Ball width, circumference of the ball, and so on. Using these 19 foot features they form a feature vector, and they define the measure of similarity of two data sets by the Euclidean distance between two feature vectors. We used the concept of this foot feature vector to represent the last features.

3 PROPOSED METHODS

The final goal of this research is to obtain the best fitting last for a specific person's foot either by picking up from last stock, or by manufacturing one. The figure 1 shows the flowchart of the whole steps. We show in this paper the area bounded by a thick curved line. In this process there are two newly developed methods; extraction of design information from a shoe last, and quantitative shape comparison method of two lasts based on their design methods. We use the design methods of last developed by Kagami Co., Ltd., because the empirical knowledge derived from manufacturing shoe lasts for a long time is systematically organized in the form easily available for implementing on a computer.

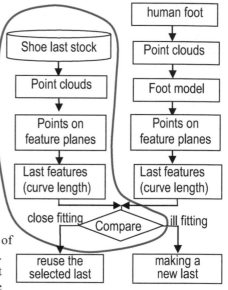

Figure 1 An overall view of determining the closest fitting last to a specific foot. The area enclosed by thick curve is what this paper covers.

3.1 Terminology

First of all we define some technical terms used in this paper.

Table 1 Technical terms to describe specific concept used in last design steps

Term	Definition
Base section	Principal vertical and longitudinal intersection of a last when put on a horizontal plane. Important points (Landmarks) for defining the shape of a last are defined on this plane (Figure 2).
Landmarks	Points in the base section defined through design process of a last. These landmarks show the position of important cross-section (Feature section) of a last (Figure 6).

feature section	A cross-sectional plane across the base section (Figure 3). On each feature section appears the distinctive cross-sectional shape (points) of a last. Each feature section has its own name such as J.
last feature	Characteristics of distinctive cross-section (feature section) Primary last feature is defined as the length of outline of contour curve, and secondary last feature is the shape
Design feature	Total information of the last taken out from the original foot form

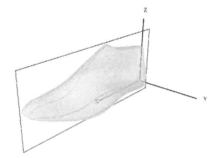

Figure 2 Base section.

Figure 3 Feature section (J)

Figure 4 Points on the Base section

Figure 5 Points on a feature section (face J)

3.2   Extraction steps of last features

At first we obtained a point cloud of a last with 3D scanner to extract the feature sections. Since the coordinate values of a point cloud depends on the coordinates system set by the equipment, its direction and position change slightly from setting to setting, so the base section must be fixed according to the shape itself. As a pre-process, we set the coordinates system and alignment by the Principal Component Analysis of the whole point cloud data.

The position of a feature section is determined by its angle to the base section, upper landmarks and sole landmarks taken on a base section (Figure 6) according to Kagami's design method. As shown in Figure 6, we calculated point A, B, and AVB and divided the point cloud of the base section into upper, sole, and the other part. Each landmark is defined by the ratio of a certain curvilinear distance from particular points. In fact upper landmarks are developed from point NOB on tiptoe and sole ones also from point OO on the center of weight. At this time it is only OO that can be determined among landmarks, so we calculate dimensions of the original foot and solve simultaneous equations to get other unknown landmark positions.

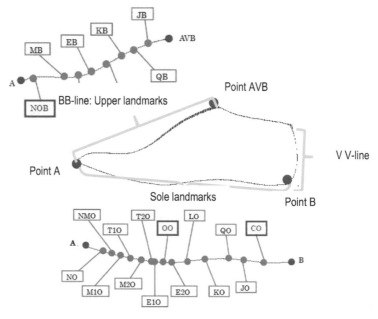

Figure 6 Landmarks on the Base section are divided into three parts;
Sole landmarks, Upper landmarks, and landmarks on VV-line

Since there are many feature sections and landmarks, they are identified by adding an extra character; upper as B and sole as O. For example, upper landmarks of J-section are represented as JB. There are seven feature sections which pass along only sole landmarks such as M1 and T2 and eight ones which pass along both upper and sole landmarks such as E and Q. Using them, we extracted the feature sections as shown in Figure 7.

3.3   Comparison method of last features and similarity

The last features are defined in Kagami's design method as the length of a curve approximating points on the feature section. Considering that not only curve length but the cross sectional shape would be important, we have made two comparison methods of last shape in order to verify whether comparison is effective enough only with the last features; one comparing the above mentioned curve length, and the other the difference of the cross sectional shapes of feature sections. We also implemented Adamek's method (Adamek, 2003) for our reference. Since we aligned the coordinates of the point cloud, we can compare the same feature section from two lasts such as J-sections and M-sections as they are. We especially focus the seven important feature sections (M-T-E-L-K-Q-J) (Figure 7).

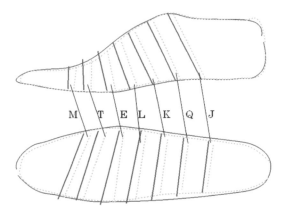

Figure 7 Comparison of positions of the feature
sections between last A (solid line)
and last B (dashed line)

Each section has upper and sole parts, so that 14 last features were obtained from one last. The comparison methods are

· Method 1: Comparison by the vectors of the last features

The 14 last features are combined into one as a 14-dimensional vector, and it is defined as a feature vector. Since a feature vector is defined for individual last, the comparison is easy. Let Euclid distance of the two vectors be the similarity measure.

Method 2: Comparison by discrepancy area

The point clouds obtained in the same feature section of two lasts are superposed. The point cloud is represented by some points at regular intervals, and made are triangles which connects the points (Figure 8). The total sum of triangle areas is approximated as discrepancy area between the point clouds of the feature section and the sum of all discrepancy area of feature sections is defined as the similarity measure.

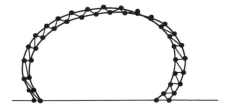

Figure 8 Calculation of the area discrepancies by triangle train

Method 3: Comparison using the shape dissimilarity calculation method of Adamek.

Calculation method of the of dissimilarity measure between the curvilinear figures is formulized in the paper of Adamek (Adamek et al. 2003).

$$D(A,B) = c \cdot |\overline{d}| \cdot \sigma / Cir^2_{min} \qquad (1)$$

A and B are the figures to be compared, D is the dissimilarity measure, c is an arbitrary constant and $|\overline{d}|$ is mean distance between figures.
Cir_{min} is the circumference length of the shorter one and σ is the standard deviation of the distance between them.

4   EXPERIMENTS AND RESULTS

We have conducted comparative experiments using point clouds data of shoe lasts obtained by the 3D scanner INFOOT by I-Ware Laboratory Co., Ltd., Osaka, Japan. We used three last data A, B, and C with similar whole lengths. The figure 9 shows plan form and side view of superposed last pairs. As you see A is largest in both whole length and width and has an almost straight line at shank part, while B and C are quite similar with each other in side view, though C seems slightly narrower in width in plan form. The figure 7 shows the positions of extracted feature planes of A and B. Compared with the difference of total foot lengths the difference of the positions of extracted feature planes seems larger. This may be caused by the inaccuracy of extraction of point OO basing position. This problem should be investigated further in the future. The figure 10 shows last features, the outlines of the cross-sections of each feature plane M, E, K, and J.

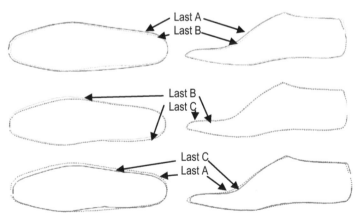

Figure 9   Plain views and side views of
superposed contour lines of last A , B and C

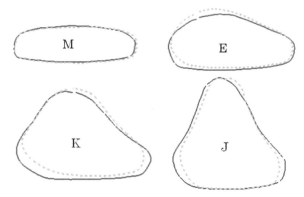

Figure 10   Comparison by superposition of two lasts;
A (Solid line) and B (dashed line)

Then we applied comparison methods described in 3.3 to these three last, and the results are in Table 2. All methods says that B and C are most similar, and, A and B least similar. Table 3 shows the result of weighting three times at EB+EO plane, and twice at JB+JO. However the order of similarity remained unchanged. We have obtained the reasonable results, however, we found that we have to investigate further the most effective comparison method.

Table 2 Similarity without weight

Dissimilarities	Lasts compared		
	A&B	B&C	C&A
Curve length	32.1	20.7	25.6
Area descrpancy	2934	2192	2710
Adamek's	8.4	5.3	8.0

Table 3 Similarity with weight

Dissimilarities	Lasts compared		
	A&B	B&C	C&A
Curve length	33.2	23.6	27.2
Area descrpancy	4313	3297	3806
Adamek's	10.4	7.64	11.0

5 CONCLUSIONS

We proposed a method to extract last features which are reflected into last shape on its design process. The extracted last features are formed into a feature vector and is used to compare the similarity between two shoe lasts. The last feature is originally obtained from a foot shape through the design process of a shoe last. Therefore using the proposed comparison method we can be able to select the most suitable shoe last among a last stock using the proposed comparison methods. However the validation cannot be done at this stage.

For further studies, the validation method must be developed as well as computerizing the steps from foot to design features of a last.

ACKNOWLEDGMENTS

The authors would like to thank Mr. Fusao Kagami, the President of Kagami Manufacturing Co. Ltd, and Mr. Takeshi Kagami for showing us their valuable information on design process. This research was supported by Grant-in-Aid for Scientific Research KAKENHI 23560151.

REFERENCES

Adamek, T., N. O'Connor : Efficient Contour-based Shape Representation and Matching, Proc. 5th ACM SIGMM international workshop on Multimedia information retrieval (MIR '03), pp. 138 - 143, 2003

Grimmer, R., B. Eskofier, H. Schlarb, J. Hornegger : Comparison and classification of 3D objects surface point clouds on the example of feet, Machine Vision and Applications, 22, 235-243, 2011

Kagami, F.: Manufacturing leather shoes and validation of their fitness to feet, Leather and footwear, 112, Tokyo Metropolitan Leather Technology Center (in Japanese), 2000

Luximon, A., Y. Luximon : Shoe-last design innovation for better shoe fitting, Computers in Industry, 60, pp. 621-628, Elsevier, 2009

Niu, J., Z., Li, G. Salvendy : Mathematical Methods for Shape Analysis and Form Comparison in 3D Anthropometry: A Literature Review, Digital Human Modeling, HCII2007, LNCS 4561, pp. 161-170, Springer-Verlag, 2007

CHAPTER 58

Design and Assessment of Digital Drawing System for Children Gesture Movement with Visual Interface Feedback

Fong-Gong Wu, Ya-San Fong, Chien-Hsu Chen

National Cheng Kung University
Tainan, Taiwan
fonggong@mail.ncku.edu.tw

ABSTRACT

We take the intuition creation as a starting point in this research. First, the situation was investigated when child use traditional tools to draw pictures. The function of drawing system for children was defined according to spot observation, questionnaire interviews, opinions of specialized teachers. Second, by using the Usability Engineering the children's demand was discussed when they operate different pointing devices. In order to control drawing function, the result from transformation and movement by the hand signal makes the invisible control interface of pointing device the most suitable document. Third, combination of vision sign rule and sign language and discussion with experts construct the gesture movement image. Finally, carrying on operation of drawing system by gesture movement and visual interface feedback and using technique of experiment confirm the system performance criteria. Also, we look for system design specifications and the correcting directions. At the same time, the subjective feeling of participants was thoroughly investigated. In the future, we provide the reference for company to develop the children drawing system.

Keywords: intuition, drawing, pointing device, gesture movement

1   INTRODUCTION

In terms of digital drawing, whether it is the conventional Paintbrush or the multi-functioned professional drawing software such as Corel Painter, Illustrator, and Photoshop etc., are almost omnipotent with the Wacom digital drawing boards or tablet PCs. However, the strong combination of soft and hardware is not necessary compatible with the intuition and usability of children's creations. Problems include complicated level interface design, small key arrangements, excessive function operation, and unrecognizable function icons etc.

This research employs the convenience and intuition of tablet PCs as the drawing platform, observing the drawing behavior of children, transferring it into an intuitive operation directional device control panel design. This is combined with the intuitive and fluently controlled digital paintbrush, analyzing the development of the current drawing systems, considering the thinking and cognitive behaviors, developing a sketching system suitable for children that is as easy to use as the conventional paper and pens, but also include the digital advantages of saving, communicating, and editing.

Children start expressing themselves from birth with manners that could be traced. Drawing, being their way of self-expressing, the features in children's drawings are closely related to their physical and psychological developments. Hence, their drawings change with their development phases.

Garder (1980, 1982) presented the U tendency theory of child art development, claiming that preschool children are like artists, they create freely and express vividly. However, for children in their latent period and teenagers, the free, vivid, and unique features decrease with age. Children's drawing activity is to express the objective emotions and ideas, it is an expression of reception and art creation.

Eisner(1972) pointed out that the development of art is not the natural outcome of growth and development, but a process affected by children's experiences. Children's art abilities are mostly a learnt function. Hence, knowing children's growth and development, and understanding the psychological features and physical abilities of children's creations in order to develop creative activities and expand children's conceptual abilities and depth to avoid disappointment and a decrease of interest due to unsatisfactory work, has always been the direction for the art education of children.

For children, drawing is a process of creative thinking. During the process, children integrate and revise personal impressions, experience, and imagination to create more possibilities. Drawing is a process of seeing-moving-seeing (Schon, 1992). The visual focuses on the image, at the same time they observe, fantasize, imagine, or remember past ideas. Therefore, visual plays an important part in the process of drawing, and excessive visual transformation could distract attention (Palmer, 1993). The current digital drawing system employs mostly digital boards or computer mouse as the input device, combining with drawing software of visual icon interface to proceed with varies digital drawing work. However, the visual icon interface control mechanism often causes obstacles in children's drawing process.

2   METHODS

2.1   Understanding the Problems

We have employed tablet PCs and Paintbrush, which is currently the most suitable drawing system for children. After instructions, the children were asked to operate the system for 15 minutes, drawing colored pictures on the topic of the zoo. Questionnaire interviews were done with the seven preschool children focusing on their feelings after the drawing activity, in comparison with the conventional paper and pens. As a whole, over 75% of the children felt that computer aided drawing was better and more interesting than the conventional drawing medium. However, on the difficulty of operation and learning, the conventional medium was easier. From this, we can state that the current digital drawing system is disruptive for the creative process of children.

Other than that, we employed participants' operation video of 15 minutes or more and their behavior and movement photos, under the company of their parents, proceeding a retrospective observational analysis of the operational movement features and the first questionnaire interview. The function of the questionnaire is to assist, through guiding suggestions, the children were encouraged to recall the experience focusing on operational features, aiming to obtain all the possible operational requirements. The major requirements include the following four items: simple and understandable icon interface, easier control movements, easier direction control, and an undisruptive key interruption.

Secondly, able to point accurately, easy to learn, moving with speed and mobility, intuitive and simple operation, suitable for children's maturity level, operational feedback compatible with the icon interface, and a horizontal drawing tool arrangement takes 49.98% of the total requirement. As for the radial tool arrangement, sliding control, and the choice of hue by point-and-click on the color chart occupies only 2.38%, presumably due to the distraction the radial tool arrangement causes by occupying too much of the drawing surface. As for the sliding operation and the choice of hue by clicking on the color chart, children are affected by the habit of operating a mouse and needs even more time to adapt to the intuitive absolute displacement of the tablet PC and digital paintbrush. This causes inaccuracy in the clicking of color charts and the clicking, pressing, and sliding of the digital paintbrush on the control panel.

Most children stated that the current direction device (the commonly used direction device is a computer mouse) are too big in size, therefore it is necessary to put this factor into consideration while designing a materialized control device for the left hand. Also, most children agree to use their left hand to choose drawing tools, stating that using both hands to control and eliminating unnecessary searching and point-clicking tasks should increase the fluency of the drawing process. Special attention should be paid to the fact that computer informational devices used by children often consist of image and games, which is very different to the document work used by adults. This causes extra attention on the performance of control devices under a large amount of high speed movements.

Other than that, in terms of choosing drawing tools, amongst radial, vertical, and horizontal arrangements, horizontal came to the top of the acceptance scale with a 5.95% approval. Hence, in the following design factor weighing comparison, horizontal arrangement is used as the representative and the other two eliminated. Same for the choosing of hues, the turning manner, with a top approval rate of 3.57% will be used as the representative for the choosing of hues.

2.2   Design Elements and Development

The purpose during this stage is to organize the operational requirements of children's drawing tasks discovered by usability engineering, in order to establish the design and evaluation criteria for the designing process, the selecting of design ideas, and the construction of functional models.

While observing the children from the preschool of National Cheng-Kung University during their art classes, the following behavior were found through their use of conventional color markers, crayons, and water colors: (1) When the right hand is holding a marker, the left hand holds the marker cap. (2) When the right hand is holding a pencil or crayon, the left hand holds the eraser. (3) When the left hand is holding a cap or eraser, flipping, turning, or pressing often occurs. (4) When choosing a color, the finger slides pass all the colors before making a choice. (5) When using paint, the right hand holds the brush while the left hand holds the palette. (6) While drawing, the paper moves according to the change of drawing posture.

Since the factor weight concluded by the usability engineering settings could only represent the importance of each factor valued by the participants, but not the relationship between the factors, during this stage, we will employ the ISM law to categorize item properties, transforming complicated design suggestions into circulating structures.

Using the ISM structure diagram, designers could clearly see the connection between varies factors. The simple to learn and use features in the structure diagram (increasing the simplicity of direction control, learning, movement control, accuracy of point and click, left hand tool choosing control, and less interference with the keys) are the main objectives of the design (design exists). These objectives could be reached by handling external features or technical program settings. These external features include horizontal arrangement tools, rotation hue selection, integration of operational feedback and icon interface, simple icon interface, fast moving speed and flexibility, intuitive and simple operations, and small in size. The handling of the above features needs to be compatible with the left hand operational habits, motion, and comfort. Understanding of children's physical features is also important to reach the goal of designing a drawing device for children.

According to the relation analysis from ISM, the size of the drawing device, the simplicity and intuitive operational manner, the moving speed, the harmony between the feedback of the drawing software interface operation and the icon interface, the simple and understandable icon interface, and the arrangement and manners of drawing tool choices are the eight major factors that makes up the design entrance of the first stage. In the development of concept design, ingenious fingers and used instead of the tardy wrist, forming an ingenious and fast operational feature. The size of the directional drawing device is being decreased in order to decrease the operational load of the users. The goal is to operate with the hands under a static situation, in order to decrease the physical load of users, completing the above goals. The following four concept designs all consist of the above features as shown in table 1.

Table 1. Conceptual Design

Conceptual Design	Design Features	Strength	Weakness
1	This concept is based on the finger tip tactile sense, using the posture change and movement of the five fingers to change between functions, making the directional device interface invisible.	(1) No specific control interface. (2) Light in weight. (3) Hands are at a natural and rest position. (4) Simple to learn and use.	(1) The support would be difficult to design.
2	This concept is based on playing with the device itself. Users flip and device and slide their fingers to change functions, making it fun to operate with.	(1) Simple operation (2) Simple to learn and use.	(1) Low work performance . (2) Large in size.
3	This concept offers support for the side and the center of the palm to avoid the weak carpal tunnel. It employs the most flexible three fingers and the thumb to operate the device.	(1) Perfect write support (2) High speed switch of functions.	(1) A low range of human engineering adaption.
4	This concept uses the flexible fingers to control the sliding of the cross section sliding tunnel, adjusting parameters, and using the stronger elbow to turn and switch between functions.	(1) Hand left in a natural position. (2) Rational work sharing.	(1) Lack of support for the wrist. (2) Complicated operation.

As a whole, three out of the four concepts are stronger than the current directional drawing device (digit pens). Amongst those, concept 1 achieved the highest adaption level of 1.845 while case 3 achieved a 1.571, and case 4 achieved only 1.047. Hence, this research would employ concept 1 as the top priority design, basing the analysis and research of the hand posture operation transformation on concept 1.

According to the previous observation on the common drawing tools used by children, the questionnaire and interview results, the suggestion from experienced teachers of children, and the usability engineering analysis, we were able to conclude the choice of tools in the children's drawing system, along with the editing of parameters, constructing three fundamental functions of drawing: The choosing of paintbrushes, the adjustment of hues, and the movement of paper, as shown in chart 1.

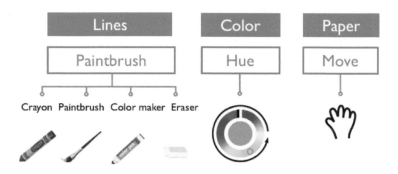

Figure 1 The drawing factors of systems

2.3 Experiment Design

The experiments are grouped into: (1) Hand gesture image design and categorizing – by employing focus groups and brainstorming, producing a large amount of hand gesture images (focus group members being teachers with at least two years of child drawing teaching experience, people who are familiar with drawing software, and people who possess an understanding of drawing system design). Literature review on the procedures of icon design was done to understand the principles of symbol design before hand gesture symbols are categorized accordingly and evaluations are done as a whole. (2) Final hand gesture image sample choice – The hand gesture image samples concluded from the focus groups are made into evaluation questionnaires (hand gesture images are grouped into three items including the choice of paintbrush, the adjustment of hue, and the moving of the paper). Preschool children evaluate the hand gesture images under each item, then leaving it to professionals to categorize, select, and compare in order to produce the final experiment sample. (3) Functional operation intuition experiment – This experiment is to test the drawing operation performance of the three types. The visual interface input type is used as the control group (touch pen and mouse) to compare the fluency and error rate between hand gesture interface operation and visual image interface, at the same time testing whether the hand gesture operational manner achieves user expectancy. During the experiment, participants were asked to follow instructions to choose functions using both hand gestures and the touch pen, in order to simulate the actual condition of the specific position tasks and drawing function adjustments during drawing process.

3. RESULTS AND ANALYSIS

3.1 Function Operation Intuition Performance

From the ANOVA analysis of the experiment data results we can see that significant results were obtained from the paintbrush change experiment, the hue adjustment experiment, and the paper shifting experiment. In both the paintbrush change experiment and the paper shifting experiment, the significant value were far less than 0.05, showing clearly significant results in all tested events between the control and experiment groups.

From the ANOVA analysis we can see that significant differences were obtained between all tested events between the control and experiment groups. Also, from the statistic data, we can see that in most experiments, hand gesture controls take a longer time than visual icon operations, with an exception of the paper shifting experiment, in which using hands to directly move the drawing screen is much simpler than the visual point and click steps. In the general experiments, hand gesture control also took a shorter average operation time than visual icon operations. From this, we could see that even though in the single task operation experiments, hand gesture operation takes a longer time to adapt to due to the learning of the task each gesture represents, hand gesture operation presented better stability in the general experiments. The inference being that a child's experience in mouse and touch pen operation would affect her operation reaction time. Children with no experience are often interrupted by the cursor and are often unable to accurately point and click on the desired task, which results in longer operation time. Hence, the standard deviation showed in each task experiment is much higher than hand gesture operations.

Also, in the Tukey and Scheffe paint brush change performance experiment, it was found that in the visual icon click operation, touch pen and mouse are being grouped together while in the hand gesture control, they are independent. From this, we can see that in the paintbrush change experiment, the touch pen and mouse showed similar operation times. However, hand gesture control showed a significant difference as during hand gesture control, participants use the flipping between the palm and the back of the hand to select paintbrushes, hence time is needed to wait for the program to change brushes. The program is unable to adjust according to the participant reactions. Therefore waiting time often occurs which delays the operation time.

3.2 Error Rate

In the single function (change of paintbrushes, adjustment of hues, shifting of paper) experiments, 6 questions were present. In the general operation experiments, 12 questions were present. The number of questions answered and the error rate were calculated and analyzed by ANOVA. All items showed significant results (sig<0.05). The significant values are listed as the following respectively. Change of paintbrush: 0.001; Adjustment of hue: 0.021; Shifting of paper; 0.000; General operation experiment: 0.036.

The hand gesture control manner presented a lower performance in terms of error rate than touch pen and mouse operation in both single function and general operation experiments. From this, we can see that hand gesture control presented a lower operation performance than the other two operation manners (results obtained from the function operation intuition performance experiment), but better stability in terms of operation. The average error rate were lower than 30%, which is two thirds of all questions correct.

3.3   Subjective Rating

After the experiments, participants were asked to grade the visual icon interface and hand gesture control interface systems, and the operation manners of mouse, touch pen, and hand gesture. The average data results show that as a whole, touch pen operation interrupts tasks and causes finger and wrist fatigue more than the other two. The possible reasons being that the distance between visual and drawing interface is closer when using a touch pen, resulting in visual oscillation during point and click tasks, causing interruption. Also, hand operation movement development for children during this stage has not been completed, which decreases the accuracy and speed of the movement. Hence, comparing to the small range point and click and moving motions when using a touch pen, the hand fatigue level is lower during hand gesture controls of a larger range.

The ANOVA analysis showed significant levels of 0.124 for the subjective comparison of wrist and finger fatigue levels. The value is larger than 0.05, which is insignificant. This means the score for the three operation manners are very close and that hand fatigue levels would not be affected by the different operation manners. The analysis results show that the average score for the three operation manners are the same.

In the subjective reaction meter, in general, the hand gesture control scored higher than mouse and touch pen controls. Aside from subjective comparisons – Which tool is easier to use (usability), which tool is easier to learn (training ability), and as a whole, which device is better these three items showed a slightly lower score. These three items are closely related to proficiency and prediction. The presumed reasons being that comparing to the frequently used icon interface system, the participants are less familiar with the hand gesture control system, which still has room for improvement even after practice. On the contrary, hand gesture control received a much higher score in the items of "Which tool is more interesting", "Which tool do you like better", "Picture abundance", and visual attention during operation process. Participants pointed out that it had never occurred to them that drawing system tools could be operated by hands. Even though it takes more time to learn in the beginning, it is interesting, fun, and they wanted to try different colors and brushes during the drawing process with a longer concentration time and persistence.

In the ANOVA analysis, the question "Which tool causes less fatigue (comfort)" showed a significance of 0.339, which is higher than 0.05 and did not reach significant level. This means that the scores for the three operation manners are very close. From the above line graph of the average score from the subjective reaction meter, we can see that the comfort level is mouse > hand gesture control > touch pens. This is related to the fatigue level in the physical meter, in which touch pen > hand gesture control > mouse. The rest of the subjective items showed a significant value of far less than 0.05, which is clearly significant. From this we can see a very positive subjective evaluation of hand gesture control.

Combining the above results, hand gesture control showed a less satisfactory performance results. However, there is a great room for improvement after practicing, learning, and technique improvement. On the contrary, in subjective evaluation, it received a higher evaluation than the visual icon interface control. Hence, it is feasible to apply hand gesture control interface system on the digital drawing of children.

4.　DISCUSSION

The function operation intuition experiment results show no significant results in the single task operation for the hand gesture control interface. From this we can see that participants are less flexible in terms of function transformation. However, with more practice, more accurate expectations could be met with the system operation (Jordan, 1995), increasing the operation flexibility. In the general experiment, hand gesture control achieved better performance. Therefore, we presume that participants could intuitively operate all function. This conforms to Fitts' law, the further away the target, the longer the moving time, and the smaller the target, the longer the moving time. Intuition and fluency in the drawing process are the goals of this research, and the speed of operation is one of the indexes of these goals.

The subjective evaluation results show that even though in the questions "Which tool is easier to use" and "Which tool is easier to learn", the hand gesture operation received a slightly lower evaluation, but there is a great room for improvement after practice. Also, in questions "Which tool is more fluent to use", "Which tool attracts you to continue playing", "Visual feedback is sufficient for the required information level", and "Visual attention on the interface during operation" participants evaluated the hand gesture control system higher than the visual icon system. This results explains that compared to visual icon, the hand gesture control is more compatible with the interface, more intuitive, and easier to memorize. This matches

As a whole, on either performance measurement or subjective evaluation, hand gesture control operation obtained a higher satisfaction level than visual icon interface.

5.　CONCLUSION

The goal of this research is to discuss the possibility of using hand gesture control interface input to replace graphic user interface (GUI) to improve performance, lower error rate and avoid interruption during the creative process due to visual search. (1) This gesture control digital drawing system operates intuitively:

The general operation intuition experiment explains the intuitiveness of this system. Also, in the participant subjective evaluation, participants pointed out that gesture control is more intuitive than the graphic user interface. Mechanisms that operate intuitively increase satisfaction, cuts down on excessive information, and effectively increases fluency in a drawing task. (2) This gesture control digital drawing system shows better control performance than the graphic user interface system. (3) This gesture control digital drawing system possesses control fluency. (4) This gesture control digital drawing system obtained an excellent subjective evaluation from users. The result states that the gesture system not only showed effectiveness in the drawing function control, but also achieved a higher satisfactory level than the graphic user system. School teachers also agree that drawing with hands increases hand and brain coordination, which is helpful for the learning and development of children. Integrating the above results and suggestions, it is clear that the possibility is high to replace the graphic point and click mechanism with the gesture control mechanism.

ACKNOWLEDGMENTS

The authors would like to thank the National Science Council of the Republic of China for financially supporting this research under Contract No. NSC 98-2221-E-006 -086 -MY3.

REFERENCES

Byong, K. K., and S. Y Hyun,1997. Finger Mouse and Gesture Recognition System as a New Human Computer Interface, *Comput.& Graphics* 21(5): 555-561.

Egloff, T. H. 2004. Edutainment: a case study of interactive cd-rom playsets. *Comput. Entertain., 2*(1), 13-13.

Eisner, E. W. (1972). Educating artistic vision. NYC : Harper & Row.

Gagne,R.M. 1974. Essentials of learning for instruction. New York: Dryden press,.

Hummels, C., K. C. Overbeeke, and S. Klooster. 2007. Move to Get Moved: a search for methods, tools and knowledge to design for expressive and rich movement-based interaction. *Personal Ubiquitous Comput.,* 11(8): 677-690.

Ishii, H., and B Ullmer. 1997. Tangible bits: towards seamless interfaces between people, bits and atoms. Paper presented at the Proceedings of the SIGCHI conference on Human factors in computing systems.

Kang,H., W. C. Lee, and K. Jung. 2004. Recognition-based Gesture Spotting in Video Games, *Pattern Recognition Letters*, 25: 1701-1714.

Palmer, J., C., T. Ames and D. T. Lindsey 1993. Measuring the Effect of Attention on Simple Visual Search. *Journal of Experimental Psychology-Human Perception and Performance,* 19 (1), 108-130

Schon, D. A.. 1992. Kinds of Seeing and their Functions in Designing. *Design Studies,* 13(2): 135--156

Shaw, C. D.. 1997. THRED: a two-handed design system *Multimedia Systems,* 5(2): 126-139.

CHAPTER 59

An Evaluation Tool the Subject / the Pointing Device Pair

Frédéric VELLA, Damien SAUZIN, Nadine VIGOUROUX

Paul Sabatier University
IRIT laboratory
118, route de Narbonnne
31062 Toulouse Cedex 9
France

ABSTRACT

The paper aim is presented a tool for evaluate the pointing devices / a subject pair. For that, the analysis tool is disposed to parameters of the literature. Thanks to its, we can studied the pointing devices using. We illustrated an evaluation with Wiimote. We chose the Wiimote to the absence in the literature for use with accelerometers as a pointing system. Directions on the analysis of the Wiimote, the use of accelerometers in two simultaneous problems appear on the diagonal movements. This result is important to interact with a WIMP interface.

Keywords: Subject, pointing device, Fitts law, wiimote

1   INTRODUCTION

The variety of interaction techniques currently offered innovation in the field of information technology and communication allows us to consider the design tools replacement or rehabilitation to promote independence and increase the quality of life for people in situations of disability. However, these interaction technologies are not accessible to all (cannot make a motion to move the cursor to a pointing device, inability to perform validation by a click of the pointing device, etc.).

To interact with an interactive system (office applications, environmental control, etc.) the disabled person with upper limb disorders needs to have an interaction device adapted to these functional capabilities. The severity of functional impairment of the upper motor disabled person forced the choice of pointing device. Figure 1 provides some pointing devices (guide finger, trackball, switch, eye tracking) in the residual upper limb motor function. They are less mobile, more devices are operated with other body parts that remain functional such as eyes ("eye tracker"), head ("head tracker"), etc..

Furthermore, the long use of these devices causes fatigues motor and attention for disabled motor (Vella and Vigouroux 2007). This is sometimes due to an inadequate a) between the subject and device and b) and non compliance with requests by users.

Degree of impairment of the upper

Figure 1 Pointing devices according to degree of impairment of the upper

We begin this article with a background on the parameters and Psychophysics models for the evaluations of devices. We report the concepts selected for the design of the Model Processing Human (MPH) platform. Then we describe the MPH architecture and the data files record. Finally we report experimentation with Wiimote which used MPH platform.

2   BACKGROUND

2.1   Parameters to evaluate the device

The following variables (MacKenzie, Kauppinen, Silfverberg, 2001) used to analyze parts of a global movement acquisition. For clarity, we define the axis of the task: it is the right starting from the point of origin of a moving and passing through the center of the target. (In Figure 2 the axis of the task is the dotted lines).

From the axis of the task and the direction of movement, MacKenzie has defined four variables and in addition it has identified three other variables that depend on the distance to the axis of displacement of the task.

Target Re-entry (TRE) (Figure 2 a): number of input and output on the target before you click on the target.

Task Axis Crossing (TAC) (Figure 2 b) number of the displacement which intersects the axis of the task.

Movement Direction Change (MDC) (Figure 2 c) MDC are direction changes depending on the determined distance from the axis.

Orthogonal Direction Change (ODC) (Figure 2 d) The ODC is changing direction from the perpendicular to the axis of the task.

Movement Variability (MV) = square root (Sum $((y_i - y)^2 / n-1)$) with n numbers of records. MV = 0 is a test performed to perfection. (Standard deviation of the average distance to the axis).

Movement Error (ME) = Sum $(| y_i |) / n$ (Axis 1 dimension) Average distance to the axis of the task.

Movement offset (MO) = Sum $(y_i) / n$ defines the tendency to go above or below the axis of the task.

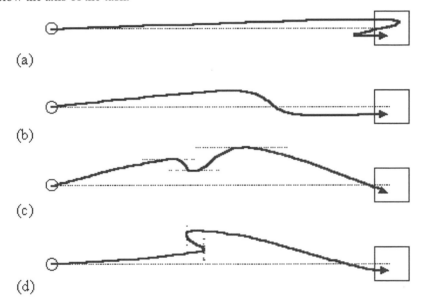

(a)

(b)

(c)

(d)

Figure 2   Variation of cursor movements borrowed from (MacKenzie, Kauppinen, Silfverberg, 2001)

The others parameters allowed to compare the devices: the speed, the accuracy, the movement times, the acceleration, etc..

We notice that there are many parameters to evaluate the devices. However, these parameters do not allow studying human behavior on the use of a device.

2.2   Psychophysics models

The psychophysics models allowed studies human behavior and predicting the time to make a task (here the task is hit a target). We choice the Fitts` law (Fitts,

1954). The validity of this law was shown by (Card, English, Burr, 1978) in tasks of hitting targets on a computer screen. It has certainly subjected to various evolutions (Mackenzie, 1992) but today it remains a tool for comparing pointing techniques. The formula (1), is related to the movement time prediction (MT) of a target to another. It depends on three parameters of regression a and b, empirically determined and index of difficulty (ID). The ID parameter (2) depends on: W_j is the width of the target; A_{ij} is the distance between two targets i and j.

$$MT = a + b \times ID \quad (1) \qquad\qquad ID = \log_2(A_{ij}/W_i + 1) \quad (2)$$

We are interested by Fitts' law because it takes into account the movement preferential according to the targets directions. This allows us to study: the pointing devices using and the human behavior. For this study, we are calculated the linear regression according to MT and ID. Then, we are calculated the integral under the linear regression. The movement preferential is depended of the integral value, more the value is small more the movement is preferential.

2.3 Devices evaluation

The literatures often mention comparative studies on the pointing devices usability. We show that the pointing devices such as mouse and / or trackball, remained the most accurate and faster to new technologies like touch screen (Sasangohar MacKenzie and Scott, 2009), Wiimote (Natapov, Castellucci, and MacKenzie, 2009) and eye tracking (Miniota, 2000). However, these studies use a lot of parameters to evaluate the devices. These parameters are different from one study to another. Nevertheless, we want compared the subject / the pointing device pair. What parameters are responding to this study?

3 MPH PLATFORM

Several platforms have been developed ((Kabbash, MacKenzie, Buxton, 1993), (Ritter, Baxter, Jones, Young, 2000]). It allowed the experiments around the Fitts' law adapted to meet targets on screen. For example, the application "A web-based test of Fitt's" aims to allow "testers" to estimate their own motor skills and illustrate (for classes and demonstrations) the principle of this law.

Nevertheless, the vast majority of these platforms are designed ad hoc for a specific experiment, or as a demonstrator.

Our experimentation platform aims to be more general: it includes several exercises to model laws and model human systems whether disabled or invalid.

3.1 Architecture

We designed a platform to study motor behavior. This platform called MPH (Platform Model Human Processor) (http://www.irit.fr/MPH), The MPH platform meets two major requirements:

- The first is to enable disabled people not to move the place of experimentation. Indeed, the experimental conditions are often better at home because of the suitability of their positions. To meet this objective, the MPH platform is based on a client / server architecture (Figure 2) to perform exercises "in situ" in the subject (client side), and retrieve various data (profiles of subjects, results of the exercises) to laboratory (server side);
- The second is to allow the pooling of different experiments carried out for comparison among French and international scientific communities.

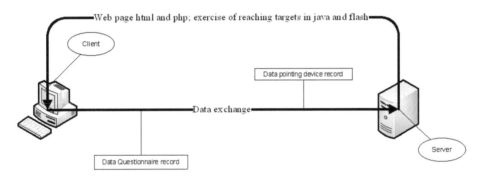

Figure 3 Architecture of MPH Platform

3.1.1 The client part

Each exercise deployed on the platform respect the following sequence:
- The subject used a web browser of choice (such as Internet Explorer, Firefox, Safari, Opera) and enters the URL of the exercise;
- Next, the subject has two possibilities: either he refuses consent and he is directed to the home page, or he accepts the consent and he is directed to the instructions of the exercise. When the subject has finished reading them, he replies to a questionnaire describing their profile, knowing that his anonymity is respected. The subject is identified by a number that increments as and when a new topic passes the test. When the subject submits the form, he is realizing the exercise;
- When the subject finished the experiment, he is directed to a final assessment questionnaire. This allows us to get the feel of the subject relative to the test.

3.1.2 The server part

We have an MPH part in our laboratory. This part called server saves the interaction traces of the different experimentations controlled through the experiment design. We can analysis the data files MPH with ECSD tool. We integrated in ECSD the parameters found in the literature.

The visualization tool (ECSD) parameters are available to compare devices (time required to reach the target error rate, accuracy of target attainment, study the trajectory of the pointing cursor, the index of difficulty according to Fitts 'law, etc..) and according to different cardinal directions, and various models of Fitts' law (linear regression modeling of full frame) to study the difficulty of using a pointing device.

The visualization tool has a friendly interface that allows you to select parameters according to the subjects, types of devices, the size and distance of targets. The display of these parameters can make decisions and perform a classification.

3.2 Data file

The MPH application data is stored in trace files. These data are saved in a file named as follows:

"UserID" + "v_ (valid) or h_ (handicap)" + "TypePériphérique" + ". Txt"
(Eg 168h_wiimote.txt).

The file structure is a list of different configurations for the same distance from the same target and target size. Each of these configurations includes a description of events sent by the device and targets.

3.2.1 Description of mouse events

Moving (Move) pointer and the status of the mouse: pressure (clickDown) and relaxation (clickUp), are in the format (time (time in ms), x (x-axis coordinates) and y (coordinate axis)).

Move: time = 5048, x = 300, y = 296
ClickDown: time = 5206, x = 300, y = 296
ClickUp: time = 5356, x = 296, y = 296

3.2.2 Description of target

Configurations give the description of target size (size) and distance (distance). These two variables are in pixels. Time (time in ms) is the time when the configuration of the targets appear on the screen.

Setting: time = 0, size = 32, range = 80

This configuration is followed by a list of targets called Button. Each is numbered (here from 1 to n) and has the x and y coordinates of the display position.

Button 1: x = 284 y = 188

...

Button n: x = 288 y = 288

At the elementary task of achieving targets, we also record the events of the latter. They are of two kinds: Pressure (Press Button) and relaxation. (Button release). These two events of the target are the same size as mouse events. The gap of time between the two types of events (mouse and target) is explained by the processing time of each.

Each target has two states: when we click on it is either active (between Button and Button press release the active state is marked by: Click on the active target) or is disabled (nothing is noted between Button and press Button release).

Move: time = 2297, x = 304, y = 308

ClickDown: time = 2584, x = 304, y = 308

Button press 9: time = 2585 x = 304, y = 308

Click on the active target

Release button 9: time = 2736, x = 304, y = 308

ClickUp: time = 2801, x = 304, y = 308

4   ILLUSTRATION WITH WIIMOTE

We chose the Wiimote for the absence in the literature for use with accelerometers as the pointing device. Our hypothesis for this evaluation is the use of the Wiimote for pointing.

For this he was asked 12 valid to make a exercise of application MPH in two dimensions.

4.1   Progress

For all sessions, the user is comfortably seated at the computer, in a lit environment. The pointing device is installed in the position that the subject wishes. From there, we start the session.

The sessions take place in two stages: the first to find the calibration of the device. This calibration is used to define the sensitivity of the pointing device. It allows a user to have an optimum use of the Wiimote.

Once the calibration set, the user performs the exercise MPH, it opens a browser (Chrome) and between the following address in the browser: http://www.irit.fr/MPH. On the page that appears, the subject must select the link "Experimentation of Fitts' law in two dimensions." The page set is displayed on the screen, the subject read the instructions and it has two options. Either he refuses consent and he returned to the home page, or he accepts the consent and the experiment continues. A login page appears if the user is at its first session, he joined several elements describing the profile. Following the first questionnaire, it gets an ID that will be noted for the following sessions. If he had an identifier is that it is the second day so the information is already known and it passes to the next step.

Following the first questionnaire, if the user wishes, he clicks the button launched.

Targets appear (Figure 3); the size and distance of the first targets are randomly chosen. The subject place the cursor over the central target and prepares to click on the external target is going to turn as quickly and as accurately as possible.

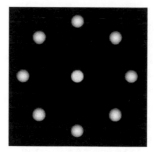

Figure 4 Exercise Fitts 2D, 9 targets with a center. The eight targets are arranged in a circular on the cardinal points (North, Northwest, etc ...). Four distances from the center of the target center and the center of Target (D = 40, 80, 160, 240 pixels) 4 target size (W = 16, 24, 32, 40 pixels)

The first target is always the central target, when it is the target to reach the user can make a pause, after this attack, target outdoor lights. After clicking this target, the subject must again return to the central target.

For each block the subject must click on the 8 targets outside before moving to the next block. Once the session is finished, click on the topic and send a new questionnaire appears. The questionnaire allows for the views and perceptions about the opposite experience.

4.2 Results

At first the results obtained here on accuracy, the integral of the curve MT / Fitts' law and the speed of cursor movements.

Accuracy is a stable parameter (Figure 3) in any direction but the result shows the integral around a difficulty in a southwesterly direction (Figure 4). The speed of user movement (Figure 5) is greatly reduced in the same direction.

So there is a direction that users do not seem well mastered using the Wiimote.

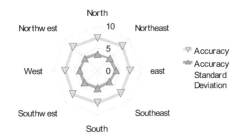

Figure 5 Accuracy according to directions

Figure 6 Preferential movement according to directions

Figure 7 Speed according to directions

5   CONCLUSIONS AND FUTURE WORK

From this background, we have developed a tool for analyzing data to assess the subject / device pair. In making this assessment, we established a protocol of experimentation. The study on the Wiimote with able bodied people shows the large amount of data and variables to analyze. We concluded that the Wiimote accelerometer as the scoring system is not the most efficient device.

In perspectives, it would be interesting to compare with other devices the results of the Wiimote to determine a ranking of features based on their results. The addition of experience with people with disabilities could help to identify a link with a classification, which can help people with disabilities to choose a device.

Fatigue is very important for people with disabilities, should therefore look at the behavior of parameters on motor fatigue.

REFERENCES

Vella F, Vigouroux N. 2007. Layout keyboard and motor fatigue: first experimental results. *AMSE journal, Association for the Advancement of Modelling and Simulation Techniques in Enterprises (AMSE)*, avril 2007, Vol. 67: 22-31.

MacKenzie, I. S., Kauppinen, T., & Silfverberg, M. 2001. Accuracy measures for evaluating computer pointing devices. *Proceedings of the ACM Conference on Human Factors in Computing Systems - CHI 2001,* New York: ACM: 9-16..

Card, S. K., English W. K., Burr B. J. 1978. Evaluation of mouse, rate-controlled isometric joystick, step keys, and text keys for text selection on a CRT. *Ergonomics,* 21: 601-613.

MacKenzie I. S. 1992. Fitts' law as a research and design tool in human-computer interaction, *Human-Computer Interaction,* 7, Moscow: 91-139.

Fitts P.M. 1954. The information capacity of the human motor system in controlling the amplitude of the movement, *Journal of experimental psychology* 47: 381-391.

Sasangohar F., MacKenzie I. S., & Scott S. D.2009, Evaluation of mouse and touch input for a tabletop display using Fitts' reciprocal tapping task, *Proceedings of the 53rd Annual Meeting of the Human Factors and Ergonomics Society* – HFES 2009, Santa Monica, CA: Human Factors and Ergonomics Society, 2009: 839-843.

Natapov D., Castellucci S. J., and MacKenzie I. S.2009, ISO 9241-9 evaluation of video game controllers, *Proceedings of Graphics Interface 2009.* Toronto: Canadian Information Processing Society: 223-230

Miniotas, D. 2000. Application of Fitts' Law to Eye Gaze Interaction, *CHI '00. ACM Press* (2000): 339 – 340.

Kabbash P., MacKenzie I. S.; Buxton W. 1993. Human performance using computer input devices in the preferred and non-preferred hands. *Proceedings of the ACM Conference on Human Factors in Computing Systems*, New York: 474-481.

Ritter F. E., Baxter G. D., Jones G., Young R M., 2000. Supporting cognitive models as users. *ACM Transactions on Computer-Human Interaction (TOCHI) archive*, Volume 7, Issue 2, Part 2: 141 – 173.

CHAPTER 60

Effects of Seat and Handgrips Adjustments on a Hand Bike Vehicle. An Ergonomic and Aerodynamic Study for a Quantitative Assessment of Paralympics Athlete's Performance

Marco Mazzola, Giuseppe Andreoni, Gabriele Campanardi, Fiammetta Costa, Giuseppe Gibertini, Donato Grassi, Maximiliano Romero

Dipartimento di Industrial Design, Arts and Communication (INDACO),
Dipartimento di Ingegneria Aerospaziale
Politecnico di Milano, Milano, Italy

ABSTRACT

The aim of this paper is to present a research study on a Handbike vehicle assessment for London 2012 Paralympics competition.

The main assumption of this study is that the athlete's posture can affect directly both the ergonomics and the aerodynamics of the performance, and the objective is to determine a method to perform a quantitative analysis to understand the way in which these important factors should be considered for the optimal performance of the athlete. To investigate this hypothesis, an Italian Paralympics athletes has been experimentally evaluated at LyPhE (Laboratory of Physical Ergonomics) and at GVPM (the Wind Tunnel) of Politecnico di Milano.

The methodology proposed in this study requires a first ergonomic evaluation conducted through a kinematic to understand how specific vehicle adjustment could affect the performance. The Aerodynamic assessments is based on the measurement for each selected condition, of the aerodynamic resistance coefficient in a wind tunnel.

Results of this work demonstrate that the ergonomics benefits and the aerodynamics improvements present an opposite behavior. In the ergonomics analysis there are two important evidences: the first one is that the increase of the handgrip length influences negatively the ergonomic scores in all the different postures; the second one regards the inclination of the backseat; the more the athlete posture is vertical, the more the ergonomic index present a decrease in the discomfort perception. On the contrary, the vertical position of the trunk affects negatively the aerodynamic resistance coefficient, and this result is coherent with the observation that, an increasing of the wind exposed surface produces higher values of the aerodynamic resistance coefficient. Also results about the length of the handgrip are in contrast with the ergonomic analysis. In fact, the longer handgrips seem to be more efficient for the aerodynamic analysis.

Results about the seat adjustment are coherent with the initial hypothesis that ergonomics and aerodynamics present opposite impacts. The vertical position of the trunk requires less anti-gravity movements that are not recommended for human performance and increase the effort level and consequently the resistance in races. The most surprising result is the high relevance of the handgrip impact to the whole performances for both the analysis. As a conclusion, the optimal setup of the handbike seems to be related to the type of each single race; for longer races the ergonomics adjustments seems to be preferable while for races with high speed, the aerodynamics factors are definitely more important.

INTRODUCTION

Hand biking is an Olympic discipline where athletes drive a cycle with hands propulsion. Although the interest and diffusion of this sport is increasing in the last years, there is still a lack of quantitative studies regarding the vehicles assessment in terms of ergonomics and aerodynamics.

The proposed mixed method approach integrates biomechanical analysis and aerodynamic studies of the human-vehicle system into a research strategy to provide a comprehensive understanding of the analyzed phenomena. The aim is to find the best compromise between postural comfort and aerodynamic resistance. The biomechanical analysis was conducted at the instrumental laboratory of human motion analysis and the aerodynamic study was performed in the wind tunnel of Politecnico di Milano.

This research is focused on a single athlete in response of the need to produce specific information to improve the vehicle assessment for London 2012 Paralympics competition.

METHODS

We defined the biomechanical protocol and performed the acquisitions and evaluation with the MMGA index of discomfort (Andreoni et al., 2009).

Upper-body kinematics was recorded through a six cameras optoelectronic system while the subject performed the cycling task with several different configurations of the handbike. The cameras were placed so that a volume of about 3 x 2 x 2 m was covered. A set of 32 passive and reflective markers, placed on the subject's body surface, were used for the kinematic computation (Schmidt et al., 1999). Calibration procedures were carried out before each experimental session, which was repeated three times,

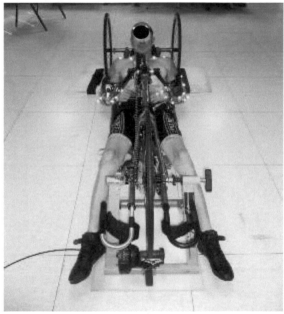

Figure 1: the subject with the 32 marker placed for the kinematic computation

The variables considered for the study were: wrist and elbow flex-extension; shoulder flex-extension, intra-extra rotation and abd-adduction; trunk flex-extension, rotation and lateral bending. The kinematics analysis is composed by the computation of the joints centers' trajectories of the trunk, left and right shoulders and left and right elbows and the flexion-extension angles of the elbows and shoulders joints. This is due to the fact that this is the most relevant movement for the analyzed task (Arnet et al., 2011).

The MMGA index has been calculated for each configuration, considering quantitative computation of the joints motion, a coefficient of discomfort and the mass of the involved body districts. The task performed with the athlete's own handbike has been considered as the reference value. The other configurations were obtained by using an adjustable handbike. As experimental conditions, the backseat inclination in 5 different positions, the handlebar at 4 different heights and 2 different lengths of the hand grips levers have been considered. Each one of the different regulations has been tested as a single variable with a total number of 32 experimental conditions.

The aerodynamics tests have been carried out in the aerodynamic test-section of the Large wind Tunnel of Politecnico di Milano (the GVPM). The test-section size is 4m x 3.84m assuring a very low wind tunnel blockage. The flow velocity can be set from 3 to 55 m/s. For the present activity the velocity has been set equal to 13.9 m/s (corresponding to 50 km/h) that is a typical value for this kind of competitions. The support structure was fixed on the wind tunnel balance and located below the ground plane except for: the two short struts that held the rear part of the bike frame, the jaws that held the rear wheels and a roller over which the front wheel was free to spin. The athlete was cycling during the tests.

The aerodynamic resistance, measured by the balance, is usually reported in terms of "drag area" (usually denoted with SCx), i.e. as the ratio of the drag force and the dynamic pressure. In this form the results are more comparable with other data obtained at different speed or different atmospheric conditions.

Fig 2. The athlete's handbike inside the wind tunnel.

Two series of wind tunnel tests have been carried out: a first series of tests to investigate the effects of same bike details and a second series to test the effects of different athlete positions. For the first series the athlete used his own bike while in the second series was necessary to use a different bike that allowed for different position settings (although it was less refined on the aerodynamic point of view).

RESULTS

As shown in Figure 3 the ergonomic analysis demonstrates three evidences:

1 – The handlebar height does not influence the ergonomics of the hand bike gesture

2 – The backseat inclination has a direct relation with the ergonomic perception; the more the seat is vertical, the more the MMGA index present lower values of discomfort. This result is confirmed by the subjective evaluation of the athlete.

3 – The handgrip length influences the ergonomic perception with an unexpected evidence.

EI	HG	HB	BS
31,06	L1	m1	s4
33,9	L1	m1	s3
33,19	L1	m1	s1
37,8	L1	m1	s2
43,21	L1	m1	s5
52,8			cecch

HB	HG	BS	EI
m1	L1	s3	**33,9**
m2	L1	s3	**34,8**
m3	L1	s3	**36,3**
m4	L1	s3	**37,8**
m4	L2	s3	**47,3**
m3	L2	s3	**49,1**
m1	L2	s3	**51,1**
m2	L2	s3	**51,2**

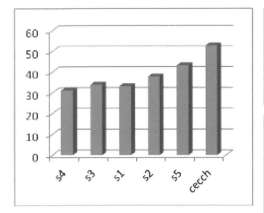

EI: Ergonomic discomfort Index
HB: Handle Bar Height
HG: Handle Grip Length
BS: Backseat Inclination
Cecch: reference test with athlete's handbike

L1: Hand Grip length = 170mm
L2: Hand Grip length = 175 mm
m: Handlebar height, with 1 high to 4 low
s3: backseat reference inclination
s1: backseat lower intermediate inclination
s2: backseat higher intermediate inclination
s4: backseat higher inclination
s5: backseat lower inclination

Figure 3: Results of the Ergonomic Analysis. On the left: effect of the Backseat inclination on the final score. On the right: effect of the Handle Grip length on the final score.

Therefore it is possible to affirm that, from an ergonomic point of view, the athlete is suggested to use the short hand grip levers and to assume a more vertical backseat position to improve his comfort during the challenge.

The biomechanical analysis provided useful information for the following step leading us to select the three intermediate backseat inclinations (s1, s2, s3) and to fix the handlebar height at the lower level (m4) for the test in the aerodynamic study.

The wind tunnel results are summarized by the table 1. A first evidence is that the handbike drag is remarkably lower than the traditional bicycle drag (referring to the whole of bike and athlete). For traditional bicycle, in case of positions and articles optimized for time trial competitions, values of SCx from 0.2 to 0.3 m^2 are normally obtained (Defraeye et al., 2010) and, as a matter of fact, values close to 0.2 are obtained only with thin (and not too tall) athletes. The values around 0.18 m^2 obtained in this series of tests (with an athlete that is not particularly thin) demonstrates an advantage even larger than the 10% reported by Gross et al. (1983) for the recumbent bicycle.

More in details, it can be observed that:

1 – the fairings did not produced appreciable advantages except for the foot fairing that shows a not negligible drag reduction

2 – the removal of the rear-view mirror produces a sensible improvement

3 – the lens wheels has to be used only in the rear position where they produce a remarkable advantage, according to what happens with the traditional bicycle (Gibertini et Al., 2008)

N.	SCx [m²]	Hand-bike	Front wheel	Rear wheel	Foot fairing	Frame fairing	Rear view mirror	BS	HB
1	0.180	A	spokes	spokes			X	s3.	L1
2	0.178	A	spokes	spokes	X		X	s3	L1
3	0.181	A	spokes	spokes		X	X	s3	L1
4	0.175	A	spokes	spokes				s3	L1
5	0.183	A	lens	spokes			X	s3	L1
6	0.166	A	lens	lens			X	s3	L1
7	0.196	B	spokes	spokes				s2	L1
8	0.192	B	spokes	spokes				s1	L1
9	0.190	B	spokes	spokes				s3	L1
10	0.184	B	spokes	spokes				s3	L2
11	0.185	B	spokes	spokes				s1	L2

Tab 1. Summary of wind tunnel test results.

Considering the tests on the athlete positions, a first observation is that, for the corresponding position, the settable handbike is less aerodynamic than the first one. Thus, the position tests results has to be considered in terms of reciprocal differences. Looking again the table 1, it is clear that the best results have been obtained with s3 and s1 backseat inclination (both with the longer hand grip). In fact, the differences between these two positions are inside the uncertainty interval.

Finally, the results about the hand grip length effect are in counter-trend respect to the kinematic results confirming that the choose of the best configuration comes from a balance between the different needs (Belluye et al., 2001)

DISCUSSION AND CONCLUSION

Results of ergonomic and aerodynamic tests are discordant. In the ergonomics analysis there are two important evidences: the first one is that the increase of the handgrip length influences negatively the ergonomic scores in all the different postures; the second one regards the inclination of the backseat; the more the athlete posture is vertical, the more the ergonomic index present a decrease in the discomfort perception. On the contrary, the vertical position of the trunk affects negatively the aerodynamic resistance coefficient, and this result is coherent with the observation that, an increasing of the wind exposed surface produces higher values of the aerodynamic resistance coefficient. Also results about the length of the handgrip are in contrast with the ergonomic analysis. In fact, the longer handgrips seem to be more efficient for the aerodynamic analysis.

The best compromise can be achieved with the seatback adjusted in the intermediate inclination and the shorter hand grip levers. In specific situation such as long competitions, where the postural comfort has to be privileged, the seatback should be put in higher intermediate inclination. If the aerodynamic issue is more relevant, for example in short chrono trials, the seatback is suggested to remain in the intermediate inclination and longer hand grip levers should be chosen.

In addition it is possible to state that the elimination of the mirror determinates a considerable advantage and that lenticular wheels have to be used only in the back where they give remarkable advantages. More work has to be done to evaluate the effects of applied especially those for the feet which showed slight advantages.

Further analysis on the effective benefits for energy expenditure in this new assessment are required to validate ergonomic data definitely. (Maki et al., 1995). Future developments comprehend the evaluation of a large number of users to go beyond the optimization for a specific athlete in order to apply the results to consumer products.

AKNOWLEDGEMENTS

The authors want to acknowledge hereby INAIL support to the research.

REFERENCES

[1] Andreoni, G., Mazzola, M., Ciani, O., Zambetti, M., Romero, M., Costa, F., Preatoni, E. (2009), Method for Movement and Gesture Assessment (MMGA) in ergonomics. In: Proceedings of the 2nd International Conference HCI/ICDHM, San Diego (USA), July 19-24, 2009, 591-598.

[2] Schmidt R, Disselhorst-Klug C, Silny J, Rau G. *A marker-based measurement procedure for unconstrained wrist and elbow motions.* J Biomech. 1999 Jun;32(6):615-21.

[3] Arnet U, van Drongelen S, van der Woude LH, Veeger DH. *Shoulder load during handcycling at different incline and speed conditions.* Clin Biomech (Bristol, Avon). 2011 Aug 8.

[4] Defraeye T., Blocken B., Koninckx E., Hespel P., Carmeliet J., *Aerodynamic study of different cyclist positions:CFD analysis and full-scale wind-tunnel tests.* Journal of Biomechanics 2010 May; 43(7):1262-1268

[5] Gross A.C., Kyle C.R., Malewicki D.J., *The Aerodynamics of Human-powered Land Vehicles.* Scientific American. 1983 Dec; 249(6): 142-52.

[6] Gibertini G., Grassi D., *Cycling aerodynamics*, in H. Nørstrud (ed.), Sport Aerodynamics, CISM Courses and Lectures, vol.506, 2008, Springer, Wien, pp. 23–47.

[7] Belluye N, Cid M, *Approche biomécanique du cyclisme moderne, données de la littérature.* Science & Sports 2001 Apr; 16(2):71–87.

[8] Maki KC, Langbein WE, Reid-Lokos C. *Energy cost and locomotive economy of handbike and rowcycle propulsion by persons with spinal cord injury.* J Rehabil Res Dev. 1995 May;32(2):170-8.

CHAPTER 61

Ergonomics and Design as Tools on the Risk Management

Débora Ferro, Laura Martins

Universidade Federal de Pernambuco
Caruaru-PE/ Recife-PE, Brasil
dtferro@gmail.com, laurabm@folha.rec.br

ABSTRACT

This paper discusses the use of Ergonomics and of Design as tools to preserve the physical and mental integrity of workers. As a case study, an industry in the Northeast of Brazil was investigated, in which activities are undertaken that involve various risks to do with fires and explosions. The behavioral intentions of employees in simulated emergency escape situations were investigated. Evidence was found of the importance of an interdisciplinary approach when risks are being managed as an alternative to avoid accidents, thus promoting safety in industrial plants.

Keywords: ergonomics, design, escape routes, risk management

1   INTRODUCTION

Every day, all round the planet, thousands of individuals perform their work activities in different environments and are submitted to very different levels of risk. Understanding work in its breadth and perceiving the dangers to which workers are exposed is important. However, knowing how to act in situations where the risks are beyond one´s control is essential and will make all the difference to the bottom line when all the losses caused by an accident are identified.

For DUL and WEERDMEESTER (2004), one of the primary factors in interface design focuses on the fact that the designer needs to know the user. Thus, considering the user of a system starting from the design phase is one of the main objectives of Ergonomics.

This paper argues that it is possible to avoid distortions in understanding information by being familiar with users' behavioral intentions. Therefore, the activity of ergonomists who are knowledgeable about the physical, psychological and cognitive dimensions of human beings is essential.

Another partner in this process is the designer, who stamps elements in his/her designs that satisfactorily meet the demands set by users. It is understood that by considering users' behavioral intentions during the design phase enables a more efficient system to be drawn up, that may avoid information being distorted and, consequently, to be better understood.

2   ABOUT BEHAVIORAL INTENTION

It is known that human error can be established before decisions are made [perception errors], while information is being processed [decision errors], or, moreover, when action is being taken [action errors] (Iida, 2005). To KALSHER & WILLIAMS (2006), it is only through behavior that the effectiveness of a sign-post can be measured. However, still according to these authors, the best indication of whether an individual will comply with the instructions is his intention to comply with them.

Thus, it is understood that in a research study in which it is not feasible to dimension behavior, an acceptable solution is to look at the context in which actions are made, by considering cognitive processes, as well as the intentions involved.

In research, there are several situations where it is not possible to measure real behavior. This may occur, for example, when:

- Ethical and safety questions do not allow participants to be exposed to real risks;
- The constructed scenario resembles the real one, but safety is not preserved;
- The costs of time and effort are not compatible with the resources of the research.

For WOGALTER (2006), when there are difficulties in conducting certain types of tests of behavior, it is advisable to use behavioral intentions as a method of measurement.

Investigating behavioral intentions consists of questioning the participants on one or more actions that they would perform in specific situations. The questions may be related to giving a warning about a product or to a particular risk in the environment.

It is worth noting that although the errors identified relate to the decision making phase this is immediately before the level of action. Thus, for designers and ergonomists, understanding users' behavioral intentions makes it possible to have a more precise and distortion-free approach to design.

This research was conducted based on identifying and analyzing the behavioral intentions [or expected behavior] of possible users of an Information System, in a processing plant located in the Northeast of Brazil. The expected behavior in emergency situations where escape routes would be used was reported on.

3   SETTING THE LIMITS OF THE RESEARCH PROBLEM

Although, in Brazil, there are Norms governing the use of colors in the workplace, during the research, no specific guidelines were found for the graphical configuration and exhibition of the informational content of safety signs and warnings for the industry. Thus, in practice, the development of the design of the signage manual is at the discretion of the manufacturer who normally is not familiar with the context of specific use of the material that he produces and markets.

It is known that particular aspects of a reader [such as their level of education, beliefs, values and repertoire] may prevent communication being fully established. Therefore, what needs to be avoided is that personal aspects of individuals interfere in the communication process.

Moreover, there is always a series of circumstances that create conditions such that mistakes and accidents occur. These factors, in general, originate in the system and not from the user. The use of escape routes - the object of investigation of this research - is just one of many situations in which signs are needed to enable the individual to make safe and effective decisions in the shortest possible time.

Given the importance of industrial signage in the prevention of industrial accidents, it is essential to develop information systems that enable users to proceed correctly when they are faced with risk situations. It is believed that, by using a design consistent with the demands raised, it is possible to prevent a minor incident from becoming a tragedy of major proportions.

4   METHODOLOGICAL PROCEDURES ADOPTED

The objective of this research was to become familiar with and analyze the behavioral intentions shown by users of a system of industrial signage in relation to accidents involving fires, leaks and explosions. Two analyzes were conducted in parallel: The first was about investigating the expected behavior of users of the system and the second concerned analyzing the vulnerability of some points of the plant under study.

By means of questionnaires, it was possible:

- To map the areas of the company occupied by the employees;
- To identify the users' perception about the environment in which they work;
- To identify the expected behavior by users of the system in escape situations.

To determine the sample [a population of 113 people], a confidence level of 95% [02 standard deviations] and a maximum permitted error of 7.5% were assumed, resulting in a sample of 69 participants. The questionnaires were administered individually and the participants did not have to identify themselves. The interviews were recorded on a digital recorder. Data were transcribed and tabulated for later analysis.

Thereafter, points of the plant were selected, and accidents were simulated in accordance with typical data from the literature. For the simulations, use was made of ALOHA [Areal Location of Hazardous Atmospheres] software, which is recognized by EPA [Environmental Protection Agency]. By using this software enabled calculations to be made of the thermal impact, shock waves, as well as the toxic cloud generated in the situations proposed. To do so, the engineers assigned values typically found in processing industries [petrochemicals, power plants, chemicals] to the input data [temperature, pressure, flow, pipe diameter and length].

Finally, the graphic simulations of the routes cited by staff were crossed with the graphs resulting from the vulnerability analysis, and errors related to safety procedures were mapped. By doing so, it became possible to support future projects related to security planning in the industry.

5   WHAT DO USERS INTEND TO DO?

The questionnaire applied consisted of objective and subjective questions, grouped into:
- Personal Data;
- Data on the activities of the respondent in the company;
- Data on the perception of safety in the workplace;
- Data on the expected behavior in situations of escape.

The Section on 'Data on the perception of safety in the workplace' showed how users perceive the environment in which they are inserted, as to their own safety and the procedures in case of emergency. Participants mentioned places which they believed to be the safest and most insecure in the plant.

The 03 places listed as the safest ones by the majority of the respondents are shown in Figure 1 with numbering in green. In the same Figure, the 03 places cited as the most insecure appear in red.

Figure 1 also indicates the dimensions of an accident in the place considered as the most vulnerable, according to the analysis of the engineering team. Within 60 seconds after the start of the fire, the red area indicates the potential for deaths within a radius of 43 meters; the orange area indicates second degree burns within a radius of 70 meters; the yellow area shows where pains will be felt – within a radius of about 92 meters. As a result, it was seen that an accident occurring at the location indicated would reach proportions that would affect the areas close to those where there is the highest concentration of people during the day.

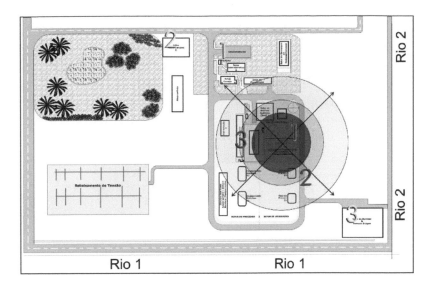

Figure 1 Places mentioned by respondents as safest (in green) and most insecure (in red) of the plant and simulation of accident with leaks in pipes feeding on sector of storage and transfer of toxic and flammable products. (From the author.)

The section on 'Data on the expected behavior in an escape situation' enabled the expected behavior in emergency situations to be identified, as regards the choice of escape routes. Respondents were asked about what routes they intended to take both in the case of not the location of the accident, and if they knew exactly where it had occurred.

In the first situation, i.e. when unaware of the exact location of the accident, 20 different routes were cited, when they were asked to indicate the route to be taken.

As to the second circumstance, what were proposed, one by one, were 07 different locations in the plant where an accident might occur. Routes they quoted [from the workstation to a safe location] were drawn on the company's floor plan. In exactly the same way, even in cases where the accident sites were specified, different routes and/ or action are mentioned, as can be seen in the following examples:

Figure 2 represents an example of how the overlaps were made. In this case, the simulation consists of an accident in pressure vessels. The red area shows the potential for deaths in 60 seconds, the orange one is where there will be second degree burns and the yellow one where pains will be felt.

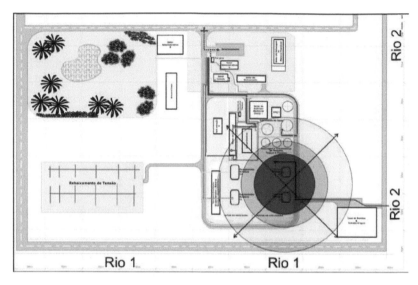

Figure 2 Jet Fire on scenario of leak in pipes feeding of pressure vessels. (From the author.)

Figure 3 simulates an accident in the flammable fuel Depot. The red area indicates the potential for death in 60 seconds and has a radius of 225 meters. The orange area covers a radius of 315 meters and indicates that people who are there will suffer second degree burns in about 60 seconds. The yellow area [which has a radius of 490 meters] indicates that people there will feel pains in 60 seconds. As can be seen, the accident in this case is of such proportions that any chosen route would be adversely affected.

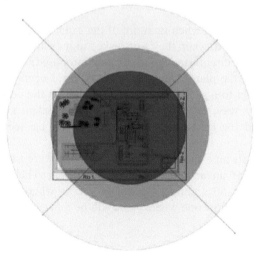

Figure 3 Simulation of accident in fuel tank flammable. Overlay traced routes. (From the author.)

For the Pressure Vessels, scenarios of a Vapor Cloud Explosion were generated. Figure 4 below shows the limit of the cloud stretches out for nearly 200 meters at its farthest point, thus reaching all the buildings of the plant except the Administrative Sector 2. Throughout the yellow area, injuries occur to people and damage to the physical environment. Glass would be easily shattered.

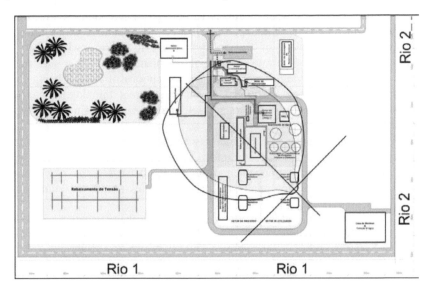

Figure 4 Vapor cloud explosions on the scenario of leaks in pipes feeding of pressure vessels. Overlay of traced routes. (From the author.)

6   CONSIDERATIONS ABOUT THE EXPERIMENT

By means of this experiment, it was found that various measures [both preventive and corrective ones] need to be taken immediately. The overlay of the graphics on the routes cited by the interviewees clearly reveals the severity of possible accidents. The number of routes mentioned indicates incompatibility between the mental scripts formed by the interviewees. It is seen that there is a lack of cohesion among the solutions that it is envisaged would be adopted, even when the location of the accident is not specified.

It was observed that although the company proposes a default behavior for evacuation situations from the area, for each situation suggested to users, new answers were presented. In some cases, the behavioral intentions disregarded the recommendations of the safety team. This shows that people still do not seem to have a fixed mental model of a set of actions to be performed in emergency circumstances.

Behavioral intentions varied not only from person to person, but also in descriptions by the same person, when asked about different situations. The variations in the statements by one and the same person when asked about different situations can be considered a positive reflection of the use of the cognitive process. Nevertheless, the variation in behavioral intentions between different individuals as to the same risk situation indicates incoherence. The lack of recycling, as well as the absence of practical simulations of how to proceed can be important factors, in this case.

The users' lack of knowledge about the characteristics of the system may lead them into error, especially when it comes to emergency situations where time is limited and the circumstances, unexpected. There were situations where the route cited was towards the place of the accident, only to comply with what was determined in training, on how to get out of the plant, indicating a stiff mental configuration when resolving a problem. The perceptions of risk as being lower than they actually are may be related to this incoherent thinking. There were respondents who were unable to accurately determine even one point of risk in the plant.

It was clear that the case of the industry under study goes beyond the mere adequacy of the signage and holding training events. Moreover, problems relating to the very layout and physical arrangement of the plant are listed.

As a way to solve and/or minimize the deficiencies found, the study points adopting measures such as:

- Standardizing safety instructions relating to the evacuation of the area and escape routes;
- Making training more suitable;
- Conducting practical Simulations of risk situations;
- Evaluations and periodic reycling events;
- Disseminating knowledge of the risks existing in the company and the dimensions of a possible accident;
- Designing alternative escape routes [e.g. on the river, by boat];
- Reviewing layout | physical arrangement;
- Making the signage fit the new project.

7   CONCLUSIONS AND FINAL REMARKS

The development of the industry, closely allied to technological developments, has become widely known about over the last thirty or so years. However, although the production processes have become ever more automated, what cannot be forgotten is the presence and activity of the human being - a figure actively involved in this context and, paradoxically, many times overlooked during the design of projects.

Industrial growth and the latent need to preserve human integrity motivate the conduct of this study. It stresses the need for projects that address users' limitations and the demands imposed by the system, by using ergonomics and design as tools to achieve this end by means of preserving workers' physical and mental integrity.

Although the level of design interventions has not been reached, many gains have been obtained from the research. The standard questionnaire proposed has made it possible to understand the mental model of the users of the system. Thus, a tool for risk management has been generated, based on the concept that the "intention of compliance" of an action is the best indicator of whether an individual will comply with the instructions (Kalsher & Williams, 2006; Wogalter, 2006).

The limitations of this tool are known, especially with regard to its high level of subjectivity, which makes it subject to the inferences of the mental map of each individual analyzed. However, despite these limitations, it is understood that the tool has advantages that make its application pertinent, especially in the preventive sense. In addition to the factors mentioned based on WOGALTER [2006], there is also the low financial investment as well as less time and effort bring spent by participants.

Despite the uncertainties cited, this tool makes it possible:
- To identify gaps in users' perception;
- To identify and anticipate possible negative actions before they become larger;
- To identify what can go wrong and how it can go wrong, thus preventing the occurrence of human errors in interactions with the system;
- To guide training based on the deficiencies investigated and the profile of the users;
- To assess the effectiveness of training already given;
- To draw up action strategies consistent with the target audience;
- To promote mental models of the individuals, thus avoiding cognitive overload, as quoted by MARTINS and MORAES (2002);
- To generate design requirements for the designer of the information system, based on the users.

The importance of integrating the designer with professionals in other fields of knowledge is emphasized. In this study, the analysis of behavioral intentions gained prominence when the consequences of failure to comply with certain actions were known. Similarly, the vulnerability of the system has become even more significant when the possible losses of human lives were identified.

It is known that preventing accidents or, at least, mitigating their consequences, is of public interest and affects both the general population and industries and government institutions. The data provided here and the methodological procedures adopted can be applied in other investigations, thus serving as support for research and sources of reference aimed at designing industrial signage, or even in future projects in the sphere of safety at work, and seek to prevent accidents.

Finally, it is understood that the design of information systems is at a higher level than the mere formulation of warning signs, thus indicating the need for professionals to design with consistency and with respect to users' limitations and the demands imposed by the design, by managing risks.

REFERENCES

ALOHA: Areal Locations of Hazardous Atmospheres. 2006. *Computer-Aided Management of Emergency Operations*. Manual do Usuário. Washington.

Dul, J. and Weerdmeester, B. 2004. Ergonomia Prática. Tradução: Itiro Iida. 2.ed. São Paulo: Edgard Blücher Ltda.

Kalsher, M.J. and Williams, K. J. 2006. Behavioral Compliance: Theory, Methodology, and Results. In. *Handbook of Warnings*. Mahwah: Lawrence Erlbaum Associates.

Matlin, M.W. 2004. Psicologia Cognitiva. Tradução de Stella Machado. 5. ed. LTC.

Moraes, A.(Org.). 2002. Avisos, Advertências e Projeto de Sinalização: *Ergodesign Informacional*. Rio de Janeiro: UsEr.

Normas Regulamentadoras. 2005. Proteção Contra Incêndios. *NR-23*. Ministério do Trabalho. Brasil.

Ramos, D. 2005. Extintores de Incêndio: *Uma Análise Ergonômica das Seqüências Pictóricas de Procedimento com Enfoque na Cognição*. Graduation thesis. Recife: UFPE.

Wogalter, M. C. 2006. Communication- Human Information Proceeding (C-HIP) Model. In: *Handbook of Warnings*. Nova Jersey: Lawrence Erlbaum Associates.

Wogalter, M. S., Kalsher, M. J. and Rashid, R. 1999. Effects of Signal Word and Source Attribution on Judgments of Warning Credibility and Compliance Likelihood. In. *International Journal of Industrial Ergonomics*. 24: 185-192.

Wogalter, M. S. and Usher M. 1999. Effects of Concurrent Cognitive Task Loading on Warning Compliance Behavior. *Proceedings of the Human Factors and Ergonomics Society*, 43: 106-110.

CHAPTER 62

Design a Force Platform for Measuring Center of Pressure (COP) Signal

Pei-Der Sue[1], Cheng-Wei Huang[1], Yuan-Jang Jiang[1], Jiann-Shing Shieh[1], Bernard C. Jiang[2]

[1]Department of Mechanical Engineering
[2]Department of Industrial Engineering & Management
Yuan Ze University, Taiwan
jsshieh@saturn.yzu.edu.tw

ABSTRACT

The aim of this study is to design a low cost and portable measuring device of center of pressure (COP) for data collecting and analysis, which is called center of pressure and complexity monitoring system (CPCMS). Firstly, we presented details about design concepts, functions of measurement component, and signal analysis of multi-scale entropy (MSE) algorithm. In order to verify the reproducibility of the measurement system, we conducted a series of experiments, including static and dynamic tests. Secondly, in order to prove that our system can estimate the different sizes of the different sways, we designed four sway tests to produce different sway displacement. We use traditional and MSE analysis methods to analyze the COP signal received by CPCMS and compare with the commercial product of Advanced Mechanical Technology Incorporation (AMTI) system. The results indicated that CPCMS can receive the same results with the AMTI system in comparison with these four sway tests under the statistical analysis. In conclusion, a low cost and portable system (i.e., CPCMS) for measuring and analysis COP signal has been designed in this paper which is allowed us to quantify the dynamic property of human's balance in order to assess the risk of falling among elderly people via large population size more handy and conveniently in the near future.

Keywords: falling, noninvasive physiological signs, center of pressure, multi-scale entropy

1   INTRODUCTION

Falling is a common accident among the elderly and it often causes great injuries. It most frequently occurs during walking and is associated with the chronic deterioration in the neuromuscular and sensory systems, as well as with ankle muscular weakness and lower endurance of these muscles to fatigue (Gefen, 2001). Falling is consequently a major cause of morbidity and mortality in the U.S. Several factors can contribute to injuries resulting from tripping, slipping and falling during locomotion, such as surface condition, transitions and the degree of walkway evenness (Schieb, 1995). Recently, biomechanical analysis has been applied to develop treatment programs for many lower extremity orthopedic and neurological injuries (Cooper et al., 1995).

Human body signals are very complex and are difficult to distinguish the difference in their characteristics. To analyze the data, which includes heartbeat (Costa et al., 2002; Yang et al., 2003), electroencephalography (EEG) and depth of anesthesia, requires use the non-linear theory such as approximate entropy (Bruhn et al., 2000; Pincus, 1991), sample entropy (Richman and Moorman, 2000) and multi-scale entropy (MSE) (Costa et al., 2003). Human body signals can be quantified by these non-linear theories to achieve the purpose of signal interpretation. In this study we chose MSE to analysis the human body center of pressure (COP) signals, measured from our balance measure system, and compared them with the commercial COP measurement system.

2   COP MEASUREMENT DEVICE

We introduce the structure of balance measurement device as shown in Figure 1.This device established by two systems, the pressure measurement platform and the data receiving systems. When the subject stands on the balance measurement device, the pressure measurement platform can receive the raw COP signal. Then the data receiving system can convert the raw COP analogy signal into digital signal for use in a computer or display device. We designed a high-resolution data receiving system, which used a 16-bit A/D card to receive the signal to provide accurate signal data. We named our system as Center of Pressure and Complexity Monitor System (CPCMS).

2.1   Pressure Measurement Platform Design

The pressure measurement platform is design to receive the raw data of the balance signal (Figure 2). Because the signals of the load cell are very small and have some noise, we needed to enlarge the signal and filter the noise in the raw voltage data. Therefore, the pressure measurement platform system includes 5 parts; load cell sensor, Wheatstone bridge, filter circuit, amplifier circuit, and calibrate circuit. The following are the brief introductions for each design.

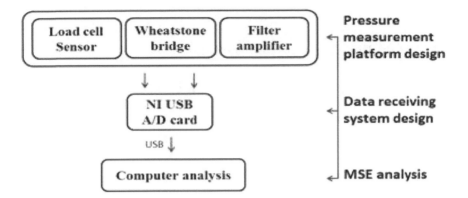

Figure 1 The center of pressure and complexity monitor system (CPCMS) design flow chart

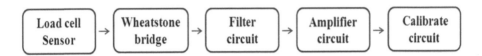

Figure 2 The pressure measurement platform design flow chart

Figure 3 Electronic weight scale

The pressure sensing principle of the electronic weight scale (Figure 3) is the same as the strain gauge. When the electronic weight scale is under pressure, load cells in the scale will produce deformation. The resistance will change because of the deformation. But, the resistance change is difficult to measure, so we collect the voltage change instead of resistance change. We use this principle as the basis for the measurement of body balance signals. The load cells are embedded in the four corners of the scale. Each load has 3 signal lines and we used Wheatstone bridge circuit to collect the output voltage, which had been changed due to pressure.

2.2 Data Receiving System Design

This data receiving system uses 16 bits analog to digital (A/D) card to receive the voltage signal from the pressure measurement platform. After the A/D card converts the voltage signal into a digital signal, we then use Borland C++ Builder (BCB) to write the program to receive the signal. Finally, we wrote MATLAB program to analyze the receiving data. The design flow chart is shown in Figure 4.

Figure 4 High resolution data receiving system design flow chart.

In the high resolution data receiving system, we chose NI (National instruments) USB-6212 16-bit A/D card as our analogy to digital device. This device has 16 analog input channels, 400 k S/s sampling rate, and 16-bit resolution. In his paper we used 4 analog input channels to receive the signal from pressure measurement platform, and sent the digital signals to a computer through USB high-speed data streams. We used the BCB to write the data receiving interface shown in Figure 5. This program can receive the digital signals converted from the A/D card and save the raw data.

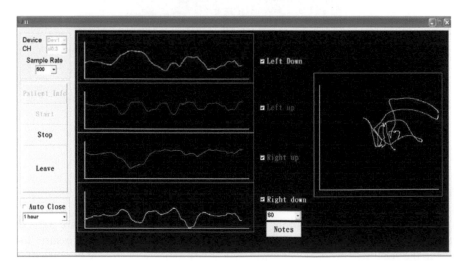

Figure 5 Data receiving program

3   ANALYSIS ALGORITHMS

Using the COP measurement apparatus to obtain the body COP data, we use empirical mode decomposition (EMD) for detrending the data. The EMD method is iterative signal processing algorithm which decomposes the intrinsic components (IMFs) from signals by iterative sifting processes. Then, the MSE is used to evaluate the body COP. MSE analysis is a new method of measuring the complexity of finite length time series. Entropy-based algorithms for measuring the complexity of physiologic time series have been widely used. They have proved to be useful in discriminating between healthy and disease slates, although some results may generate misleading conclusions. This computational tool can be applied both to physical and physiologic data sets, and can be used with a variety of measures of entropy.

4   MEASURING INSTRUMENT

In order to prove that CPCMS can measure the same displacement as the commercial system, the following test will use two systems to measure the COP signal. Because the commercial system, AMTI, can eliminate the weight on the platform, we chose this system as our comparison. We placed CPCMS on the top of the AMTI, therefore we could obtain the same displacement. We compared the difference between CPCMS and the commercial one (AMTI).

4.1   AMTI

AMTI Biomechanics Platforms simultaneously measure the three force components along the XYZ axes and the three moment components about the XYZ axes. The forces and moments are measured by strain gages attached to proprietary load cells near the four corners of the platform. The gages form six Wheatstone bridges having four active arms each with eight or more gages per bridge. Three of the output signals are proportional to the forces parallel to the three axes and the other three outputs are proportional to the moments about the three axes.

4.2   Human Body COP Signal Test

Figure 6 show that if the subjects are standing on the top of the two measuring systems, the two systems can measure the sway displacement at the same time. We also spread the displacement into x and y directions (Figure 7) and analyzed the data by complexity MSE analysis methods. We used these concepts to design the following experiment.

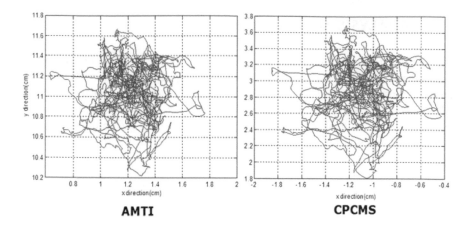

Figure 6 The displacement figure measured by AMTI and CPCMS

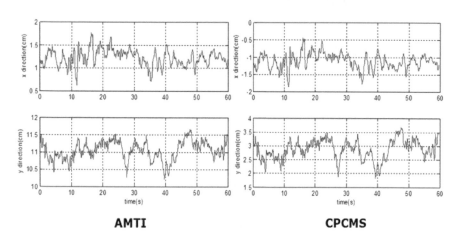

Figure 7 X and Y direction displacement figure measured by AMTI and CPCMS

5 EXPERIMENTAL DESIGN

In order to prove that CPCMS can estimate the different sizes of the different sways, we designed four sway tests to produce different sway displacement. The sampling rates of the two systems are both 50Hz, measured for 60s. The subjects were 20 young adults, aged between 20-25 years old. The four steps of the experimental are as follows (the Schematic diagram show in Figure 8):

Step 1. The subjects open their eyes standing on the balance measurement system and measured for one minute (sway 1).

Step 2. The subjects close their eyes standing on the balance measurement system and measured for one minute (sway 2).

Step 3. The subjects open their eyes standing on the water cushion placed on the balance measurement system and measured for one minute (sway 3).

Step 4. The subjects close their eyes standing on the water cushion placed on the balance measurement system and measured for one minute (sway 4).

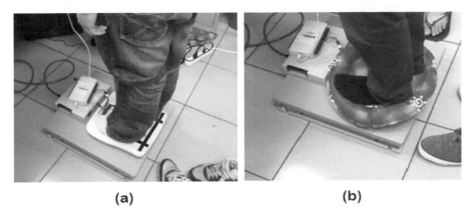

(a)   (b)

Figure 8 (a) The subjects open and close their eyes standing on the CPCMS system and AMTI system (b) Subjects open and close eyes standing on the water cushion placed on CPCMS and AMTI system

6 ANALYSIS RESULT

We applied the MSE to evaluate the COP complexity in all subjects. We used EMD for detrending the data. Because CPCMS's sensitivity is less than AMTI system, there are some minor changes we cannot receive (Figure 9). Therefore the typical results from EMD will be different in all IMF. Figure 10 shows the typical results from EMD.

The previous study (Yang, 2010) shows that IMF2 plus IMF3 can clearly show the change of human body COP complexity. Therefore we chose IMF2 plus IMF3 to do MSE analysis. We calculated the area under the curve as the quantification value of MSE analysis. An MSE analysis results example is shown in Figure 11.

According to the AMTI system analysis results (Table 1), the complexity has no obvious change between "eyes open" and "eye closed" tests in both X and Y directions. But, we found that the P_c and P_d values in the Y direction are less than 0.05, which means the complexity obviously changed before and after using the water cushion in the "eyes open" and "eye closed" tests. We think the reason is that the subjects are young adults which closed their eyes will not cause problems in maintaining balance. Therefore the complexity has no obvious change between eyes open and closed in both X and Y directions. But if subjects stand on the water cushion, they will try to keep balance to prevent falling down. This behavior will cause the complexity to reduce. The CPCMS has the same result with AMTI according to statistical analysis which the P_c and P_d values in the Y direction are also less than 0.05.

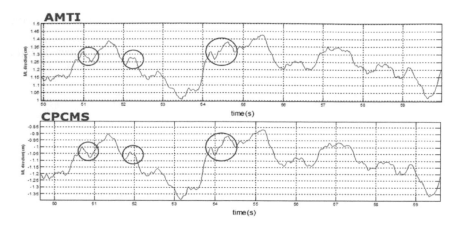

Figure 9 The minor changes between two system

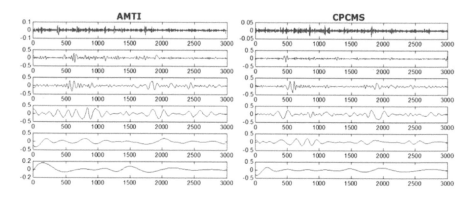

Figure 10 Two typical results from EMD measured by two systems

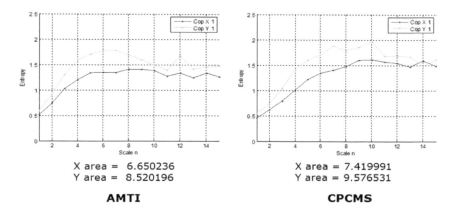

X area = 6.650236
Y area = 8.520196

AMTI

X area = 7.419991
Y area = 9.576531

CPCMS

Figure 11 MSE analysis results figure

Table 1 The MSE analysis result

	Standing directly on the platform		Standing on the water cushion		P_a	P_b	P_c	P_d
AMTI	Eyes open (sway1)	Eyes closed (sway2)	Eyes open (sway3)	Eyes closed (sway4)				
MSE(X)	4.02±1.43	4.04±1.82	4.22±0.98	4.31±1.19	0.969	0.795	0.609	0.582
MSE(Y)	5.64±1.39	6.27±1.12	4.73±1.35	5.00±1.42	0.123	0.541	0.042	0.003
CPCMS								
MSE(X)	5.40±1.10	4.54±1.60	5.76±1.28	5.40±1.48	0.055	0.416	0.346	0.086
MSE(Y)	6.88±1.48	6.91±1.74	5.85±1.56	5.67±1.57	0.953	0.718	0.039	0.023

P_a: p-value for standing directly on the platform between EO-test and EC-test.

P_b: p-value for stand on the water cushion between EO-test and EC-test.

P_c: p-value for open eyes standing directly on the platform and standing on the water cushion.

P_d: p-value for close eyes standing directly on the platform and standing on the water cushion.

7   CONCLUSION

From the analysis results, CPCMS has the same results as the AMTI system. The COP signal is similar to the AMTI system so the different sway displacement can be determined by CPCMS. However, it is less sensitive than the AMTI system, which causes CPCMS to miss some high frequency signals in the large sways. In the MSE analysis, the two systems have the same results at P_c and P_d values in the Y direction. Also, the complexity obviously did not change between "eyes open" and "eye closed" tests in both X and Y directions.

8   FUTURE WORKS

Next step we will try to use the new AD chip ADS8342 which resolution is also 16 bits, but it is cheaper than the NI A/D card. Hope we can let whole CPCMS system cost down. Beside we want to draw the COP data on a portable screen. Therefore we will write data receiving program in an APP for Android and put in tablet PC for transmitting data by Bluetooth or Wi-Fi. Then, we will use ensemble empirical mode decomposition (EEMD) (Wu and Huang, 2009) or multivariate empirical mode decomposition (MEMD) (Rehman and Mandic, 2010) to detrend the data. We hope these methods can obtain the similar values as the very expensive commercial AMTI system.

REFERENCES

Bruhn, J., H. Ropcke, and A. Hoeft 2000. Approximate Entropy as an Electroencephalographic Measure of Anesthetic Drug Effect During Desflurance Anesthesia. *Anesthesiology,* 92: 715-26.

Costa, M., A. L. Goldberger, and C. K. Peng 2002. Multiscale Entropy Analysis of Complex Physiologic Time Series. *Physical Review Letters,* 89.068102.

Costa, M., C. K. Peng, A. L. Goldberger, and J. M. Hausdorff 2003. Multiscale entropy analysis of human gait dynamics. *Physica A,* 330:53-60.

Cooper, R. A., D. P. VanSickle, R. N. Robertson, M. L. Boninger and G. J. Ensminger 1995. A Method for Analyzing Center of Pressure During Manual Wheelchair Propulsion. *IEEE,* vol. 3, no. 4.

Gefen, A. 2001. Simulations of Foot Stability During Gait Characteristic of Ankle Dorsiflexor Weakness in the Elderly. *IEEE,* S 1534-4320(01)11421-X.

Pincus, S. M. 1991. Approximate Entropy as a Measure of System Complexity. *PNAS,* 88:2297-2301.

Rehman, N. and D. P. Mandic 2010. Multivariate Empirical Mode Decomposition. *Proceedings of the Royal Society,* A vol. 466, no. 2117, pp. 1291-1302.

Richman, J. S. and J. R. Moorman 2000. Physiological time-series analysis using approximate entropy and sample entropy. *Am J Physiol Heart Circ Physiol,* 278:H2039-H2049.

Schieb, D. A. 1995. Walkway Surface Heights and Ground Reaction Forces. *IEEE,* 0-7803-:2083-2/95

Wu, Z. and N. E. Huang 2009. Ensemble empirical mode decomposition: A noise-assisted data analysis method. *Adv. Adaptive Data Anal,* vol. 1, pp. 1–41.

Yang, A. C. C., S. S. Hseu, H. W. Yien, A. L. Goldbesger, and C. K. Peng 2003. Linguistic Analysis of Human Heartbeat Using Frequency and Rank Order. *Physical Review Letters,* 90.108103

Yang, W. H. 2010. Data Mining on Physiological Signal - Integrate Dissimilarity Approach and Signal Reconstruction for Complexity Analysis. *PHD thesis, Yuan Ze University.*

CHAPTER 63

Ergonomics and Inclusivity Present in the Visual Communication Design Project

Fernando Moreira da Silva

CIAUD, Technical University of Lisbon
Portugal
fms.fautl@gmail.com

ABSTRACT

This paper presents the results of a research project overlapping Ergonomics, Visual Communication Design, Printed Colour and Inclusive Design. The target group was older people, with the aim to develop of a set of research-based ageing-centered communication design recommendations for printed material. The project was divided in two phases: in the first one, we developed a literature review in the areas of color, older people issues and vision common diseases, and communication design; in the second phase we implemented an experiment to measure the different color experiences of the participants in sample groups (three in UK and other three in Portugal). For the work with the sample groups, we used printed material, to find out the colors one should use in analogical communication material, being aware of the color contrast importance (foreground *versus* background) and the difficulties experienced by older people in what concerns readability and legibility.

The experimental phase was composed of two parts: in the first part we didn't measure de reactions of the participants, only using a video camera to report the sessions; in the second part we used eye-tracking, in a way to help us to understand the eyes movements and the efforts of each participant to read the cards.

The results of this research project were significant: we verified the outputs of the relevant literature in the different areas which were studied, as well as we identified constancy in the findings of the experiment, bringing new information for the knowledge in the area. This allowed us to develop a set of recommendations mainly for the graphic designers, when they have to design a new communicational project in a way this project can achieve vision comfort, readability and legibility, for everyone, including, this way, older people, among the different users.

Keywords: ergonomics, inclusive design, color, older people, understandability

1 INTRODUCTION

"... the interaction between the user and the product is one of the primary concerns of the product design process." (Openshaw et al, 2006)

People are central to system functioning; they have an important, often taken for granted, role coordinating processes, as well as task-related roles. Ergonomics, as a human or social science, has as its subject matter the organized goal-directed activity of people in the world. Bolton (2007) argues that dignity is "...overwhelmingly presented as meaning people are worth something as human beings, that it is something that should be respected and not taken advantage of and that the maintenance of human dignity is a core contributor to a stable 'moral order.'" According to Fisk (2004), the design products have to consider the cognitive and perceptual capabilities and limitations of these consumers, and also have to provide the optimal training, select the appropriate input and output devices and structure the interface to ensure a usable system. Since older adults do not represent a homogeneous group any product that will be designed for them should take this guidelines into consideration since the beginning of the project and after that has to go through some testing to ensure their likely successful application.

The EU and most other developed countries have identified population ageing as one of the key economic and social challenges to be faced.

Most of us are aware that in an ideal world, inclusive products and services, designed with an ergonomic approach would be standard and not the exception.

Collaboration between users, designers and producers from the very beginning of the project until its completion, would be the key for success. However, it still isn't like this. Meanwhile, "critical" users who could supply the creative stimulus that designers need are still out of the process.

To work in the Graphic Design and Visual Communication area one needs to have knowledge of different techniques and how to manipulate them. Despite the knowledge in this professional circle, there is a gap in knowledge of Colour and Inclusive Design, and how to use ergonomics to develop communicational products.

An ergonomic approach is effective for a number of reasons:

• It is mainly based on the systems approach. Very effective is macro-ergonomics – the ergonomic approach to big systems;

• Ergonomists use very different professional skills (engineering, psychology, health, toxicology etc.);

• Ergonomics uses sensitive subjective methods (studies of sensation, discomfort and feelings) which enable the detection of weak physiological changes in different tissues of the human body earlier than most objective methods. It enables the development of pathological changes to be avoided;

• Ergonomics uses a participatory approach that respects the knowledge and experience of the user of ergonomic interventions. This approach enables the use of the wide vague knowledge of the user. (Kristjuhan, 2010)

This paper summarizes the content and output of an 18 months scoping research study concerned with ergonomics, inclusive design, color and visual communication in analogical support for older people. The output from this project is the initial stage of color guidance and recommendations for designers, older people and community care professionals.

To work in Visual Communication one needs to have knowledge of different areas and techniques, and how to manipulate them. However, there is a gap in knowledge of Inclusive Design and Colour. Most of the studies that have been implemented until now addressing the research topic focus on the use of color and text in digital displays, not in analogical supports.

The research object of this study is the overlap between Ergonomics, Inclusive Design, Visual Communication Design, and Printed Color. The main objective is to introduce color as a variable of great importance in Visual Communication, in an Ergonomics and Inclusive Design perspective, having as a target older people. Until now, little research has been carried out on how colors in form and background relationship affect older people. If text does not have sufficient contrast compared to its background, people will have problems.

An inevitable outcome of the ageing population is an increase in the difficulty of reading and understanding visual printed messages (analogical support) due to the lack of information about color, mainly for communicational designers, in an ergonomics and inclusive design perspective.

This study is aimed at addressing this lack of evidence-based data. As well as a literature search, the project has involved an experiment phase using sample groups of older people between 65 and 87, in Marple (UK) and in Cascais (Portugal).

The purpose of the sample groups was to test and validate information gathered during the literature review and to generate other findings.

The results obtained from this study are summarized in this paper.

2 LITERATURE REVIEW

The visual and non-verbal systems operate relatively without a "tutor" in our society, at least in comparison with language. Visual communication is a primary system, which is located at the same expression level as is verbal language. In terms of human development, the visual system of signs occurs earlier than language itself. In terms of complexity, one can find visual interpretation more complex than verbal interpretation, because of the absence of a conventional sign system and the formalization of a training protocol. Visual communication could also be considered primary because the observer has to learn how to control better, and independently, the visual interpretative function.

Finally, visual communication is neither derivative nor peripheral to verbal language, and consequently the designation of visual communication as secondary, tertiary or of a "superstructure" built in verbal language is not right. In Paivio notion (1971), a more appropriated model would be built of a dual codification, saying that visual and verbal information are codified and uncodified by perceptive and cognitive separated systems. A system is considered visual/pictorial and manipulates simultaneously the elements of imagination; the other one is linguistic and propositional and operates in sequence.

2.1 Visual impairment and older people

As the population is growing and ageing, the proportion of older people is expected to increase. One of the effects that growing older has on vision is that, on average, less light falls on the retina, and there is less tolerance to glare. Loss in the fovea affects visual acuity and color perception and general loss of vision across the whole visual field.

In the UK, loss of sight most frequently happens later in life and as part of the ageing process. Studies (Evans et al, 2002) suggest that more than 12% of people over 75 have some sight loss. Visual acuity is reduced by 10% for 60-69 year olds, 30% for 70-79 year olds and 35% in the over-80s. In the UK, 65,000 people are diagnosed with low vision each year (Morris, 1999).

Vision is one of the primary senses and serious or complete loss of sight also has a major impact on a person's ability to communicate effectively and function independently (Jones, 2007). Lakowski and Drance (1979) found that a large number of patients with ocular hypertension showed acquired color vision losses. These losses were particularly in the blue-green part of the spectrum, the called tritan defects (Kelly, 1993). They seemed to precede nerve fiber bundle defects in the visual field (Drance et al, 1981). The loss of chromatic sensitivity in the short wavelength part of the spectrum in glaucoma was confirmed by others. Other eye diseases, such as low-grade type 2 diabetic retinopathy (Friström, 1998) and moderate cataract (Friström & Lundh, 2000) have been shown to affect peripheral color contrast. Central color contrast sensitivity is affected by macular degeneration, even with early age-related maculopathy (Frennesson et al, 1995).

Macular degeneration, one of the most common causes of visual impairment, makes close up tasks such as reading more difficult. Retinitis pigmentosa, macular degeneration and cataracts can result in problems or discomfort with glare and bright lights or backgrounds. With age, changes to the eye increase sensitivity to glare, difficulty of adapting to changing light levels, and make contrast and color harder to discern. A good color use for visual communication, as well as helping to improve visual performance, it may also increase general well-being and health.

The effect of sight ageing is partially attributed to the yellowing of the retina, lens and vitreous humor yellow with age causing colors such as violet, blue and green to be filtered out, reducing the contrast sensitivity of the eye (Kelly, 1993) and increasing the requirement for light in older subjects (Bix, 1998). The effect of age on measurement legibility is further compounded by a reduced ability of the iris to dilate, under all light conditions.

2.2 Color contrast

Color appearance is closely related to a constant property of a surface, its spectral reflectance, it is not simply determined by the composition of reflected light (Weiskrantz et al, 2007). The relative reflectances of adjacent surfaces at any specific wavelength of light determine the relative intensities of light reflected from them at that wavelength, regardless of the spectral composition of the light illuminating them. Ratios of cone signals are approximately invariant under changes of illumination (Foster & Nascimento, 1994). If the reflectance properties of one surface are known, then the properties of all other surfaces seen under the same illuminant can be inferred on the basis of these cone ratios or color contrasts (this is the essence of Land's "retinex" color constancy algorithm) (Land & McCann, 1971). The earliest mechanisms of color constancy, our ability to recognize objects' colors regardless of the color of light illuminating them, occur in striate cortex. So, people without striate cortex, like some older people, will lose color contrast processing. They should behave as if objects change color as the light illuminating them changes, their apparent color depending on the wavelengths of light they reflect, rather than the properties of their surface material (Weiskrantz et al, 2007).

The influence of contrast in reading and legibility is important not only because text of a wide range of contrasts is encountered in the environment but also because many ocular conditions lower the effective contrast of the reading stimulus. Most studies of the role of contrast in reading, however, have treated only the luminance dimension. In general, reading is found to be fastest when the luminance difference between text and background is maximal. Lippert (1986) reported that legibility of briefly presented digits depended on the color difference between the digits and the background. While Lippert systematically varied luminance and chromatic differences between text and background, he used a range of character sizes that was near the acuity limit of the chromatic-contrast-sensitivity function. Tinker and Paterson (1928) found the legibility of colored inks on differently colored papers to depend primarily on the luminance difference between the text and the background, but the range of conditions that they could examine was limited by the nature of their stimulus medium. Legge and Rubin (1985) demonstrated that for observers with normal vision the luminance contrast and the background luminance (or the text luminance when the background was dark) determine reading rate regardless of the color of the text (Knoblauch, 1991).

Many eye conditions that result in low vision also result in color vision deficiencies. While the most common congenital color defects produce little in the way of performance limitations in and of themselves, the additional loss of sensory information that arises from a color defect for an already compromised visual system could have greater consequences.

From their study, and in a way to compensate the losses of older vision, Santa-Rosa and Fernandes (2012) recommended some guidelines, such as: enhancement of figure-ground segregation, reduction in clutter, prohibition on using short-wavelength colors as blue, increasing of the font size, diminishing of menu options and items. It was also recommended to increase readability: black and thicker letters in a white ground.

When creating a composition, either something freeform, or a more text based layout, a determination for the final impact of the whole presentation needs to be identified.

The dominant element may be classified as either "contrast dominant" or "value dominant." Designs that evidence contrast dominance or value dominance are then sub-divided into low, moderate, and high contrast, or light, medium, and dark value categories. The choice of colors will enhance or minimize the overall impact.

If the proximity between the neighboring hues is less apparent when you squint, the overall composition has a displays lower contrast level; if the overall composition appears light, it has a light value. Conversely, if distinctions between hues are very apparent, the contrast is high, and if the overall composition appears dark, the value level is dark. Understanding how the relationships between the colors of a chosen palette will affect the final outcome of an overall composition is integral to mastering the use of color.

Every visual presentation involves figure-ground relationships. This relationship between a subject (or figure) and its surrounding field (ground) will evidence a level of contrast; the more an object contrasts with its surrounds, the more visible it becomes. When we create visuals that are intended to be read, offering the viewer enough contrast between the background and the text is important.

The human eye requires contrasts for visibility and legibility. Contrast creates visual interest and helps deliver accurate information. Colors that are close in value tend to blur together, and their borders "melt." This is especially important when you want readable text on a colored background.

Colors of contrasting values stand out from each other. On a blue background, yellow jumps out at the reader. When the color value has enough contrast compared to the background it is easy to read; but too much contrast or the use of complementary colors is taking the idea of contrast too far. Colors will appear to "vibrate" and will create legibility problems and give your poor reader a headache.

The present work aims to disseminate simple, comprehensible recommendations for choosing colors that work effectively for all.

2.3 How does impaired vision affect colour perception?

Arditi, A. (1999a) designed three basic guidelines for making effective color choices that work for nearly everyone. Following the guidelines are explanations of the three perceptual attributes of color - hue, value (lightness) and saturation (chroma) - as they are used by vision scientists. However, this study, as most of the research implemented till now, was developed for digital displays, not for printed material.

Partial sight, aging and congenital color deficits all produce changes in perception that reduce the visual effectiveness of certain color combinations.

Two colors that contrast sharply to someone with normal vision may be far less distinguishable to someone with a visual disorder. It is important to appreciate that it is the contrast of colors one against another that makes them more or less discernible rather than the individual colors themselves.

Color deficiencies associated with partial sight and congenital deficiencies make it difficult to discriminate between colors of similar hue.

Lightness, like hue, is a perceptual attribute that cannot be computed from physical measurements alone. It is the most important attribute in making contrast more effective. With color deficits, the ability to discriminate colors on the basis of lightness is reduced. Designers can help to compensate for these deficits by making colors differ more dramatically in all three attributes.

3 METHODOLOGY

The study started with a literature review of the relevant material. The literature review was exploratory qualitative. Then, we produced the State of the Art.

From the theoretical contextualization, we were able to draw a hypothesis:

To produce a more inclusive design project when designing visual communication analogical products, designers must be aware of the issues related with color and text legibility, due to the reading problems experienced by many users, like older people.

The Project overlaps different areas and knowledge, among which: Ergonomics; Inclusive design; Visual Communication; Colour contrasts and color measurement; Light sources (natural and artificial); The evaluation of the proprieties of the light sources and the influence of the surfaces; The evaluation of the color aspects inside visual communication area; Legibility and the obstacles to reading; Older people and visual limitations.

The work with experts in the area of Ergonomics, Colour, Inclusive Design, Visual Communication and Older People, as well as with the users and the associations of people with impairments, was also fundamental for the study.

For the second stage of this research project, we decided to develop a direct field work with the users, i.e., an experiment (active research) using sample groups of older people with the same gender composition and general characteristics.

In order to allow a detailed exploration and the handling of complex and diverse information, a qualitative method was chosen. The process involved development of tools to work with the groups, especially printed color material always relating front and background colors.

The search was carried out in several occasions (rounds), using the same material in similar lighting conditions and the same distances. During the experiment sessions, the sample groups of older people who took part in this study had the light up, on average, 900 lux (illuminance), as recommended by O'Neill (2003) in a study conducted by a Research Group for Inclusive Environments at the University of Reading, in terms of having a "Good" color rendering. The examples were very simple and designed to be readily understood. They were written in plain English for the UK sample groups and in Portuguese for the Portuguese groups. For this reason no technical terms were required when participants were asked to read the different words and sentences. The words and sentences acted as forms, using different color schemes, in form/background relationship.

Every sample group was formed by 8 people: 5 female persons and three men, with ages comprised between 65 and 85, all in sight normal conditions for people in this range of age, only with aged vision, but with no specific sight diseases.

Several messages, in different color contrasts were printed on A4 format cards for a total of 24. All messages were created using 48 point Myriad. The lettering size was chosen having in mind that the group members would be placed at a two meters distance from the cards. For the color production, we used the Pantone Matching System (PMS) Colors. The colors used in this research are colors that the researcher found more appropriated for the research aim and the target group. For this experiment, the inside level of light was measured by an A.W. Sperry light meter. Information for each group included the gender, age, requirement for eye wear. The results for color blindness were recorded as "normal" for all group members. To be selected for each of the sample groups, the members couldn't have any eye disease, only older vision.

After informed consent had been obtained, and the subjects' visual acuity and ability to perceive color had been tested, each one was seated in front of a researcher holding the A4 cards, with a 2 meters distance between them, and asked to read the text written in each card, where object (text) and background had different color combinations.

There were also cards with different types of lettering and spacing (letter and line spacing), but using always the same letter dimensions. We also wanted to test the level of legibility and the eventual experienced difficulty, as well as the level of eye comfort and color contrast.

The experiment comprised two phases:

- In phase 1 we didn't measure de reactions of the participants; we only used a video camera to report the sessions.
- In the second phase we used eye-tracking, in a way to help us to understand the eyes movements and the efforts made by each participant to read the cards.

After the implementation of the experiment with the sample groups of older people, we could achieve findings, which were confronted with the drawn hypothesis. We were able to verify that not only we had proved the hypothesis but also we had amplified the initial knowledge with a contribution for the study area.

4 CONCLUSIONS

With the sample groups, the research team verified some conclusions from the literature review and found some solutions for an inclusive approach in visual communication design, using color in analogical material, for older people.

The different rounds of tests allowed the research team to identify the main problems with color use for visual communication for older people.

After the experiment we were able to draw some recommendations:

- good legibility helps all users, but for people with low vision the issue is crucial for reading text;

- the text and background color combination should have high contrast;
- a clear open typeface (font) should be used for text;
- the characters must be of good proportions with clear character shapes;
- text should not be placed over a background image or over a patterned background;
- white or yellow type on black or a dark color is more legible;
- small type and very bold type tend to blur for some people, reducing legibility;
- avoid shades of blue, green and violet for conveying information since they are problematic for older users;
- use no more than five colors when coding information;
- be sure the elements have a *contrasting color value* unless you want the elements to just blur together;
- excessive use of colors can be distracting;
- when using colors, one must have in mind that older individuals have a harder time distinguishing between colors in the cooler range - blues and greens particularly;
- some individuals are colorblind and find it difficult to distinguish between red and green;
- color is not appropriate as the sole differentiating feature between different elements - they should vary in other design features as well;
- varying the value of colors (the lightness or darkness) by at least two levels will enable most people to differentiate between the colors.

ACKNOWLEDGMENTS

The author would like to acknowledge SURFACE –Inclusive Design research Centre – the University of Salford, UK, where the project was developed and Professor Marcus Ormerod for his committed and helpful supervision; FCT – Science and Technology Foundation, Portugal, who funded this research; CIAUD, that supported the participation in the AHFE 2012 Conference.

REFERENCES

Arditi, A., 1999, "Effective color contrast and low vision" in B. Rosenthal and R. Cole (Eds.) "Functional Assessment of Low Vision". Mosby. St. Louis. pp 129-135.

Bix, L., 1998, "The effect of subject age on legibility". Unpublished Master's Thesis, Michigan State University, East Lansing.

Bolton, S. (Ed), 2007, "Dimensions of Dignity at Work", London, Elsevier, p.7.

Charman W.N., 1991, "Limits on visual performance set by the eye's optics and the retinal cone mosaic". In: J.J. Kulikowski, V. Walsh and I.J. Murray, Editors, Vision and Visual Dysfunction Limits of Vision vol. 5, Macmillan, UK, pp. 81–96.

Drance, S.M. et al, 1981, "Acquired colour vision changes in glaucoma". Arch Ophthalmol 99. pp 829–831.

Evans, J.R. et al 2002, "Prevalence of visual impairment in people aged 75 years and older in Britain: results of the MRC Trial of Assessment and Management of Older People in the Community". British Journal of Ophthalmology, 86. pp 795-800.

Fisk, A.D., 2004, "Designing for older adults: Principles and creative human factors approaches", pp.97, 108, 148.

Foster, D.H. & Nascimento, S.M.C., 1994, Proc R Soc London Ser B 257. pp 115–121.

Frennesson, C. et al, 1995, "Colour contrast sensitivity in patients with soft drusen, an early stage of ARM". Doc Ophthalmol 90. Blackwell Publishing. pp 377–386.

Friström, B. & Lundh, B.L., 2000, "Colour contrast sensitivity in cataract and pseudophakia". Acta Ophthalmologica Scandinavica. 78(5). Blackwell Publishing. pp. 506-511.

Friström, B., 1998, "Peripheral and central colour contrast sensitivity in diabetes". Acta Ophthalmol Scand 76. Blackwell Publishing. pp 541–545.

Helena H. & Näsänen R., 2003, "Effects of luminance and colour contrast on the search of information on display devices", DISPDP, 24, no4-5, pp. 167-178.Oxford, Elsevier.

Jones, R. & Trigg, R., 2007, "Dementia and serious sight loss". Bath: Research Institute for the Care of the Elderly, St Martin's Hospital.

Kelly, M., 1993, "Visual impairment in the elderly and its impact on their daily lives". Unpublished PhD Dissertation, Texas Woman's University.

Knoblauch, K. et al, 1991, "Effects of chromatic and luminance contrast on reading". J. Opt. Soc. Am. A. 8(2), p 428.

Kristjuhan, Ü., 2010, "Decreasing aging velocity in industry workers", Ann NY Acad Sci, 1197.

Lakowski, R., Drance, S.M., 1979, "Acquired dyschromatopsias: the earliest functional losses in glaucoma". Doc Ophthalmol Proc Ser. 1979, 19, p159.

Land, E.H. & McCann, J.J., 1971, J Opt Soc Am, 61, pp1–11.

Legge, G.E., et al, 1985, "Psychophysics of reading". I. Normal vision. Vision Research, 25, pp.239-252.

Lippert, T.M., 1986, "Color difference prediction of legibility performance for CRT raster imagery", in "Digest of the Society for Information Display" (Society for Information Display, Playa Del Rey, California, 17, pp. 86–89.

Morris, C., 1999, "Visual impairments and problems with perception". Journal of Dementia and Care, Nov/Dec, pp.26-28.

O'Neill, L., 2003, "Lighting the homes of people with sight loss". The University of Reading.

Openshaw S., Taylor, E., 2006, "Ergonomics and Design - A Reference Guide", Allsteel Inc.

Pavio, A., 1971, "Imagery and Verbal Processes". NY: Holt, Rinehart and Winston.

Santa-Rosa J.G., Fernandes H., 2012, "Application and analysis of the affinities diagram on the examination of usability problems among older adults", Work (Reading, Mass.), 41, ISSN: 1875-9270

Tinker, M.A. & Paterson, D.G., 1931, "Studies of typographical factors influencing speed of reading". VII. "Variations in color of print and background". Journal of Applied Psychology 15, pp. 471–479.

Weiskrantz, L. et al, 2007, "Color contrast processing in human striate cortex". USA: PNAS.

CHAPTER 64

Glazed Tile Surfaces as Perceptual Cues in Space Recognition and Urban Design

Carla Lobo, Fernando Moreira da Silva

Faculty of Architecture, Technical University of Lisbon; CIAUD – Research Centre in Architecture, Urban Planning and Design
Lisbon, Portugal
carla.a.lobo@gmail.com

ABSTRACT

The tasks of recognition and categorisation of the environment depend on the mental processes of organising the information conveyed by the available stimuli in the visual space. While walking around the city, we guide ourselves through these personal mind maps created on a basis of meaningful perceptual / spatial / emotional / socio-cultural features.

Ceramic claddings, by they meaningful emotional and sensory features and tactile qualities, can help to increase readability acting as important factors in the *wayfinding* and *wayshowing* processes. This perceptual characteristics of ceramic claddings may not be detected consciously, however, the experience and the information which has been processed and stored, allow us to interpret this as clues and use them in the understanding and organisation of the space

Keywords: Glazed tiles, perceptive cues, colour, gloss, texture, user

1   CODING AND DECODING THE ENVIRONMENT

Movement characterises the way of perceiving and being in the human space. Movement (of the eyes, head, shoulders and body) induces consecutive perceptual variations of the urban space, the results of the linking of light and the observational conditions, whether the observer is immobile or moving, will determine their grasp and understanding of the space. Gibson (1986) considers that there are two types of vision which are crucial to the understanding of space: ambient vision, in which the observer is static; and ambulatory vision, where the observer moves within the space, constructing their perception of the space from the stimuli received from different observation points.

The tasks of recognition and categorisation of the environment depend on the mental processes of organising the information conveyed by the available stimuli in the visual space (Friedman & Thompson 1976). Whilst mobile, the inhabitant is an incessant explorer of the space, in pursuing his objectives; in this exploratory process, information is harvested from the stimuli available, and sequent encoding and storage, which is combined with data retrieved from memory (Golledge 1999; Allen 1999; Mollerup 2005), that will be used in the processes of recognition and orientation. The apprehension and comprehension of the space are a result of the relation that the body establishes with the space through the senses and proprioception. Golledge (1999) draws attention to the relevance of this cognitive information in spatial navigation decisions, further emphasising the weight of personal cognitions in the behaviour when faced with way-finding tasks. We gather information from the environment through sensory channels, relating it to previous significant processed data, (Allen1999; Mollerup 2005). At a glance, as stated by Fei-Fei *et al* (2007), a variety of global perceptual features are represented, allowing a categorization of the environment (Torralba & Oliva a*pud* Fei-Fei *et al* 2007). While walking through the city, we guide ourselves through these personal mind maps created on a basis of meaningful perceptual/spatial/emotional/socio-cultural features.

1.1   Livable Urban Landscapes

Diversity in the urban landscape is considered as a vital quality (Humphrey 1980; Lancaster 1996; Lynch 2002). Over stimulating environments can lead to crowding – decrease of object's perception by the proximity of multiple similar objects (van de Berg *et al* 2007) - affecting our ability to identify basic information in visual fields.

Size, shape, colour and location are part of the mental process of objects and space organization (Swirnoff 2003; Allen 1999). The uniqueness of each of these factors, and the way in which they combine in the space determine the user's efficiency, as they promote environment recognition and categorization (Friedman & Thompson 1976).

Colour draws our attention, and is a relevant feature in emotional memory building, promoting the establishment of centres of interest in the visual field. The use of colour in urban design, associating meaningful emotional and sensory features to the tactile qualities of the materials, can help to increase readability acting as important factors in the way-finding and way-showing processes.

As stated by van de Berg *et al* (2007), hue and saturation variations, as they cause less clutter, are more suitable for visualizing and understanding information than variations in size and orientation grids. Colour and chromatic diversity, enhanced by light variations are considered positive elements in our space image composition, not only for their emotional and psychological value, but also by its sensory capabilities that can alter our spatial perception. According to Frey et al (2008) when colour information is present, users with no visual impairment tend to look more often to those locations, particularly when hue variations are used in surface segmentation rather than lightness variations.

The experimentation of modern urban spaces is almost exclusively dependent on the visual experience, however, the blind, partially blind or the visually impaired do not have the possibility to participate and enjoy these experiences (Roberts 2004). For these users, stimuli are perceived through tactile, auditory, olfactory, and proprioceptive information (Allen 1999) and the structuring of the space, reason why the existence of the diverse sensorial stimuli in the environment is considered as a crucial factor in the design of inclusive and accessible habitats for a greater number of users. In this context, the creation of structures that are referential in way-showing systems, and way-finding facilitators, can be enhanced by the visual and tactile quality of the materials, and in the way they are perceived by the users

An ergonomic use of colour, textures and graphic patterns can highlight details, break monotony by introducing rhythm and proportion, increase spatial readability by differentiating volumes, establishing figure/ground clarification, hierarchize spaces (Porter 1982; Manhke1996) guide the traveller, improving way-finding and way-showing tasks (Mollerup 2005).

1.2   Ceramic Claddings – Urban Surfaces

As Rapoport (2002) stated, constructed environments are used in an active way by people, in their process of acculturation, and familiarization. The structuring of spaces from combining the familiar, the new, and the unexpected, allows the subject/explorer to create orientation systems, to optimise and facilitate the implementation of tasks and attain objectives. Surfaces are interfaces between users and objects (Manzini, 1993), through which, their surface qualities determine the selection of available stimuli in the visual space. Elements differentiated by colour, texture and form, or those that relate directly with the specific objectives of the subject, will be the focal points of attention, the elements at the top of the selection hierarchy.

Ceramic glazed tiles are architectural skins that solve problems related to space, light, and acoustics, protecting from the elements, making surfaces more interesting with the introduction of colour, gloss and texture without losing the cultural, sensory and symbolic meaning to the "observer / user". Due to their material characteristics, glazed ceramics are constantly changing in appearance, and a simple alteration in the weather or observation conditions (point of view, angle of incidence of light, and time of day) provide different perceptions of the same shape, the same pattern, the same colour or texture.

610

Figure 1   Colour variation in azulejo's claddings – glaze evenness and surface qualities as modifiers.

The perceptual characteristics of ceramic claddings may not be detected consciously, however, the experience and the information which has been processed and stored, allows us to interpret the clues and use them in the understanding and organisation of the space.

2   GLAZED TILES PERCEPTUAL FEATURES AS REFERENCE CUES

1.1   Colour

The possible chromatic variety with ceramic glazes, combined with the possibility of introducing graphic elements (patterns, compositions graphics), allows the construction of graphic layouts on the surfaces, which can significantly define the modelling and understanding of the space. As it is associated to glossiness, the chromatic palette of glazed tiles ranges from relatively dull colours to the brightest, and from the least saturated to the most saturated, in contrast to the architectural paint palettes that usually exclude dark and saturated colours. The glossiness of ceramic glazes creates points of brilliance and reflections that minimise the impact of the application of saturated dull colours on large surfaces, as well as dematerialising the architectural mass, making their application possible in extensive areas. Even if you have similar hues in contiguous façades, the glazed tile's perceptive variability is recognized as a reference, helping us when moving.

These characteristics differentiate the ceramic surfaces from the other finishes on the buildings, accentuating them from the rest by their visual character and physical permanence, and their perceptive impermanence.

Figure 2 Chromatic intensity, and variety as a detachment factor. The glossiness of the glaze allows its intensity

1.2   Texture

Recognition and categorization cognitive processes make use of visual data to help on explorer relative positioning. Gradients of scaling and optical mixture of colours, textures, patterns and laying grid can indicate the users' position regarding the building. Proximity or distance leads to different visual patterns. These stimuli changes, as long as they are understood as stimuli modifications and not as new stimuli, can contribute to a better understanding of the environment: representing shape modification, change of direction or time elapsed through colour and reflection variations, progression or location in a mentally designed route. As Todd *et al* (2007) stated, texture scaling is one of the primary sources of information to understand three-dimensional shapes.

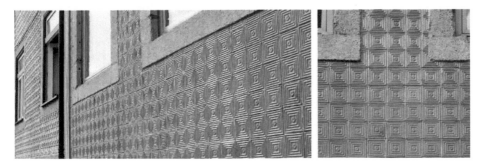

Figure 3 Texture gradient as distance and depth indicator.

The way we perceive the ceramic tile cladding patterns, their texture or the grid created by the tile laying, is fundamental to the perception of the form of the building, and for referencing our position in relation to that surface. If these features are clearly perceived, the distance is proximal and the viewing angle is close to 90°. If texture, pattern and grid merge into homogeneous colour, the viewer is at a greater distance, or the viewing angle is smaller, or both.

1.3   Gloss

Gloss is one of the extrinsic characteristics of glazed claddings, and confers a visual mutability unique to these products.

The reflections caused by the incidence of light on its surface, which can vary with the angle of vision and with the sun inclination, provide specific differences in brightness, allowing movement references for those with impaired vision or colour blindness, without creating visual uneasiness.

The shifts in specular reflection, associated to brightness variations on the surface, can help to identify the relative position of the viewer and building, in regard to the sun, thus corroborating its spatial orientation.

The levelling of surface qualities (optical mixture of colour and texture in a uniform application), when viewed from a certain distance, is contradicted by the specular reflections which remain visible until comparatively further away, they also help to highlight the surrounding tiled surfaces (Lobo & Moreira da Silva 2010).

Figure 4 Gloss as a differentiation factor.

In situations of low luminosity or under chromatic light sources that unify the colour differences by contamination, where the perception of colour and/or graphic patterns is strongly influenced, glossiness enables a differentiation between surfaces, aiding in the recognition of buildings, and their segregation in relation to the surrounding urban fabric (Davidoff 1991; Lobo & Moreira da Silva 2010).

2   CONTRIBUTION TO THE OPERATING QUALITY OF SPACES

Some building regulations recommend the creation of visual contrast between adjacent surfaces, as a means of improving identification and navigation (Dalke *et al* n.d.), and efficiency and safety for the target groups (Bright, Cook & Harris, 2010). These contrasts may be achieved through the effects of light and colour, as well as through the use of contrasting colours, luminosity, and the surface quality of the materials. In the case of blind users, the visually impaired, the partially sighted, and the elderly, the sensitivity to chromatic contrasts, specifically the hue, is less than the sensibility to luminosity variations (Mollerup 2005), which makes much of the provided information based on chromatic contrasts difficult to understand or even illegible. For colour-blind users, contrasts of luminosity and surface quality of materials may be the only clue to distinguish two surfaces. Tile varied chromatic palette in terms of hue, intensity and luminosity, as well as its (enduring) chromatic stability, allows spatial solutions to be designed, which improve navigation for the elderly and partially sighted. The possibility of using more saturated, and darker colours on large expanses, without becoming monotonous, or with excessive negative visual impact, allows the creation of long lasting contrasts of colour and luminosity with adjacent surfaces, whether they are light or dark, bearing in mind that it will remain stable for a long time, thus increasing the safety of the user when using the space.

The problem of using glossy surfaces on buildings is the possibility that it may cause glare. It has been proven that there is no direct relation between measured glossiness and perceived glossiness, with regard to the designations of glazes (glossy, satin or matt), and these may not correspond to what is perceived in the real context.

Both "quantitative" and perceptive measurements made by the author, have led to the conclusion that in the case of tiles being applied to exterior surfaces, subjected to the action of rain and wind, and the deposit of surface dust, the levels of perceived and measured glossiness, semi-matt | glossy on the comparative NCS scale, and between 5% and 60 % in units of glossiness measured with a glossmeter, are within the parameters indicated by Mollerup (2005) for the use of glossy exterior surfaces (maximum value 60% gloss). In outdoor spaces, the deposit of fine particles of dust on the surface contributes to a decrease in reflection and specular highlights, minimising the probability of dazzling, which is a characteristic of reflective surfaces.

The reflections caused by the incidence of light on the surface, which are variable depending upon angle of vision and sun inclination, provide specific differences in luminosity and specular highlights, differentiating it from adjacent ones (Lobo & Moreira da Silva 2010), creating movement references for users who are colour blind or with impaired vision without visual discomfort. In the case of sequential façades, even if the chromatic contrast is reduced, the perceptive variability of the tiled surfaces is recognised as a reference, aiding navigation.

It should also be noted that in spite of its reflective characteristics, the risk of colliding with a tile-clad surface is small when compared with glass. The affordances of glazed tile claddings generate a system of multisensory information that prevents the creation of dangerous situations (Gibson 1986), which are inherent in openings or façades in achromatic glass. As noted by Lobo & Moreira da Silva (2010), this differentiation "allows the design of efficient relationships of form/background in exterior public spaces, by taking advantage of chromatic and gloss contrasts between glazed tile claddings – background – and fixed and moveable urban equipment– forms".

It should also be noted that in cases of low luminosity, or at greater observation distances, the greater reflective capacity of glazed tile surfaces makes them stand out from other rendered or painted surfaces (Lobo 2006), creating reference points for perceptive structuring of the space (Lobo & Moreira da Silva 2010).

Moreover, the differences between the glossiness of the façade and the shadows of the openings increase the legibility and identification of the building's characteristics (Cook *et al* s.d.), facilitating the tasks of navigation and recognition of the important functional elements of buildings, such as access to the interior, warning of danger, changes of direction, framing a building's limits, or the demarcation of public service entrances and dwellings. (Lobo & Moreira da Silva 2010:12).

Due to its plasticity prior to firing, the surface of the ceramic body can be modified, making reliefs and textures possible; by associating surface qualities to colour and brightness the tile's ergonomic potential is increased as a solution for cladding and tactile and visual communication (Braille, letters in relief, pictograms, orientation guides). This eliminates the need for communication physical supports, thereby freeing the circulatory space, and increasing the safety of disabled users, and minimising visual crowding. The various information elements are distinguished by contrasts, clarifying the relations of form and background, providing greater visibility at a distance, broadening the range of people that benefit from its use as a cladding material whilst incorporating signage and aesthetic enjoyment.

3   FINAL CONSIDERATIONS

Due to their visual, tactile, cultural, and symbolic attributes, tiles stand out from the generality of cladding materials and urban equipment, facilitating their recognition and memorisation, offering the user/explorer of urban spaces a set of visual, tactile and proprioceptive references enabling them to construct an emotional and three dimensional space, thus facilitating the tasks of recognition and spatial orientation.

Due to the chromatic, reflective and tactile nature of the material, as well as the differentiation at a graphic level, the presence of tiles in inhabited spaces distinguishes them from other components of the landscape, becoming what Lynch (2002) termed a landmark. Their uniqueness and socio-cultural significance corroborate the visual attraction inherent in ceramic claddings, enabling them to be identified and referenced by subjects from different cultural backgrounds.

The visual variety and diversity that we defend, should be potentialised by the application of glazed tiles surfaces in urban spaces, are perceptible both at the scale of a town square and a street, providing distinct stimuli depending on the observation variables (distance and angle of observation, quality and angle of light incidence), and the extrinsic qualities of the tile (quality of the glaze, pattern or textures).

Its proven longevity (structural, chromatic and brightness) allows an environmentally responsible use, without losing visual or tactile qualities during its lifetime.

Ceramic claddings urban performance is enhanced by the presence of other materials; functional performance can be increased, and visually they can turn into an integration element of the building in the environment, or a detachment factor.

Colour schemes, as well as tactile features can contribute to spatial readability, and as way-finding coordinates. The sensory-perceptual and spatial features of this material, qualify it as an improving element in spatial legibility and readability, allowing us to characterize, organize, and hierarchize objects and spaces.

Ceramic glazed tiles can be more than protective architectural skins; through the interaction of light and colour they can act as an information finding system. Incorporating perceptual features analysis, and users experience information related to environment and materials, can be effective tools in the design process of ceramic glazed tiles, as well as in urban design.

ACKNOWLEDGMENTS

The present study as been developed with the support from CIAUD - Research Center in Architecture, Urban Planning and Design, Faculty of Architecture, TU Lisbon.

REFERENCES

Allen, G. 1999, Spatial Abilities, Cognitive Maps, and Wayfinding: bases for individual differences in spatial cognition and behavior. In Golledge, R. (ed.). *Wayfinding behavior: cognitive mapping and other spatial processes,* pp. 46-80. Baltimore, The Johns Hopkins University Press.

Bright, K & Cook, G 2010, *The Colour, Light and Contrast Manual: Designing and Managing Inclusive Built Environments*, Wiley-Blackwell, Chichester.
1st chapter online media.wiley.com/product_data/excerpt/45/14051950/1405195045.pdf

Cook, G Bright, K, Yohannes, I, Dalke, H, Camgoz, N, *Colour and Lighting for Intermodal Transport Environments.* (internet) Available at
http://www.wiley.com/legacy/wileychi/brightandcook/references.html. Retrieved 22.02.2011

Dalke, H, Conduit,G. Conduit, B. Corso, A. n.d., *Measurement for a more visible world: colour contrast and visual.* (internet). Available at:
http://www.tcm.phy.cam.ac.uk/~gjc29/Papers/ DalkeConduitConduitCorso09.pdf
Retrieved 22.02.2011

616

Fei-Fei, L., Iyer, A., Koch, C., & Perona, P. (2007). What do we perceive in a glance of a real-world scene? *Journal of Vision*, 7(1):10, 1–29, http://journalofvision.org/7/1/10/, doi:10.1167/7.1.10. Retrieved 26.12.2010.

Frey, H.-P., Honey, C., & König, P. (2008). What's color got to do with it? The influence of color on visual attention in different categories. *Journal of Vision*, 8(14):6, 1–17, Available at http://journalofvision.org/8/14/6/, doi:10.1167/8.14.6. Retrieved 26.12.2010.

Friedman, S & Thompson, S 1976, Colour, competence, and cognition: notes towards a psychology of environmental colour. In Porter, T & Mikellides, B (ed.) 1976, *Colour for Architecture*, Studio Vista, London.

Gibson, J. 1986, *The ecological approach to visual perception*. London: Lawrence Erlbaum Associates.

Golledge, R. (ed.) 1999, *Wayfinding behavior: cognitive mapping and other spatial processes*. Baltimore, The John Hopkins University Press.

Humphrey, N 1980, Natural Aesthetics. In Mikellides, Byron (ed.). *Architecture for the People*. London, Studio Vista.

Lancaster, M 1996, *Colourscape*, Academy Editions, London.

Lynch, K 2002, A *imagem da cidade*, Edições 70, Lisboa.

Lobo, C & Moreira da Silva, F 2010, Interacção da luz e corn as superficies como factores ergonómicos no design urbano: o azulejo como concretização. Silva, J Paschoarelli, L Moreira da Silva, F (Orgs.) *Design ergonômico: estudos e aplicações*, PPGDesign – FAAC – Universidade Estadual Paulista, Bauru. (CD-Rom).

Manzini, E 1993, *A Matéria da invenção*, CPD, Lisboa.

Mollerup, P 2005, *Wayshowing - A Guide to Environmental Signage: Principles & Practices*, Lars Muller Publishers, Baden.

Porter,T 1982, *Architectural Color*, Whitney Library of Design, New York.

Rapoport, A 2005, 'Spatial organization and the built environment', in: T Ingold, (ed.) , *Companion encyclopedia of anthropology*, Routledge, London, pp.460-502.

Roberts, M 2004, *In sight - A guide to design with low vision in mind: Examining the notion of inclusive design, exploring the subject within a commercial and social context*, Rotovision, Mies.

Swirnoff, L 2003, *Dimensional color* – Second Edition, W W Norton & Company, New York.

Todd, JT Thaler, L Dijkstra, TMH Koenderink, JJ & Kappers, AML 2007, The effects of viewing angle, camera angle, and sign of surface curvature on the perception of three dimensional shape from texture. *Journal of Vision*, 7(12):9, 1–16. Available at http://journalofvision.org/7/12/9/, doi:10.1167/7.12.9. Retrieved 26.12.2010.

Van den Berg, R., Roerdink, J. B. T. M., & Cornelissen, F. W. (2007). On the generality of crowding: Visual crowding in size, saturation, and hue compared to orientation. *Journal of Vision*, 7(2):14, 1–11, Available at http://journalofvision.org/7/2/14/, adoi:10.1167/7.2.14. Retrieved 26.12.2010.

CHAPTER 65

Breast Design: The Role of Ergonomic Underwear during Lifetime

Filipe, A. B., Montagna, G., Carvalho, C, F. Moreira da Silva
CIAUD, Technical University of Lisbon
Portugal
anabrigida84@gmail.com

ABSTRACT

The image of women changes considerably in different stages of her life, from childhood to maturity (Leonard, R. 2008), so her chest also undergoes changes that drive women to seek proper lingerie. In order to reach an ergonomic adjustment of the body to the bra it's necessary to understand if the stages of women's lives are connected with those of her chest. In addition there are several types of bras, suitable for every situation that may change the visual appearance of the breast and the appearance of clothing, due to its technical characteristics.

Ergonomics in a bra is felt by the user in a very intimate way due to the direct result on her body. If a woman repeatedly the wrong size bra, there will be repercussions on her body. In terms of clothing, ergonomics is intrinsically linked with the comfort or the lack of it, and due to the development of the bra, also the female body has changed according to fashion's demands. Today, a good bra is associated to what can be trasmitted as a good ergonomic comfort and this is the fundamental premise for the welfare of the women's chest.

Keywords: Lingerie Design, Ergonomics, Women's breast, market segments

1. INTRODUCTION

With the previous development of a research project regarding Lingerie, its visual image and its market - consumers and brands, it was possible to understand what was not yet studied about bras and breasts. This particular area within the

Fashion Design has vast research possibilities and is therefore a relevant study from different perspectives. Among the conclusions within this project, it was concluded that women experience different stages of her chest, and that these are in agreement with her different stages of life - childhood, adolescence, reproductive years, perimenopause, menopause, active retirement and old age (Leonard,R.2008).

The women's chest has a character that goes far beyond is physical appearance. Not being a matter of consensus, the various categorizations of breast formats demonstrate the singularity of such an important part of the female body and makes the design of bras a complex search for comfort, pleasure and welfare for each user.

Of all the features that covers the bra, the comfort was, in the research study conducted, the most appreciated, crucial for the purchase and the one that is intrinsically connected with satisfaction.

Figure 1 Bra types, acessed 16 Dec. 2011,
< http://intimateapparelconsultancy.blogspot.com>

2. FROM THE CORSET TO TODAY'S BRA

The feminine silhouette has undergone continuous changes in ways that shaped appearance according to the trends of the time. In history, there has been times when more emphasis was given to the chest, other to the waist or back, etc. (Bressler, Newman, Proctor 1999; Boueri 2008).

Fashion forced women to possess a wasp waist designed to use tight corsets. This was not a healthy practice for women - it caused reproductive deformations and drastically narrowed down the ribs within the abdominal area. The women who did not wear a corset were not considered worthy and was so offensive as to leaving the house without gloves and hat (Boueri 2008).

The characteristics of the corset changed according to seasons, they were burly objects, cruel, sometimes comfortable, but always seen with a dense, sensual and erotic nature. The corset was considered the most intimate part of women until the early twentieth century, and, in this century its use both fit as underwear our outdoors. It is still considered an object of provocation and continues to be associated with sexual fantasies and eroticism in the year 2000 (Bressler, Newman, Proctor 1999; Piveaut 2005).

Name	Age	Size of waist	Reduce to	To wear corset
Nelly G.	15	20 inches	16 inches	Night and Day
Helen Vogler	12	21 inches	15 inches	Day
G. Van de M.	14	19 inches	13 inches	Night and Day
V.G.	13	22 inches	12 inches, if possible	Night and Day
Alice M.	17	16 inches	14 inches	Night and Day
Cora S.	16	18 inches	13 inches	Night and Day

Figure 2 Table of Medical Measurements waist with ages.
< http://intimateapparelconsultancy.blogspot.com>

In 1550 Catherine de Medicis contributed to limit the width of women. An adult woman had to draw up a narrow waist of 10 inches (about 25 cm) while breasts should be enhanced. For centuries the women's body had an artificial modeling. Even with the beginning of the Belle Epoque, the corset was one of the most desired objects raising the breast, reducing the waist and giving more curves on the hips. But with sports, the corset had to adapt, and so the the sports corset was created, which provided some freedom of movement and consequently an increase in women's health.

In the twentieth century, the corset had changed but its genesis wasn't affect. Thereafter, elastic materials such as Lycra gave the body an ergonomic comfort (Bressler, Newman, Proctor 1999). The deconstruction of the corset helped other pieces to develop, such the basque, the bustier, the girdle and the bra itself. The brassiere, essential part of lingerie, has shown many different shapes over time such as the bandeau of Balconette of décoletté, etc. (Bressler, Newman, Proctor 1999).

The variety of bras at that time allowed the enhancement, the uplift, the breast reduction and the smoothing (Bressler, Newman, Proctor 1999). Considered the most important element of the western women wardrobe, based on the fundamentals of Design, it's now a part of the women's daily life.

The bra arrived when the corset was put aside, allowing women to move, to breathe and express herself. So more than freedom to the women's body it also brought a freedom of thoughts and attitudes (Berry 2006). The creation of the bra was based on the necessity to use a support for the chest with the shoulder straps and two cups to fit the breasts. The aim was to produce a comfortable piece, ergonomic, resistant and healthier than the corset, according to the changing needs of society (Yu, Ng 2006).

The bra has changed the body visually, deceiving in size and shape the desired chest (Bressler, Newman, Proctor 1999).

Also, practical improvements began to emerge. Charles de Bevoise devoted to the creation of more than twenty models of bras with silk, worked embroideries and

620

brocades to meet the new need. At this point the bra suited women, contrary to what happens today. Among other styles a corset was created for the chest which was supported by the shoulders.

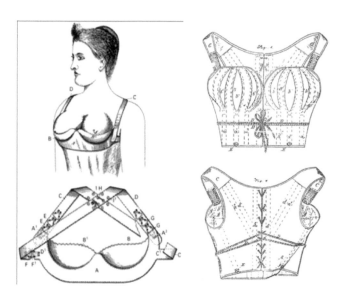

Figure 3 Image of Marie Tucek's bra (Left) and Luman Chapman's bra (Right)
< Farrell-Beck, J., Gau, C. (2002). Upfit - The Bra in America. Philadelphia, University of Pennsylvania Press.>

The bra began to arouse curiosity and its design gained commercial value when Mary Phelps Jacob sold her patent to Warner Brothers Corset Company.

According to Berry (2006), the bra Kestos reflected different types of breasts and was adapted to each women's size. All these details meant that women were becoming more feminine and devote more attention to her own individual beauty (Yu, Ng 2006), Fields 2007). However, secret dominated when a bra was bought due to shame.

The developments of each decade made the bra suit different types of clothing. The advances in technology and the industrial revolution made possible the birth of bras with cups adapted to the chest and back. The income, fine silks, satins, taffetas, cotton, tulle and later nylon bras were the characteristics that confer resistance, lightness, flexibility (Farrell-Beck, Gau 2002).

In the post Second World War a diversity in the types of bras began to be explored and women could choose what that suited to her chest and style: push-up bra, strapless bra, spiral-stitched bra, front closure bra, etc. (Berry 2006).

In the '50s the bra had more attention than ever. The major reason was the need for women to show their more pronounced curves and her large and sumptuous breasts could be shown by a more low-cut dress and fair enough to the body (Berry 2006; Farrell-Beck, Gau 2002).

The bra was alluding to the comfort and safety but maintained the features of an attractive piece.

Bras are created for all type of occasions and effects, and visual effects, the Demi bra or Half bra, minimizer bra to reduce the women's chest at least one size. In 1960 Louise Poirier, crieted the push up bra and the plunge bra for a lingerie company in Canada with distinct Wonderbra push-up .

In the 80's the bra brook with its main function. The pieces of lingerie, were always lower than the outer parts, however, now they garments themselves, without being hidden or being complemented (Fields, 2007), as Jean-Paul Gautier was inspired in underwear to create outwear. The main responsible for such change was the singer Madonna. In the film Desperately Seeking Susan, she gives emphasis to her black-lace bra (Berry 2006) and makes her Blond Ambition tour with a conic bra with theatrical character.

In the 90's women seek superior comfort and so their chest defined the type of bra that defined the body (Berry 2006; Fields 2007).

Since the appearance of the brassiere in the late nineteenth century, its basic construction has not changed substantially. What exists today are versions of the bras that have long been developed, applying a function and significance that attracts consumers to the updates, the relationship with the body, innovation and re-design (Yu, Ng 2006).

As the years passed, the bra became one of the most complex pieces of lingerie ever created. Towards the end of the 90's it's could be composed by to 43 components which turned it into a high-tech product (Bressler, Newman, Proctor 1999). The intimate products have the capacity to improve the life of the user, help to increase demand of the product they are buying and the different features that it may have (Vicentini, Castilho 2008).

3. ERGONOMIC COMFORT OF BRA

Handled daily, women wear bras with a natural and unquestionable need, which makes it a very important element and builder of her visual image. Sports, daily or special occasions, the bra will affect the level of comfort, health, posture and physical performance of the user. In this sense, the size and comfort are critical to the success of a bra (Mintel 1997).

Physiologically speaking, women do not have both breasts symmetrical and there aren´t two women with exactly the same shape. Furthermore, it is considered that the breast has six different phases, by means of their life cycle. The physical changes in her figure from the fact that going through puberty, hormonal changes, whether temporary as in the menstrual cycle or pregnancy, or permanent menopause, react and gradually change the chest. So, it is not stable throughout life, due to its behavior, historical data, and nature, and all these factors contribute to the change. (Hart and Dewsnap 2000).

If we compare the seven stages of a women's life - childhood, adolescence, reproductive years, perimenopause, menopause, active retirement and old age (Leonard, R. 2008), with the six phases of the breast, adolescence, pregnancy,

622

menopause (Yu , W., Ng., SP. 2006), etc. we can understand that it's possible to interconnect them. For this reason, we can adjust the different stages of a women's life to certain bras. Due to its technical characteristics, it can improve in a more assertive relationship between the brassiere and the user.

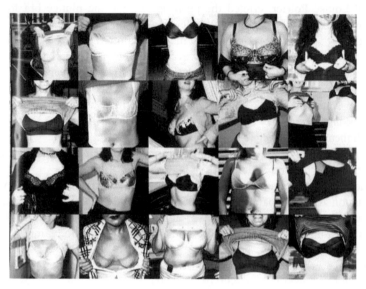

Figure 4 Image of photographer Harri Peccionotti - women´s bra
<Campbell, E., Cicolini, A. (2000). Inside out - underwear and style in the UK. London, Black Dog Publishing Limited.>

In order to meet the size of the existing breasts, there is a wide variety of choice in bra size allowing greater user satisfaction in different styles (Dewsnap and Hart 2000). No other product offers variety such as the bra (Bressler, Newman, Proctor 1999), which in practice improves the quality of choice.

There are physiological factors, psychosocial, economic, and functional (Richards and Sturman 1977; Damhorst, Miller, Michelamn 1999), that also "depend on the culture, affiliations with other groups, race, tribe, caste, language, location and also membership of other groups such as the case of friends, social classes, religious groups, sport groups and economic groups."[1]

Fashion trends and psychosocial factors determine the shape and size of the bra for a season, changing the chest by the imposition of Fashion. Although the intimate clothing are invisible, it defines the outwear shapes that part from the inside, this means, lingerie (Dewsnap and Hart 2000). In contrast with the outdoor clothing, very little is known about the behavior of the user because it is essentially a category of very intimate and personal clothing (Dewsnap and Hart 2000).

---

[1] Damhorst, Miller, Michelman (1999). The Meanings of Dress. New York, Fairchild Publication.

Banavage and Koff (1998) found that low self-esteem is associated with dissatisfaction with the breast size, when there is a large discrepancy between the ideal size and actual size. In this sense a bra can affect self-image, which helps to understand consumer behavior.

The complex physical nature of the bra and its composition ensures that the physiological behavior of women when choosing a bra is different from the kind of choice of other outdoor clothing products. Dewsnap and Hart (2000) consider that the variables of the decision process exploit some stages. The involvement and self-knowledge are considered variables that mediate the behavior of consumers and act on the stages of buying a bra. It was found that the bras have a high profile engaging in the product. Thus, bras are seen by the consumer with in following characteristics:
- Delight in the added value of the product;
- Strong link between the product and personality;
- High symbolic value of the product.

When a consumer finds a bra that appreciates, she has as tendency a similar or even repeated one of a different color of a from another color. This trend can be explained by the following reasons:
- Power of purchasing;
- Collection with great choice of patterns and colors;
- Bra that will be useful on several occasions.

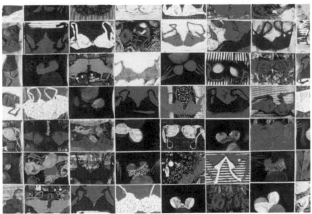

Figure 5 Image of painting bras
< Cox, C. (2000). Lingerie - A Lexicon of Style. London..>

The comfort is associated with daily use and women are able to be uncomfortable with a bra and yet use it because there is compensation (Hart and Dewsnap 2000). The evaluation of the bra is not made only in purchase but in daily use, based on the attributes that it gives the user and the self-image it transmits.

Another study that describes interviews with women who are selling lingerie, aims to understand the type of purchase, the use of lingerie that is acquired and the changes in consumers shapes when experiencing different identities.

With the analysis carried out, it is concluded that the lingerie is bought and used for the thrill. The consumers feel their own bodies, with minor changes that encourage them to build their self-image, which is created for their social identity. The importance of lingerie for the interviewed in this study reveals that this type of clothing allows them to control their "modern femininity" and expressed a deep social side.

Jantzen, C., P. Østergaard, C. Vieira (2006), support the duality of lingerie to show how a symbol and a tool for women's identity can create a personal level of satisfaction, pleasure and comfort.

Defining comfort is complex, but if it is fragmented and analyzed, it is possible to make a subjective evaluation - sensory and tactile comfort, psychological comfort or esthetic, and an objective overview – Thermal and physiologic comfort as well as ergonomic comfort.

The ergonomic comfort relates to body movement by a more objective evaluation because it analyses the relationship of the product construction with the body. In the perception of ergonomic comfort it's necessary to take into account that there is a very close relationship with the sensory comfort. This is due mechanical parameters that reveal a good ergonomic design of clothing, the two are intimately related to the way of dressing by cutting, sewing, modeling, and anthropometric tables used for this purpose. A good performance of movements is always coupled with the use of suitable materials for this purpose and intimate clothing in the proportion of the body. This has to be evaluated so that the product used is the most appropriate for those who wear it.

4. RESEARCH GOALS

An investigation in Fashion Design, focusing on the study of lingerie may have a direct benefit to society and result in the improvement of living individual conditions. The contact with women analyzing their real needs and requirements is the base structure carried out in this investigation.

The aim is then to add scientific knowledge to a specific area of fashion that understands the necessities of the women's chest throughout her life.

It is also outlined the aim to achieve a strong connection with the industry, where collaboration between companies of Lingerie Corseterie with academic studies is enriched.

So, understanding the life stages of a women's breast and needs beyond the stage of the chest, may provide higher quality products, reflecting the needs of consumers of a specific stage. The search for coordination between breast and bra can help encountering shapes that are best suited. The objective is to create an incentive to settle and regulate shapes through each stadium. Again the proximity to the industry can help to validate this correspondence and implement it with in the market.

To achieve this, manage brands and materials that are already focused to reach segment markets with specific needs, will allow to understand the best way to adapt women's bras and breasts through every stage. This domination of the sector may also, point out market segments with potential to develop industrial activity, contributing to a richer business. However, the interpretation of the evolutionary line of the sector Lingerie / Corseterie in the global market, may depend on a strategy plan with a more realistic view of the market needs.

5. METHODOLOGY

The methodology focuses initially on gathering information on the contribution of design to adapt the products of lingerie consumers' needs. A study was developed and analyzed and data concerning their needs and general requirements. Combining still experience and expertise of specialists in the area, will range from designers, teachers, medical specialists and ergonomists, designed to combine a sustainable basis for differing viewpoints. Itch one can define the needs of consumers who, according to this study are more focused on the characteristic of comfort, well-being of the chest and the harmonious relationship between the user and bra. Besides the comfort features such as touch, durability and quality, a large weight on purchase and daily use is noticed.

After comparing the results, with experts, simplify the heterogeny view that a woman has on her chest and her bra. Thirty women in each of the different phases were questioned, in order to deepen the technical features, combine the ergonomics of the bra and describe the level of comfort and quality. A more specific knowledge concerning levels of comfort of the handles, the rim, the brackets and their own materials result in the qualification levels of comfort and usability in the user's relationship with her bra. With the features most frequently analyzed, a product can be achieved that meets a certain stage of women's lives, helping to ergonomically improve the relationship with the bra and contributing to an improved quality of life. In order to carry out this process, it's intended to ensure the use of size charts, anthropometric tables of different brands that work different market segments and understand that changes may be considered in order to adequated to the reality of the consumers.

To this end, this research will be supported by a company in the sector, which designs and manufactures bras, investing in innovation and new product development, using the ergonomics of the design as a tool that defines market segments supported by a women's point a view.

6. EXPECTED RESULTS

A synonym for success in this research is to achieve the objectives set, contributing to the enrichment of the area. With the division of the bra, just studying an ergonomic perspective, it is expected that we can understand the complexity of a fashion product with very specific functions but also understand the logic of correspondence between bras and their users.

It is intended to reach the applicability of results for industry by creating a focus for segment and withdraw the generalization of products for all different stages.

Encourage partnerships between companies and academic research and knowledge, to result in a partnership to improve quality of products developed, and greater scope to satisfy the real needs of women.

Conclude that the evolution of the bra contributed to the adaptation of each type of chest and of each phase. The bra is at the center of the study as clothing of architectural construction that can perform various functions and improve women's self-image regardless of their life stage.

It is also intended that the knowledge achieved from this study fosters greater scientific objectivity, contributing to this specific area of fashion design as a complete product to be created, modeled and produced.

ACKNOWLEDGMENT

The authors would like to express their gratefulness for support from CIAUD – Research Centre in Architecture, Urban Planning and Design, which guarantee the sucess of this research.

REFERENCES

Berry, C. (2006). Hoorah for the Bra - A perky peek at the history oh the brassiere. New York, Stewart, Tabori & Chang.

Bressler, K., Newman, K., Proctor, G. (1999). Un Siglo de lenceria. Arrigorriaga, Status Ediciones, S.L.

Farrell-Beck, J., Gau, C. (2002). Upfit - The Bra in America. Philadelphia, University of Pennsylvania Press.

Fields, J. (2007). An Intimate Affair - Women, Lingerie, and Sexuality. Berkeley, University of California Press.

Hart, C., Dewsnap, C. (2000). "An exploratory study of the consumer decision process for intimate apparel." Journal of Fashion Marketing and Management Vol. 5(Henry Stewart Publications): 108 - 119.

Jantzen, C., Østergaard P., Vieira C. (2006). "Becoming a 'Woman to the Backbone' - Lingerie consumption and the experience of feminine identity." Journal of Consumer Culture Vol 6(2)(SAGE Publications): 177–202.

Leonard, R. (2008). As sete idades da mulher Barcelona, Editorial Presença

Vicentini, C., Castilho, K. (2008). Design de moda - diversos olhares. São Paulo, Estação das letras e da cores.

Yalom, M. (1997). A history of the breast. New York, Ballantine Books.

Yu, W., Ng., S-P. (2006). Innovation and techonology of women's intimate apparel. T. T. Institute. Cambridge, Woodhead Publishing Limited.

CHAPTER 66

The Importance of Integrating Perceived Affordances and Hazard Perception in Package Design

Hande Ayanoğlu[1,2], Emília Duarte[2,3,4], Paulo Noriega[2,3], Luís Teixeira[2,3] and Francisco Rebelo[2,3]

[1] IDEAS, Department of Industrial Design, Environment and History, Second University of Naples, Aversa(CE), Italy
[2] Ergonomics Laboratory – FMH – Technical University of Lisbon, Cruz Quebrada-Dafundo, Portugal
[3] CIPER – Interdisciplinary Center for the Study of Human Performance, Technical University of Lisbon, Cruz Quebrada-Dafundo, Portugal
[4] UNIDCOM/IADE – Institute of Arts, Design and Marketing, Lisbon, Portugal
[handeayanoglu@yahoo.com; emilia.duarte@iade.pt; pnoriega@fmh.utl.pt; lmteixeira@fmh.utl.pt; frebelo@fmh.utl.pt]

ABSTRACT

Since it is not always possible to design-out hazards from some packages, injuries are, unfortunately, a common outcome while handling packages with hazardous contents (e.g., chemicals) at home. This can be due to inadequate assumptions about the packages' content hazard level. Thus, to increase safety, it is fundamental that users perceive quickly and accurately the nature of the hazard and the hazard level associated with the product that they are handling. Generally, warning labels are used for such goal, since they provide a method to convey the required risk information, however, people do not always read labels, which creates a potential health and safety problem. One possible approach to overcome this problem can be the use of perceived affordances to convey adequate action possibilities to the users and, at the same time, convey the adequate hazard nature and level, which can promote cautionary behaviors. In this context, based on a literature review, this paper's main objective is to challenge experts engaged in package design to explore variables such as affordances, as well as hazard perception, in order to design safer and more effective packages for consumer products.

Keywords: Package Design, Safety, Affordances, Hazard Perception

1. INTRODUCTION

Effective packaging is an important concern in safety. Each year many people are injured while using, sometimes improperly, products at home. This is particularly evident for cases of containers for hazardous substances (e.g., household cleaning products or pesticides). This can be due, in part, to the imperceptible features of the different package designs. Furthermore, many users are unaware of the hazards often encountered in the products for domestic use and/or are misinformed about the seriousness of hazards and their consequences (e.g., Leonard and Wogalter 2000). If the user is unaware of the existence of hazards associated to a product, or if perceives that the product is less hazardous than it actually is, the user may pay less attention to safety concerns. In addition, for products to be used in the home context, no formal training in safety procedures is generally possible.

Not surprisingly, packaging has been an important issue of research for fields such as marketing and design (graphic and product design), because the package is a key factor in the purchase decision (Bloch 1995). For instance, food and beverage packages have been analyzed to understand the relations between tastes and shapes (e.g., Smets and Overbeeke 1995; Wang, Chou and Sun 2009). Other studies assessed if the proportions of packages' shapes (i.e., ratio) can affect the consumers' purchase preferences (e.g., Raghubir and Greenleaf 2006).

However, most of the studies about packages, in Human Factors and Ergonomics' (HFE) literature, have focused on the shape, the color and the warnings' symbols, or the packages' labels, and whether these features generate a perception of hazard and risk when using the products (e.g., Serig 2001; Smith-Jackson and Wogalter 2000; Wogalter and Laughery 2001). As suggested by Wogalter, Laughery and Barfield (2001), the containers' features (i.e., shape) evoke different levels of hazard perception on the users, thus resulting in different precautionary behaviors (e.g., likelihood to read the label). The opening of packages is another subject frequently studied for safety reasons (e.g., Caner and Pascall 2010; Galley, Elton and Haines 2005; Langley et al. 2005; Winder et al. 2002; Yoxall et al. 2006). Furthermore, many studies have been conducted to improve the design of warnings on packages (e.g., Hammond et al. 2006), leading either to an increased size of the warnings, to an improved positioning of such warnings (e.g., Wogalter et al. 1996) or, in some cases, to the introduction of picture-based warnings on labels (e.g., cigarettes' packages).

However, and although there is extensive work done about package design, little attention has been given to the packages' ability to induce safe behaviors, without relying only on warnings for that purpose. Furthermore, increasing safety mainly through the use of warnings encompasses a certain probability for failure, since the hazardous nature of some products is not always clear and obvious (as in the case of some consumer products) and people do not always read warning labels (e.g., Laughery and Wogalter 1997). One way to potentially increase safety is by considering the packages' affordances as a mean to induce a cautionary behavior (i.e., read the label).

The idea of affordances has its roots in the work of Gibson (1986) however, for this study we also consider the Norman's approach (1999), more commonly used in the field of design, which noted that affordances are the products' properties (actual and perceived) that suggest, invite, or prohibit certain behaviors. Several features of the package (e.g., shape, handle, bottleneck, opening, textures and colors) can provide users with information (nonverbal), which they can use to evaluate the situational characteristics, the nature and the extent of the product's hazard, as well as to support their behavioral decisions.

In this context, this article discusses affordances that can be used to design safer packages, for hazardous chemical products in the home context, and that can effectively communicate the adequate hazard level in other words, promote cautionary behavior. We begin with a brief approach to package design. Next, we present an overview of affordances theories to help explain how package's features can influence the users' behavior. The hazard perception, as a factor that influences behavior, is also discussed.

This paper does not intend to be an exhaustive review but, rather, it intends to highlight some aspects that support our argument that package design, for consumer products, could be improved by taking into consideration variables such as affordances as well as hazard perception. Thus, we expect that researchers and experts engaged in package design can benefit from this discussion and enhance their knowledge about this subject.

2. PACKAGE DESIGN

Packages have been in continuous use since very early times and are presented under different forms, features and technologies. A product's package is usually the first thing that the users interact with before actually using the product, therefore, its design needs to be effective.

Initially, a package's main purpose was to protect its content, although nowadays, besides protecting and conserving the content, packages are an important form of selling a product and serve as an interface to deliver safety information to the user. Additionally, packages can transmit some of the utilitarian functions of the product, which can be directly obvious from their appearance, such as the handle indicating that the product is portable (Creusen 2005). Furthermore, a package can be an important value for the people who are looking for an aesthetical appeal.

However, occasionally, users may get the wrong impression because of the packages' appearance. Bloch (1995) states that the ideal product does not need to be beautiful, but the form of the ideal product would be the most comprehensible and usable. A product's appearance communicates messages, and Bloch (1995) denotes that consumers may use the product's appearance for categorization. So, when a package's design is difficult to categorize based on its appearance, users might not perceive it the way they should.

Due to conventions and past experiences, people seem to form expectations and make inferences just by looking at the package of the product. Hence, in what regards safety, unless users have conventions or experiences about a certain package design, the package's design needs to provide the users with adequate signals that indicate potential hazardousness of the content, in order for them to act cautiously when handling the product. In other words, package design can be used proactively and provide the users with a certain impression about the product's function, leading to a safe behavior. As Laughery (1993) suggests, in the design process, designers should do their best in determining what safety knowledge requirements exist for the user and whether those requirements are met through the *a priori* experiences of the user or through safety communications to the user.

3. AFFORDANCES

The notion of affordances was originally coined by the perceptual psychologist James J. Gibson, who depicted the relationship between an actor and its environment. An important fact about affordances of the environment is that they are in a sense objective, real and physical, unlike values and meanings that are supposed to be subjective, phenomenal and mental (Gibson 1986). According to Gibson's theory, affordances are a part of nature and they do not have to be seen, known, or desirable. Since they are a part of nature, the actor-environment mutuality, i.e., the actor and the environment are an inseparable pair (McGrenere and Ho 2000). Gibson's work focuses on direct perception, a form of perception that does not require cognitive mediation by an actor. Direct perception is possible when there is an affordance and there is information in the environment that uniquely specifies that affordance (McGrenere and Ho 2000) and a possibility of action.

On the contrary, Norman (1988), in his book "The Design of Everyday Things", states that affordances are perceived and as actual properties of a thing, primarily the fundamental properties that determine how that thing could be used. According to Norman (1999), we should not confuse affordances with perceived affordances nor with conventions. For Norman, designers can invent new real and perceived affordances, but they cannot so readily change established social conventions. With respect to this, designers should be able to understand the differences between real affordances, perceived affordances and conventions, and then exploit that knowledge in order to design an effective package that encompasses them all.

Gibson (1986) points out that affordances are not dependent on perception; they exit whether or not they are perceived. Norman's viewpoint on perception is in conflict with Gibson's direct perception, since Norman believes that affordances result from the users' mental interpretation of things and themselves, which are based on their previous knowledge and experience. Following Norman's interpretation, the information implying the usage of an object is regarded as an affordance, regardless whether the actual affordance exists or not. To Gibson, affordances are the action possibility of objects with reference to the physical condition of the user, while in Norman's interpretation, it is the perceived information with reference to the mental and perceptual capabilities of the user (You and Chen 2007).

McGrenere and Ho (2000) state that Gibson and Norman seem to have, at first glance, similar definitions about affordances. Gibson intended an affordance to mean an opportunity for action possibility that is available in the environment for an individual, independent of the individual's ability to perceive this possibility. In Norman's position, affordances are perceived properties. He attempted to clarify the misuse of the term affordances in design practice and literature, and used "perceived affordances" to differentiate his version from Gibson's concept (Norman 1999).

Perceived affordances, as opportunities for action, depend both on the individuals' characteristics such as action capabilities, beliefs, previous experiences/familiarity, as well as on conventions, and also incorporates both concepts of self-efficacy and relevance, as factors that affect motivation. Self-efficacy refers to both the individuals' beliefs about how effective the behavior will be at accomplishing a given goal and how effective the individual is at doing that behavior. Perceived relevance is related with the evaluation of the trade-off between the costs associated to a given behavior and the granted benefits.

In the field of Warnings and Risk Communication, Ayres, Wood and Schmidt (1998), cited by Riley (2006), suggests that perceived affordances (i.e., whether or not an action can be completed successfully) can be a relevant factor of influence over behavior, namely compliance with warnings, maybe even stronger than the risk perception. According to this perspective, behavior is much more goal-oriented rather than avoidance-driven. These authors also argue that, because the perception of affordances takes place pre-consciously, or automatically, warnings might not be able to affect behavior as intended. So, in this sense, situations in which the degree of safety conveyed by the affordance is greater than the actual safety should be examined because they can involve greater risk for users.

4. HAZARD PERCEPTION

Research in Warnings and Risk Communication literature identify a number of factors that affect motivation to comply as well as behavioral compliance (e.g., Riley 2006), namely hazard perception. In this topic, we review the hazard perception concept, in order to highlight it's relation to affordances and its role in behavioral compliance.

Hazards are everywhere. To Laughery and Wogalter (1997), hazard is a set of circumstances that can result in injury, illness, or property damage. Products such as detergents, drain cleaners and bleach are considered hazardous because they have the potential to cause severe injuries and are potentially fatal if used improperly. However, the user might not perceive them as hazardous because, as stated by Ridley (2004), hazard identification varies from person to person, depending, among other things, on their experiences (previous knowledge), attitude to risks and familiarity with the process.

We are referring to perceived hazard that describes peoples' beliefs about dangers associated with a product or a situation and encompasses individual knowledge that is accepted as true, although it might be untrue (Laughery and Wogalter 1997). Also, perceived hazard is closely tied to the injury's expected severity level. The greater the potential injury, the more hazardous the product is perceived to be and people will be more cautious when handling it (Wogalter and Laughery 2001), hence the converse is also true.

In certain circumstances, such as when people are familiarized with the product or situation, it may be required that the message be made more persuasive to be able to surpass the existing (sometimes incorrect or inadequate) beliefs. Wogalter and Laughery (2001) define familiarity as beliefs which are formed from past similar experiences that are stored in a person's memory. According to Ortiz, Resnick and Kengskool (2000), users can become familiar with a product through experience, advertising, and other interactions. They can be familiar with several aspects of the product, such as its use, composition and hazards. Turner (2005) denotes that there can be "good" or "bad" affordances but in some cases, familiarity operates. Therefore, affordances with the help of familiarity might lead users to believe they are in the possession of all the knowledge needed for an effective and safe use. In cases where there is no familiarity with a product, the context of use might have enough effect on hazard perception. Furthermore, people might also assume that a different, or new, but similar product presents similar hazards, or is to be used in a similar way, which may not be true.

The association between shapes and hazardousness can have an impact on hazard perception. According to Wogalter, Laughery and Barfield (2001), some package shapes might serve as a cue to the type and extent of the hazard of the product, therefore, people may make assumptions of the potential danger just looking at the appearance of the package design. However, although people form an impression by seeing the product (Creusen 2005), if packages present misleading information that indicate that it is safe when, in fact, it is not, a safety problem may take place. Thus, product safety problems may be caused by misperceived features such as materials, shapes, sizes, colors, textures, etc.

Situational variables, such as characteristics of the situation in which the product is being used, are also important to determine how the user will interact with the product. For example, normally, to keep safe while using a hazardous product, some actions are required (e.g., to use protective equipment) and usually there are costs associated with it. These costs, also referred to as cost of compliance, may include time, effort, discomfort or even financial expense. But, when people perceive that costs of compliance are higher than the benefits, they are less likely to comply (Wogalter and Laughery 2001). Increasing the perceived hazard can motivate the user to be more careful (Wogalter, Brems and Martin 1993), improving the odds of compliance.

Another important aspect is associated to the designers' problem of overestimating what people know and what people might be able to perceive, which can result in poorly designed solutions. Leonard and Wogalter (2000) connote that ergonomic development of products commonly used by consumers includes the necessity for users to be informed about hazards associated with those products. Moreover, we believe that without the users input in the product's design (User Centered Design's approach) there may be solutions that fail to meet the users' needs.

In summary, two important factors for safety are the physical characteristics of the packages and the extent to which the limitations of use and the hazards associated with the products are recognized by the users.

5. IMPROVING PACKAGE DESIGN

Some packages, independently of their content and inherent hazardousness (e.g., water vs. bleach) will, most likely, share affordances that are perceived as critical for their basic use (e.g., portability, openability, drinkability). From the affordances, users may perceive that they may pour the content over a surface, drink it, and/or decide to throw it away. However, depending on the nature of the content, some of these actions can be unsafe. This means that, in what regards the action possibilities with impact on the users' safety, affordances can work in two different levels; one related to conveying action possibilities that are desirable and the other related to action possibilities that are undesirable. Thus, weakening the affordance that conveys undesirable action possibilities seems the most effective strategy to ensure safety. However, in some cases, this is not feasible, since some affordances are intrinsically associated with each other (e.g., pouring and drinking abilities). One possibility could be to weakening the affordance, by turn it less desirable, pleasurable or easier to follow. For example, in order to discourage drinkability without interfering with the action of pouring the content over a surface, a solution could be to design a bottleneck with a prismatic shape (i.e., hexagonal), with a coarse-texture, or to cover it with a repulsive material. Thus, for instance, a package with poison as content must not afford to be drinkable, but must afford an action that alerts a person to "pay attention, check the warning".

A different approach is to use affordances to communicate an adequate level of hazard and, therefore, promote a cautionary behavior (i.e., read the warning before use). However, designers must ensure that the hazard level being conveyed is adequate, i.e., it must not provide inefficient or over exaggerated information. If the situation seems safer than it actually is, the users might not take the appropriate protective measures. On the contrary, if users overestimate the hazardous level, they may decide not to use the product. In both cases, affordances are biasing the users' behavior, which might result in a safety issue.

Nevertheless, in spite of the approach taken, the affordances should be quite visible to the users since, as pointed by Norman (1988), the most important principle of design is visibility, so the product can convey the correct message in order to be easy to use (Gaver 1991; McGrenere and Ho 2000). Furthermore, in what regards package design, there are well established conventions that determine most of the packages' features which in turn, by a repeated exposure, are incorporated in the expectations of the users that are familiarized with the package. Whenever it is possible, package design should also cast on the common psychological processes, experiences and cultural background of the users to create clear affordances, resulting in a simpler, effective, pleasurable, and a safer interaction with packages.

6. CONCLUSION

Products that are capable of inflicting injuries on their users should be able to communicate the nature of the hazard, convey the adequate hazard level, as well as to inform the users about the action possibilities (i.e., affordances). Such descriptions, if adequate, should provide users with an accurate picture of the situation, and provide them with appropriate information needed to make rational safety-related decisions.

In this context, the purpose of this paper was not to present a detailed review of theories on affordances, or hazard perception, but rather to outline a few ideas about how to use these concepts in package design in order to improve human safety. For such discussion we have considered the case of hazardous products for domestic use (e.g., liquid chemical products).

In package design, the use of warnings and other safety information is usually what is recommended by researchers in order to avoid or reduce the injuries related with the hazardous products. Unfortunately, it is known that warnings frequently fails to prevent injuries.

Perceived affordances are suggested as another perspective to communicate hazard to users while handling packages. By considering the relationship between affordances and the users' cultural constraints, conventions, beliefs and experiences, which affects the users' hazard perception, it can be recommended safety communication that would be appropriate and the approach that would be feasible to designers.

These considerations should alert experts engaged in package design to the need of examining, in a systematic manner, how perceived affordances and hazard perception interact and to what extent they affect behavioral compliance and they promote a safer behavior. It will also be a challenge to establish the role of the packages' features, such as shape, color, textures, among others in this framework.

ACKNOWLEDGMENTS

This study was developed in the scope of a research project (PTDC/PSI-PCO/100148/2008) funded by the Portuguese Science Foundation (FCT).

REFERENCES

Ayres, T J, C T Wood, and R A Schmidt. 1998. "Risk perception and behavioral choice." *International Journal of Cognitive Ergonomics* 2 (1-2): 35-52.

Bloch, P.H. 1995. "Seeking the ideal form: product design and consumer response." *The Journal of Marketing* 59 (3): 16–29.

Caner, C., and M.A. Pascall. 2010. "Consumer complaints and accidents related to food packaging." *Packaging Technology and Science* 23 (7): 413–422.

Creusen, MEH. 2005. "The different roles of product appearance in consumer choice." *Journal of product* (22): 63-81.

Galley, Magdalen, Edward Elton, and Victoria Haines. 2005. Packaging: a box of delights or a can of worms? The contribution of ergonomics to the usability, safety and semantics of packaging. In *FaraPack Briefing 2005 New Technologies for Innovative Packaging.*

Gaver, William W. 1991. Technology affordances. In *Proceedings of the SIGCHI conference on Human factors in computing systems: Reaching through technology*, 79-84. New York, NY, USA: ACM.

Gibson, J J. 1986. *The Ecological Approach to Visual Perception.* New York: Taylor & Francis.

Hammond, D, G T Fong, A McNeill, R Borland, and K M Cummings. 2006. "Effectiveness of cigarette warning labels in informing smokers about the risks of smoking: findings from the International Tobacco Control (ITC) Four Country Survey." *Tobacco control* 15 Suppl 3 (June): iii19-25.

Langley, J., R. Janson, J. Wearn, and A. Yoxall. 2005. "'Inclusive' design for containers: improving openability." *Packaging Technology and Science* 18 (6) (November): 285-293.

Laughery, KR. 1993. "Everybody knows or do they." *Ergonomics in Design: The Quarterly of* (1959).

Laughery, KR, and M.S. Wogalter. 1997. Warnings and risk perception. In *Handbook of human factors and ergonomics*, ed. Gavriel Salvendy, 2:1174–1197. 2nd ed. Wiley) New York.

Leonard, S D, and M S Wogalter. 2000. "What you don't know can hurt you: household products and events." *Accident; analysis and prevention* 32 (3) (May): 383-8.

McGrenere, J., and W. Ho. 2000. Affordances: Clarifying and evolving a concept. In *Graphics Interface*, 179–186. Citeseer.

Norman, DA. 1999. "Affordance, conventions, and design." *interactions*: 38-42.

Norman, Donald A. 1988. The Design of Everyday Things. New York: Doubleday.

Ortiz, Julio, M.L. Resnick, and Khokiat Kengskool. 2000. The effects of familiarity and risk perception on workplace warning compliance. In *Proceedings of the Human Factors and Ergonomics Society Annual Meeting*, 44:826–829. SAGE Publications.

Raghubir, Priya, and Eric A. Greenleaf. 2006. "Ratios in proportion: what should the shape of the package be?" *Journal of Marketing* 70 (April): 95-107.

Ridley, John. 2004. *Health and Safety in Brief.* Ed. S Zhang. *SME Mining Engineering Handbook, American Institute.* 3rd ed. Elsevier Ltd.

Riley, DM. 2006. Beliefs, attitudes, and motivation. In *Handbook of warnings*, ed. M. S. Wogalter, 289-300. Mahwah, NJ: Lawrence Erlbaum Associates.

Serig, E.M. 2001. The influence of container shape and color cues on consumer produckt risk perception and precautionary intent. In *Human Factors Perspective on Warnings Volume 2: selections from Human Factors and Ergonomics Society annual meetings, 1994-2000*, ed. Michael S. Wogalter, Stephen L. Young, and Kenneth R. Laughery, 185-188. Human Factors and Ergonomics Society.

Smets, G.J.F., and C.J. Overbeeke. 1995. "Expressing tastes in packages." *Design Studies* 16 (3): 349–365.

Smith-Jackson, T.L., and M.S. Wogalter. 2000. Users' hazard perceptions of warning components: An examination of colors and symbols. In *Proceedings of the Human Factors and Ergonomics Society Annual Meeting*, 44:6–55. SAGE Publications.

Turner, Phil. 2005. "Affordance as context." *Interacting with Computers* 17 (6) (December): 787-800.

Wang, R.W.Y., M.C. Chou, and C.H. Sun. 2009. Research on Taste Synesthesia Induced by the Shape of Food Package Bottles. In *17th Congress of International Ergonomics Association*.

Winder, Belinda, Keith Ridgway, Amy Nelson, and James Baldwin. 2002. "Food and drink packaging: who is complaining and who should be complaining." *Applied Ergonomics* 33 (5) (September): 433–438.

Wogalter, M S, A.B. Magurno, K.L. Scott, and D.A. Dietrich. 1996. Facilitating information acquisition for over-the-counter drugs using supplemental labels. In *Proceedings of the Human Factors and Ergonomics Society Annual Meeting*, 40:732–736. SAGE Publications.

Wogalter, M S, and Kenneth R. Laughery. 2001. Warnings. Ed. Waldemar Karwowski. *International encyclopedia of ergonomics and human factors, Volume 2*. Taylor & Francis.

Wogalter, M.S., Kenneth R. Laughery, and David A. Barfield. 2001. Effect of container shape on hazard perceptions. In *Human Factors Perspective on Warnings Volume 2 selections from Human Factors and Ergonomics Society annual meetings, 1994-2000*, 231-235.

Wogalter, Michael S., Douglas J. Brems, and Elaine G. Martin. 1993. "Risk perception of common consumer products: Judgments of accident frequency and precautionary intent." *Journal of Safety Research* 24 (2): 97–106.

You, H, and K Chen. 2007. "Applications of affordance and semantics in product design." *Design Studies* 28 (1) (January): 23-38.

Yoxall, A., R Janson, SR Bradbury, J Langley, J Wearn, and S Hayes. 2006. "Openability: producing design limits for consumer packaging." *Packaging Technology and Science* 19 (4): 219–225. doi:10.1002/pts.

CHAPTER 67

A Case Study of Creating Public Space for People with Disabilities

Joanna Bartnicka[1], Agnieszka Kowalska-Styczeń[2]

[1]Institute for Engineering of Production
[2]Institute for Economics and Informatics
Silesian University of Technology
41-800 Zabrze, ul. Roosevelta 26-28, Poland

ABSTRACT

An example of solutions addressed to people with disabilities increasing their self-reliance and capability for mobility through the building is presented in the article. The solutions take into account such aspects of creating public space as: information system based on maps and plates; signs of selected areas of the building, color arrangement, signs in vertical communication routs.

There is described the project PA-UNIV with non-discriminatory solutions. A methodology of proceeding for develop the solutions creating the friendly public space consists of four main stages. The first stage is identification the needs of people with disabilities including physical, sensory, mentally and cultural disability. The second stage includes audit of accessibility based on special evaluations tables. The outcomes of the audit are the basis for elaborating a technical project of adjustment of building space to the needs of people with disabilities and the last stage is implementation of designed solutions.

The object of the project was university building in Zabrze, Poland, which was being renovated then. The authors focused a special attention on the organisational issues of project related with transfer of information and way of cooperation between participators: architects – investor – construction manager – designers of accessibility solutions.

638

Keywords: disabled persons, public space, university, accessibility, mobility, innovation, design for all

1 INTRODUCTION

One of the most important aspects of people's life is their development and related to it education. Meanwhile the situation of people with disabilities in both education and the labor market is particularly difficult and requires extensive corrective action. Statistics indicate that persons with disabilities registered in the Institution: District Labour Office (in Polish PUP) in one of the developed cities in Poland are usually characterized by low levels of education. Nearly 80% of them are not qualified or have low qualifications. It constitutes a serious barrier to entry into the competitive labour market. Undoubtedly, this situation translates in high unemployment among this community.

There are many barriers to social inclusion of persons with disabilities, including training on the same level as the healthy persons. They are the following:

- physical (architectural) barriers associated with the inadequacy of infrastructure of school (university) buildings and open spaces for people with disabilities,
- cultural barriers associated with the perception of disability as something negative and undesirable, and revealing in such reactions of society, as avoidance of contact and the inability to freely retain in the company of people with disabilities,
- social barriers associated with: low social awareness about the needs of people with disabilities, and thus prejudice about their intellectual capacity as well as the negative attitudes that result in reduced participation by persons with disabilities in society, including the performance of important social roles,
- barriers in the process of learning associated with maladjustment of teaching methodology to the possibilities of perception of people with specific types of disability, as well as the lack of teaching tools or technologies, which would support learning.

Taking into account the needs of disabled people in area of developing solutions to increase the availability of public space and access to services, including educational services, the authors of this article have undertaken the research on non-discriminatory solutions, which can be implemented in schools and other public building as well as in open spaces. The original method of audit of availability of public buildings was developed in frame of research, which is based on a documented analysis and assessment of the accessibility of information, and the degree of adaptation of buildings and its equipment for the needs of all people, including people with disabilities and the elderly. This method was used to analyze the availability of three public buildings, including the building of the university. In addition to the audit method the modern and innovative solutions for increasing the availability of public space was being developed. An example of such solution is a system, which consists of audio, visual and tactile signs. The project of such system was implemented in the building of the Silesian University of Technology in Zabrze and is described in this article.

2 THE PROJECT OF ADJUSTMENT OF THE UNIVERSITY BUILDING TO PEOPLE WITH DISABILITIES (PA-UNIV)

2.1 PROJECT OBJECTIVES AND ASSUMPTIONS

The project (PA-UNIV) aimed at developing the non-discriminatory solutions for the modernized building of public university. The authors (Universal Designers - see Fig. 1) define the concept of non-discriminatory solutions as the solutions that are adjusted for all persons, including persons with disabilities.

The building consists of five storeys: ground floor, which includes a cloakroom, a library and reading room, the ground floor and three floors contain classrooms: seminar rooms, computer rooms and consulting rooms.

The main assumptions of the PA-UNIV were:

1. The solutions in area of availability concern the open spaces of the building, ie corridors and main hall.
2. Project takes into account basic needs of users in the passageways of the building and surrounding buildings in terms of: mobility, orientation, accessibility to information and security.

The project scope included the following list of tasks:

1. Elaboration of information system based on maps and plates;
2. Elaboration of signs system of selected areas of the building;
3. Elaboration of color scheme and color arrangement;
4. Elaboration of signs system in vertical passageways.

The work was carried out with using the following methods and tools:

– field investigations, including photographic recording and metrology measurement of in the building during renovation work,
– analysis of the architectural documentation for the modernized building,
– expert meetings with architects, construction manager and contractors,
– computer simulation using CAD systems,
– simulation of colour vision (Color Blindness Tests).

2.2 ANALYSIS OF THE ORGANIZATION AND INFORMATION FLOW WITHIN THE PA-UNIV

An important issue of the PA-UNIV was the cooperation and communications between the entities involved in the investment process. Apart from the Universal Designers there were the following units: the investor, main contractor, a team of architects, a contractor who implemented the PA-UNIV and the beneficiary (university).

Figure 1 illustrates process map, which presents in simplified way the process of investment with highlighted stage of PA-UNIV.

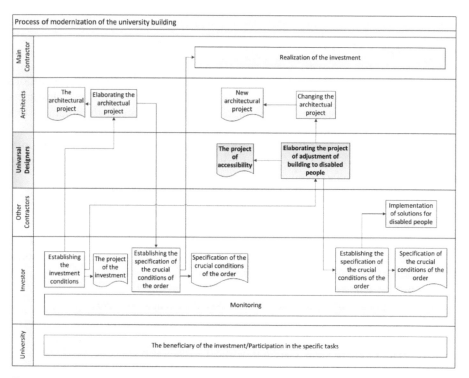

Fig. 1 The process map of Investment

The point of beginning the PA-UNIV in the schedule of investments is important. It is located here in the middle of the construction works after preparing the complete set of architectural documentation. This situation forced the need of making changes in selected elements of architectural project by the team of architects, in particular: changing color of the walls and floors, changing some finishing elements such as floor tiles and the way their tiling. On the other hand, the degree of advancement of the modernization was a huge barrier, which blocked the intervention in selected elements of the infrastructure and colors within the building. It was a significant problem, which hindered the implementation of the principles of accessibility.

3 CASE STUDY OF SOLUTIONS FOR PEOPLE WITH DISABILITIES

The case study describes the following solutions in area of adapting the university building to the needs of people with disabilities:
- information system based on maps and plates;
- signs of selected areas of the building,
- color arrangement,
- signs in vertical communication routs.

Information system based on maps and tables.

Taking into account the needs for orientation and accessibility to information, a project of installation of specialized maps and plates adjusted to blind and color-blind people was designed (Kowalska – Styczeń, Bartnicka and Bevilacqua, 2010). The PA-UNIV included the following types of plates: main map installed in the main hall, which shows the vertical structure of the building, the maps on each floor, which contain a floor plan and the plates at the door, which contain the room number and a simple plan of reaching the nearest exit. Each plates and maps are embedded in special terminals, which have been designed according to ergonomic criteria. The size of the terminals and the way of their installing enables easy access to them by people with different heights or people moving on wheelchairs as well as easy visual or tactile reading by visually impaired or blind people. The construction of the terminals is made according to criteria of safety. They do not contain sharp edges or protruding elements. The construction of terminals is shown in Figure 2.

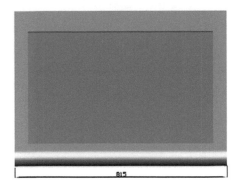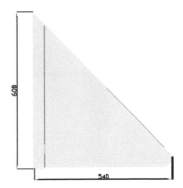

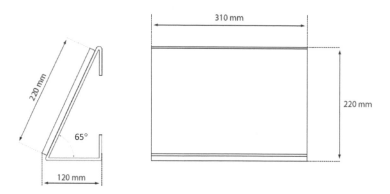

Fig. 2 The construction of terminals for maps and plates

Each plate consists of two parts: a bottom part "sighted part" (clear sharp colours for people with good and weak vision, also suitable for colour-blind people) and a top transparent and raised part made of Plexiglas. The top transparent part contains plans with additional information and a legend in braille. Figure 3 illustrates the examples and the way of installing the terminals with plates.

Fig. 3 The examples of plates installed in the building

The signs of selected areas of the building.
The main points of reference for self-contained moving around the building are maps and information plates. Therefore these elements should be exposed in full view, and access to them after entering to the building should be easy and intuitive. The last notice is particularly important for blind and weak vision people as well as

for deaf people, who are not able to establish directly contact with other persons in order to obtain specific information. This problem was solved by selecting the specific communication zones and using the tactile signs (guiding tiles), which demarcate the pathways to the terminals with maps. These zones are the main hall and the areas with staircase and elevators. Particularly the pathways determine:

- how to move from the main entrance of the building to the terminal with the general plan of the building, which is located next to the elevator (Fig. 4a),
- how to navigate the route from the elevator to the terminal with a floor plan (Fig. 4b).

Fig. 4 The examples of specific pathways in the building

The plates have a different texture and special grooves, which help blind people to navigate.

The selection of color for guiding tiles was conditioned by the color of the floor, architecture and general color arrangement of the whole space in the building. The tiles are in color contrasting with the rest of the parquet floor indicating in intuitive way how to move.

Color arrangement.

The studies on the color selection were based on assumptions of wayfinding methods (Chien-Hsiung, Wen-Chih and Wen-Te, 2009; Dalke, Littlea and Niemanna, et al. 2006; Helvacioglu and Olguntürk, 2011; Holscher, Meilinger and Vrachliotis, et al. 2006).

The project of color arrangement aimed at:

- improvement of orientation within the building by selecting a specific color for the each storey (a storey is associated with a particular color), for

644

example, yellow is the storey with the library and reading room, the purple is the storey with specialized laboratories, etc.,
–   improvement of the aesthetics of space - the colors of storeys create a coherent set of color resembling the rainbow. In addition color of the certain storey is associated with the color of equipment, such as furniture or plates.

The individual shades of colors were chosen in such a way so as to be distinguishable by color-blind people with disorders:
–   Deuteranopia: color deficiency affecting red-green hue discrimination
–   Protanopia: mild color vision defect in which an altered spectral sensitivity of red retinal receptors (closer to green receptor response) results in poor red-green hue discrimination.
–   Tritanopia: hereditary color vision deficiency affecting blue-yellow hue discrimination.

The Figure 5 shows how to paint walls with color. On gray background is painted the 10cm wide strip in the selected color for storey at a distance of 170cm from the floor. To strengthen the effect of contrast the color strip should be separated from the gray walls by white narrow stripes with a width of 2.5 cm. The color strip on a gray background enables on one hand the identification of color storey, on the other hand, it does not bedim the space. The element integrated with certain storey is the plate with the number of floor, which possess the same color scheme as the strip. The plate with number is located opposite to the elevator and staircase. This location makes the information about the number of storey readily noticeable.

Fig. 5 The way of color arrangement: a project and after implementation

The signs in vertical communication routes.

The vertical communication routes consist of staircases and elevators.

The staircases are one of the main routes in the system of evacuation of the building. Therefore they should include such technical and organizational solutions that will ensure safety in moving people, including people who are blind, with low vision or people with mobility impairments.

There was proposed a solution based on a system of horizontal tactile signs, whose purpose is to warn people (especially people who are blind and with low vision) before the changes of levels. It was pointed out that those signs should be installed only at the top of staircase. In addition these plates must have a high resistance to abrasion and skid.

The selection of tactile tiles took into account the architecture, arrangement and color in the space of the building. There was chosen the tiles with a contrasting color with respect to the color of the floor (see Fig. 6a). Tactile tiles create a strip with a width of 40 cm, 30 cm from the edge of the stairs.

Adaptation of elevators for the disabled people is one of the elements affecting the availability of the whole building.

Developed guidelines take into account the conditions for the adaptation of lifts for people with disabilities, where it was highlighted the following aspects in terms of accessibility:

1. Accessibility of tactile information panel in the hallway, which is located on the wall next to the elevator and in the elevator cab. It is proposed to distinguish the "0" which indicates the storey with exit from the building and simultaneously is the reference to the other buttons on panel (Fig. 6b).
2. Accessibility of audio and visual information in the corridor in front of the elevator and inside the elevator cab. There was drew attention to the audio and visual information that indicates the arrival of the elevator, the direction of moving (up, down) and the number of storey.
3. Accessibility of elevator space for people moving with wheelchairs by installing handrails on both sides of the cab and mirrors.

Fig. 6 The examples of adjustment communication routs to people with disabilities

CONCLUSIONS

The university building, which is the subject of this study, is the first building in Poland, with the mixed implemented solutions in terms of accessibility. The basic premise of this study was to find solutions that are complex and proper for all people. It has been proven that improving the mobility and accessibility of information can be based on simple solutions, such as correct selection of colors or finishing elements.

A very important issue in the process of investment was the organization and ability to cooperate with dispersed groups of participants with varied disciplines. Undoubtedly, the inclusion the PA-UNIV into the process of modernization of the building was too late, which resulted in problems of integration of the architectural project with PA-UNIV. Both projects should have proceeded parallel from the moment of beginning the investment.

REFERENCES

Kowalska – Styczeń A., J. Bartnicka and Ch. Bevilacqua. 2010. Innovative solutions for adjustment of city area to disabled persons and the aged with using computer techniques, w (red.) Khalid H., Hedge A, Ahram T. Z.: Advances in Ergonomics Modeling and Usability Evaluation, wyd.: CRC Press / Taylor & Francis, Ltd. 3rd International Conference on Applied Human Factors and Ergonomics (AHFE), July 17-20, 2010 Miami, Florida, USA.

Chien-Hsiung Ch., Ch.Wen-Chih and Ch. Wen-Te. 2009. Gender differences in relation to wayfinding strategies, navigational support design, and wayfinding task difficulty, Journal of Environmental Psychology 29 (2009) 220–226.

Dalke H., J. Littlea and E. Niemanna, et al. 2006. Colour and lighting in hospital design, Optics & Laser Technology 38 (2006) 343–365.

Helvacioglu E. and N. Olguntürk. 2011. Colour contribution to children's wayfinding in school environments, Optics &Laser Technology 43 (2011) 410–419.

Holscher Ch., T. Meilinger and G. Vrachliotis, et al. 2006. Up the down staircase: Wayfinding strategies in multi-level buildings, Journal of Environmental Psychology 26 (2006) 284–299.

CHAPTER 68

The Use of Reverse Engineering Development of Cranial Prosthesis for Thermal Spray

Catapan, M., Okimoto, M., Paredes, R.

Federal University of Parana
Curitiba-PR, BRAZIL
marciocatapan@ufpr.br

ABSTRACT

Patients with craniofacial defects or resulting from injury can do cranioplasty. Through of prosthesis, patients who do this surgery, get protection and aesthetic restoration. Thus, the manufacturing of these prostheses has been made subject of various studies, where innovations like new biomaterials, improved techniques for image acquisition, has evolved considerably in recent years. Due to the complexity of its forms generates iterations in the development process, where there are repetitions for changes in shape of the prosthesis during the stages of manufacturing. However, the objective of this study is to propose the establishment of a procedure that consists of image capture from the defective region of the patient, until to obtain the final product. All this monitored by 3D scanning methods, where one can evaluate both the mold and the part model, during the manufacturing steps, thus avoiding the iterations of design and process. Thus, one can acquire a cranial prosthesis quickly and according to the 3D model, acquired early in the process.

Keywords: cranial prosthesis, thermal spray, reverse engineering, systematic design

1   INTRODUCTION

Due to the great engineering developments in recent years, investments in research and support for health care demands have generated benefits the ability to play highly complex geometry, typical of anatomical geometries. Within the theme called cranial prosthesis, there is a great evolution of surgical procedures and, in parallel, equipment and materials used.

Several studies have been developed in order to find solutions to find out which materials and methods that can be used for the development and manufacture of cranial implants, because they know they are scarce, making the cost of these prostheses are high.

In this context it is inserted in the intention of this work. Who will propose a new method for the manufacture of cranial prosthesis, by thermal spraying, and also using reverse engineering principles, in order to achieve higher speed and confiability in the process, from the image capture to manufacture the prosthesis.

2   THE CONTEXT OF THE PROBLEM

Bone defects of the skull often occur by tissue loss related to trauma or tumor treatment. Bone loss in the total thickness of the skull, the cranioplasty is the main objective of protecting the brain and corrects a cosmetic deformity extremely apparent (YAREMCHUCK, 2006).

Currently in cranioplasty, the use of reverse engineering (RE) is used at the start of the procedure for obtaining three-dimensional model of the prosthesis. That is, from the patient's head with the problem, the use of ER makes you able to obtain this model through technologies such as computed tomography and magnetic resonance imaging. However, it would be able to use other resources, such as 3D scanner. The scanner is a device for capturing images because they have a laser scanning device, which captures the image point by point and with the aid of computer programs, transforms them into the shape of the product surface.

2.1   Materials Used in Cranial Prosthesis

Rotaru et al. (2006) mentioned that planning for a cranioplasty, the choice of biomaterial is a very important subject. Certainly the material chosen will direct much of the planning, as in the manufacturing process, the surgical technique, adjustment and attachment to the bone.

Eppley (2003) mentions that the Ti is the material of almost all fixation devices, and in some cases the entire prosthesis. Some acrylic and polyester biomaterials are being studied in cranioplasty, however, authors like Lai, Sittitavornwong and Waite (2011) cite some cosmetic problems that make these materials are not the best for this application. Also, the bioceramic materials are studied for the replacement of Ti, since they have high biocompatibility. However, Staffa et al. (2011) report that these materials have low porosity and not promote migration of osteoblasts thus not achieving integration of the implant and bone regeneration inefficiently.

2.2   Ways To Get If Cranial Prosthesis

There are several ways to manufacture cranial prostheses, such as Rapid Prototyping (RP), processes forming, Machining and foundry. The rapid prototyping is a multidisciplinary technique, involving resources from the field of reverse engineering, design, biomaterials and medical. An important factor for the successful application of RP is the choice of technology and adequate material. These choices influence the ability of the process, comprising: coating and manufacturing tolerances. Furthermore, attention should be paid to the fact that direct manufacturing of implants by RP techniques be limited to that process biomaterials. However, the focus of this technology is still in the surgical planning. Thus, one must make another prosthesis, with all the necessary requirements. This can cause errors during replication to another prosthesis.

2.3   Possible Problems in Cranial Prosthesis Construction

An example of the sequence to the cranial prosthesis construction, the casting process is illustrated in Figure 1. This can be noted that steps shown possible that there can be dimensional problems.

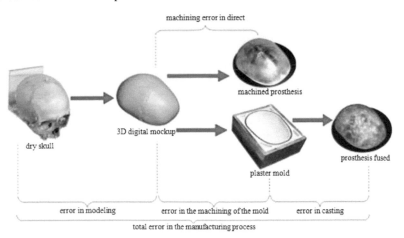

Figure 1 Stages of the fabrication of devices, where it is possible for errors.

In addition to obtaining the image of the skull, early in the development of the prosthesis, a possible problem is the difficulty of auditing the sequence of process operations. Thus, it increases the risk of error in obtaining the final dimension of the product, wasting time and materials to continue in possible operations. There are ways to avoid the problems presented in Figure 1, working with contact measurement technique, like the CMM (coordinate measuring machine), and without contact, such as laser and optical scanning.

In studies by Ambrogio et al. (2005), a reverse engineering technique has proposed to manufacture an implant ankle. This product was chosen for two main reasons: First, it requires a high degree of customization and second, the need for low cost is not strategic. Unfortunately this work was not possible to use scanning the ankle, for lack of use. But the idea is quite valid in this article.

3   METHODOLOGY PROPOSAL

This sections is responsible to propose a methodology that is based with the opportunity to add some technologies such as scanning of parts and thermal spray, in the process of manufacture of cranial implants. Thus there a new method for the construction of this type of artifact. This methodology will be detailed showing all the steps and their assigned tasks, to form the cranial prosthesis by thermal spray, see Figure 2.

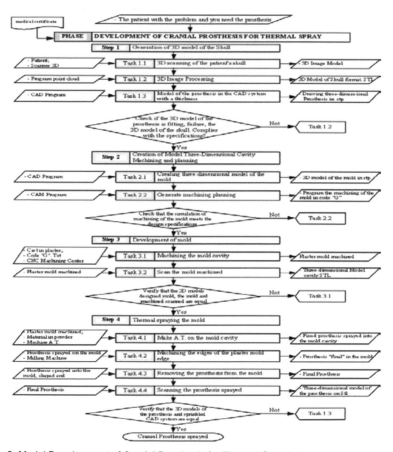

Figure 2 Model Development of Cranial Prosthesis for Thermal Spraying

Importantly, as input to the top of the methodological procedure proposed first need to be patient with the problem head. Since this article has the principles of research, which will not be considered the study of a living being and, yes, a dry skull.

3.1   Step 1: Generation of 3D model of the skull

Some tasks that make up this step are explained below.

3.1.1   Task 1.1: Scanning 3D Skull Patient

The dry skull to be used in this work is the same as studied by Lajarin (2008), see Figure 3a. The Figure 3b is 3D scanner to scan the patient's skull. This equipment is available in UFPR Laboratory of Ergonomics and Usability. Once done scanning, it generates the 3D model. But this image is polluted because there are many points that are not representative, which requires the need for image processing. This task will be described under 1.2.

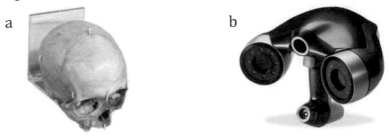

a   b

Figure 3 – (a) Image of Skull dry; (b) Scanner 3D – ZScanner model

3.1.2. Task 1.2: Treatment of 3D

This task transforms the image obtained from the scanning, which still contains some problems such as smoothing and is still not in a format where there is a CAD system. One of the problems that can occur with the image acquired in task 1.1, stick with the different way the product scanned. Since impurities generated in the scan, it doens't only captures the proposed product, but also nearby regions of the product.

But what stands out most in this task is: how to get the three-dimensional model of the prosthesis, through the skull and 3D scanner?

This paper proposes how to solve this is to scan the entire skull of the patient. Thus, when working with all the head, including the region of the fault, it is possible to extend the surfaces to be able to fill the region fails the skull. In this case, since the three-dimensional model of the skull and the other model without failure overlap the two and removing the surface which has been added. This will be the outer surface of the cranial prosthesis. Of course, this procedure is possible after the image processing with the aid of specific programs for this application.

3.1.3. Task 1.3: Model of the prosthesis with a thickness

This task transforms the image on the task treated 1.2, which is in the form of surface to volume. That is, adding a thickness similar to the three-dimensional model of the skull of the prosthesis.

In studies as Lajarim (2008), it was found that implants of titanium have thicknesses of about 1.5 mm. How does the work proposed to work with niobium and for theoretical purposes, this thickness is taken as a reference for this task.

3.1.4. Checks tasks 1.1 to 1.3

After the three-dimensional model of the prosthesis is necessary to verify whether the 3D model generated in the task 1.3 is consistent with the failure model of the skull scanned in task 1.1.

According to Kasprzak et al (2011), the distance between the skull and the graft should be about 0.25 mm. To make such a simulation is necessary to verify in an atmosphere of mounting CAD programs, the prosthesis models superimposed on the skull.

According to the methodology proposed, if the verification of the distance is in accordance with the specification cited in the previous paragraph should continue with the process of obtaining the fabrication of prosthesis. That is, proceed to Step 2. Otherwise, should return to Task 1.2.

3.2. Step 2: Creation of three-dimensional model of the mold cavity and machining planning

As shown in the flowchart illustrated in Figure 2, this step has as input a CAD program, the three-dimensional model of the prosthesis in the STL format and another CAM program.

3.2.1. Task 2.1: Create three-dimensional model of the mold

The three-dimensional model proposed mold has a length of 210 mm, 180 mm wide and 70mm thick. Having as a starting point the model in STL extension, the first step in the construction of the mold is used the exterior surface of the prosthesis and subtracting the CAD program. Important to ensure a perfect fit of the cranial prosthesis in surgery, may be used a method for the flat region of failure. Thus improves the fit of the prosthesis.

For a better view of the mold, the material is drawn that will ensure a level surface of the prosthesis, called a sacrificial material with approximately 5mm. Figure 4 illustrates a section showing the situation of positioning the gypsum block for machining.

Thus, there is a three-dimensional model in STL format template that will later be prepared for the machining and after thermal spraying.

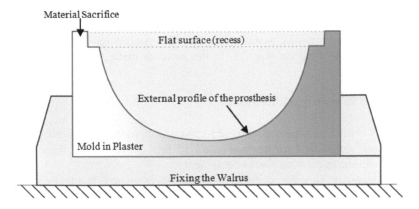

Figure 4 Cutaway illustration of the positioning block of plaster machining

3.2.2. Task 2.2: Generate the planning of mold machining

This task is to plan the machining of the mold using a CAM program. Necessary information is to inform what the CNC machine to be used. This planning can generate machining of the mold which is plasterboard. The end result of this task is the code "G" of this planning machining of the mold, with the CAM program. This code can be TXT.

3.2.3. Verification of tasks 2.1 and 2.2

Once obtained the three-dimensional model of the mold and machining planning is necessary to check if the machining meets design specifications. If complying with the specifications, it should proceed to the next step. If you have any impediment, should return to Task 2.2.

3.3. Step 3: Development of such

As shown in the flowchart in Figure 2, this step has as input the plaster model to be machined, the machining simulation program format txt and machining center to manufacture the mold. Also, as proposed in this work, it is necessary to have the 3D scanner model that was used in task 1.1. All these features have already been mentioned in previous paragraphs.

3.3.1. Task 3.1: Creation of the mold

In this paper does not proposed insertion into a living being, soon the sterility of the mold material it is proposed for the future works. For this reason, the chosen material is plaster. Even in the liquid form, the gypsum is poured into an aluminum mold holder. After drying the material, is cast as a block, ready to be machined to the shape of the cavity. Thus, it can be inserted into the CNC machine, the task 2.1, fixing it by a clamp to the machining of the mold. Thus, it becomes possible thermal spray into the mold cavity as illustrated in Figure 5.

However, there may be some errors in this machining, leaving the mold designed differently than the task 2.1. In this case, is made to the process of 3D scanner.

3.3.2. Task 3.2: Scan machined mold

The task is proposed that the resource to be used to verify your dimensional 3D scanner is the same used in previous tasks of this process. Thus, there are no problems with the acquisition /use of equipment.

As the scanning time for complex parts, it is estimated that takes about 15 minutes with the machine, about an additional two hours to image processing to obtain three-dimensional model that has been scanned.

3.3.3. Verification of tasks 3.1 and 3.2

With the three-dimensional model in STL, intends to make a comparison with the 3D model of the mold, coming from the task 2.1. These two designs must have the same shape, respecting the tolerance of the equipment. If the models are similar, you can go to step 4. Otherwise, you must return to the task 3.1.

3.4. Step 4: Thermal spray the mold

As shown in the flowchart illustrated in Figure 4, this step has as input the plaster model machined, the biomaterial powder to be sprayed and thermal spray equipment.

3.4.1. Task 4.1: Making the mold machined thermal spray

It is proposed herein that the mold machined is secured by means of a clamp in a spray chamber. Other settings need to be defined as: what is the thermal spraying process is used, that the biomaterial is used in place and, if the conditions for the thermal spray treatment; parameters of the thermal spraying process, the conditions of the substrate, in this case, the mold itself will be machined. With these variables defined, you can complete this task successfully. To have a reference to these variables were examined in UFPR available resources, which are discussed below.

The Figure 5a illustrates how the task will be performed. The numbers are: (1) The pistol in the process of thermal spray flame; (2) The biomaterial being sprayed; (3) The mold cavity, which has the negative of the three-dimensional model of the cranial prosthesis; (4) The mold; (5) A vise for clamping the process of thermal spraying. This produces the result shown in Figure 5b. That is, the cranial prosthesis sprayed in the plaster mold.

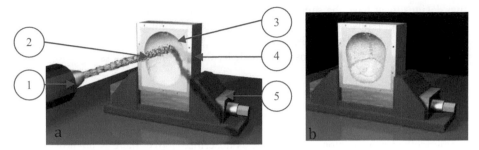

Figure 5 - (a) Illustration of the thermal spray process plaster mold, (b) Illustration of the cranial prosthesis sprayed into the mold.

The process parameters are used for thermal spraying of the Marino (2008), but serve as a reference. Parameters will be defined by more precise experiments in future work.

3.4.2.   T ask 4.2: Machining the edges of the cast

As shown in task 2.1, and illustrated in figure 5 of this work, the mold has an additional layer is designed on the edge part, called sacrificial material. This has to be removed to ensure that they can leave this flat region of the prosthesis. To achieve this requirement, we propose in this paper that this region is machined.

As it is a flat region, can use a conventional cutter, or even the CNC machining center described in the task 2.2 of this work.

3.4.3. Task 4.3: Remove the prosthesis from the mold

Here it is proposed that the layer removed is sprayed through a liquid solution, which causes gypsum dissolves, leaving the prosthesis with its final shape. Thus reducing or eliminating possible problems with the finished product, if the prosthesis is forcibly withdrawn.

3. 4.4. Task 4.4: Scanning the prosthesis sprayed

There may be problems with the product of 3.2 task, since it is sprayed metal material in multiple layers for high temperature, pressure and speed. So you can use the same 3D scanner to check this dimension.Thus, the same procedure described in 1.1 and 3.2 tasks for scanning the prosthesis. It is suggested here, it is tested at both the inner surface exteriore to verify the thickness of the prosthesis.

Scanned the prosthesis in STL format, can be compared with the same model obtained in task 1.1.

3. 4.5. Checks tasks 4.1 to 4.4

With the three-dimensional model in STL sprayed prosthesis, it is proposed to make a comparison with the 3D model of the part, coming from the task 1.1. These two designs must have the same shape. If so, it is arguable that the prosthesis is in its final form, ready for use in living after going through a proper cleaning. Otherwise, you must return to the task 1.3.

Thus, there is the cranial prosthesis made by thermal spray process.

4   CONCLUSIONS

The methodology proposed, where the goal is to obtain a cranial prosthesis with a new method was developed. However, as described above, there is unknowns as parameters spray plaster on curved surfaces, and other details need to be cleared for this research.

Thus, experiments on thermal spray plasters with curved shapes, analysis of porosity for better osseointegration and biocompatibility testing of niobium metal sprayed, are proposed for future work.

However, we can conclude that this study achieved the goal of this work, where you can add a new method for the development of cranial prostheses.

REFERENCES

Ambrogio, G., Napoli, L., Filice, F., Gagliari, Muzzupappa, M. Application of incremental forming process for high customized product manufacturing. Journal Materials Processing Technology 162-163, 2005, p.156-162

Eppley, B. L. Alloplastic Cranioplasty, Operative Techniques in Plastic and Reconstructive Surgery, v. 9, n. 1, p. 16-22, 2003.

Hara, T., Farias, C. A. S. A., Costa, M. J. M., Cruz, R. J. L.; ISSN 2177-1235, Cranioplasty, parietal versus custom prosthesis, Journal of Plastic Surgery, 2011. Pag. 32 – 36

Kasprzak, P., Tomaszewski, G., Wrobel-Wis´niewska, G., Zawirski, M., Polypropylene–polyester cranial prostheses prepared with CAD/CAM technology. Report of first 15 cases. Journal Clinical Neurology and Neurosurgery. 2011. Pag. 311–315

Lai, J. B., Sittitavornwong, S., Waite, P., D. Computer-Assited Designed and Computer-Assited Manufactured Polyetheretherketone Prosthesis for Complex Front-Orbito-Temporal Defect. Journal Oral Maxillofac Surg. 2011.

Lajarin, S. F., Dimensional evaluation of implant tailored to cranioploastia, Dissertation of UFPR – Brazil - 2008

Marino, C. Obtaining niobium coatings deposited by thermal spray to protect the marine corrosion. Dissertation. UFPR.Curitiba, Brazil. 2008.

Rotaru, H.; Baciut, M.; Stan, H.; Bran, S.; Chezan, H.; Josif, A.; Tomescu, M.; Kim, S. G.; Rotaru, A.; Baciut, G. Silicone rubber mould cast polyethylmethacrylate-hydroxyapatite plate used for repairing a large skull defect, Journal of Cranio-Maxillofacial Surgery, v. 34, p. 242–246, 2006.

Staffa, G., Barbanera, A., Faiola, A., Fricia M., Limoni, P., Mottaran R., Zanotti B., Stefini, R.. Custom made bioceramic implants in complex and large cranial reconstruction: A two-year follow-up. Journal of Cranio-Maxill-Facial Surgery. 2011.

Yaremchuck MJ. Acquired cranial bone deformities. In: Mathes SJ, Hentz VR, eds. Plastic surgery. 2nd ed. Philadelphia: Elsevier; 2006. p.547-62.

CHAPTER **69**

Usability Evaluation of Touchscreen Phone in Emergency Context

Fernanda Pozza, Vanessa Roncalio, Cristiana Miranda, Maria Lucia Okimoto

Federal University of Paraná - UFPR
Curitiba, Brazil
fepozza@gmail.com

ABSTRACT

Emergencies are always unexpected, causing changes in the emotional state of the individual, such as tension and stress. In this situation, a simple task such as making a phone call can prove difficult, considering that the user may not be familiar with the device. Also, interfaces are often not designed for emergency use. This paper reports on a pilot test of the usability of a touchscreen phone in an emergency context, which consisted of three phases: pre-test questionnaire, test and post-test questionnaire. Thus, it was possible to obtain information about the usability of the device and the user´s response in a simulated emergency situation.

Keywords: emergency context, usability, touchscreen phone

1   INTRODUCTION

According to Paulheim *et al.* (2009), in their study "Improving Usability of Integrated Emergency Response System," the usability in an emergency situation is gaining more attention from researchers. He also highlights that the unpreparedness of designers and users to this issue is due to the fact that most devices are not designed considering their use during a future emergency scenario. Consequently, when emergencies occur, the users are not very familiar with a particular interface to respond in a very short time, considering that he or she is under stress and out of their normal responses to the environment (Paulheim et al., 2009).

It is known that emergency situations increase the probability of human errors. The severity of effects, in this case, can lead to serious consequences to human life. Thus, in order to contribute with usability in emergencies, this study is based on the usability test of a touchscreen phone in a simulated emergency setting.

Over the past few years have been brought to market devices with different capabilities and features. Such devices involve increasingly complex tasks, requiring more cognitive ability of the user. Thereby, the interfaces play a key role in communication between human and machine. The touch screen phones belong to this category of devices.

Santos and Maciel (2010) states that cell phones today are designed to not only make calls but to be "mobile multimedia terminals." Thus, they offer more features, storage and processing. Most of the phones interfaces are composed by visual elements on the screen alongside with pressing or sliding buttons. These are the means by which it is possible to interact with the device. Therefore, it is important that the interface is designed to optimize the product's performance.

According to the Brazilian Standard NBR 9241-11 (2002) usability is the ability of a product to meet certain goals to specific users with effectiveness, efficiency and satisfaction, considering the specific context of use. Hence, the context is an important factor in usability studies, because according to the NBR 9241-11 (2002), the usability can vary significantly in different usage situations.

In order to evaluate the usability of a touchscreen phone, Nokia model 5530, this pilot test was set for the users 'to make a call', during a simulated emergency context. Were considered in the experiment, the emotional and behavioral aspects of the user and the interface characteristics of the cell phone. However, this article deals only with issues related to the interface of the device being tested.

2   METHODOLOGICAL PROCEDURES

2.1   Description of the tasks

Seven participants (n = 7) were chosen for this study, considering only that they did not own the same model of the phone being tested. Each participant was instructed to 'make an emergency call' from a locked screen phone.

The tasks to be completed by the participants included: to unlock the phone screen (Task 1), to activate the phone's keypad (Task 2); to dial the given number (Task 3), to initiate the call (Task 4), as illustrated in Figure 1.

Figure 1 Tasks to be performed on the phone´s interface during the test

2.2   Description of the room and equipment

Two rooms were prepared for the test isolated from one another. In the first room (Room 1), each participant had their heart rate measured (initial frequency). In this room the participant also answered a pre-test questionnaire about stress perception, and a post-test questionnaire in order to enquire the participant's impressions about the device and the test itself.

The test was conducted in Room 2, equipped with: a laptop connected to speakers and the phone (test instrument) on a table, two video cameras, one facing the participant's face and torso and the second directed to the phone screen to assist in the counting of 'touches' the heart rate monitor adjusted on each participant with the digital display positioned on the table. Thus, it was possible to compare the initial heart rate (before testing), intermediate (during test) and final (right after the experiment) of each individual. Besides the equipment, there was an observer next to the participant counting the number of 'touches' and another observer further away marking the time for the completion of the task.

The experiment was performed in three steps, as Figure 2:

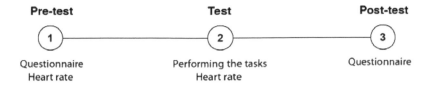

Figure 2 Phases of the test

2.3   Emergency context

An emergency event causes an immediate response in any person, generating increased levels of stress. To simulate the emergency context in this study we opted for an approximation of the person´s emotional state in a real emergency situation. Because there are ethical principles that limit the level of stress the test participant is submitted.

For the induction of stress in the laboratory is commonly used sensory stimuli such as noise or exposure to extreme temperatures or psychological stressors that include demanding cognitive tasks and social assessment (Bollini, Walker, 2002).

We tried to induce some significant levels of stress by combining the following factors: psychological suggestion triggered by a story told to the participant, by the need to perform the task in the shortest time and with the disturbance caused by a sudden and loud siren (71.9 decibels , 0.8 Hz). Each individual heard the following story: 'you and a friend are in an isolated location. Suddenly, your friend passes out. You try to call for help using your own cell phone when you realize its battery is discharged. Your only option is to use your friend´s touch screen phone. Every minute is crucial. You need to act fast! Pick up the phone, unlock it and call 0000-0000'. The total time for performing the test should not exceed two minutes, being this time set from the time of the expert (10 seconds).

The best known test for induction of stress in the laboratory is the Trier Social Stress Test - TSST (Kirschbaum, Pirke, Hellhammer, 1993), designed to cause anxiety and whose protocol has been adapted to induce stress in other studies. The TSST has an anticipatory period in which the participant is preparing for a speech to be presented to two or three evaluators, besides other monitored tasks.

However it was not intended to replicate the TSST procedures in this study. Instead of creating the anticipatory period reported in TSST the present experiment was set to be unpredictable for the participant, as would occur in a real emergency. In this case, the participant was informed of the task a few seconds before executing it.

2.4   Usability metrics

To evaluate the usability of the touchscreen phone were applied these metrics proposed by Tullis and Albert (2008): task success, the time taken to perform it and the efficiency.

For task success the authors mean its completeness, with ease or not (Tullis, Albert, 2008). In this study were considered the time each participant took from the start of the first task to the completion of the last task. According to Albert and Tullis (2008), the efficiency can be measured by both the time and effort. This comprises the cognitive and physical aspects and it can be defined as the user´s actions or steps to complete the action. (Tullis, Albert, 2008). For this metric, we took into account the number of 'touches' and the time of each participant in relation to the number of 'touches' and time of an expert, considering only individuals who have succeeded in all tasks. Thus, the number of ´touches´ was counted until the individual accomplish the tasks. For this study, we consider 'touch' any attempt to unlock the phone and make the call, whether giving quick touches with fingertips, pressing or sliding them over the phone or on the screen.

2.5   Independent variables

The pre-test questionnaire reported on the emotional aspects, therefore, not to be covered in this article. In the post-test questionnaire, were asked questions related to opinions of the individual in relation to the test and the device being tested. Thus, Figure 3 highlights (in gray) such issues, proposed on a semantic differential scale.

I FULLY AGREE								I FULLY AGREE
I WOULD BUY THIS PRODUCT								I WOULDN'T BUY THIS PRODUCT
I APPROVE THE LOOK OF THIS PRODUCT								I DON'T APPROVE THE LOOK OF THIS PRODUCT
I CONSIDER THIS PRODUCT VERY COMPLEX TO USE								I CONSIDER THIS PRODUCT EASY TO USE
I THINK I WOULD NEED INSTRUCTIONS TO USE IT								I THINK I WOULD USE THIS PRODUCT WITHOUT TRAINING
I THINK OTHER PEOPLE WILL ALSO HAVE DIFFICULTY IN USING THIS PRODUCT								I THINK OTHER PEOPLE WILL NOT HAVE PROBLEMS IN USING THIS PRODUCT
I FEEL CONFIDENT USING THIS PRODUCT								I DON'T FEEL CONFIDENT USING THIS PRODUCT
I THINK THE INITIAL SCREEN ICONS ARE EASY TO IDENTIFY								I THINK THE INITIAL SCREEN ICONS ARE CONFUSING
IT WAS EASY TO IDENTIFY THE UNLOCK BUTTON								IT WAS DIFFICULT TO IDENTIFY THE UNLOCK BUTTON
I WAS ABLE TO MAKE A CALL WITH EASE								I COULDN'T MAKE A CALL WITH EASE
I CONSIDER THIS PRODUCT EASY TO USE IN AN EMERGENCY								I CONSIDER THIS PRODUCT COMPLICATED TO USE IN AN EMERGENCY
I WAS NERVOUS DURING THE TEST								I STAYED CALM DURING THE TEST

Figure 3 semantic differential scale applied in the post-test questionnaire

In open-ended questions were asked the following questions:
• How do you think it should be the process to unlock the screen of the phone tested?
• What did you think of the emergency simulation?
• What did you think of this touchscreen phone?

3   RESULTS

3.1   Participants

The group aged between 22 and 49-years-old, 40% between 22 and 25 and 60% between 32 and 49. Among the participants there was only one woman. Four participants had Nokia cell phones; however they were different models of the device being tested. The other three participants had different brands: Foston, Samsung and Gradient.

3.2   Influence of emergency context

The emergency context in the proposed experiment was emphasized by these elements: a short story told to the participant, by the loud siren during testing and by the urgency to complete the tasks in the shortest possible time. The verification of the influence of these elements can be seen by comparing the initial heart rate with the both the intermediate and final heart rate, as shown in Table 1.

Table 1 Comparison of initial, intermediate and final heart rates (beats per minute)

Participant	Initial	Intermediate	Final
1	85	89	76
2	86	112	97
3	93	98	95
4	95	112	99
5	85	101	94
6	94	89	90
7	85	105	103

It can be noted that with the exception of participant 6, all other participants showed a higher heart rate during the test. Therefore, we believe that these individuals were influenced by the emergency resources used - history, loud siren and timing.

3.3   Success and task time

Two participants (2 and 5) failed to complete the task "to unlock screen" because they did not find the unlocking button of the device. This action demanded more time for all participants. The task time for each participant, as well as the number of 'touches' made by them can be viewed and compared to the time and 'touches' of the expert, as shown in Table 2.

Table 2 Total time and number of touches of each participant

Participants	Time to accomplish all tasks	Number of touches
Expert	10"	11
1	60"	38
2	50"	42
3	36"	8
4	82"	49
5	36"	38
6	75"	43
7	31"	14

It is noted that the lowest time achieved by the participants was 31 seconds while the expert took 10 seconds. In relation to the number of touches, the third participant was able to make the call under the expert time. However, he used the auto redial instead of following the specified tasks. Therefore, the participant closest to the number of touches of the expert was number seven, with 14 touches.

3.4   Efficiency

To measure the efficiency were taken only participants who were successful in all tasks, as suggested by Tullis and Albert (2008). So in order to validate the pilot test were compared, first, the time for each participant (PT) at the time of the expert (ET), considering ET / PT, as shown in Table 3. In a second phase, were considered the number of touches of the participants (PN) and the number of touches of the expert (EN), EN / PN, as shown in Table 4.

Table 3 Measured efficiency considering the time to complete all tasks.

Participants	Efficiency (time)
1	0,16
3	0,27
4	0,12
6	0,13
7	0,32

Table 4 Measured efficiency considering the effort to complete all tasks.

Participants	Efficiency (effort)
1	0,28
4	0,22
6	0,25
7	0,78

Considering that the third participant made the call using the auto redial instead of dialing the number, his results were not included in the calculation of the efficiency and effort. Following the formulas described and considering the time and effort of the expert, the efficiency rate in both cases equals to 1. Observing the Tables 3 and 4, it can be argued that the participant 7 came closest to the index of the expert, with 0.32 for time and 0.78 for effort. Therefore, none of the subjects achieved the efficiency ratio taken as the parameter.

664

3.5   Results of the independent variables

The post-test questionnaire allowed the collection of additional data. The responses obtained with the semantic differential were coincident in some aspects and discordant in others, as shown in Figure 4.

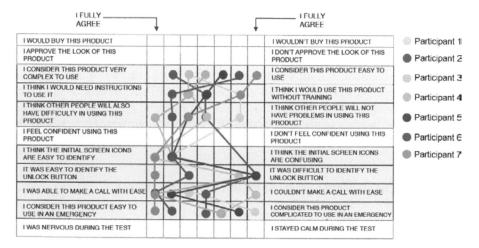

Figure 4 Results of the semantic differential scale

By analysing the responses certain inconsistencies became evident. Participants did not consider the device too complex and most of them said it was easy to use it, while most of them agreed partially that they need instructions to use it. They also agreed fully or partially that others will also have difficulty using the product. All of them agreed that the figures on the initial screen are easy to understand. Five participants found it difficult to identify the unlocking button of the device, while the call was considered an easy task for most of them. However, only two participants found the device easy to use in an emergency.

When questioned about how the screen unlocking process should be, we obtained the following: five participants suggested that the unlocking should be integrated into the touchscreen display; one participant said it should be automatic, and another said it did not need to change at all. By comparing these responses to the results of time and effort it can be noted that there is an inconsistency between their performance and opinions.

About the emergency simulation test, the opinions given by the participants are as follows (Table 5):

Table 5 Participants' opinions about the test.

Participants	Comments about the test
1	"Well, a little concentration was required"
2	"Interesting"
3	"It enabled an experience that really could happen"
4	""I really got the impression of an emergency"
5	"Very interesting"
6	"Interesting"
7	"Simple and practical"

Asked about their views on the mobile device tested, the participants pointed out that (Table 6):

Table 6 Participants' opinions about the device being tested.

Participants	Comments about the phone
1	"The touch screen display is a bit weak, it failed sometimes"
2	"Good and practical"
3	"An aesthetically beautiful device, but it took some time to figure out how to unlock it"
4	"Okay, I've had worse gadgets"
5	"Apart from the unlocking button the phone is apparently good"
6	"Better than my own phone"
7	"With minimal tools for the user"

CONCLUSION

The proposed method of usability in emergencies consisted of an exploratory study of the context of use. It could be inferred that the emergency context altered the performance of users, with the addition of elements in the environment such as a sudden and loud noise through siren and a timing device. The perception of the influence of these factors can be attributed to the increased heart rate during and after the test. However, a comparative study needs to be done in a context of stress free environment. Thus, one cannot state with certainty to in what degree the emergency context caused tension and stress in participants.

From the techniques applied (questionnaires and test) and metrics used (success of tasks, time and efficiency), were gathered sufficient information to assess the usability of a touchscreen phone, considering the proposed tasks in an emergency context.

In the post-test questionnaire, were obtained inconsistent responses. It is believed that this can be mitigated by asking more objective and specific questions.

Regarding the usability of the touchscreen phone tested, the main difficulty for the participants was to unlock the screen to access their features. This was emphasized in the post-test questionnaire. Such difficulty was due to the lack of signaling and information about this action in the device. From this, it is believed that the manufacturers assume that the phone users have read the manual and therefore know how to use all the phone features. But the manufactures does not consider that the same device might be used in an emergency by someone other than the owner of the phone. To be able to access the features of the mobile phone can be decisive in an emergency, so the question is: are the products designed for an emergency situation?

Thus, knowing the importance that mobile phones have gained in our lives it is recommended that there should be more specific guidelines for usability considering different contexts of use.

REFERENCES

Bollini, Anna M.; Walker, Elaine F. *Efficacy of a laboratory stressor*: failure to replicate the Trier Social Stress Test. Accessed February 5, 2012, http://www.jasnh.com/c2.htm.

Associação Brasileira de Normas Técnicas - ABNT. *NBR 9241-11*: requisitos ergonômicos para trabalho de escritórios com computadores - parte 11 - orientações sobre usabilidade. Rio de Janeiro: ABNT, 2002.

Kirchbaum, Clemens; Pirke, Karl-Marlin; Hellhammer, Dirk H.; The Trier Social Stress Test: a tool for investigating psychobiological stress responses in a laboratory setting. *Neuropsychobiology*, 28:76-81, 1993.

Paulheim, H., Döweling, S., Tso-Sutter, K.H.L., Probst, F., and Ziegert,T. Improving Usability of integrated emergency response systems: the SoKNOS approach. In *Proceedings of GI Jahrestagung*. 2009, 1435-1349.

Santos, Robson; Maciel, Francimar. Estilos de interação em interfaces para dispositivos móveis. *Ação Ergonômica* 5: 21-27. Rio de Janeiro, PUC-Rio, 2010.

Tullis, Tom; Albert, Bill. *Measuring user experience*: collecting, analyzing, and presenting usability metrics. Massachusetts: Morgan Kaufman, 2008.

Author Index

T - #0293 - 071024 - C688 - 234/156/30 - PB - 9780367381134 - Gloss Lamination